MOSPOWER
Applications
Handbook

Editor in Chief

Rudy Severns
Senior Staff Engineer

Assistant Editor

Jack Armijos
Senior Applications Engineer

Siliconix
incorporated

Santa Clara, California 95054

Library of Congress Cataloging in Publication Data
Main entry under title:

MOSPOWER applications handbook.

    Bibliography: p.
    Includes index.
    1. Metal oxide semiconductor field-effect transistors--
Handbooks, manuals, etc. 2. Power transistors--Handbooks,
manuals, etc. I. Severns, Rudolf P. II. Armijos.
Jack, 1949–   . III. Title: M.O.S.P.O.W.E.R. applica-
tions handbook.
TK7871.95.M67 1984   621 31'7   84-14140
ISBN 0-930519-00-0

Editorial production/supervision: Robin Berliner
Interior design: Robin Berliner
Cover design: Thomas Washburn Associates

Printed in the United States of America
10 9 8 7 6 5 4 3 2 1

ISBN 0-930519-00-0

Contributing Authors

Edward R. Abramczyk
Mark Alexander
George Allen
Dr. Richard H. Baker
Dr. Richard Blanchard
Gordon (Ed) Bloom
Dr. Dan Y. Chen
Robert R. Cordell
C. F. Der
Phil Ekstrom
Alton Eris

Z. D. Fang
W. E. Frank
David Giandomenico
James Harnden
Barry Harvey
Michael Herrick
John G. Kassakian
Steve Kent
David Lau
Dr. Fred C. Lee
David Mele

Dr. R. D. Middlebrook
Ramesh Oruganti
Ed Oxner
George Riehm
William Roehr
Gary Rothrock
Ray Ruble
Rudy Severns
P. D. Wesel
Richard Williams

Acknowledgement

This book is the result of the efforts of many people. We have elected not to identify every contribution by scattering the authors' names throughout the text. Rather we have grouped them here for convenience.

The following is an alphabetical listing of the authors and their contributions:

Contributing Authors	Sections
Edward R. Abramczyk	2.9
Sr. Design Engineer	2.9.1
Siliconix incorporated	
Santa Clara, CA	
Mark Alexander	2.9
Applications Engineer	2.9.1
Siliconix incorporated	2.12
Santa Clara, CA	3.1
	4.2
	5.1
	5.2
	5.6
	5.6.2
	6.6.4
	6.10
	6.11
	6.14.2
	A.1
	A.2
George Allen	7.1
Testing Engineer	
Siliconix incorporated	
Santa Clara, CA	

Dr. Richard H. Baker	6.1.1
Gould Inc.	
Andover, MA	
Dr. Richard Blanchard	1.3
Vice-President of Design and	2.9
Advanced Technology Development	2.9.1
Siliconix incorporated	2.11
Santa Clara, CA	4.2
	5.6
	5.6.2
	7.1
Gordon (Ed) Bloom	6.1.2
E. & J. Bloom Associates	6.1.3
Arcadia, CA	
Dr. Dan Y. Chen	5.3.4
Assistant Professor of Engineering	5.5
Virginia Polytechnic Institute	6.7
and State University	6.8
Blacksburg, VA	6.9
Robert R. Cordell	6.6.3
Tinton Falls, NJ	
C. F. Der	6.2
Westinghouse Electric Corp.	
Sykesville, MD	
Phil Ekstrom	6.12
Chief Scientist	
Northwest Marine Technology	
Shaw Island, WA	

Alton Eris   6.1.2
Litton Industries
Woodland Hills, CA

Z. D. Fang   6.9
Virginia Polytechnic Institute
and State University
Blacksburg, VA

W. E. Frank   6.2
Westinghouse Electric Corp.
Sykesville, MD

David Giandomenico   5.3.1
Graduate Student
University of California
Berkeley, CA

James Harnden   6.12
Finell Systems   6.14.1
San Jose, CA   6.14.2

Barry Harvey   6.1.5
Advanced Micro Devices
Santa Clara, CA

Michael Herrick   6.1.6
New York, NY

John G. Kassakian   5.3.2
Laboratory for Electromagnetic   5.3.3
and Electronic Systems
Massachusetts Institute of Technology
Cambridge, MA

Steve Kent   7.2
Cypress Semiconductor   7.4
San Jose, CA

David Lau   5.3.3
Student
Massachusetts Institute of Technology
Cambridge, MA

Dr. Fred C. Lee   2.13.1
Professor of Engineering   5.3.4
Virginia Polytechnic Institute   5.5
and State University   6.7
Blacksburg, VA   6.8
    6.9

Dave Mele   6.1.6
MacLeod Labs, Inc.
San Jose, CA

Dr. R. D. Middlebrook   2.2
Professor of Engineering
California Institute of Technology
Pasadena, CA

Ramesh Oruganti   2.13.1
Massachusetts Institute of Technology
Cambridge, MA

Ed Oxner   1.1
Staff Engineer   1.2
Siliconix incorporated   2.2.1
Santa Clara, CA   2.3
    2.4
    2.5
    2.6
    2.7
    2.10
    2.13.2
    3.2
    5.3
    5.6.1
    A.3
    A.4
    A.5

George Riehm   6.1.4
NEC Electronics
Sunnyvale, CA

William Roehr   6.3
Applications Manager   6.6.1
General Semiconductor Industries   6.6.2
Tempe, AZ

Gary Rothrock   6.13
Fairchild
Sunnyvale, CA

Ray Ruble   2.8
Sr. Applications Engineer   6.5
Siliconix incorporated
Santa Clara, CA

Rudy Severns   Preface
Sr. Staff Engineer   2.1
Siliconix incorporated   2.13.2
Santa Clara, CA   4.1
    5.3
    5.4
    5.6
    5.7
    6.1.3
    6.1.4

P. D. Wesel   6.8
Virginia Polytechnic Institute
and State University
Blacksburg, VA

Richard Williams   5.2
IC Design Engineer
Siliconix incorporated
Santa Clara, CA

A book is a creation not only of the authors but also of the support people who made the finished book a reality and without whose help the book would never have been printed. Special thanks are due to Susan Scott for her endless hours of editing for English (engineers really can't spell!); to Ray Lubow, Ray Ruble, and Ed Oxner for technical editing; to Pam Lewis, Susan Hamel, and Linda Cosner for their seemingly unlimited patience in typing and retyping and retyping…the manuscript; and to Robin Berliner who took the manuscript and the drawings (which most closely resembled chicken scratchings) and turned them into a beautifully finished text.

A substantial portion of the credit for the book belongs to Jack Armijos who did the real grunt work of collating material, hounding the authors and editors to finish their work, and in general, seeing to it that the project kept moving.

Rudy Severns
July, 1984

Table of Contents

Chapter 5   Practical Design Considerations   **Page**

Chapter 6   Applications Information

Chapter 7   MOSFET Testing and Reliability

Chapter 8   Appendices

Preface

Since the introduction of the first practical power MOSFETs, in October 1975 by Siliconix, these devices have undergone major performance improvements and are now widely accepted and used in power electronics equipment.

Along with improvements in the devices, an understanding of how to use these devices in practical circuits has gradually evolved. Like any new semiconductor device, the process of understanding has been slow and the progress uneven. Even though our knowledge of these devices is far from complete, much has been learned which would be helpful to circuit designers. Unfortunately, this information has, until now, been scattered through a wide variety of publications or in some cases was unpublished.

The purpose of this handbook is to solve this problem by providing a source of detailed, up-to-date information on device characteristics and circuit applications.

This is an ambitious goal, and total success in the first edition is not possible. What is available here is essentially a snapshot in time. This book is intended to be a beginning, not an ending, and subject to revision and updating as our understanding grows. If the book is to continue to grow and become the central reference work for MOSFET applications, it is vital that you, the reader, send us your comments, criticisms, and contributions. If you do this and we are diligent in correcting and expanding the text, future editions of this handbook should provide a rich source of information for the user of power MOSFETs.

In this first edition, approximately 25 percent of the material is new, and a number of important issues are treated here in detail for the first time. Another 35 percent of the material is a rewrite and expansion on earlier work with an emphasis on putting the information in context by interrelating the various subjects. The remaining material is a collection of application examples taken from a variety of industries.

The book is divided into four sections. Chapters 1 through 3 discuss the basic operation of MOSFETs. Chapters 4 and 5 deal with the practical problems of using MOSFETs from a device point of view. Chapter 6 is a collection of applications examples selected to demonstrate how to use MOSFETs from a circuit point of view. Chapter 7 provides an introduction to the testing and reliability of MOSFETs.

The intent throughout is to provide a balance of basic theory and practical design information since it is our belief that this leads to the best designs. The balance should make this book useful to technicians while still providing plenty of meat for the advanced worker.

Rudy Severns
July 1984

1.1 General Remarks

Introduction

There are two basic field-effect transistors (FETs): the junction FET (JFET) and the *metal-oxide semiconductor* FET (MOSFET). Both have played important roles in modern electronics. The JFET has found wide application in such cases as high-impedance transducers (scope probes, smoke detectors, etc.), and the MOSFET in an ever-expanding role in integrated circuits where CMOS (*Complementary* MOS) is perhaps the most well-known.

All of these applications are best described as *small-signal* where signal levels are measured in a few volts and where the relative power levels range from pico- to micro-watts. Such FETs are microscopically small with the potential of placing thousands of them upon a single semiconductor chip.

This handbook is about MOSFETs—*power* MOSFETs—and they are not microscopic. Before the reader plunges into the text, it is wise to review the differences between a small-signal FET and a power MOSFET. Also, the reader should identify *our* definition of power since this handbook is for power applications.

FETs and MOSFETs—How They Differ

Junction FETs (JFETs) and small-signal MOSFETs differ in their cross-section as well as in their operation. Both are majority-carrier transistors in that they do not rely upon the mechanism familiar to the users of bipolar transistors: *injection, diffusion,* and *collection.* While the bipolar transistor needs an injection of base current to initiate *transistor action,* the FET is controlled by a gate-to-source potential. Current, hence power, is not a requirement for *FET action.*

Before embarking on the differences between JFETs and MOSFETs, note that operationally there are three classes of FETs easily identified in Figure 1: depletion-mode JFETs, depletion-mode MOSFETs, and enhancement-mode MOSFETs.

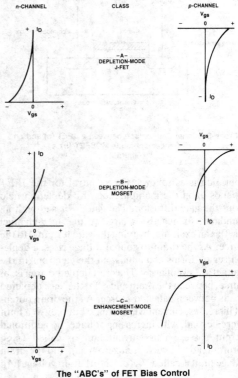

The "ABC's" of FET Bias Control
Figure 1

Class A is defined as the depletion-mode class; all JFETs perform in this class. Class B is also defined as being in the depletion-mode class, but, as can be seen from the figure, it can support enhancement current. Class C is strictly enhancement. MOSFETs perform in either Class B or Class C. No JFET is capable of performance in either of these latter classes because it would require forward biasing of the gate junction—a condition strongly discouraged.

The JFET and the MOSFET have fundamental fabrication differences which are easily visualized from the cross-sectional view in Figure 2. Here the JFET has a *diffused* gate electrode, but the MOSFET has a gate electrode electrically isolated from the channel. In this illustration, the MOSFET is a Class C MOSFET—more commonly known as an *enhancement-mode* MOSFET.

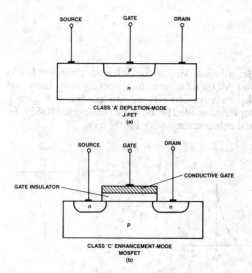

Cross-Sectional Comparison between a JFET (a) and an Enhancement-Mode MOSFET (b)
Figure 2

Operation is as follows. Gate control for the JFET depends upon the manipulation of a depletion field about the gate diffusion. The gate is biased with an abundance of negative charge (using an *n*-channel JFET as an example) by tying the gate terminal to the source. As the drain-to-source voltage rises, a depletion region forms and almost all the electrons in the channel are swept away. The most depleted area is, of course, between the drain and gate; the least depleted area is between gate and source. As the drain potential increases, current will increase but only to a limit where beyond, with increasing voltage, no additional current passes. This current limit is called I_{DSS} (saturation drain current at zero bias), and the drain voltage limit is called V_p (pinch-off drain voltage). In

Figure 3, the output characteristics for a typical *n*-channel JFET are shown with I_{DSS} and V_p clearly identified.

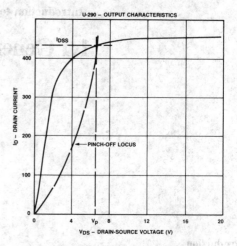

Typical Output Characteristics of a JFET Showing the Saturation Drain Current, I_{DSS}, and the Pinch-Off Voltage, V_p
Figure 3

If the JFET gate is forward biased (positively for an *n*-channel JFET and negatively for a *p*-channel JFET), the *pn*-junction formed by the gate diffusion would be in current conduction. Two things occur: first, substantial gate current would flow resulting in a lowering of the gate input resistance; and, second, a slight increase in drain current would be observed. The latter results from both the contribution of gate current as well as from some further reduction of the depletion field which increases the available channel current.

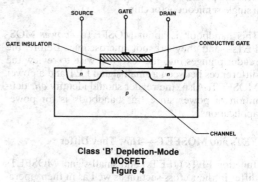

Class 'B' Depletion-Mode MOSFET
Figure 4

Gate control for the MOSFET differs from that of the JFET. A Class B depletion-mode MOSFET, shown in Figure 4, may be compared with a Class C enhancement-mode MOSFET shown earlier in Figure 2. Note in the Class B MOSFET there is a channel connecting the source and drain. Although gate control is essentially the same for both, operationally there is a difference. For an *n*-channel depletion-mode (Class B) MOSFET, a negative potential upon the gate will deplete the electrons within the channel (like charges repel, unlike charges attract), and drain current will be proportional to the gate potential. In fact, the output characteristics of a Class B MOSFET are remarkably similar to those of a Class A JFET as shown in Figure 3. If, however, a positive potential is applied to the gate, free electrons within the *p*-body will accumulate under the oxide and existing *n*-channel inverting the *p*-region. The enlarged *n*-channel allows additional drain current to flow over and above that current flow at zero bias. Thus, the depletion-mode MOSFET takes on the characteristics of an enhancement-mode MOSFET.

Conversely, the Class C enhancement-mode MOSFET has no current flow until *inversion* occurs. Consequently, for an *n*-channel enhancement-mode MOSFET, a positive gate potential is required. Closer examination of Figure 1C shows an offset before conduction which is called the threshold voltage (V_T). Before drain current flows (before *inversion*), an electric field must be established closely dependent upon such variables as the thickness of the gate insulation (silicon dioxide), the doping of the body (which, in turn, controls the availability of free electrons), and the gate material itself—whether metal or polysilicon. Threshold voltages generally lie between 1 and 6 volts.

MOSFETs and MOSPOWER—The Successful Solution

Both the Class B (depletion-mode) MOSFET and the Class C (enhancement-mode) MOSFETs shown in Figures 2 and 4, respectively, were originally developed for small signal applications requiring, at most, a few milliamperes of drain current with attendant power requirements of a few hundred milliwatts. There were several interrelated problems inhibiting high-power performance that were eventually solved by the introduction of the vertical *double-diffused* MOSFET structure. As the reader continues through this handbook, a clearer understanding of MOSPOWER theory, design, construction, and applications will evolve.

POWER: Its Definition Used in This Handbook

World-wide, the semiconductor industry has established a 1 watt power dissipation to set apart small-signal transistors from power transistors. At the time of this writing (1984), it is premature to establish an upper limit although we can, without embarassment, suggest that the upper limit may not be totally the responsibility of the semiconductor itself but more probably that of the package. Today packages, such as the ubiquitous TO3 (TO-204), limit performance to 250 watts. New packages just now emerging show promise to 500 watts.

1

1.2 A Little History

The power MOSFET, as a practical commercial device, has been available since 1976. However, the history and attempts to produce power FETs go back much further. It may come as a surprise, but Shockley, Bardeen, and Brattain were actually trying to fabricate a field-effect transistor when they stumbled—yes, *stumbled*—upon the bipolar transistor. Actually the field-effect transistor preceded the invention of the bipolar transistor by nearly 20 years! Although some try to contest who actually made the discovery, the U. S. Patent Office credits the invention to Dr. Julius Lilienfeld—the date of his patent? 1930! Unfortunately, the good doctor did not build the field-effect transistor (or FET), for in his day semiconductor material was not sufficiently advanced to be of much use. After Shockley and his crew developed their Nobel-prize winning bipolar transistor, Shockley went on to develop the FET. He announced the FET to the world in his classic paper [1] where he offered a general theory of operation and his predictions of performance.

But Shockley's work was still 20 years from the introduction of a successful *power* FET although he was very close to the eventual understanding of what it would take to build a power FET. Many followed Dr. Shockley, making numerous experimental devices and hoping for power, but they all failed—and they all failed for the same reason. The path these unsuccessful experimenters followed was generally to parallel many cells. What they disregarded, forgot or tossed off as inconsequential was that paralleling for additional current handling may drop the channel resistance, but the parasitic elements increase at a higher ratio. As a consequence, these early power FETs were self-defeating and offered excessively high parasitic capacitances for the little gain in performance.

Undoubtedly, the breakthrough came in 1959 when a paper authored by Wegener [2] identified the fundamental failing of the classic approach and offered a remarkable solution. Wegener's solution was simplicity itself. He proposed that rather than build *planar* FETs, we could achieve higher current densities if we built *cylindrical* junction FETs. Within a few years of this remarkable discovery, a plethora of published papers appeared, scattered through several different technical journals which announced the successful fabrication of moderate-power, field-effect transistors. Notable among these early researchers were Teszner, Zuleeg and Nishizawa [3]. The latter scientist is well-known for his Static Induction Transistor (SIT) which made its commercial entry during the mid-1970's in Japan.

Unfortunately, their time had not yet come, and little was heard of these power FETs. A few would appear, undergo some preliminary evaluation, possibly attain some publicity, and then quickly fade into oblivion. Part of the problem was the difficulty in manufacture and the problem of needing two power sources—one for drain voltage and the other for gate voltage (they were depletion mode devices). Finally, without support engineering, they simply never caught on. Consequently, although occasionally discussed, power FETs remained dormant until 1976.

In 1976, Siliconix announced the world's first commercially-available power *MOS*FET in volume production with the registered name *MOSPOWER*.

Nearly simultaneously with this announcement by Siliconix, Hitachi of Japan announced the availability of complementary pairs of power MOSFETs ... and the race was on. Unlike previous attempts at achieving power, these devices were first, MOSFETs—*Metal Oxide Semiconductor Field-Effect Transistors*—and second, they utilized a novel technology called *double-diffusion,* which will be further explained in a later chapter.

The first Siliconix power MOSFET was of novel vertical construction, utilizing a four-level semiconductor process and an easily identifiable anisotropic-etched V-groove shown in Figure 1. Hitachi, on the other hand, followed the classic planar design with source, gate, and drain accessible on the top surface of the chip.

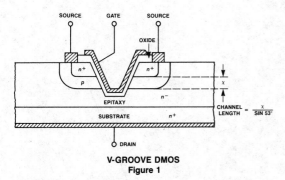

V-GROOVE DMOS
Figure 1

As the power MOSFET technology improved and competitors appeared, the technology evolved into what is now the standard—a vertical, planar four-layer semiconductor process commonly called DMOS or double-diffused MOS, shown in Figure 2.

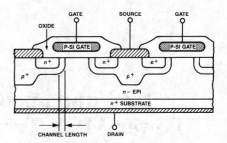

Planar Vertical Double-Diffused Power MOSFET
Figure 2

It is important for the reader to note that despite the various trade names (TMOS, HEXFET, SIPMOS), the basic operation and construction of the power MOSFET are, for all practical intent, identical.

References

[1] Shockley, W., "The Unipolar 'Field Effect' Transistor," *Proc. IRE 40* November 1952) pp. 1365–76.

[2] Wegener, H.A.R., "The Cylindrical Field-Effect Transistor," IEEE Trans. Electron Devices, Volume 6 1959, pp. 442–449.

[3] Teszner, S. and R. Giquel, "Gridistor—A New Field-Effect Device," *Proc. IEEE 52* (1964) pp. 1502–13.

[4] Zuleeg, R., "Multi-Channel Field-Effect Transistor Theory and Experiment," *Solid State Electronics* 10 (1967) pp. 559–76.

1

1.3 Fundamental Limitations of Power MOS Transistor Performance

The goal of power MOS transistor suppliers is to produce devices with the best performance and the lowest manufacturing cost. When die manufacture alone is considered (i.e., the cost of assembly and test are ignored) this goal becomes the production of devices with the lowest on-resistance per unit area for a given breakdown voltage. While one might conclude that this statement is obvious, the method for obtaining this performance is not. There are many practical considerations to be made before the "optimum" design for even a single voltage is obtained. Various engineering approaches have led to a large number of device designs in the growing power MOS transistor field.

Before discussing the practical issues faced when optimizing device performance, it is appropriate to identify the theoretical maximum performance per unit area as a standard against which various manufacturer's devices can be measured. This limit, shown in Figure 1, is obtained when the product of the resistance of a block of silicon and its surface area are plotted as a fraction of its breakdown voltage. This "normalization procedure" assumes that 100 percent of the surface of the silicon is injecting carriers and that 100 percent of the theoretical breakdown voltage of the silicon is obtained. Both of these assumptions can never be achieved in real devices, but it is possible to obtain accurate estimates of the limits imposed by practical considerations.

The utilization of the transistor chip surface may be examined in two steps. First, the percentage of the surface area available for the active device is determined. The relationship between the chip area and

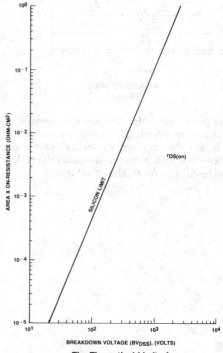

**The Theoretical Limit of
MOS Power Transistor Performance [1]
Figure 1**

the potential active chip area may be derived from knowledge of the width of the region at the perimeter of the chip required for device termination, the number and size of the bonding pads, and the spacings and tolerances required for the active area.

A graph of this relationship consists of the three regions outlined below for a device with source and gate bonding pads on the surface.

Region I: The chip size is dominated by the need for the two bonding pads placed far enough apart to allow for bonding.

Region II: Minimum size bonding pads and minimum conductor widths are sufficient for the current densities involved.

Region III: The current density increases to a level that requires either greater than minimum conductor widths and bonding pad sizes.

Figure 2 shows the percentage utilization versus chip area for MOSPOWER transistors with two bonding pads and a nominal width for the region at the edge of the chip. For a chip 20 mils on a side (500 μm on a side), the surface utilization is below 40 percent, but rises rapidly to approximately 80 percent for a chip 40 mils on a side. If a manufacturer chooses to eliminate the area of the chip dedicated to the source pad and to "bond over active area," the chip utilization is improved by another 5 to 7 percent over the utilization factor shown in Figure 2.

The surface geometry and layout dimensions chosen by a device manufacturer also impacts the device efficiency. The surface geometry requires optimization to obtain the greatest amount of injecting source perimeter per unit area. Figure 3 compares the layout efficiency of various surface geometries. While the geometries differ in efficiency by only a few percent, the "square-on-hex" pattern is most efficient. The width of the gate region must also be optimized.

Figure 4 shows the cross section of a contemporary MOSPOWER transistor. If the gate width is too large, surface utilization is poor; while if it is too small, the JFET formed by the body region results in an increase in device resistance. Optimum gate width is a function of the resistivity of the drain region and decreases as the voltage rating of the device decreases.

The voltage rating of the MOSPOWER transistor also sets limits on device performance. At high voltages, devices are designed to obtain a breakdown

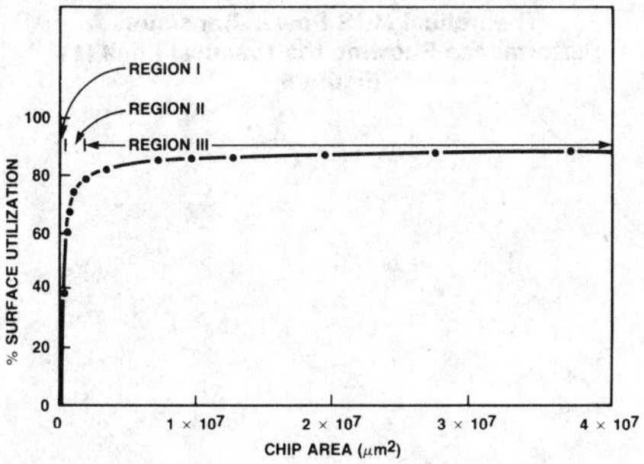

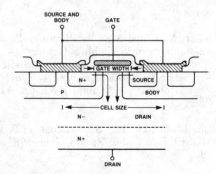

Cross Section of a Power MOS Transistor
Figure 4

**Surface Utilization vs. Chip Area
for MOS Power Transistors
Figure 2**

	LINEAR GEOMETRY	SQUARE ON SQUARE GRID	CIRCLE ON SQUARE GRID	HEXAGON ON SQUARE GRID	SQUARE ON HEXAGONAL GRID	CIRCLE ON HEXAGONAL GRID	HEXAGON ON HEXAGONAL GRID
SOURCE GEOMETRY AND GRID							
UNIT CELL							
COEFFICIENT G FOR CELLULAR GEOMETRIES	NOT APPLICABLE	1.0	0.8862	0.9306	1.0746	0.9523	1.0

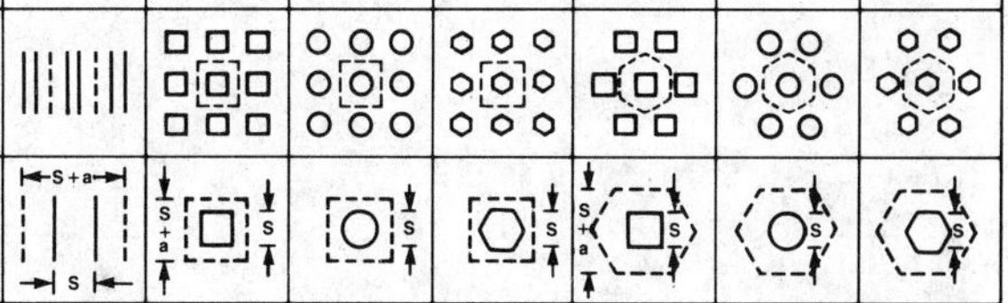

**The Efficiency of Various MOS Power
Transistor Surface Geometries [2]
Figure 3**

voltage close to the theoretical minimum value set by the silicon. As the voltage is increased, it becomes difficult to obtain greater than 80 percent of the theoretical maximum. As the voltage rating decreases, another limit is encountered. The decrease in gate width and other device dimensions cannot continue indefinitely as the voltage rating decreases. The requirements for low resistance ohmic contacts to the source and body regions set a minimum source and body contact dimension, while the lateral diffusion of the body region beneath the gate sets another minimum dimension. These minimum dimensions establish a limit for low voltage devices.

When all of the limits discussed in this section are imposed on a power MOS transistor, the "practical limit" of Figure 5 results. This figure accounts for all of the overhead requirements for edge terminations, bonding pads, device dimensions, surface efficiencies, and voltage effects discussed. (This figure assumes an 85 percent surface utilization value.) The performance of various contemporary devices are also included on this figure for information.

References

[1] Adler, M.S. and S.R. Westbrook, "Power Semiconductor Switching Devices — A Comparison Based on Inductive Switching," IEEE *Transactions on Electron Devices*, ED-29, No. 6, June 1982, pp. 947-952.

[2] Chi, M. and C. Hu, "Some Issues on Power MOSFETs," PESC Record, June 1982, p. 392.

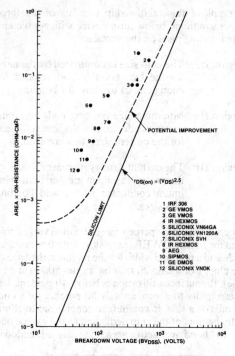

Theoretical MOS Power Transistor Performance Showing the Practical Limit [1]
Figure 5

2.1 Principles of MOSFET Operation

One of the attractive features of MOSFETs is the ease with which the basic principles of operation can be understood. In many ways, the MOSFET is a solid state equivalent to a vacuum tube and, at least for first order behavior, is as easy to understand.

A simplified model will be used to explain how a MOSFET works. The model chosen does not represent how practical devices are actually built, but it does operate in the same manner and makes the explanation easier. This discussion will address an *n* channel device. A *p* channel device would be just the opposite, with *n* and *p* regions interchanged. The basic operation is, however, still the same.

Figure 1a shows an *npn* junction structure. In many ways, this structure is identical to a bipolar junction transistor; the differences arise from the connections made to this basic structure. As shown in Figure 1b, for a MOSFET (Metal-Oxide-Semiconductor Field Effect Transistor), three ohmic contacts and an insulated capacitor plate are added to the *npn* structure. As long as the potential between body and gate is not positive, this device is essentially two diodes back-to-back (Figure 1c), and only a small junction leakage current will flow if a + or − potential is applied between the *n* region ohmic contacts. In this state, the device is OFF.

The *p* region is doped so that there are more holes than electrons. This is, by definition, what makes it a *p* region. Even though the holes outnumber the electrons, there are still plenty of free electrons in the *p* region, and if the gate is made positive with respect to the body, some of these electrons will be attracted

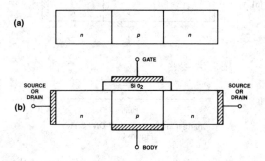

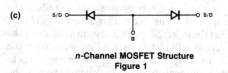

n-Channel MOSFET Structure
Figure 1

to the gate structure, as shown in Figure 2a. Due to the presence of the gate oxide insulator, the gate metal and the body semiconductor form a capacitor which accumulates charge. Even if the potential between the gate and body is small (0-3V), the electron density in the body side of the capacitor will be less than the hole density and the device acts like two diodes back-to-back with a moderate increase in leakage current.

As the gate to body potential is increased, however, the charge density in the body, immediately adjacent to the gate oxide, will increase to the point where the electron density exceeds the hole density, and a portion of the body region (the channel) inverts to become *n* rather than *p*. This is shown in Figure 2b. The semiconductor structure is now *n-n-n* and has become simply a silicon resistor through which current can flow easily in either direction (Figure 2c). This is a variable resistor in which the resistance of the channel is controlled by the potential from gate to body.

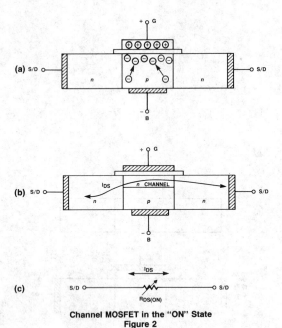

Channel MOSFET in the "ON" State
Figure 2

There is a limit to the minimum resistance of the channel. As the gate-body potential is increased, more charge collects in the channel region. This charge acts like an electrostatic shield to reduce the field in the rest of the body. This action very closely resembles the effect of space charge in a vacuum tube which acts to reduce the electron emission from the cathode. In a MOSFET, this "space charge" acts to limit the additional charge in the channel so that the channel resistance quickly reaches its minimum value.

This charge scenario can be used to explain the $R_{DS(on)}$ versus V_{GS} characteristic shown in Figure 3 which is typical of all power MOSFETs. Region I corresponds to the condition when the accumulative charge is not sufficient to cause an inversion. Region II corresponds to the condition where sufficient charge is present to invert a portion of the *p* region, forming the channel, but not enough that the "space charge" effect is important. Region III corresponds to the charge limited condition where $R_{DS(on)}$ does not change appreciably as the gate-body potential is raised.

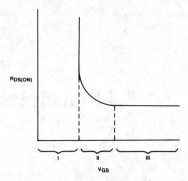

MOSFET Turn-On Characteristic
Figure 3

At low values of drain-source voltage (V_{DS}), the MOSFET, when turned ON, acts much like a normal resistor. However, as V_{DS} is increased, $R_{DS(on)}$ becomes a function of V_{DS} as illustrated in Figure 4. For high values of V_{DS} ($> 10\text{-}20$ V), the MOSFET is no longer a resistor but acts as a very good current source if the gate-body voltage (V_{GS}) is fixed. The transition from resistor to current source will depend on V_{GS}, as indicated. The actual behavior in the resistor region will vary depending on the design compromises adapted for a particular device.

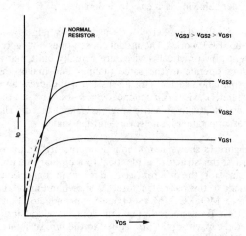

Comparison Between a MOSFET and a Normal Resistor
Figure 4

For example, some power MOSFETs have a linear $R_{DS(on)}/V_{DS}$ curve, like a normal resistor. Other designs, where a minimum $R_{DS(on)}$ at low V_{DS} is desired, will have a curved characteristic as indicated by the dashed line in Figure 4.

Both of these effects are caused by a narrowing or "pinch off" of the channel as V_{DS} is increased. Figure 5 illustrates the channel narrowing or pinch off.

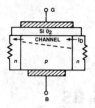

Channel Pinchoff in a MOSFET
Figure 5

For a variety of reasons, a practical power MOSFET does not have four terminals. The normal practice is to connect the body directly to one of the n regions as shown in Figure 6a. When this is done, the n region to which the body is connected is defined to be the source connection. The other n region becomes the drain connection. To turn the device ON, the gate is made positive with respect to the source.

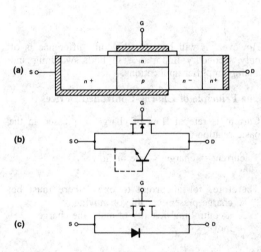

The Body Diode in a MOSFET
Figure 6

The body-source connection places a short between the base and the emitter of the parasitic *npn* BJT as shown in Figure 6b. The base-collector junction is, however, still present so that the equivalent circuit for the MOSFET is a MOSFET in parallel with a diode as shown in Figure 6c.

For most applications, this equivalent circuit is adequate, but in some cases, a more accurate model is needed. Inherent in the structure of the MOSFET are parasitic resistances and capacitances. These parasitic elements are shown in Figure 7a and the equivalent circuit in Figure 7b.

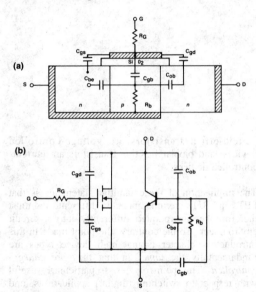

Parasitic Elements Present in a MOSFET
Figure 7

The effect of these parasitic elements on the device and circuit behavior is discussed in Sections 2.2, 3.1, 3.2, 5.4, and 5.5.

2.2 FETs and BJTs as Charge-Controlled Devices

"Field-effect transistors are voltage-controlled devices, and bipolar junction transistors are current-controlled devices."

The implication of these familiar statements is that FETs and BJTs are fundamentally different, and must therefore be incorporated differently into a circuit environment. To the contrary, the point made in this introductory chapter is that both device types are fundamentally the *same,* in that they are *charge-controlled.* This point of view is particularly useful with respect to switching (digital) applications, and even more important for high-power switching applications such as switched-mode power conversion. The main conclusion is that, as switches, FETs must be driven just as "hard", both ON and OFF, as BJTs in order to achieve comparable switching speeds.

In the broad sense of this chapter, *all low-frequency electronic devices are charge-controlled,* including that old-fashioned device, the TUBE. The term "low-frequency" here means that "transit-time" devices, such as the travelling-wave amplifier and the klystron, are excluded.

There are, of course, many differences between the various types of charge-controlled devices, differences which are of greater or lesser importance depending upon the application, and most of the balance of this book is concerned with these differences. Only one of these will be discussed in this chapter, namely, the requirement for a steady state input current in order to maintain an ACTIVE or ON condition in the BJT; the absence of this requirement in the FET is the principal reason why the FET is commonly thought to be fundamentally different from the BJT.

However, it will be seen that this difference is of only secondary importance in both switching and analog amplifier applications.

The Principle of Charge-Controlled Devices

Current is rate of flow of charge, expressed in the basic relation

$$\text{current} = \text{charge} \times \text{rate of flow}$$

Therefore, for a current to exist, there must be:
1. charge present capable of moving
2. a controlling quantity to make the charge move

In the simple device represented in Fig. 1, there is a "channel" terminated at two contacts by which the device may be connected through a switch to an external circuit containing an emf potentially capable of controlling the circuit current. If the channel is an insulator, there is no mobile charge available, so no current flows even though there is a controlling quantity available when the switch is closed.

If the channel is a conductor or an *n*-type semiconductor, as represented in the corresponding structure of Fig. 2, there are two types of charge originally present: positive charge on the crystal lattice ions, and negative electron charge. The channel is electrically neutral, and the total negative charge equals the total positive charge.

When the external circuit switch is closed, the emf sets up an electric field in the channel which exerts a force on both charge types. However, only the

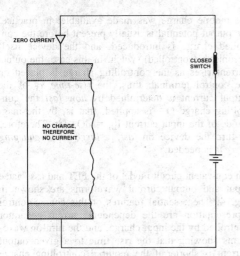

A channel containing no mobile carriers (an insulator) can carry no current

Figure 1

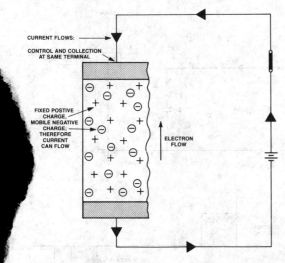

A channel containing mobile carriers (a conductor) is able to carry current

Figure 2

negative electron charge is mobile and capable of carrying current; the positive charge is constrained to the fixed lattice ions. In the diagram of Fig. 2, a (clockwise) flow of electrons results, constituting a counterclockwise steady state (DC) current in which the electrons in the channel may be visualized as moving continuously past the fixed positive charges while retaining the same density distribution; that is, the channel is still essentially electrically neutral.

In the two-terminal device of Fig. 2, the controlled current is collected at the same terminal at which the

controlling quantity is applied. The essence of a *three-terminal active device* such as the FET or the BJT is the *separation of control and collection functions* so that control is exercised at an input terminal and the resulting controlled current is collected at an output terminal.

The FET

Consider again the device of Fig. 1 containing the insulating channel. How can mobile charge be introduced into the channel? That is, how can the device be converted into an electronically operated switch in which current collected from the channel can be controlled by a third terminal?

Charge can be induced in a previously insulating channel by induction across a dielectric. Suppose such a dielectric is introduced in a layer between a third (control) terminal and the side of the channel, as in Fig. 3(a). The control terminal, the dielectric, and the channel connected to the lower terminal constitute a capacitance: if a positive charge Q^+ is inserted at the control terminal, an equal negative charge Q^- is induced in the channel on the other side of the dielectric. This negative charge is capable of carrying current, so that if a positive potential is applied to the upper channel terminal by closing the output circuit switch, a current flows as in Fig. 3(b). This is the principle of the FET, in which the control terminal is designated the gate, and the lower and upper channel terminals are the source and drain, respectively.

Quantitatively, the device turn-on process from the OFF into the ACTIVE mode is as follows. A charge Q^+ is placed on the gate by an input current $I_{in} = dQ^+/dt$; the gate-channel capacitance is thereby charged up and an equal and opposite charge $Q^- = -Q^+$ of electrons is induced in the channel. If a positive potential is placed on the drain, an output current results that is given by $I_{out} = -Q^-/\tau_t$, where Q^- is the *charge in transit,* and τ_t is the *mean transit time* of the charge Q^- that is in transit between the source and drain.

The most significant features of this process are:
(1) To establish mobile charge in the channel, an input transient current (capacitive charging current) is required.
(2) After a steady state is established, the *controlling* charge Q^+ remains static (there is no more input current), but the *controlled* charge Q^- is in continuous motion clockwise around the output circuit, constituting a steady state (DC) output current. Thus, even though *the total charge in transit in the channel remains constant at Q^-, the individual negative charges are constantly being replaced,* entering the channel at the

2

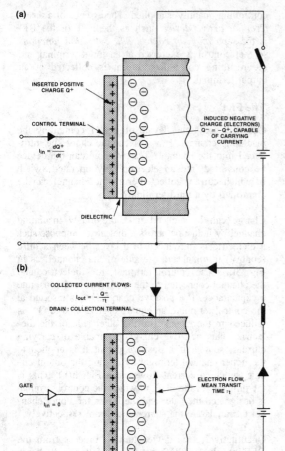

(a)

INSERTED POSITIVE CHARGE Q⁺

CONTROL TERMINAL

$I_{in} = \dfrac{dQ^+}{dt}$

INDUCED NEGATIVE CHARGE (ELECTRONS) $Q^- = -Q^+$, CAPABLE OF CARRYING CURRENT

DIELECTRIC

(b)

COLLECTED CURRENT FLOWS:

$I_{out} = -\dfrac{Q^-}{\tau_t}$

DRAIN : COLLECTION TERMINAL

GATE

$I_{in} = 0$

ELECTRON FLOW, MEAN TRANSIT TIME τ_t

SOURCE

Principle of a charge-controlled device (FET): (a) An insulating channel can have mobile carriers induced in it by a control charge inserted at a control (input) terminal (gate); (b) the induced channel charge is capable of carrying (output) current when a potential is applied to the collection terminal (drain).
Figure 3

source and leaving at the drain after a mean transit time τ_t.

(3) The magnitude of the output *current* $I_{out} = -Q^-/\tau_t$ is determined by the controlling *charge* $Q^+ = -Q^-$.

These three statements describe the principle of a charge-controlled device, in which a current in an output circuit is controlled by the charge placed on a control terminal.

For clarity in the above explanation of the control and collection functions, the potential at the output terminal was asssumed to be applied by a switch after

the mobile charge was made available; in practice, the output potential is usually present before the controlled charge is introduced, and the device itself becomes a (controlled) switch. In this case, the output current rises as the controlling charge is placed on the control terminal: thus, the *rise-time* τ_r of the output current is determined by *how fast* the controlling charge Q^+ is applied, that is, by the magnitude of the input current $I_{in} = dQ^+/dt$. Therefore, to turn the device on *fast*, a *large input charging current* is needed.

An equivalent circuit model of the FET and associated input and output current waveforms are shown in Fig. 4. The essential features of this charge-control representation are the dependent current generator controlled by the input charge, and the turn-on waveforms showing that the rise time to a given output current is shorter if the required controlling charge is applied faster, requiring a larger transient input current (for a shorter time).

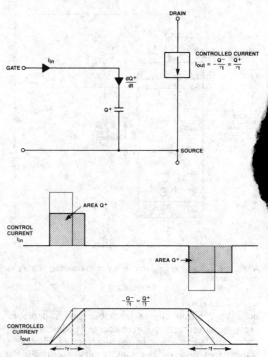

Basic equivalent circuit model of a charge-controlled device (FET). The same control charge inserted or extracted over a shorter time lowers the rise or fall time, respectively.
Figure 4

A steady state, or DC, output current continues to flow as long as the controlling charge remains (statically) at the control terminal. Equally important as turn-on is the reverse, turn-off, requirement: to *stop* the output current, the controlling charge must be *extracted*, requiring a *reverse gate current*. To turn the device off *fast*, which means a short fall time τ_f, the controlling charge must be withdrawn *fast*, requiring a *large* gate extraction current $I_{in} = -dQ^+/dt$. The turn-off waveforms are also shown in Fig. 4.

The essential charge-control quantitative relations may be summarized as:

$$I_{in} = \frac{dQ^+}{dt}$$

$$Q^- = -Q^+$$

$$I_{out} = -\frac{Q^-}{\tau_t}$$

The charge-control model represented by Fig. 4 emphasizes that, to realize the potential fast speed of the FET, the gate drive source must be *low impedance* and capable of both *supplying* and *sinking* the required transient insertion and extraction control charge currents.

The input capacitance shown in the model of Fig. 4 merely accounts for the fact that a voltage appears at the input: it is a nonlinear capacitance, and so the input voltage is a nonlinear function of the control charge. However, the charge-control description of the device operation makes it clear that it is the input *charge* that does the work, and the input *voltage* is of strictly secondary importance.

The BJT

The charge-control description of the BJT almost exactly parallels that of the FET, with appropriate change of terminology.

In an *npn* junction transistor, the channel is the base *p* region, and the lower and upper contacts are the emitter and collector *n* regions, respectively. A conducting contact on the side of the base region becomes the third, control terminal, as shown in Fig. 5(a).

An input current $I_{in} = dQ^+/dt$ introduces positive charge (holes) into the base region, which induces an equal negative charge Q^- of electrons to be injected into the base region from the emitter. This negative charge is capable of carrying current so that, if a positive potential is applied to the collector terminal, an output current $I_{out} = -Q^-/\tau_t$ flows where, again, $Q^- = -Q^+$ is the total charge in transit through the base region, and τ_t is the mean transit time between emitter and collector.

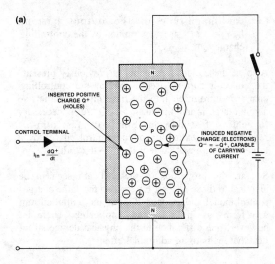

(a)

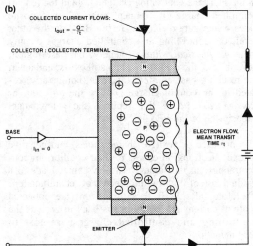

(b)

Principle of a charge-controlled device (BJT): (a) Insertion of control charge (holes) at the base terminal induces injection of electrons into the base region (channel) from the emitter; (b) the induced electron base charge is capable of carrying (output) current when a potential is applied to the collection terminal (collector).

Figure 5

The most significant features of the BJT are identical with those of the FET:

(1) To establish mobile charge in the base, an input transient current $I_{in} = dQ^+/dt$ is required.

(2) After a steady state is established, the controlling charge Q^+ remains static, but the controlled charge Q^- is in continuous motion around the output circuit constituting a steady state DC output current. Thus, even though the total charge in transit in the base remains constant at Q^-, the individual negative charges are constantly being replaced.

(3) The magnitude of the output current $I_{out} = -Q^-/\tau_t$ is determined by the controlling charge $Q^+ = -Q^-$.

Also, if the collector potential is previously present, the collector current rises as fast as the controlling charge is introduced into the base input terminal, so again a large input current $I_{in} = dQ^+/dt$ is necessary to achieve fast turn-on. Similarly, fast turn-off requires a fast extraction of the control charge by a large negative input current $I_{in} = -dQ^+/dt$.

So far, the properties of the BJT are seen to be identical to those of the FET, and so the same charge-control model and associated input and output current waveforms as in Fig. 4 are applicable. There is, however, one difference in the physical device structure that leads to an additional effect.

In the FET, the controlling charge Q^+ and the (equal) controlled charge Q^- remain separated on opposite sides of the gate dielectric. In the BJT, on the other hand, the controlling and controlled charges occupy the *same physical volume,* namely, the base region; nevertheless, the one-to-one ratio still exists, and again the resulting base-emitter voltage is (nonlinearly) related to the control charge Q^+ by the *diffusion* or *storage capacitance* that replaces the gate-channel capacitance of the FET in Fig. 4.

The presence of positive charge (holes) and negative charge (electrons) in excess of their equilibrium concentrations, in the same physical volume, leads to a gradual loss of both by the process of recombination. Therefore, after a BJT is turned on, the collected (output) current gradually decays back to zero as the controlling and controlled charges are lost to recombination.

Conversely, if the collected current is to be prevented from decaying, the recombination charge loss must be replaced: thus, a steady state (DC) *maintenance* control current must be provided to keep feeding in control charge Q^+ at the same rate that it is lost by recombination. This loss rate is Q^+/τ, where τ is the *mean lifetime* of the hole-electron pairs in the BJT base region.

To account for recombination loss, the charge-control model and associated current waveforms of Fig. 4 may be augmented, as in Fig. 6, to represent an additional feature of the BJT:

(4) To maintain the original DC output current at turn-on, a maintenance input DC current must be provided, represented in the model by the current through a resistance in parallel with the storage capacitance.

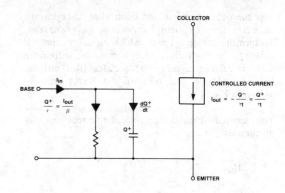

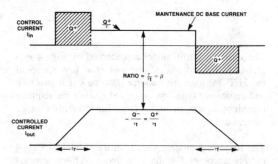

Basic equivalent circuit model of a charge-controlled device (BJT). A maintenance DC base current I_{out}/β is required to make up for the hole-electron recombination loss of the stored charge.

Figure 6

The essential charge control quantitative relations are correspondingly augmented to become

$$I_{in} = \frac{dQ^+}{dt} + \frac{Q^+}{\tau}$$

$$Q^- = -Q^+$$

$$I_{out} = -\frac{Q^-}{\tau_t}$$

In the steady state condition in the ACTIVE region, only the DC maintenance component of the input current remains, and the ratio of the DC output to DC input current is

$$\left.\frac{I_{out}}{I_{in}}\right|_{DC} = \frac{\tau}{\tau_t} \equiv \beta \text{ or } h_{FE}$$

where β and h_{FE} are the familiar symbols for the "current gain" of the BJT.

It is clear that from a switching point of view the FET and the BJT are identical in principle. Both are charge-controlled devices whose input is capacitive, and to make the device operate fast the input capacitance must be charged and discharged fast, that is, the drive must be capable of sourcing and sinking sufficiently large transient currents.

In contrast, the most obvious difference between the BJT and the FET, the presence of DC input current in the BJT, is irrelevant as far as the switching properties are concerned. It may be noted that the same conclusion applies to the BJT operated as an analog amplifier: for high frequencies the input admittance is dominated by the diffusion susceptance, and the β is irrelevant; only for low frequencies, down to and including the DC bias conditions, is the input admittance dominated by the β-determined conductance. From this viewpoint, an "ideal" BJT would have an infinite β, as essentially does the FET.

Charge-Controlled Devices as Digital or Power Switches

All the above discussion has been related to how a charge-controlled device is driven from the OFF to the ACTIVE region, and back, in order to emphasize the underlying simplicity of the device nature. In practice, when operated as a switch, either in digital signal processing or in switched-mode power processing applications, the device is driven from OFF through the ACTIVE to the ON or saturated condition, when additional effects come into play. Nevertheless, the charge control description continues to be useful and, qualitatively at least, the additional effects can be accounted for by superposition.

When a device is used as a switch, the initial turn-on phase is the same as already described: when the collection potential is already present on the output terminal, the channel charge Q^- begins contributing output current as the input current $I_{in} = dQ^+/dt$ induces the corresponding charge $-Q^- = Q^+$. A load carrying the output current causes the output terminal collection potential to fall. If this potential remains high enough to continue to collect all of the channel charge after the input current ceases, the device is turned on into the ACTIVE region as already discussed.

However, if the output terminal potential falls sufficiently low that it ceases to collect all the induced channel charge before the input current ceases, then the saturated ON condition results in which the output current is limited to a value determined by the output circuit. Since input current continues beyond the point

at which all the induced charge can be collected, the device is being *overdriven* and *excess* or *uncollected* charge is present in the channel. The waveforms are shown in Fig. 7, in which the saturated ON condition replaces the final ACTIVE condition as a result of input current being present beyond the time needed for saturation.

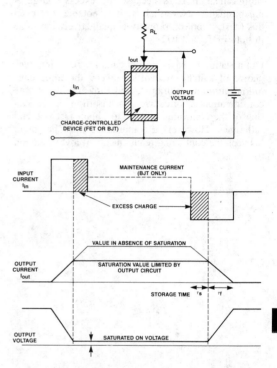

Charge-controlled device operated as a switch: overdrive is needed to cause output saturation, resulting in excess stored charge that must be withdrawn before the switch begins to turn off (causing the storage time).
Figure 7

In the saturated ON condition, the channel contains two components of charge: the charge necessary to contribute the saturated output current as if the device were in the ACTIVE condition, plus the excess charge. From a practical point of view, there must always be some margin of excess charge to ensure that saturation is maintained, and also, the larger the excess charge the smaller is the saturation voltage drop along the channel, which is a desirable circuit property in the ON condition.

If the device is a BJT, there must be a suitable steady state input current to maintain not only the active but also a suitable excess channel charge.

When a charge-controlled device is to be turned OFF, a qualitatively new effect occurs when the initial condition is saturated. As also shown in Fig. 7, the turn-off input current must first extract the excess charge before the output current begins to fall. This interval is known as the *storage time* τ_S, familiar in BJT devices, and the greater the excess charge the longer the storage time for a given turn-off input current. However, a quantity analogous to storage time also exists in FETs, although quantitatively it is much shorter than in BJTs because the excess charge is much smaller. Nevertheless, it is notable that even this (charge-control) property is qualitatively the same in both FETs and BJTs.

Optimization of base drive circuits for BJTs involves choice of suitable tradeoffs between the three conditions turn-on, saturation, and turn-off. The prime consideration is, of course, fast insertion and extraction of the controlling charge to obtain fast rise and fall times. However, a compromise must be found between the conflicting requirements of low saturation ON voltage drop (large excess charge) and short storage time (small excess charge). "Proportional base drive," in which the ON maintenance base current is made proportional to the output current (by a positive feedback transformer connection) is often used to alleviate this conflict.

The FET is "easier" to drive than the BJT, not because one is "voltage-controlled" and the other "current-controlled", but because in the FET a maintenance ON gate current is not required, and the "storage time" is much smaller, so the design tradeoff between ON drop and storage time is essentially eliminated without the need for proportional drive. Basically, the FET and BJT are both charge-controlled, and both must be driven from a low-impedance source capable of both sourcing and sinking sufficient current to provide the desired turn-on and turn-off times.

2.2.1 Charge Transfer Characteristics

Introduction

Casual readers might question beginning the study of hybrid power FET circuits by examining the charge transfer characteristics. Indeed, some readers unfamiliar with power FETs might erroneously believe that no energy, ergo no charge, is needed to operate a power FET. Furthermore, these same readers might anticipate that the coupling of a high gate impedance power FET to a bipolar transistor or thyristor would obviate any need for an intermediate power driver other than, perhaps, a level shifter. Under some conditions, these might be proper assumptions, but first examine exactly how to turn ON a power FET. Surprisingly, it is similar to that of a bipolar transistor since both the power FET and the power bipolar transistor *are charge-coupled devices!*

Although the power FET and the power bipolar transistor are both charge-coupled devices, much of the similarity stops there. Operation of a power bipolar transistor is principally concerned with *current* gain whereas the power FET is, in effect, a *voltage-controlled* current source. If, indeed, this is true then how can a power FET be called a charge-coupled transistor?

The answer lies in the way a power FET is turned ON. More properly it is identified as an enhancement-mode MOSFET. This section concentrates on the operation of an *n*-channel, enhancement-mode power MOSFET remembering that the *p*-channel enhancement-mode power MOSFET obeys all the same laws but in complement. Any enhancement-mode MOSFET requires a finite gate voltage before conduction occurs. The level at which it conducts is called the *threshold voltage,* V_T. As the gate voltage begins

to rise (more positively for an *n*-channel MOSFET and more negatively for a *p*-channel MOSFET), two events begin: As the gate potential rises, channel inversion begins (see Figure 1), and the input capacitor, C_{gs}, begins to charge. In reality, the second event precedes the first; the order is reversed here for emphasis since the first event is recognized without explanation. The current necessary to charge this input capacitor is computed using the following equation.

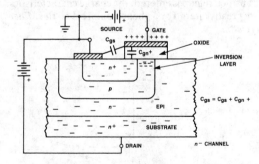

Positive Gate Bias Inserts the p-Region Immediately Beneath the Oxide to Effect Majority-Carrier Current Flow
Figure 1

$$i = C \frac{\partial v}{\partial t} \qquad (1)$$

where, ∂v is the gate-to-source voltage, V_{gs}
    ∂t is the rise time of the incoming signal
    C is the input capacitance, C_{gs}, prior to turn-ON.

but

$$Q = C \, \partial v \tag{2}$$

so combining Equations (1) and (2) yields

$$Q = i \, \partial t \tag{3}$$

where, Q is the charge necessary to raise the gate to V_{gs}.

What is unique about equation (3) is that it is easily measured! However, before looking into the mechanics of how to measure charge, note that the discussion up to this point only involves the gate charge prior to turn-ON [see equation (1)]. The entire switching cycle must be examined. The portion thus far is known as the *subthreshold region*.

(a) The Dynamic Characteristics of Gate Charge

Up to this point, the power MOSFET is OFF. The input capacitor, C_{gs}, has a charge, Q, so the voltage across this capacitor is equal to the threshold voltage, V_T, of the MOSFET. At this point, it might seem that only a small additional charge is necessary to drive the power MOSFET into action, that is, into conduction. When this occurs, the MOSFET is in the dynamic region–the transition region that demands attention.

Generally, whenever power MOSFETs are used as switches, they operate in what is commonly called the common-source mode as shown in Figure 2. As this transition region is entered, the charge characteristics are greatly altered by a phenomenon called the *Miller effect*.

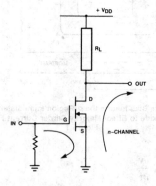

Common-Source Configuration. Source is "Common" Both to Input Circuit as Well as Output Circuit
Figure 2

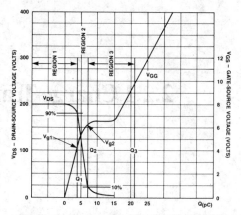

Gate Charge Characteristics Showing the 3 Regions
Figure 3

(a.1) Driving the power MOSFET into Conduction

In Figure 3, the complete *charge-transfer characteristics* of a typical vertical planar double-diffused (VDMOS) MOSFET, region 1 is the area previously discussed, the sub-threshold region.

Two very noticeable events occur in region 2. First, in time, the drain-to-source voltage begins to drop; second, what appears to be an incredible increase in input capacity, C_{gs}, occurs as the fast-rising gate potential apparently stalls [see equation (2)]. What is happening is simply the Miller effect brought about through the interaction of the feedback, or gate-to-drain capacitance, C_{gd} (more commonly recognized in FET terminology as *Reverse Transfer Capacitance*, C_{rss}), and the common-source forward voltage gain, A_V.

$$C_{in} = C_{gs} + C_{gd}(1 - A_V) \tag{4}$$

where
$$A_V = \frac{\partial V_{ds}}{\partial V_{gs}}$$

Remember that the quantity, ∂V_{ds}, is, for an n-channel MOSFET, a *negative* value, and for a p-channel MOSFET, the quantity, V_{gs}, is a negative value. Thus in either case, this equation is calculated as (1 + *voltage gain*).

Thus, either considerably more time is needed (assuming that the charging current, i, is a constant), or, to keep time, ∂t, a constant considerably more current is needed. One thing must be kept in mind: the execution of equation (4) is not easy! If the charge-transfer characteristics for the particular power MOSFET are available, *use them!* Poised to move into the conduction or *transient region,* first take time to see how to use these charge-transfer characteristics.

(a.2) Using the Charge-Transfer Characteristics

It takes a finite charge to energize a capacitor and that charge, Q, can be calculated using equation (2). If equation (2) is merged with equation (3), a constant-current source will provide a linear rate of increasing voltage with time, ∂t, as seen in Figure 4 as well as in region 1 of Figure 3. If the capacitance changes, for example, or if it increases because of the Miller effect then with a fixed charging current, i, the charge time, t, increases. Additionally, the effect of increasing capacitance (Miller effect) upon the voltage, ∂V, can be examined. Where, with a fixed capacitance the rate of change was linear, now there is an abrupt slowing. All of these effects may be seen by careful examination of Figure 3.

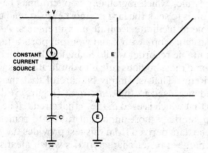

A Linear Charging Rate Results When a Capacitor is Charged from a Constant-Current Source
Figure 4

To determine the amount of *energy* necessary to move a power MOSFET from a normally OFF state to a normally ON state, the difference in charge provides the answer.

$$W = 1/2\,(\partial Q)\,(\partial V) \tag{5}$$

Using data extracted directly from Figure 3, we can calculate, with precision, the input capacitance states in all three regions: the subthreshold region (region 1), the turn-ON region (region 2), as well as after V_{SAT} has been passed (region 3).

From equation (2) calculate the capacitance in the subthreshold region remembering that at the beginning of our initial turn-ON charge, $V_{gs} = 0$, which simply means that $\partial V = \partial V_{gs}$.

$$C = \frac{Q}{V_{gs}} \tag{6}$$

Next the Miller effect must be considered: the apparent flattening of the charging cycle. Since the calculation of input capacitance is, from equation (2), merely the slope of the charge curve, it appears to be a tremendous increase in input capacity. Fortunately, equation (4) can be bypassed; C_{in} is easily determined by an obvious extension of equation (2).

$$C_{in} = \frac{\partial Q}{\partial V} = \frac{Q2 - Q1}{V_{gs2} - V_{gs1}} \tag{7}$$

This value of C_{in} obtained from equation (7) equals that value that might have been obtained from equation (4) after a lot of labor. Finally consider the calculation of input capacitance for region 3, and an equation similar to equation (7):

$$C_{in} = \frac{Q3 - Q2}{V_{gs3} - V_{gs2}} \tag{8}$$

Whether these equations are used to calculate the various input capacities or simply observed from Figure 3, it is clear that C_{in} of the subthreshold region (region 1) differs from the C_{in} of region 3—the former being somewhat lower. The reason for this is that capacitances within a power MOSFET—as with *any* FET—are depletion-field dependent. Being depletion-field dependent, they are voltage dependent as is evident from Figure 5. Since in the subthreshold region a power MOSFET is OFF, the drain voltage is at the rail, but when driven into *full* conduction, the voltage across the MOSFET is simply its V_{SAT}, at worse only a few volts.

(a.2.1) Calculating Switching Times from Charge Transfer

These charge-transfer characteristics can also be used to determine switching times with remarkable precision. Aside from turn-ON and turn-OFF switching

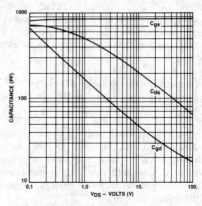

Voltage-Dependent Depletion Areas Also Affect Depletion-Dependent Capacitances, C_{ds} and C_{gd}
Figure 5

times, the *delay* times must be considered. The first delay is that time required for the input capacity to charge to the threshold voltage at which turn-ON begins. In actual practice, turn-ON time (or more properly *rise time*) is generally specified as that interval between the 90% and 10% rise time, so region 1 really extends a bit further than the threshold voltage and, conversely, region 3 begins at the 10% level. A careful study of Figure 3 shows this quite plainly. The second delay is at turn-OFF when the input capacitor must be discharged to the voltage where the channel begins to revert to the non-conducting state. This latter state is akin to storage-time delay caused, not by minority carriers as might be anticipated with a power bipolar transistor, but by *overcharging* the input capacitor when the power MOSFET switch was turned-ON.

There are three equations which can be used with the charge-transfer characteristics as depicted in Figure 3 to determine the three switching times: turn-ON delay, $t_{d(on)}$; rise time, t_r; and turn-OFF delay, $t_{d(off)}$. A fourth, fall time, t_f, cannot be calculated using the charge-transfer characteristics of Figure 3.

$$t_{d(on)} = \frac{Q1}{V_{g1}} \; R_{gen} \; \ell n \; \frac{V_{GG}}{V_{GG} - V_{g1}} \qquad (9)$$

$$t_r = \frac{Q2 - Q1}{V_{g2} - V_{g1}} \; R_{gen} \; \ell n \; \frac{V_{GG} - V_{g1}}{V_{GG} - V_{g2}} \qquad (10)$$

$$t_{d(off)} = \frac{Q3 - Q2}{V_{GG} - V_{g2}} \; R_{gen} \; \ell n \; \frac{V_{GG}}{V_{g2}} \qquad (11)$$

where the first two terms within each equation, $\frac{\partial Q}{\partial V_g}$ R_{gen}, represent the classic RC time constant where R_{gen} is the driving impedance as seen *from the gate terminal*.[1]

The correlation of calculated switching time using these equations with that of measured data can be precise if data is taken carefully from the charge-transfer characteristics such as has been done in Figure 3.

Before addressing the so-called problem of fall time, t_f, and why it cannot be calculated from the data

---

[1]Occasionally data sheets with erroneous generator impedance listed with the switching time parameters are discovered. If a data sheet offers a switching time test circuit, check whether the gate shunting resistor is considered in parallel with the generator impedance, viz., the correct value of

$$R_{gen} = \frac{R1 \times R2}{R1 + R2}$$

contained in Figure 3, the method used to generate the charge-transfer characteristics must be considered.

(a.2.2) Developing the Charge-Transfer Characteristics

In Section (a.2), the charge-transfer characteristics were derived by observing the gate voltage with respect to time with a constant-current drive. That, in effect, is how the characteristics shown in Figure 3 are developed. Achieving the drain voltage characteristics involves nothing more than a simple monitor on the drain terminal of the power MOSFET under test. The circuit for this characterization is shown in Figure 6. Although admittedly simple, a few rules must be observed if these will be used to evaluate the charge characteristics. The first rule is to make absolutely sure that the base line—the abscissa—is in easy-to-divide integers. If the current source is fractional, for example 4.7 mA, then the sweep rate should be such that the product (i•t) is 5 nC/cm and not 4.5 nC/cm which would occur at a sweep speed of 1 μs/cm. The second rule, which regrettably is often times beyond control, is to seek the largest, most easy to read chart possible to facilitate accurate measurements. A digitized oscilloscope with x and y output ports attached to a suitable recorder would be highly desirable. The third and most obvious rule is to duplicate the switching circuit. That is, simply be careful that the load impedance used in the *test* circuit (Figure 6) duplicates that which is used in the actual circuit. The final and potentially most important rule is to remember that the data derived from this test provides the turn-ON charge-transfer characteristics which means—as seen in Section (a.2)—that they only provide turn-ON delay, rise time, and turn-OFF delay [equations (9), (10), and (11)]. To calculate fall time, a new switching model is needed.

(a.2.3) Developing a Switching Model for Fall Time

Already many commercially available power MOSFET data sheets include the charge transfer characteristics

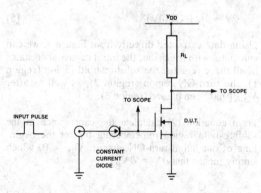

Circuit for Monitoring Charge Transfer Characteristics
Figure 6

in graphical form similar to what is shown in Figure 3. Offering such a graph is not particularly difficult for most vendors. For the most part, they can anticipate most *turn-ON* switching configurations based on such power MOSFET parameters as *Breakdown Voltage,* BV_{DSS}; *ON-Resistance,* $r_{DS(on)}$; and *Operating Drain Current,* I_D. When a power MOSFET is turned ON, the action is well defined: the fast-dropping ON-resistance eventually crowbars the output capacitance of the MOSFET reducing the voltage drop to V_{SAT}. All of this action can be reasonably well-defined on the vendor's data sheet. What he cannot do is to predict turn-OFF since this depends to a large measure on the type and size of the load and whether the load is resistive, reactive, or both. Consequently, the burden of developing a Fall-Time Charge-Transfer Characteristic lies with the end user and not with the supplier. The difference between turn-ON and fall time is shown in Figure 7. Note that this figure is elementary and does not represent a true model of the power MOSFET. The reader is encouraged to review Section 2.13 for specific modeling information.

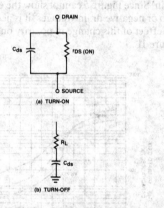

During turn-ON (a), C_{ds} begins fully charged and is discharged through the low $r_{DS(ON)}$ of the MOSFET. However, at turn-OFF C_{ds} begins fully discharged and charges to the rail through the load.

Figure 7

(a.2.4) Developing the Fall-Time Characteristics

In Section (a.2.2) and the accompanying Figure 6, the gate of the power MOSFET is charged from a constant-current source. In the initial pre-drive state, prior to any attempt at turning ON the power MOSFET, the input capacitor, C_{gs}, is uncharged, and the gate voltage is zero. On the other hand, the drain-gate capacitance, C_{gd}, has the full drain-gate potential, V_{gd}, impressed across it. When considering the fall time, quite the opposite is true. The input capacitance, C_{gs}, is charged—perhaps overcharged—and the gate-drain capacitance, C_{gd}, rather than having the full drain-gate potential, enters the fall region with a very small potential which is discussed in Section (a.3).

It should be obvious to the reader that to develop a fall time characteristic, the current (charge) must be controlled as it *leaves* C_{gs}, in other words, as the input capacitor is discharged. The circuit might look much like that in Figure 8. This circuit, however, presents a very fundamental problem: how to configure the circuit shown in Figure 6 (the *turn-ON* charge-transfer characteristics) to that shown here (the *turn-OFF* charge-transfer characteristics). The solution is simply to switch between the two using the circuit shown in Figure 9. Instead of a turn-ON charging diode (our constant current source) limiting the current from the pulse generator (Figure 6), the input capacitor of our power MOSFET is charged from a pre-set voltage. An analog switch is used to switch the circuit configuration from that of a turn-ON charging network to a turn-OFF discharging network. The finite ON-resistance inherent in the analog switch is not significant since a *constant-current* source or load represents a very high resistance. What is important is to have the voltage drop across these constant-current diodes—and in particular, the discharge diode—as low as possible to ensure that the gate charge truly drops to as near zero as practical.

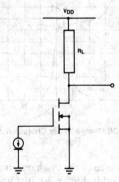

Simple Turn-OFF Charge Transfer Circuit
Figure 8

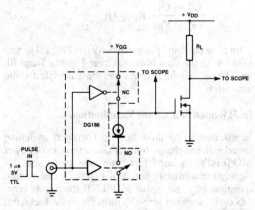

Automatic Turn-ON, Turn-OFF Charge Transfer Test Circuit
Figure 9

(a.2.5) Calculating the Fall-Time

The results obtained from testing a power MOSFET in the circuit just described (Figure 9) can provide any of three views depending on how the viewing oscilloscope is synchronized. It can be set to view only the rise-time characteristics with results similar to what was seen earlier in Figure 3, or the fall-time characteristics as shown in Figure 10 can be seen, which, incidentally, also provide the turn-OFF delay, $t_{d(off)}$, information. In the calculation of turn-OFF delay [equation (11)], data can be extracted from either Figure 3 or Figure 10. A third option would be to view the entire switching cycle: turn-ON delay, rise time, the steady ON state (which, depending upon the test conditions, could also be the turn-OFF delay), and the fall-time.

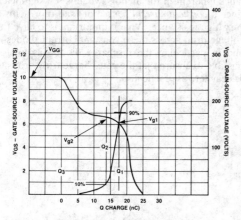

Turn-Off Charge Decay Characteristics
Figure 10

With the assistance of the discharge-transfer characteristics provided in Figure 10, it is now possible to calculate the fall-time using the following equation:

$$t_f = \frac{Q2 - Q1}{V_{g2} - V_{g1}} R_{gen} \ln \frac{V_{g2}}{V_{g1}} \qquad (12)$$

Using these four Equations—(9), (10), (11), and (12)—with the data obtained from Figures 3 and 10, all of the switching times can be calculated quite accurately.

(a.3) Region 3: A Closer Examination

In attempting to show how to correlate switching speed with the charge-transfer characteristics of power MOSFETs, an important point was by-passed. By closely examining region 3 of Figure 3, one should question why the power MOSFET, during its turn-ON cycle, does not *turn-ON fully*. Not only does it not turn-ON, but the Miller effect holds down the gate voltage build-up. It is only after what amounts to an

excruciatingly long time that V_{SAT} is finally achieved. The question that needs to be asked is: "Why? What makes this obvious slowdown in turn-ON occur?" Since this slowdown occurs after the measured (and calculated) rise time, the effect and its cause are generally disregarded.

Section (a.2.4) began by discussing the effects of turn-ON on the various parasitic capacitances paying attention to the rapidly changing voltages impressed across the gate-drain capacitance, C_{gd}. When the MOSFET is OFF, the voltage impressed across this gate-to-drain capacitor equals the full rail voltage, V_{DD}. But as the MOSFET begins to switch ON, the drain voltage drops along with the impressed voltage across this capacitor. Figure 5 shows that as this drain voltage drops, the capacitance rises dramatically. In fact, as the drain voltage slowly settles toward V_{SAT}, the drain-to-gate voltage, V_{dg}, actually *reverses polarity!* Where, at the beginning of the rise time there was a drain voltage many times higher than the gate voltage, now the gate voltage is *more positive* than the drain! Since Figure 5 cannot show the effects of either zero or negative drain voltages, it is necessary to view the effect of this change in polarity on capacitance in Figure 11.

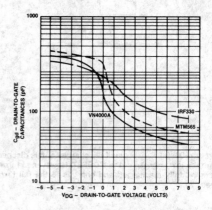

Gate-Drain Capacitance versus Drain-Gate Voltage
Figure 11

During the initial rise time (region 2 of Figure 3), there is the contribution of a gradually rising gate-to-drain capacitance, C_{gd}, on the Miller effect which, with some difficulty, could have been determined from equation (4). Now after having completed the turn-ON cycle (V_{DS} dropping to within 10% of its original value), the voltage gain, A_V, has also dropped ($\partial V_{DS}/\partial V_{GS} \rightarrow 0$), but because of the reversal of polarity across the gate-drain capacitor, its capacitance has soared to astounding heights. As a consequence, there is an extended Miller effect and slow decay of the drain-to-source voltage to V_{SAT} (region 3).

2.3 Planar MOS

When small signal MOSFETs (such as those found in FM and TV receivers) were in vogue, someone might have asked if they could be used for a higher power than the few milliwatts obtained, and the answer would have been a resounding "No!"

The small-signal MOSFET was, in hindsight, simplicity itself. The structure is shown by the *n*-channel depletion-mode device in Figure 1. If our desire was to turn-ON this MOSFET, a positive gate voltage would accomplish it. When the gate receives a positive charge, the capacitor effect forces an accumulation of electrons to migrate under the oxide. But since there is a *p*-doped channel beneath this oxide, the migration of electrons effectively inverts this region into what might be termed *p*-doped. Once inversion occurs, current can flow from source-to-drain since the entire path including both the source and the drain has *n*-accumulation throughout. Conversely, a negative gate potential depletes the region, enhancing the *p*-type characteristics of the channel and preventing any current flow aside from any leakage that exists between the pair of back-to-back *pn*-diodes over which the gate has no control.

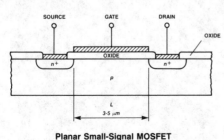

**Planar Small-Signal MOSFET
Figure 1**

The construction of a planar small-signal MOSFET reflects a unique symmetry in that both the drain and source diffusions are, for all practical purposes, identical. What actually differentiates source from drain is the biasing arrangement; the gate is generally biased with respect to the *source*.

The *p*-channel MOSFET is the exact inverse of the *n*-channel version. Where the *n*-channel MOSFET was *n*-doped, the *p*-channel is *p*-doped. All that remains identical to the *n*-channel MOSFET is the oxide, the gate structure, and of course, the metallic contacts to the gate, source, and drain. Operationally it, too, is reversed. Now a negative bias enhances current conduction through the channel.

The obvious extension combines *n*-channel and *p*-channel MOSFETs on the same substrate into what is popularly known as CMOS (complementary MOS). It is one of the foremost technologies today extending its influence from the fabrication of analog switches to microprocessors.

Why the small-signal MOSFET is a *small-signal* MOSFET and incapable of handling power can be simply explained if the basic equations for current are examined both in the linear region and in the saturation region.

In the linear region, the drain-to-source current is given by:

$$I_{DS} = \frac{W}{L}\mu C_{ox}\left[V_{DS} - \frac{V_{DS}}{2}\right]V_{DS}$$

where:

    W = device channel width
    L = device channel length
    μ = electron mobility in the inversion layer

In the saturation region, it is given by:

$$I_{DS} = \frac{\mu C_{ox} W}{2L} (V_{GS} - V_T)^2$$

The fundamental problem lies in the single term L, the effective device channel length. If a MOSFET is designed to withstand high voltages, the channel length must be increased to reduce the deleterious effects of the body-drain diode (*pn* junction) depletion region that, because of the increasing voltage gradient, moves progressively further into the channel region. If the channel length is not increased, punch-through breakdown results, causing catastrophic destruction of the MOSFET. However, careful inspection of these two equations identifies that current is *inversely* proportional to L; consequently, increasing the breakdown characteristics reduces the power handling capability!

Although this restriction was well understood, there were, nonetheless, a few attempts during the 1960's to achieve power by paralleling multiple chips. These attempts failed for many reasons: an incorrect understanding of the contribution of parasitic elements (notably capacitance), improper headers and die attach methods to maintain low thermal resistance, and prohibitive costs.

In the early 1970's, investigators began in earnest to resolve the conflict of high voltage and current in their efforts to overcome the problem. The novel technology that evolved was called *double-diffused* MOS which made possible the revolution in discrete power MOSFETs that is occurring today.

References

Richman, Paul, *MOS Field-Effect Transistors and Integrated Circuits,* New York, John Wiley and Sons, Inc., 1973.

Sigg, Hans J., et. al., "DMOS Transistor for Microwave Applications," *IEEE Trans-Electron Devices,* Vol. ED19, 1 January 1972, pp. 45–53.

2.4 DMOS Double-Diffused MOS

Realizing the fundamental limitation of conventional planar MOS—a channel length proportional to breakdown voltage but inversely proportional to current—researchers focused upon a new structure which simultaneously accomplished both high voltage and higher current. During the decade of the 1970's, the literature described progressively advanced designs using this new structure—what became known as the *double-diffused* MOS.

Simultaneous with the progressive development of the double-diffused MOS technology was the evolution of the MOS gate. Previously the gate consisted of a metal overlay on oxide, but this new technology used a layer of polycrystalline silicon as shown in cross-section in Figure 1. Among the advantages offered by this polycrystalline silicon gate structure was the ease in interconnection between cells which now could be done in diffusion rather than by the more awkward techniques of metalization and bonding. A more significant advancement was the "self-aligning" polycrystalline structure that now could be controlled by a mask rather than by using metalization. Consequently, interelectrode capacitances fell, thus affording higher frequency (faster switching) performance for the MOSFET.

A double-diffused structure consists of a sequentially introduced set of impurities into the epitaxy that for an *n*-channel MOSFET consists of first a deep *p*-

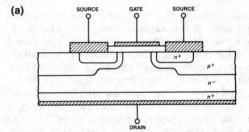

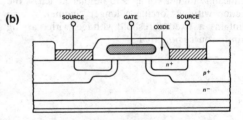

**Cross-Sectional Comparison: Metal-Gate (a)
vs. Silicon-Gate (b) DMOS
Figure 1**

doped region followed by a more heavily doped *n*-region. A typical profile showing this double-diffused structure is shown in Figure 2. The channel length is controlled by careful monitoring and control of the doping levels and the subsequent diffusion cycle which must consider both time and temperature.

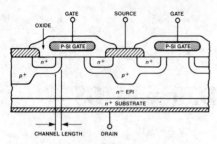

Planar Vertical Double-Diffused Power MOSFET
Figure 2

One obvious difference between the DMOS structure and the conventional planar MOS structure is the channel length, L, determined by the difference between two sequential diffusions. Unlike the conventional planar MOS where the channel length is achieved through the photolithographic process and limited to 4 to 5 μm, the DMOS channel length can be reproducibly controlled to values in the 1 to 2 μm range!

A less obvious difference, but one of paramount importance, is what happens to the *pn* depletion field that formerly forced the construction of a long channel in the conventional planar MOS. In the DMOS process, the body region (for the *n*-channel MOSFET, the *p*-doped area) is more heavily doped than the *n*-drain (epitaxy) region. Consequently, the depletion field extends further into the drain region than into the body region when a reverse bias is placed across the drain-to-body (*np*) junction. This not only allows significantly higher voltages to be placed across the junction without forcing a longer channel, it also maintains a fixed threshold voltage with varying drain-to-source potential.

These two differences allow a MOSFET that has both a short channel length and the ability to withstand a high drain potential without fear of punch-through.

DMOS transistors exhibit a difference in their drain current-to-gate voltage relationship. With the simple planar MOS structure, a square-law relationship is identified, and in the short-channel DMOS structure, the accelerating field potentials can easily reach values exceeding 10^4 V/cm. When this occurs, the drain current becomes linearly related to the gate voltage, as given by the equation:

$$I_{DS} = C_{ox}W \cdot V_{SAT} \cdot (V_{GS} - V_T) \qquad (1)$$

Furthermore, as this equation shows, drain current is no longer dependent upon channel length, L. Additionally, we can manipulate equation (1) to show that transconductance, g_m, is independent of V_{GS}.

$$g_m = \frac{dI_{DS}}{dV_{GS}} = C_{ox} \cdot V_{SAT} \qquad (2)$$

Operationally, the DMOS transistor works similarly to the planar MOS. An *n*-channel enhancement-mode DMOS transistor requires the application of a positive gate potential to cause surface inversion of the *p*-region. Once the surface has inverted, a conducting channel then bridges the source and drain-drift region allowing conduction to occur.

This fundamental double-diffused process opened the door for the fabrication of at least three basic high-power MOSFET designs: the V-groove transistor, the lateral DMOS (LDMOS), and the popular vertical DMOS structure (VMOS), usually called DMOS. All of these power MOSFETs perform similarly as we shall now examine.

2.5 V-Groove MOS

The first commercially available power MOSFET had a unique way to achieve the performance goals that the double-diffused MOS was expected to offer. It was, after a fashion, a truly double-diffused transistor despite its somewhat unconventional design. With the conventional double-diffused technology, the channel length was controlled by precise *lateral* diffusion; however, with V-groove, the control was *vertical*. It was critical to the performance of the device that a narrow channel be available; as a result, the structure's design began closely resembling the four-layer construction typical of the high-frequency transistor as we see in Figure 1. To gain access to the channel, it was necessary to anisotropically etch a groove that, once completed, could have an oxide overlay, and then have the gate metal sputtered. The finished V-groove cross-sectional view is shown in Figure 2.

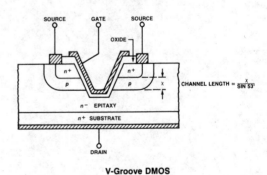

V-Groove DMOS Power MOSFET
Figure 2

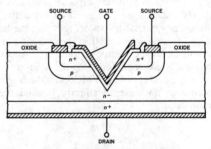

$$\text{CHANNEL LENGTH} = \frac{X}{\text{SIN } 53°}$$

V-Groove DMOS
Figure 3

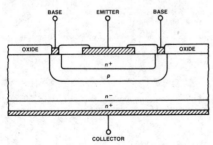

Bipolar Transistor
Figure 1

Early in the development of these devices, the V-groove was etched to a sharp trough. Once higher voltages were demanded, it was quickly recognized that this sharp-edged trough made reliable high voltage performance difficult; consequently, a truncated V-groove was developed. This new design is shown in Figure 3.

Before beginning the discussion of V-groove technology, it is necessary to recognize that silicon is a crystalline allotrope element with three well-known crystallographic planes — $<100>$, $<110>$, and $<111>$. To build a V-groove structure, the starting silicon is always $<100>$. The V-shaped groove results from what is known as preferential etching that follows the silicon crystallographic plane. Without interference, the etching process terminates when the sharp V-groove is formed; however, since a truncated etch is desired, it is critical to the fabrication that accurate timing be employed. Nevertheless, since the etch is "controlled" by the crystallographic plane of the silicon, the sides of the V-groove form a 54.7° angle with the surface and terminate at the $<111>$ plane. Because of the extreme care which is required in the manufacture of V-groove transistors, it is only used when the desired parameters cannot be achieved in any other way.

The channel region of an n-channel V-groove transistor is the p-diffusion (shown in Figure 1). To achieve electron velocity, this region must be short; it is in the design of high-frequency bipolar transistors to effectively reduce transit time. However, unlike the high-frequency bipolar transistor, reducing the length of this p-diffusion does not affect the operating breakdown voltage of the V-groove transistor. Access to this channel is accomplished using the V-groove. It is crucial to the success of the VMOS that the V-groove (truncated or not) extend slightly beyond the p-diffusion depth.

Performance of this V-groove transistor follows that of the DMOS transistor closely with the following notable exceptions:

1. If the same diffusion schedule is maintained for the V-groove structure as for the DMOS structure, the channel length is longer by the geometrical factor associated with the 54.7° angle. The difference is approximately 1.7 times the channel length of an equivalent planar DMOS transistor.
2. Electron mobility is dependent upon the crystallographic plane, and for V-groove, the $<111>$ plane at the surface (channel) of the V-groove results in a reduction of this mobility of approximately 25 percent.
3. Threshold voltages tend to be higher for the same body dopant than a comparable DMOS structure as a result of a higher oxide charge inherent in $<111>$ structures.

References

[1] Fuors, Dennis and Verma Krishna, "A Fully Implanted V-Groove Power MOSFET," *Technical Digest,* IEEE Electron Devices International Meeting, Washington, D.C., 1978, pp. 657-60.
[2] Heng, TMS, et. al., "Vertical Channel Metal–Oxide–Silicon Field Effect Transistor," Final Report, Westinghouse Research and Development Center, Pittsburgh, PA, November 1, 1976.

2.6 Lateral Power DMOS (LDMOS)

Although lateral small-signal DMOS structures appeared in the closing years of the 1960's, power devices only became available commercially in 1977.

The lateral DMOS transistor, shown in cross-section in Figure 1, has all three terminals—source, gate, and drain—on the top surface. This topology offers unusual advantages as well as several disadvantages that have limited the application of the LDMOS.

The following advantages are offered by the lateral construction:

1. It has a very low zero-temperature coefficient drain current, I_{DZ}. For example, an 8 amp, 200 volt

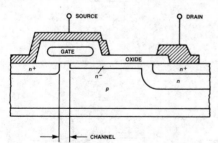

Cross-Sectional View of Idealized Lateral DMOS Structure
Figure 1

LDMOSFET can exhibit an I_{DZ} of 100 mA compared to a more conventional VDMOS I_{DZ} of 3 amps.

2. The gate-to-drain capacitance, C_{rss}, is low. A low C_{rss} offers the advantage of high switching speeds combined with high-frequency operation.

3. The drain is isolated from the case.

The disadvantages of lateral construction are:

1. It has a severe geometric constraint and an inefficient utilization of chip geometry. This constraint tends to limit the high voltage capabilities because limitations are imposed by the topology on the use of field shaping techniques.

2. As with the small-signal lateral MOSFET, the depletion region of the *pn* body-drain diode spreads laterally and thus adds to limitations in topology.

Reference

Richman, Paul, *MOS Field-Effect Transistors and Integrated Circuits,* New York, John Wiley and Sons, Inc., 1973.

2.7 Vertical DMOS

The disadvantages recognized as inherent in the lateral DMOS structure have paved the way for industry to examine the vertical DMOS FET (DMOS).

The DMOS structure is conceptually similar to the LDMOS transistor except the drain contact lies on the backside of the wafer. Consequently, current flow is more vertical than horizontal as the cross-sectional view in Figure 1 illustrates.

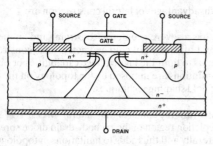

Figure 1

The advantages of the DMOS structure, when compared with the lateral (LDMOS) structure, are simply the elimination of the disadvantages outlined previously for the LDMOS. However, although we gain in chip topology utilization, we do add another topology-related limitation which becomes increasingly more pronounced as the current flow rises.

The electrical model of the vertical DMOS, although basically the same as any other DMOS structure, does offer an anomaly that distinguishes it from all other structures. As shown in Figure 2, there appears to be

a series-connected JFET in the drain because the physical relationships between adjacent p-diffusions, acting as gates, pinch the current flow in its vertical descent to the backside drain. A simplified view of this restriction is shown in Figure 3. The spacing between the adjacent p-diffusions can be crucial in the performance of a high-power (current) DMOS FET. A more detailed analysis appears in Section 2.13.

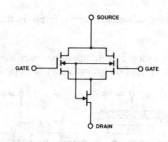

Figure 2

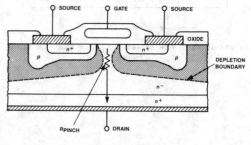

Figure 3

References

Hu, C., "A Parametric Study of Power MOSFETs," *Conference Record* PESC, San Diego, CA, 1979.

Lidow, A., T. Herman and H. Collins, "Power MOS-FET Technology," *Technical Digest,* IEEE Electron Devices International Meeting, Washington, D.C., 1979, pp. 79-83.

Lisiak, K. and J. Burger, "Optimization of Nonplanar Power MOS Transistors," *IEEE Trans-Electron Devices* 25 (1978), pp. 1229–34.

2

2.8 The Bilateral MOSFET:
A New Device Sees New Applications

For many years there has been a need for a semiconductor capable of general purpose AC control. While there have been, and continue to be, some special purpose devices available which will do part of the job, no truly general purpose device for AC control was available. This situation has just changed. Advances in the field of MOSFET technology have now produced the first truly general purpose AC semiconductor: The Bilateral MOSFET.

Until now, engineers wishing to control AC had numerous problems and few solutions. For low frequency AC switching, one could use thyristors, but they do not work at high frequencies and are hard to turn OFF. For switching both ON and OFF, one could use a bipolar transistor inside a diode bridge, but this is inefficient, cumbersome, expensive, and slow. For linear low-level applications, special attenuator and MUX ICs are available, but these function at low power levels only. For either linear or switching control of high level, high frequency AC, you were generally out of luck. Now, the bilateral MOSFET can handle all these jobs.

Bilateral MOSFET switches are a new product resulting from recent advances in power MOS technology. In terms of construction, the bilateral switch may be thought of as two *n*-channel DMOS power FETs with their source and gate leads connected in common. In fact, the first bidirectional devices were made by connecting two discrete power FETs in this fashion. The device that resulted from this pairing had such impressive characteristics that unified bilateral switches are now being manufactured.

Reprinted with permission from Electronic Design, September 6, 1984; copyright Hayden Publishing Co., Inc., 1984.

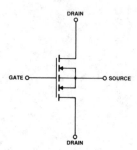

Bilateral MOSFET–Symbol
Figure 1

New bilateral switches, such as the Siliconix BLS100 family, give the designer requiring analog switching, subscriber line switching, or AC power conduction, a nearly ideal combination of characteristics which evades many of the compromises required by the approaches now in use. These *n*-channel enhancement mode devices have no inherent offset voltage and can transmit AC or DC signals with equal facility. Furthermore, the impedance in the ON-state is resistive, providing very low harmonic distortion.

In operation, when a positive voltage greater than the threshold voltage (V_T) is applied to the gate (with respect to the source), both FETs turn ON and conduct in series via their channels (see Figure 1). When the gate signal is removed, both channels turn OFF and conduction ceases. The source-to-drain diodes of the individual FETs do not conduct in reverse as they do in a unidirectional MOSFET because such conduction is blocked by the diode in the other half of the device.

Either of the diodes (depending on the drain-to-drain voltage polarity) may conduct when the device is turned ON but it will only conduct when the voltage drop across half the channel resistance exceeds the forward voltage of the diode. When one of the diodes does turn ON, it clamps the voltage across its associated half of the channel. This feature both reduces power loss in the bilateral switch at high power levels and serves as a surge protection in signal applications. Both diodes will conduct from the source to their respective drains if the source is biased above drain potential, a condition which should be avoided. Note that despite the fact that the two halves of the device are *n*-channel and operate in the enhancement mode, the resulting device is symmetrical and conducts identically in both directions.

Good gate drive for the bilateral MOSFET switch requires some new, yet not difficult techniques. The gate-source terminal pair may be viewed as a voltage-sensing capacitive load exactly as in a unidirectional MOSFET. Like the unidirectional MOSFET, input capacitance (C_{iss}) will vary with chip size, and thus, with voltage and current ratings. Bilateral devices will have approximately double the input capacitance of unidirectional devices of equal rating. The available threshold voltage (V_T) range will also be the same as for unidirectional devices and is well controlled on the bidirectional devices to assure symmetry.

There is one striking difference between the bilateral MOSFET switch and any other device (see Figure 2). When the device is ON, the source terminal (gate return) is electrically in the center of a resistive divider formed by the two gate channels, and when the device is OFF, the source is referenced one diode drop above the more negative drain terminal. In many cases it will be necessary to use an isolated gate drive in order to avoid common mode effects. This seems a fair trade for true symmetrical AC capability and may, in some cases, be a convenience.

Today, isolated gate drive presents few problems, both because of a proliferation of devices meant to accomplish this function and because FETs require so little static drive power. The designer has a choice of optical, transformer, or capacitive isolation, as well as combinations thereof. Each type of drive has its own properties. The choice of drive will depend on the system's requirements.

Transformer isolation may be either linear or switching in nature. Linear transformer drive is usually accomplished by rectifying a variable amplitude high frequency tone sent through the transformer. Transformer switching drive is accomplished by driving the transformer at two opposite polarity voltage levels: one for ON, the other for OFF. Note that proper volt-second balance must be maintained in the transformer, or the core will saturate. To prevent saturation, the ExT applied to the primary in one direction must equal the ExT applied in the opposite direction. By making the OFF voltage greater than the ON voltage, ON intervals greater than 50% of total period can be achieved. Using 40 V gate breakdown FETs, this technique allows ON intervals to reach 80% of total period while maintaining proper volt-second balance in the transformer. Automatic control of volt-second balance in the drive transformer may be achieved by placing a suitable capacitor in series with either the primary or secondary of the transformer. This will also reduce the reverse voltage stress on the gate oxide when duty cycles are below maximum.

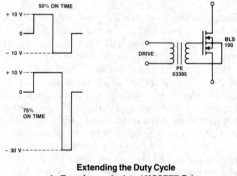

Extending the Duty Cycle
of a Transformer Isolated MOSFET Drive
Figure 3

Transformer drive, especially when using shielded transformers, offers the best common mode rejection ratio (CMRR). However, transformers tend to be bulky and expensive, and if it is necessary to build them to UL or VDE insulation requirements, performance (primarily coupling) will suffer significantly, especially at higher frequencies.

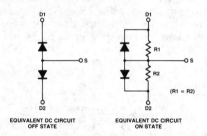

Source Reference
Figure 2

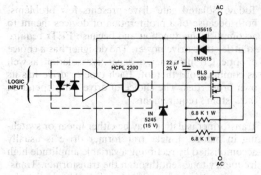

Parasitically Powered, Optically Isolated Drive
$T_r < 75$ nsec $T_f < 50$ nsec   120 VAC Switch
Figure 4

Optically coupled gate drive, likewise, may be either linear or switching in nature (see Figure 4). Generally, optical drives are smaller and less expensive than transformer drives and somewhat slower. Modern, shielded, optocouplers do, however, have good CMRRs, approaching the best transformers. Furthermore, units with sufficient voltage swing to drive a FET with switching times in the tens of nanoseconds are now available. Even using such high-speed devices, optocoupler speed remains the primary limitation on system switching speed. Sophisticated optical drives made from discrete PIN photodiodes and fast amplifiers can improve switching speed significantly when the extra speed is found to be worth the extra cost.

Linear optical drives suffer primarily from nonlinearities in the optocoupler. These nonlinearities may be evaded by several means. The most common system is executed by using two matched optocouplers, each of which serves as a local feedback path to compensate for the nonlinearities in its companion. Similar, but more elaborate, systems can be made which are adequate for precision applications. Again, the primary limitation on the small signal gain bandwidth of linear optocoupling in terms of FET drives is cost.

Capacitively isolated drives are both an old system and a very new innovation. New technology (includ-

ing special purpose ICs) has evolved making old methods more practical. Modern capacitively isolated switching drives consist of a gated oscillator driving a rectification circuit through two matched capacitors (see Figure 5). The frequency of the oscillator must be high, perhaps 100 times the desired switching frequency. ON/OFF control of the MOSFET is achieved by gating the oscillator ON and OFF. The isolator in Figure 5 has a 1 MHz oscillator frequency and works well switching signals of ≤ 10 kHz and ≤ 30 VAC. Some telecommunications companies are studying this method to control bidirectional MOSFET switches for signal switching. Linear control could be achieved similarly by amplitude modulation of the oscillator.

There are still limitations to the technique, however. CMRR is low, as would be expected. Capacitors, after all, conduct AC equally well in both directions. Presently available systems are better suited for linear controls, or for switching under tightly constrained circumstances than for high power switching systems where large common mode voltage transitions can be expected. If the load frequency being switched is too high, spurious turn-ON of the bilateral MOSFET may result. Present devices, which use a 1 MHz oscillator, can control frequencies of 5 to 10 kHz at levels suitable for telephone switching, but a 50 V_{RMS}, 20 kHz signal, for example, would make some of the present units appear very leaky. The necessity of running the oscillator at frequencies much higher than the maximum frequency of the switched device also places a constraint on either the switching frequency of the main switch or the cost of the oscillator. Faster chips can be expected to alleviate this problem in the future. High DC isolation voltages are also a problem. The dielectric materials available for capacitors included in integrated circuits now place a limit on the maximum DC standoff voltage of such capacitors at 500 volts DC. Improvements in standoff voltage can be expected to come slowly.

To recap briefly, with optically isolated drive, you can perform any drive function except very rapid switching. With transformer isolated drive, you can perform any drive function except keeping the switch in the

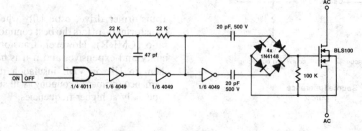

Capacitor Isolated Gate Drive Driving a Bilateral Switch
Figure 5

ON state indefinitely. Transformer drive requires the most volume. By using capacitor-isolated drive, you can reduce costs, but only if you can accept limitations on load frequency and slower switching.

One of the most obvious applications for the bilateral MOSFET is in the Venturini converter. The Venturini converter is an elegant voltage-independent, frequency-independent, bidirectional N-phase to M-phase switching converter. For example, a Venturini converter can be built to convert 3-phase 60 Hz power to 3-phase 400 Hz power bidirectionally, in a single stage, at well over 80% efficiency while regulating output voltage and input power factor. Venturini converters can also be built to convert 3-phase power to 2-phase, 5-phase, or any other number of phases. They can be built to change frequency to either a higher, lower, or variable frequency, including DC. Because Venturini converters are bidirectional, they will also function to synthesize multiphase sinusoidal power from DC or single-phase AC. They may also be used as regulators and power factor correctors. They may also be built as bidirectional synchro-to-resolver converters, or, with an extra A to D stage, as an either-to-digital converter. The Venturini converter is probably as close to being a power anything-box as any currently known device.

Until now, the Venturini converter has had two horrendous drawbacks which limited its applicability. First, an N-phase to M-phase Venturini converter requires NxM AC power switches with individually isolated drives. In practice, this means using 2(NxM) unidirectional switches, each with its own drive and protective power diode. For a 3-phase to 3-phase converter, this adds up to 18 assemblies, each consisting of a power transistor, its drive electronics, a fast power diode with the same voltage and current rating as the transistor, and an isolated driver. The resulting system is incredibly complex, and only those who truly need the multifunctional capabilities of the Venturini are willing to bear the cost.

This problem is now solved. Bilateral MOSFET switches conduct AC naturally and require a much less complex drive (though isolation is still necessary). In a 3-phase to 3-phase converter using bilateral MOSFET switches, several reductions take place: nine power components replace 36 power components; nine drivers replace 18; and drive complexity, cost, and volume are reduced.

The second drawback to the Venturini converter is the lack of optimized control algorithms. Progress in solving this problem is more likely now that a practical

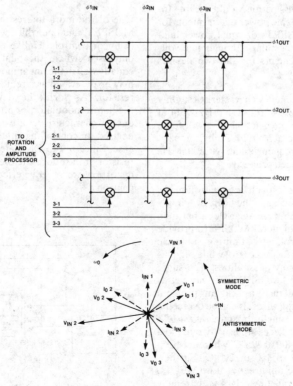

3-Phase to 3-Phase Venturini Switch Matrix
Figure 6

converter can actually be built. Existing algorithms, though limited, give a reasonable basis for beginning.

Besides the multiphase Venturini converter, there are many single phase and DC applications for the bilateral switch. Heretofore, the only AC conducting power semiconductor available was the triac, which, while it is useful in a variety of applications, has its drawbacks. Although many multi-device assemblies exist which will conduct AC power, most of these have been designed for specific instances where triacs were unsuitable and few have general applicability. Most multi-device approaches have their own shortcomings.

The most obvious problem with a triac is turning it OFF. To do this, the current through the device must be reduced below some temperature and geometry dependent minimum—either by means of external circuitry or by waiting until the AC line crosses zero. The necessity of maintaining a minimum "holding current" through a triac when the device is to stay ON usually limits the dynamic range of controllable AC signals to less than 150:1 (unless a DC bias is used). The bidirectional MOSFET, of course, may be turned OFF at any time, regardless of load current, simply by removing the gate signal. Furthermore, the bidirectional MOSFET can control AC signals of any dynamic range that a resistor of equal ON-resistance ($R_{DS(on)}$) and power dissipation would accept. Interestingly enough, total gate power requirements for triacs and bilateral MOSFETs of equal power handling capability are quite similar although gate voltage and current requirements for the two devices differ drastically.

Triacs also have poor frequency characteristics. Only selected units will phase control at 400 Hz without spurious turn-on, and no production triac will function beyond 1 kHz. Bilateral MOSFETs, by comparison, will function at frequencies above 100 MHz providing they are packaged properly.

Bilateral MOSFETs are capable of controlling as much power as a triac. Because they can switch ON and OFF many times within an AC cycle rather than twice as a triac would, bilateral MOSFETs provide much more flexible control. Surge limiting, sub-cycle load shaping, and power factor control while regulating are only three of many possibilities. The surge characteristics of bilateral MOSFETs are not yet as good as those of triacs, but this is a packaging problem caused by the chip bonding wires. It can be overcome by revised packaging. Thermally, bilateral MOSFETs are far superior to triacs. Triacs are usually rated to a junction temperature of 100 or 125°C, whereas MOSFETs are usually rated to 150 or 200°C. Further, bilateral MOSFETs, like unidirectional MOSFETs, are easy to parallel. Paralleling triacs is extremely difficult.

Bilateral MOSFETs are occasionally being used instead of triacs as pseudo-stepdown transformers to provide power for low voltage tungsten filament lamps and other similar low voltage devices. Bilateral MOSFETs appear to be superior in this service to either triacs or transformers when stepdown ratios exceed approximately 5:1. Transformers with large stepdown ratios can become difficult to wind. Triacs operated at high reduction ratios produce small, widely-spaced pulses which are not properly averaged by small filaments having comparatively short thermal time constants. Filament life may also increase because the peak-to-average power ratio is substantially reduced when power is applied every 50 μsec instead of every 8.3 msec. Electromagnetic Interface (EMI) also becomes easier to deal with because bilateral MOSFETs switch much faster than triacs. Triacs are subject to across-the-chip propagation delays when switching which means they actually switch slowly. The higher frequencies resulting from MOSFET switching can be filtered more easily using smaller filters.

The absence of across-the-chip propagation delays at turn-on in bidirectional MOSFETs also means that their di/dt characteristics are superior to those of triacs. The improvement is two or more orders of magnitude.

One of the more interesting areas for the bilateral MOSFET, and one which will reward exploration, is in linear controls. Unlike the triac, which functions only as a switch, the bilateral MOSFET functions equally well in the linear mode. In essence, the device may be described as a remotely controllable variable resistor. As such, the device may be used as a regulator, modulator, or attenuator. Because its frequency characteristics are very good, the device may be used for controlling either audio or RF with equal ease. Since in the ON-state the device virtually is a resistor, it introduces neither significant nonlinearity distortion nor reactive effects into the AC signals it controls.

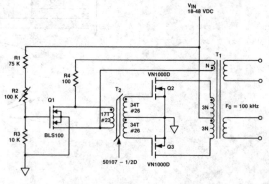

Frequency Control in a Jensen Converter
Figure 7

For example, in Jensen and similar saturating-magnetic converters, a bilateral MOSFET operating in the linear mode can be connected across a control winding on the saturating core to provide frequency regulation for the converter (see Figure 7). The floating drive, high gain, and true bidirectionality of the device mean few parts are necessary, and a simple, frequency stable, converter results. Similar remotely variable resistor controls may be used to stabilize or linearize a wide variety of other oscillator circuits as well. Amplitude stabilization of Wein bridge oscillators is an example.

Bilateral switches may also be used to tune oscillator circuits. In one current application, the devices are being used to short individual turns in an RF coil in a transmitter, providing an extremely wide tuning range without the necessity of swapping coils. When properly packaged, the devices will function in this mode at frequencies well above 100 MHz.

Another interesting control scheme dealing with magnetics and bilateral MOSFETs involves using one bilateral MOSFET device as a current source/sink to provide reset current to a pair of magnetic amplifiers used for post regulation on an auxiliary output of a switching converter (see Figure 8).

In a multi-output switchmode converter, usually only the main (highest power) output is directly regulated by the PWM IC. The smaller auxiliary outputs are regulated only by transformer coupling to the main output. When the regulation accuracy achieved that way ($\pm5\%$ at best) is not sufficient, mag amp post-regulation is a cheap, simple, reliable way to obtain closer regulation without degrading system efficiency as a 3-terminal regulator would. $\pm0.5\%$ regulation is easy using mag amps.

In operation, the two saturating-core inductors, "mag amps" L1 and L2, function as switches (high impedance unsaturated, low impedance saturated) con-trolling the flow of power from the secondary to the output. On any given half-cycle of the SMPS, one mag amp is timing out and eventually saturates, conducting power to the output. Meanwhile, the other mag amp is being reset (pulled back *out* of saturation) a controlled amount by running current (generally robbed from the output) through it backwards. The amount the reversed biased mag amp is reset determines the length of time it blocks before it conducts on the following half cycle.

Reset current in the system in Figure 8 is controlled by the bilateral MOSFET, which replaces an op-amp, a bipolar transistor current source, and two steering diodes. Because the bilateral MOSFET is an AC device, it can be connected directly across the mag amps ahead of the rectification rather than after it as is normally the case. That way it uses the leakage current through the mag amp which is timing out to reset the other mag amp.

Because it is a true AC device, it continues to function. No reset current steering diodes are necessary—neither is a separate reset current supply. The high-gain and wide gain-bandwidth product of the bilateral MOSFET allows a fast, high-gain regulator loop to be built without additional amplification. Also, because the "set" current in one mag amp is equal to the "reset" current in the opposing core, such a post regulator will function down to zero output current without requiring a dummy load.

The inherent symmetry and linearity of bilateral MOSFET switches makes them useful for switching and attenuation in both high quality audio and tele-communications circuitry. An additional benefit for these applications is the low inherent noise of the devices. It is true that junction FETs have even less noise, but most new MOSFETs (bilateral switches included) run JFETs a close second and offer much greater power handling capability as well as increased dynamic range.

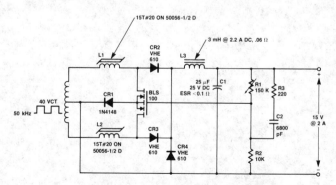

**Reset Mag Amp Post-Regulator Circuit Showing a Bilateral
MOSFET Operating in Linear Mode
Figure 8**

It should be noted, however, that a bilateral MOSFET will become nonlinear whenever the drain-to-drain ON-voltage exceeds approximatley 1.4 V peak in either direction (about 2 V_{RMS}). This happens because each of the two controllable channel resistances has a diode in parallel with it, and only one of these diodes is reverse biased at a time (see Figure 2). If the voltage drop across the half-channel that is in parallel with the forward biased diode exceeds the forward drop of that diode then the diode will conduct. This will shunt current away from that half of the channel and clamp the voltage drop across that half of the device to the V_f of the conducting diode. The other half of the device, where the diode is reverse biased, continues to function as a variable resistor. This tends to limit the amount of distortion that can be generated but will not prevent such distortion from occurring. In AC signal switching systems, this distortion may be detectable during the switching interval. Fortunately, with proper gate drive, the switching intervals can be made short enough that distortion from this cause will probably not be detectable. For high-level attenuators, the problem can be controlled by balancing a series element with a shunt element, but this will not correct the nonlinearities in impedance that would occur. This cannot, therefore, be considered good practice.

It should be noted that in view of the low $R_{DD(on)}$ of the bilateral MOSFET, the 1.4 V drop across the device required to initiate this type of distortion represents a signal better than 90 dB above OFF leakage. This figure is for a small device suitable for audio switching. In larger devices or devices selected for low leakage, the range before distortion onset becomes even larger.

In summary, a new device with a very useful and novel set of properties is now available for applications which require symmetrical conduction with low distortion or AC conduction to high power levels. The device has excellent high frequency characteristics both in terms of the currents it carries and the response rate of the device itself. These characteristics are achieved without sacrificing simplicity in the surrounding circuitry, ruggedness, or reliability. A new generation of systems requiring devices with these properties will soon be available.

References

Gardner, N.C., "Isolated Capacitor Drive," *Electronic Product Design,* August 1983.

Jensen, J.L., "An Improved Square Wave Oscillator Circuit," *Transactions of the IRE,* Vol. CT-4 (September 1957), pp. 276-279.

Mullett, C.E. and S. Smith, "Techniques for Optimizing the Design and Application of High Frequency Power Magnetics," *Proceedings,* POWERCON 7.

Venturini, M., "A New Sine Wave In, Sine Wave Out Conversion Technique Eliminates Reactive Elements," *Proceedings,* POWERCON 7.

2.9 Depletion-Mode Power MOSFETs: New Devices to Solve Old Problems

Introduction

Since their introduction in 1976, enhancement-mode power MOSFETs have enjoyed widespread acceptance by the electronics industry. These devices offer very fast switching times, linear on-resistance, and low average gate current requirements. They are used widely in applications ranging from audio-amplifiers to motor-drives and switched-mode power converters.

Since the beginnings of power MOS development, designers have concentrated mainly on perfecting enhancement-mode power devices and making them a cost-effective alternative to bipolar transistors. The result of these efforts is a wide variety of enhancement-mode parts, available from several manufacturers, with power ratings up to 250 watts.

Certain potential applications of power MOSFETs necessitate that a depletion-mode (or normally-ON) MOSFET be used, rather than an enhancement-mode (or normally-OFF) device. Depletion mode devices are not new; they have been available in the form of small-signal JFETs for a number of years. Unfortunately, the limited voltage and current ratings of most common JFETs restrict their use to low power applications where device dissipation is less than 1 watt.

Many applications could benefit from depletion-mode devices with higher ratings. Until now, such devices have not been available. Consequently, designers seeking depletion-mode devices have often had to resort to less satisfactory solutions for their particular applications.

Reprinted with permission from Electronic Design, Vol. 32, No. 13; copyright Hayden Publishing Co., Inc., 1984.

The introduction by Siliconix of the NOS100 n-channel depletion-mode power MOSFET marks a new step in the evolution of power MOS device technology. This device can withstand a drain to source voltage of up to 150 volts, and it conducts a drain current of typically 500 mA with zero applied V_{GS}. The TO-39 packaged NOS100 is a fairly low power device by present day enhancement-mode MOSFET standards (P_{DISS} = 6.25 W max.), but it is only the first member of a new family of depletion-mode products. Ultimately, Siliconix will design and manufacture depletion-mode MOSFETs in a variety of packages with power ratings extending up to 150 watts.

The majority of uses for depletion-mode devices will be those in which the characteristics of an enhancement-mode power device are not suited to the circuit at hand. These applications often require some form of logic compatible AC or DC power switch that remains ON when the logic supply fails and V_{GS} drops to zero.

Other applications may also benefit from the use of depletion-mode MOSFETs to replace existing devices in new designs. By reducing the component count in certain cases, both a savings in circuit cost and overall board space may be achieved.

This article provides an introduction to the operation and fabrication of depletion-mode power MOSFETs. Following this, two applications representative of the majority of the uses for these devices are presented. Because they possess unique char-

2

acteristics compared to enhancement-mode MOSFETs, many of the applications for depletion-mode devices will likely be unique as well. Rather than serve as replacements for enhancement-mode devices, the NOS100 and its quad version, the VQ5000, are intended more as complements to the existing range of Siliconix power MOSFETs.

Depletion Mode Versus Enhancement Mode

Because the device structure of a depletion-mode power MOSFET like the NOS100 is so similar to that of an enhancement-mode power device, one would expect their operating characteristics to be similar as well. In fact, a depletion-mode MOSFET may be thought of as an enhancement-mode device with a negative threshold voltage (V_T). Figure 1 compares the saturated I_D versus V_{GS} transfer characteristics of each device. Both types of MOSFETs exhibit a square-law relationship between saturated drain current and gate to source voltage, but the depletion-mode device conducts substantial current with $V_{GS} = 0$. This zero V_{GS} drain current is termed I_{DSS}, and is proportional to the square of the threshold voltage.

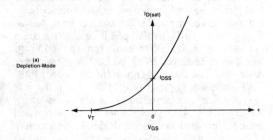

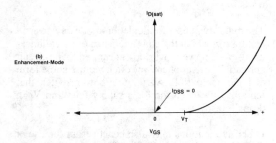

The Difference Between an *n*-Channel Depletion-Mode MOSFET and an *n*-Channel Enhancement-Mode Device, is that I_{DSS} is Substantially Greater than Zero for the Former, and Equal to Zero for the Latter
Figure 1

Unlike a JFET which is a depletion-mode only device and can accept only a single polarity of applied V_{GS}, the NOS100 can accept either a negative or positive V_{GS}. With a negative applied V_{GS}, the NOS100 works as a depletion-mode device with $I_D < I_{DSS}$. But with V_{GS} positive, the NOS100

functions as an enhancement mode device and I_D can be made larger than I_{DSS} if desired. This dual-mode operation is illustrated in Figure 2. One can see that with $V_{GS} = 0$, the NOS100 behaves as a linear resistor of just a few ohms in value when V_{DS} is reduced to a low level.

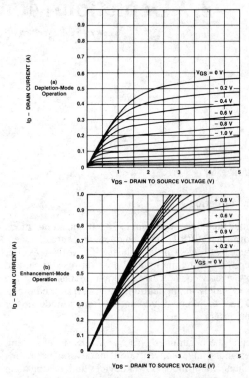

The NOS100 Functions Both as a Depletion-Mode MOSFET (a) with $V_{GS} < 0$, and an Enhancement-Mode Device (b) with $V_{GS} > 0$
Figure 2

It is important to note that the on-resistance ($R_{DS(on)}$) of a NOS100 may vary substantially from device to device with zero applied V_{GS}. This variation can be as large as 100%. Enhancement-mode devices with similar specifications, on the other hand, may show only a 20 to 30% maximum variation in $R_{DS(on)}$.

Unexpectedly, this behavior results from the differing test conditions used when measuring $R_{DS(on)}$ for each type of device. Enhancement-mode power MOSFETs generally have a threshold voltage of 2 to 4 volts, and are tested for $R_{DS(on)}$ at $V_{GS} = 10$ volts. The effective enhancement voltage ($V_{GS}-V_T$) can therefore be as large as eight volts, but is no less than six volts.

The NOS100 and other depletion-mode power MOSFETs, however, are normally tested for

RDS(on) with V_{GS} equal to zero volts. Since the typical threshold voltage variation of the NOS100 is from -2 to -4 volts, the effective enhancement voltage ($V_{GS} - V_T$) is then only $+2$ to $+4$ volts.

For most power MOSFETs in the 100 to 150 volt range, the resistance of the conducting channel comprises 25 to 40% of the total device on-resistance. The conducting channel resistance (R_{CHAN}) can be shown to be inversely proportional to the effective enhancement voltage ($V_{GS}-V_T$). Consequently, an enhancement-mode MOSFET will have an R_{CHAN} that varies only by about 30% over its 2 to 4 volt threshold range. A depletion-mode MOSFET with a -2 to -4 volt threshold, however, will exhibit an R_{CHAN} that varies by 100% over this range. It is this variation in R_{CHAN} that accounts for the wider spread in $R_{DS(on)}$ for the NOS100 and other depletion-mode power MOSFETs.

Applications

1. A two-terminal adjustable current regulator

Many analog circuits require a constant current to be generated that is independent of supply voltage. This current may be used to bias an amplifier stage, such as a differential transistor pair, or may be used to charge a timing capacitor which generates a sawtooth voltage waveform.

Two-terminal current regulator diodes such as the Siliconix CR series of devices have been available for a number of years. While these devices are very useful in certain applications, they are limited by their maximum voltage and current ratings (100 volts and 4.7 mA, respectively). Clearly, if constant currents larger than 4.7 mA are required, and at higher supply voltages, the designer must choose some form of discrete regulator circuit.

Until now, the best way to implement a high-voltage two-terminal adjustable current-regulator was with an enhancement-mode MOSFET and a bipolar transistor as shown in Figure 3. This circuit employs

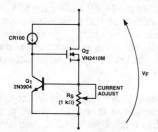

The Two-Terminal Adjustable Current-Regulator Uses Four Components Including an Expensive Current Regulator Diode
Figure 3

a standard 1 mA current-regulator diode (CR100) that supplies a constant collector current for the bipolar reference transistor Q_1. This transistor sets a constant voltage drop of one V_{BE} across resistor R_S. By varying the value of R_S, adjustable output current is obtained. This circuit works well for currents up to several hundred mA, but its sustaining voltage is only equal to the breakdown voltage of the current-regulator diode.

By connecting the gate terminal of an NOS100 back to the other end of a source degeneration resistor, as shown in Figure 4, a simple adjustable current regulator is obtained. This two-component regulator performs just as well as the circuit in Figure 3, but uses fewer components and requires less board space.

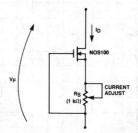

The NOS100 and One External Variable Resistor Form a Simple but very Useful Two-Terminal Current Regulator
Figure 4

Because the negative gate bias for the NOS100 is generated by the drain current flowing through R_S, one can solve analytically for the value of this resistor given the desired I_D and the two parameters V_T and I_{DSS}. To accurately determine R_S, however, both V_T and I_{DSS} must be known accurately as well. This necessitates measurement of V_T and I_{DSS} for every device, since these parameters may vary widely from device to device.

Rather than using the analytical approach to determine the required value of R_S given a desired value of I_D, a simpler method would be to replace R_S with a multi-turn trimpot. This trimpot can then be adjusted until the desired operating current is reached.

With R_S being the gate bias resistor, the maximum possible output current will be equal to I_{DSS}. This current is reached when R_S equals zero. It is not possible to obtain values of I_D in the low μA range, unless the maximum value of R_S is quite large (around $1 M \Omega$). This behavior occurs because the V_{GS} of the NOS100 is equal to $- I_D R_S$. For V_{GS} to approach V_T, progressively larger values of R_S are required. But V_{GS} can never be equal to V_T because I_D would then be equal to zero and there would be no gate bias—a paradoxical situation.

2

In most real applications for the circuit of Figure 4, the required output current will probably be in the range of a few mA to several hundred mA. The 1kΩ trimpot chosen for R_S, provides an I_D adjustment range of approximately 5 mA to 500 mA.

To illustrate the performance of the NOS100 as a current regulator, the 1 kΩ trimpot used in the circuit of Figure 4 was adjusted until I_D was equal to 30 mA at a V_F of five volts. Figure 5 shows how the output current varied when V_F was varied between 0 to 200 volts. As can be seen, the output current changed less than 1 mA (or 3.3%) resulting in a high output resistance of approximately 200 kΩ.

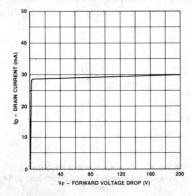

The Two-Terminal Current Regulator Using a NOS100 Performs Very Well Over an Applied Voltage Range of 0 to 200 V
Figure 5

When the current regulator is to be operated at moderate current levels with large sustaining voltages, a small clip-on heatsink should be attached to the NOS100 because the temperature coefficient of I_D is positive. This recommendation was followed for the 30 mA regulator described above.

2. A normally-ON solid-state AC switch

Solid-state AC switches that can handle reasonable amounts of power have been implemented with a variety of semiconductor devices such as bipolar transistors, SCRs, triacs and MOSFETs. Of these devices, only MOSFETs offer high breakdown voltage, low leakage, and linear on-resistance down to zero volts.

If a solid-state AC or "bilateral" switch is built using enhancement-mode MOSFETs, then the switch is by nature normally-open or normally-OFF. This means that when the gate bias voltages for the MOSFETs are lost due to failure in the gate-drive circuitry (or its power supply), the switch turns off. In many AC switching applications this characteristic behavior is acceptable.

Certain applications, particularly those in the telecom domain, often require a logic controllable AC switch with high breakdown voltage as well. The major difference here is that the switch must remain on whenever there is a failure in the gate drive circuitry or its power supply. Until now the only way of obtaining normally-ON switching action, combined with high breakdown voltage and moderate current handling capability was with electromechanical relays. The disadvantages of relays are numerous and include large drive power requirements, excessive EMI generation, finite contact lifetime and slow switching speed. Depletion-mode power MOSFETs with their normally-ON resistive characteristics provide a solution to the problem of building a normally-ON solid-state AC switch.

The VQ5000 depletion-mode MOSFET array contains four devices. These transistors can be used as the basic switching elements in a dual normally-ON solid-state AC switch shown in Figure 6. When the sources and gates of each pair of MOSFETs are connected together, two bilateral switches are formed. This switch configuration can conduct current and block voltage in both directions.

The rest of the circuitry shown in Figure 6 comprises a capacitive isolated driver stage that generates logic controllable DC bias for the two bilateral switches. Because the MOSFET's gate and source terminals are isolated by two capacitors from the logic circuitry, the AC signal being switched need not be referenced to the logic power supply ground. This is a very useful feature of the circuit.

Each half of the driver circuit in Figure 6 is identical, and its operation can be understood if the signal path is traced from the LM555 output to the gates of the MOSFETs.

The LM555 is connected as a free running oscillator and generates a rectangular waveform at approximately 100 kHz. This signal is fed to the inputs of a Siliconix D469 CMOS quad driver IC. Each pair of drivers within the D469 buffers the 555 output, and generates antiphase 100 kHz outputs that swing between ground and 12 volts. The remaining inputs of the two pairs of drivers are connected to control logic that allows the D469 outputs to be gated on or off as required.

The antiphase D469 outputs are fed into a pair of AC coupling capacitors that drive a bridge rectifier comprised of fast small-signal diodes. Both outputs of the bridge rectifier are connected to the bilateral switch, such that when the outputs of the D469 are pulsing

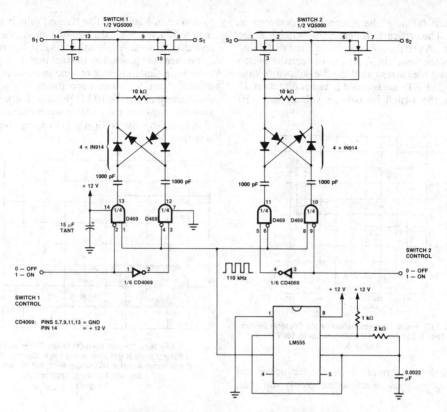

**A VQ5000 Quad Depletion-Mode MOSFET Array, a D469 Driver
and a 555 Timer Form the Basis for an Isolated, Dual Normally
On-Solid State Switch That Is Logic Compatible
Figure 6**

the V_{GS} of the MOSFETs is negative. Under these conditions the bilateral switch is OFF.

When the logic control signal gates the antiphase D469 outputs off, the output of the bridge rectifier drops to zero. Consequently, the effective gate to source capacitance of the MOSFETs discharges through the 10 kΩ resistor, and the bilateral switch turns on. The switch also turns on when the logic/driver power supply fails, since the output of the bridge rectifier is then zero as well.

Figure 7 shows the on-state characteristics of the depletion-mode bilateral switch. The vertical axis in Figure 7 is the switch current in A and the horizontal axis is the voltage drop across the switch in V. One can see that the switch on-resistance is equal to the reciprocal of the slope of the trace in Figure 7, and is approximately 7.5Ω. It should be noted that the on-resistance remains very linear through the zero-crossover region in the center of the photograph. As the voltage across the switch increases in either direction, the current increases linearly up to a point but then levels off at a

constant value. Not surprisingly, this value of limiting current is equal to the I_{DSS} of the MOSFETs. The bilateral switch thus has built in current limiting which is a useful self-protecting feature.

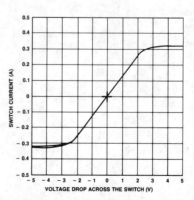

**The On-State Transfer Characteristics of the Normally-On
Bilateral Switch Show a Constant $R_{DS(on)}$ Through the Zero
Crossover Point
Figure 7**

The switch off-state characteristics are shown in Figure 8. The vertical axis here is the switch leakage current in μA and the horizontal axis is the sustaining voltage of the switch in V. As can be seen the switch leaks negligible current until the breakdown voltage of the MOSFETs is exceeded in either direction. In this case the switch breakdown voltage is $\pm$180 volts.

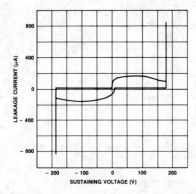

The Off-State Transfer Characteristics of the Bilateral Switch Show That It Can Block AC Voltages Up to 180V Peak
Figure 8

The turn-on and turn-off times for the driver and switch configuration described are greatly different, as shown in Figure 9. The turn-on time is approximately 50 μsec, while the turn-off time is substantially less than this value (about 10 μsec). The observed difference is explained by the fact that during switch turn-on, a finite amount of time is required for the effective gate to source capacitance ($C_{GS(eff)}$) to discharge through the 10 kΩ resistor. During switch turn-off, however, more than enough current is available to rapidly charge $C_{GS(eff)}$ to a negative voltage less than V_T.

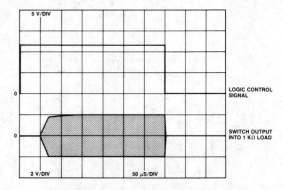

The Turn-On and Turn-Off Delay Times of the Bilateral Switch Relative to the Logic Control Signal (Top-Trace) Are Not Blindingly Fast, but Nonetheless Adequate for Many Applications
Figure 9

2.9.1 Fabrication of Depletion-Mode Power MOSFETs

Combining low voltage depletion-mode MOS technology with high voltage MOSPOWER technology has produced a new type of power device—the depletion-mode power MOS transistor. This new class of transistors represents a merging of the strengths of both of its parents. Its announcement is the result of an 18-month effort to obtain a delicate balance between the need to turn the device off when a low negative voltage is applied to its gate (i.e., to pinch off), while having a low on-resistance when no voltage is applied to the gate. The hurdles that were overcome in the development program may be understood by referring first to Figure 1. This figure shows the cross section of both a low voltage enhancement-mode and a low voltage depletion-mode MOS transistor. The threshold voltage of a MOS transistor is determined by a number of process parameters. For a given gate conductor (usually aluminum in a metal-gate process and phosphorus-doped polycrystalline silicon in a silicon-gate process), the threshold voltage is set largely by the dopant profile in the body region. The low voltage enhancement-mode transistor of Figure 1(a) is converted to a depletion-mode transistor by locally adjusting the body dopant using an ion-implantation. For the n-channel transistor of this figure, a phosphorus implant is used.

The development of a high voltage, n-channel MOSPOWER transistor requires considerably more process optimization than is shown in Figure 1. The dopant profile through the double diffused source, body and drain regions of a vertical DMOS device is shown in Figure 2. In this structure, the channel length L is determined by the difference between two sequential diffusions moving in the same direction

Reprinted with permission from Electronic Design, Vol. 32, No. 13; copyright Hayden Publishing Co., Inc., 1984.

from a common origin. The channel length L is reproducibly controlled to values in the 1 to 2 μm range. A small value of L results in both a large current per unit area and a low value for the device channel resistance.

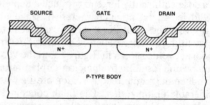

a. Enhancement-Mode

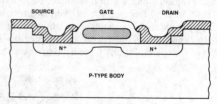

b. Depletion-Mode

The Cross Section of an NMOS Silicon Gate Transistor
Figure 1

In the DMOS power transistor structure, the body region is more heavily doped than the n-drain region as shown in Figure 2. With a reverse bias across this junction, the resulting depletion region extends further into the drain region than the body region. This behavior allows high voltages to be placed across the body-to-drain junction without significantly affecting the channel length of the DMOS transistor.

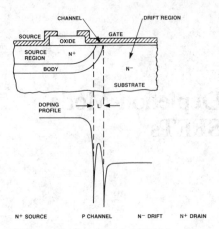

The Doping Profile in a Vertical DMOS Transistor
Figure 2

The first attempt to produce a depletion-mode power MOS transistor added only an ion-implantation step to the normal process sequence. Transistors produced in this fashion did exhibit the required depletion-mode characteristics, but suffered from two problems. First, these transistors could not be turned off with less than −10 volts from gate to source. In addition, the breakdown voltage was considerably less than desired. The use of only an ion-implantation with a dose sufficiently large to produce a depletion-mode device altered other characteristics in an unacceptable manner. Analysis of these results lead to an understanding of the problem. The lateral nature of the dopant profile shown in Figure 2 and the vertical dopant profile resulting from an ion-implantation do not produce the required body region profile without considerable optimization. The peak concentration in the body must be reduced while the remainder of the body region concentration is impacted in a minimal fashion. This balance was arrived at through an iterative procedure. Computer simulations were first used to home in on the body dopant profile. The results of these simulations were used to devise both

diffusion cycles and the specific ion-implantation dose and energy. These profiles were verified experimentally on existing device geometries and the resulting transistor characteristics were measured. This information was used as a starting point for another round of simulations and experiments.

While the iterations were taking place in the process area, device layout was started in design. The device was designed to meet a 150 volt breakdown and a 5 ohm on-resistance specification. The design was completed as the process experiments were drawing to a close, so the mask set was run using the final process sequence. The culmination of this work is shown in cross section in Figure 3 where an enhancement-mode device is compared to a depletion-mode transistor. A photograph of the device is shown in Figure 4. This device was designed for a specific application, but the process was developed so it would be applicable to the complete line of power DMOS transistors. This goal was met, and the NOS100 is the first device in a complete family of depletion-mode DMOS power transistors.

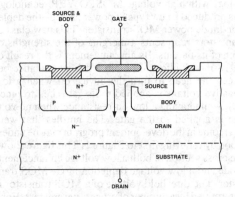

The Cross Section of a Depletion-Mode Vertical DMOS Power Transistor Shows the Conducting Channel with V_{GS} = 0V
Figure 3

Photograph of the Depletion-Mode Device (NOS100)
Figure 4

2.10 Other Structures

There are many alternative solid-state devices that can handle high voltages, high power, and respectable switching performance although perhaps not at the speed of the power MOSFET. Since the early 1950's, the Darlington pair has offered quasi-MOS-like performance for high gain and reasonably high input resistance. Their nemesis has been their poor V_{SAT} and long turn-OFF time. Thyristors, such as the ubiquitous SCR, have tremendous voltage standoff and power handling capability, but triggering and switching time have been their shortcoming. Early in the 1980's, a concentrated development effort appears to be rewarding industry with innovative power devices that, in some cases, have often encircled MOS technology and MOS performance in their designs.

One of several useful criteria for judging the effectiveness of solid-state power devices is to compare their current capability per unit area of silicon. Another criterion, of course, is simply to compare performance.

The composite MOS-bipolar transistor structure as shown in Figure 1, whether the Darlington configuration or one of several alternative configurations, has found widespread industrial acceptance. Although often touted as exhibiting ultra-low ON-resistance, much like the power bipolar transistor, their slow switching speed and attendant high switching losses (again, much like the power bipolar transistor) have severely limited their usefulness above 30 kHz.

More recent developments have focused attention on MOS-gated SCRs—variously called the Insulated-Gate Transistor (IGT), the High-Conductance MOS-FET, or Insulated-Gate Rectifier (IGR)—as shown

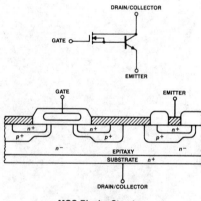

MOS-Bipolar Structure
Figure 1

in Figure 2. Although they exhibit an impressive current density, their principal shortcoming is, as with all composite MOS-bipolar structures, an excessive *storage time* or turn OFF delay.

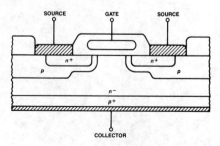

Insulated-Gate Rectifier (IGR)
Figure 2

Reference

Ghandi, S. K., *Semiconductor Power Devices,* New York, Wiley Interscience, 1977.

Schultz, Warren, "High Current FETs–A New Level of Performance," *Power Conversion International,* March 1984, pp. 43–6.

Smith, Marion, et. al., "Insulated Gate Transistors Simplify AC-Motor Speed Control," *EDN* (February 9, 1984).

2.11 Power Control with Integrated CMOS Logic and DMOS Output

Abstract

This paper examines the evolution and the present status of a new class of power integrated circuits— DMOS/CMOS or D/CMOS ICs. This type of circuit is a natural extension of the development of the discrete MOS power transistor that has occurred in the last few years. However, there are significant differences between D/CMOS ICs and either conventional MOS or bipolar circuits. The major difference is their ability to withstand and control voltages that are considerably greater than those of either circuit type. A second difference is their low power dissipation. Less obvious advantages also make D/CMOS ICs attractive in other applications. These advantages and their effect on future D/CMOS circuit trends are also discussed.

Introduction

The term "integration" as used in electronics means the incorporation of more functions in a physical unit. The trend to integrate more devices on a piece of silicon is one manifestation of this trend. In integrated circuit development, progress within a given technology is usually made in a very predictable fashion. An example of this type of progress is the storage capability of the largest commercially available RAM. Over the last fifteen years, RAM capability has doubled every one and one-half to two years from 1K in 1969 to 256 K today. Breakthroughs occur when new technology is injected into the mainstream of IC development, allowing previously unfeasible circuits to be developed. Such a breakthrough is occurring today as high current, high voltage DMOS tran-

sistors are being fabricated simultaneously with CMOS control logic to obtain DMOS/CMOS or D/CMOS power integrated circuits. This paper begins by quickly reviewing the history of DMOS or double-diffused discrete MOS transistors. Next, the development of the circuits presently available using D/CMOS technology is covered. D/CMOS circuits may be manufactured using different process sequences. The characteristics of each of the D/CMOS IC technologies are compared. Finally, the expected directions of technology and circuit development are discussed. In this section, unexploited aspects of D/CMOS devices are pointed out for specific applications.

History of Discrete Power DMOS Transistors

A detailed description of the development of contemporary power MOS transistors was given in a previous paper [1]. The flow of discrete MOS power transistor development is shown in Figure 1. As shown in this figure, a ten-year span has witnessed large multiples in both current and voltage rating. The next few years will see continued advancement in discrete power MOS transistors, but not the large increases in current and voltage ratings that have occurred in the last few years. The on-resistance characteristics of these devices restrict their use in most cases to applications requiring less than 1000 volts. Larger chip sizes are possible, but yield considerations make it more attractive to parallel die or packages to obtain higher currents. The main focus for discrete technology in the coming years will be manufacturability and economics. Large reduction

(This manuscript was prepared for and presented at Electro 84)

in price will lead to greater device acceptance. The economics of scale resulting from increases in MOS power transistor use will enable them to compete more effectively with bipolar transistors.

The most significant development in discrete DMOS devices in the last two years is the announcement of MOS-gated regenerative (i.e., latching) and conductivity modulated devices. These two similar types of devices promise to significantly enhance the current ratings of D/CMOS ICs.

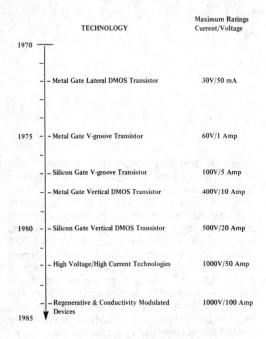

A Chronology of Discrete DMOS Power Transistor Development
Figure 1

TECHNOLOGY	Maximum Ratings Current/Voltage
Metal Gate Lateral DMOS Transistor	30V/50 mA
Metal Gate V-groove Transistor	60V/1 Amp
Silicon Gate V-groove Transistor	100V/5 Amp
Metal Gate Vertical DMOS Transistor	400V/10 Amp
Silicon Gate Vertical DMOS Transistor	500V/20 Amp
High Voltage/High Current Technologies	1000V/50 Amp
Regenerative & Conductivity Modulated Devices	1000V/100 Amp

The advent of MOS power transistors has lead to numerous comparisons between them and bipolar transistors. At first, the goal of the comparison was to show that one type of device was superior to the other for a given application. One such comparison is contained in Table 1. A more enlightened approach is to examine the characteristics of each type of transistor, and then to make a selection based on the resulting match. Each transistor type has a range of applications which match the device requirements.

The Development of ICs with Low Voltage CMOS Control and High Current, High Voltage DMOS Outputs

The comparison of characteristics presented in Table 1 compels IC designers to consider the benefits derived from the incorporation of ICs in monolithic

Table 1
A Comparison of Power MOS and Bipolar Transistor Electrical Characteristics [1]

Parameter	MOS Performance	Bipolar Performance
Input Impedance	High (109-1011 ohms)	Intermediate (103-105 ohms)
Current Gain	High (105-108)	Intermediate (101-102)
Switching Frequency	High (100-500 kHz)	Intermediate (20-80 kHz)
On Resistance	High	Low
Off Resistance	High	High
Voltage Capability	Intermediate (500 V)	High (1200 V)
Ruggedness	Excellent	Good
Cost	High	Intermediate
Maximum Operating Temperature	High (200°C)	Intermediate (150°C)

circuits. The ease with which DMOS transistors may be driven when compared to bipolar transistors is one obvious advantage. Ruggedness, particularly the absence of secondary breakdown in DMOS transistors, is another advantage. However, the voltage drop per unit area for transistors with equivalent voltage ratings is considerably greater for DMOS transistors. This disadvantage dictated that DMOS transistors be utilized first in ICs with voltage ratings beyond those ratings of bipolar transistors. The first application of D/CMOS technology was for high voltage display drivers. The development of these circuits took place in the 1977/78 time frame.

Two classes of displays—plasma and electroluminescent panels—required voltages from 80 to 200 volts for successful operation. Line density and allowable cost per driver also dictated that as many outputs as possible (in multiples of 8) be fabricated on one chip. Different drive schemes were developed to address and solve the problems presented by each panel type. This work resulted in the three different D/CMOS technologies which are the starting point for contemporary D/CMOS ICs. These three D/CMOS IC technologies are discussed in the next section.

Self Isolated D/CMOS IC Technology

In one driver configuration, the output drive transistors were required to withstand 200 volts in an open drain configuration. The drive circuitry was required to dissipate minimum power, so CMOS logic was an appropriate choice. These two requirements lead to the development of "self isolated" D/CMOS technology as shown in Figure 2. The CMOS logic in this technology operates with a power supply up to 15 volts, while the DMOS output transistors are capable of 400 volts. Using silicon-gate technology with a 4 μm feature size, the clock frequency of the logic is 5 MHz.

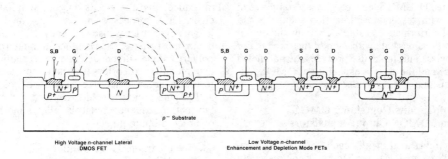

High Voltage n-channel Lateral
DMOS FET

Low Voltage n-channel
Enhancement and Depletion Mode FETs

Cross-Section of a Self-Isolated D/CMOS I.C.
Figure 2

Junction Isolated D/CMOS IC Technology

Other displays required lower voltages (60 to 120 volts) with totem-pole, n-channel outputs. These requirements were met using junction isolation to stand off the voltage between adjacent devices. DMOS transistors, because of their configuration, have a higher breakdown voltage than simultaneously fabricated bipolar transistors. A cross section of this technology is shown in Figure 3. Because of its added complexity, junction isolated technology is capable of providing a variety of additional components. The most useful of these components is the npn transistor, though pnp transistors and a variety of diodes and resistors are also available.

Junction Isolated D/CMOS Technology with Complementary Outputs

A third D/CMOS configuration was required when high voltage push-pull outputs were needed. Junction isolation was essential but the process sequence had to be modified to accommodate the high voltage p-channel device. Size constraints made it advantageous to use DMOS technology for the p-channel device. This combination of constraints led to the junction isolated complementary output structure shown in Figure 4. Both the junction isolation and the DMOS output transistors are capable of withstanding 120 to 150 volts. The low voltage CMOS logic is designed to operate in the 5 to 15 volt range.

2

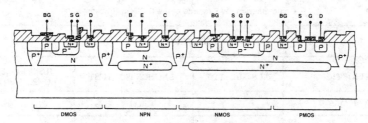

| DMOS | NPN | NMOS | PMOS |

Cross-Section of a Junction Isolated D/CMOS I.C. [1]
Figure 3

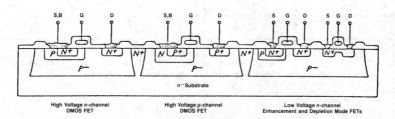

High Voltage n-channel
DMOS FET

High Voltage p-channel
DMOS FET

Low Voltage n-channel
Enhancement and Depletion Mode FETs

High Voltage CMOS Process

Cross-Section of a Junction Isolated, Complementary Output
D/CMOS I.C. [2]
Figure 4

These three D/CMOS technologies are compared in Table 2 with respect to fabrication sequence and electrical performance. Table 2 contains the output configurations possible with each of the three major technologies. This table also provides the paralleled resistance and the current capability of DMOS output transistors.

Status of Circuits Containing CMOS Logic and DMOS Output Transistors

Circuits fabricated with each of these three D/CMOS technologies are becoming commercially available. Considering the design flexibility and economics of IC manufacture, the second technology — junction isolated D/CMOS — is likely to become the most popular industrial choice. This technology has a number of advantages for both IC manufacturers and users.

Similarity to Existing Bipolar Process Technology

The process cross section shown in Figure 3 differs only in a few ways from the cross-section of a bipolar IC. The most significant of these differences are the need to include a deep, lightly doped region called a p-well for the n-channel transistors, and the requirement for a gate dielectric (usually thermally grown SiO_2), and a gate conductor. The p-well is easily

introduced and represents only an additional, well characterized process sequence. The growth of quality gate dielectrics is also well known. The gate conductor most often chosen is either aluminum or doped polycrystalline silicon. Aluminum as the interconnect metallization offers process simplification but results in larger device sizes. The use of poly silicon as the gate conductor adds a process step to the fabrication sequence, but it also allows another level of process interconnect.

Device Characteristics Similar to Those of Other Technologies

The similarity between the bipolar and the junction isolated IC fabrication leads to a second significant advantage. Many of the available circuit devices have characteristics that are similar to those of bipolar technology. The circuit designer has to become familiar with the characteristics of only a few new devices as opposed to unlearning and relearning a whole set of new characteristics.

Circuit Design Techniques Are Familiar

Design techniques using the grounded substrate configuration of the junction-isolated D/CMOS technology are also well known. A designer is not faced with learning a set of new design techniques. Only an extension of existing knowledge is required.

Table 2
Comparison of the Features of the D/CMOS Processes

D/CMOS TECHNOLOGY			
CHARACTERISTIC		**SELF ISOLATED**	**JUNCTION ISOLATED**	**JUNCTION ISOLATED WITH COMPLEMENTARY OUTPUTS**
FABRICATION	Epi Required	No	Yes	Yes
	Masking Steps	9 - 11	10 - 12	11 - 13
ELECTRICAL	Logic Voltage	5 - 15V	5 - 15V	5 - 15V
	Clock Rate	5 MHz	5 MHz	5 MHz
	Output Voltage	200 - 400V	150V	150V
	Output Configuration	Single Ended (open collector)	Single Ended, Totem Pole	Single Ended, Totem Pole, Push-Pull
	Parallel Resistance of Output Devices (200 x 200 mil chip)	3 - 5 Ω	1 - 2 Ω	2 - 4 Ω
	Total On-Chip Current Capability	2-4 amps	10-20 amps	5-10 amps

Circuits Are Similar to Other Device Families

The negative ground configuration with CMOS compatible inputs is common in the electronics industry. Little, if any, systems modification is necessary to begin taking advantage of circuits fabricated with the junction isolated D/CMOS technology.

For these reasons, the junction isolated D/CMOS technology is likely to gain industry-wide acceptance more rapidly. An example of a circuit fabricated using this technology is shown in Figure 5. This chip contains analog and digital control functions implemented using silicon-gate CMOS technology and two large output drive transistors in the totem-pole configuration. The input circuitry is TTL compatible with the low voltage section operating from a 30 volt supply. Level shift is accomplished using high voltage lateral DMOS transistors. The two large DMOS transistors are rated at 80 volts, and the large transistor supplies 12 amps while the small one supplies 8 amps. Resistors are placed in series with the two output transistors for sensing and controlling the current in each transistor. The chip size is 195 mils × 335 mils (4.9mm × 8.4mm). This junction isolated IC represents the state-of-the-art in power D/CMOS circuits.

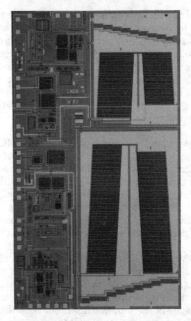

Photograph of a Driver Circuit Fabricated Using the Junction Isolated D/CMOS Technology
Figure 5

Directions for Future Development in D/CMOS Integrated Circuits

The initial enthusiasm resulting from the promise of D/CMOS technology has tended to obscure two things—problems that need addressing before greater acceptance is earned, and growth directions to fully take advantage of the promise of this new technology. Concerns in both of these areas are essential if D/CMOS technology is to realize its full potential. These issues are discussed in this section.

A major potential concern for D/CMOS circuit designers and circuit users is circuit performance in environments with considerable electrical noise. The high currents through the transistors coupled with the presence of various four layer (and hence potentially latching) structures require care in both circuit layout and circuit use. The sensitivity of the IC to latch-up can be minimized through proper circuit design, but it cannot be eliminated in conventional junction isolation. Use of dielectric isolation instead of junction isolation as shown in Figure 6 will eliminate latch-up, but this technology adds considerable cost to the circuit. The dielectric isolation technique may be applied to any of the three D/CMOS technologies, but it is probably more applicable to the two junction isolated versions.

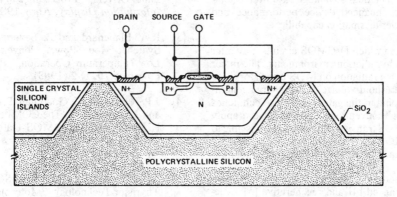

Cross-Section of a Dielectrically Isolated D/CMOS I.C. [1]
Figure 6

Careful circuit use is as important as the design of the circuit in systems with large currents flowing. For this reason, it may be necessary to have more than one ground on the chip. A logic ground should be capable of providing stable current sinking. The presence of a separate high current ground for the output devices means that some ground "bounce" may be allowable without interfering with the operation of the low voltage section of the chip.

One major hurdle facing high power D/CMOS IC manufacturers and users is the availability of packages, heat sinks, etc., that allow circuits to operate at or near their intrinsic limits. The last few years have seen continued progress in the high power package area in both plastic and hermetic packages. However, this area is key for future improvements in circuit performance and reductions in circuit costs.

The DMOS transistor's ability to operate at both high and low temperature extremes offers significant advantages for many applications. The package power dissipation limits discussed in the previous paragraph set the ultimate limit, but both the logic and the output stage of a D/CMOS are capable of operating at 150°C and above.

The positive temperature coefficient of resistance of DMOS transistors has both advantages and disadvantages. The output device resistance does increase with temperature resulting in increased power dissipation and decreased efficiency, but DMOS transistors are capable of stable operation at temperatures as high as 200°C and above. This behavior is in contrast with that of bipolar transistors which have a negative temperature coefficient. The low temperature performance of D/CMOS circuits also offers significant advantages over bipolar circuits. The speed and current handling capability of both the low voltage CMOS transistors and the output DMOS transistors increases as the temperature decreases from room temperature down to 77°K (liquid nitrogen) [3]. The dual advantages at lowered temperatures are enhanced device performance and increased power dissipation capability.

Another area in which D/CMOS circuits may offer advantages is in radiation environments. The major effect of ionizing radiation on MOS transistors is to change the threshold voltage due to charge generation and trapping in the gate dielectric. Techniques for "hardening" conventional MOS transistors apply as well to DMOS transistors. It should be possible to obtain D/CMOS integrated circuits capable of withstanding a total dose of 10^5 to 10^6 (rads silicon).

The announcement of discrete DMOS-gated conductivity enhanced [4] and regenerative [5] structures foreshadows future circuit developments. Both of these device types promise to control more power per square mil of silicon surface than DMOS transistors. The conductivity enhanced device is gate controlled during both turn-on and turn-off, while the voltage across the regenerative device must go to zero before conduction stops. It is possible to integrate both of these structures in D/CMOS ICs. However, this integration cannot proceed without some caution. The presence of excess minority carriers in the drain region may increase circuit sensitivity to latch-up. Considerable care must be taken to guarantee latch-up free operation.

Summary and Conclusions

This article has examined the evolution, present status, and future directions of D/CMOS IC technology. Three different specific technologies were discussed, and their features were compared. Based on this comparison, one D/CMOS technology, the junction isolated D/CMOS, was selected as the technology most likely to become an industry standard. One circuit that pushes junction isolated D/CMOS technology to its limits was examined as an example of what is possible. Finally, future directions for D/CMOS IC circuits were indicated by discussing both areas of concern and advantages of this technology. The theme running through this paper is the dynamic nature of this field. The promise of D/CMOS technology is being realized as both users and manufacturers begin to understand its limits.

References

[1]   R.A. Blanchard, "MOSFETS IN ARRAYS AND INTEGRATED CIRCUITS," *Proceedings of Electro 83* Paper 7/4 (1983).

[2]   R.A. Blanchard and W.G. Numann, "A High Voltage Chip Set for Use as Electroluminescent Panel Drivers," *Proceedings of the Society of Information Display*, April, 1982.

[3]   R.A. Blanchard and R. Severns, "Designing Switched-Mode Power Converters for Very Low Temperature Operation," *Proceedings of Powercon 10*, D-2 (1983).

[4]   J.P. Russell, A.M. Goodman, L.A. Goodman, and J.M. Neilson, "The COMFET — A New High Conductance MOS-Gated Device," *IEEE Electron Devices Letters*, P63, *EDL-4* No. 3 (March, 1983).

[5]   A. Pshaenich, "The MOS SCR, A New Thyristor Technology," Motorola Engineering Bulletin *EB-103* (1982).

2.12 Drain to Source Breakdown and Leakage in Power MOSFETs

An examination of almost any power MOSFET data sheet reveals values for both the drain to source breakdown voltage, BV_{DSS}, and leakage current, I_{DSS}, with zero applied gate-to-source voltage. At a cursory glance, these two parameters might seem to be unrelated until one looks at the test conditions for each. The breakdown voltage is measured at a specified leakage current and leakage current is measured at a specified voltage, often a percentage of the breakdown voltage. Specifying these two parameters separately on a data sheet is accepted practice (by many manufacturers), but is somewhat redundant.

A statement relating breakdown voltage and leakage current would certainly help to clarify matters. In the ideal case, the BV_{DSS} of a MOSFET is reached when the leakage current begins to increase drastically with a small change in applied voltage. In a properly designed and manufactured power MOSFET this effect is as shown in Figure 1. Beyond a certain voltage, the drain current suddenly begins to increase uncontrollably if not limited.

Not all power MOSFETs show such predictable behavior as that illustrated in Figure 1. An unusual effect known as "latch-back" has also been observed in certain types of MOSFETs [1] (notably metal gate V-groove devices), and is shown in Figure 2. Here normal breakdown begins as BV_{DSS} is reached, but as the leakage current rises towards a critical value, the sustaining voltage of the device suddenly drops or "latches-back" to a lower value. This latched-back breakdown voltage can be as much as 80% less than BV_{DSS}.

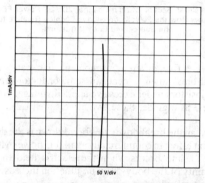

1mA/div

50 V/div

Normal Breakdown Occurs in an *n*-Channel DMOS Power FET When the Applied Drain to Source Voltage Exceeds the Device's BV_{DSS}
Figure 1

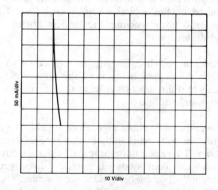

50 mA/div

10 V/div

Latchback Occurs in a Power MOSFET When the Parasitic Bipolar Transistor Begins to Conduct Collector Current
Figure 2

Both the normal and latched-back modes of break-down (Figures 1 and 2 respectively) are non-destructive if the increased leakage current is limited to a safe value. This prevents excessive power dissipation in the MOSFET, which could force the junction temperature above the maximum allowable limit and cause device destruction.

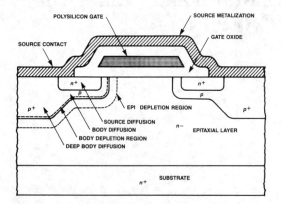

This Cross Section of an *n*-Channel VDMOS Power FET Shows How the Polysilicon Gate is Located Relative to the Source and Body Diffusions
Figure 3

There are five different phenomena that can cause normal breakdown to occur in semiconductor diodes and transistors under excessive applied voltage. Using Figure 3 as a guide, these five effects are [1,2]:

1. Avalanche breakdown; this breakdown is the dominant effect in commercially available power MOSFETs, and occurs when the electric field in the vicinity of the body-drain *pn* junction increases to a critical value.

2. Reach-through breakdown; this breakdown is a special case of avalanche breakdown which is caused by the depletion region in the n^- type epitaxial layer reaching the heavily doped n^+ substrate.

3. Punch through breakdown; this breakdown occurs when the depletion region of the reverse biased body to drain junction reaches the heavily doped n^+ source diffusion.

4. Zener breakdown; a high electric field (on the order of 10^6 V/cm) initiates this type of breakdown. There is a finite probability that electrons in their covalent bonds will be excited directly into the conduction energy band at the junction. These high electric fields can only be reached in heavily doped p^+/n^+ junctions, with resulting breakdown voltages < 6 volts. This type of breakdown is avoided in commercial power MOSFETs by

keeping the heavily doped *n* and *p* regions sufficiently far apart that there is no interaction between them.

5. Dielectric breakdown; here the electric field in a dielectric layer such as silicon dioxide (SiO_2) or silicon nitride (Si_3N_4) exceeds the dielectric strength of the material. This field causes very large currents to flow through the dielectric and can permanently alter its insulating properties. Commercially available power MOSFETs use sufficiently thick dielectric layers such that the electric fields within them never reach the critical value necessary to initiate breakdown.

Since both zener and dielectric breakdown have been eliminated as potential causes of normal breakdown in power MOSFETs, we will examine the first three in more detail. Following this section, the phenomenon of latch-back and how modern power MOSFETs are designed to avoid it is discussed. Finally the temperature dependence of breakdown voltage and leakage current is examined.

Causes of Normal Breakdown in Power MOSFETs

1. Avalanche breakdown [2]

This phenomenon, as its name implies, is a sudden avalanche of mobile carriers caused by the increased electric field present within the depletion regions at the body drain *pn* junction. Electrons or holes that enter the depletion regions acquire sufficient energy from the electric field to knock bound valence electrons out of the silicon lattice atoms in that region. If one electron or hole produces on the average less than one additional carrier then the leakage current is not increased.

If, however, one or more additional carriers are produced and these extra carriers each produce one or more additional carriers then avalanching ensues. The depletion regions around the body-drain *pn* junction must be wide enough so that the mobile carriers can gain sufficient energy from the local electric field to initiate this process.

The mathematics for describing this effect are already known from the analysis of gaseous breakdown phenomena. When this analysis is applied to the semiconductor case, the carrier multiplication factor can be derived. Basically it is [2]

$$M = \frac{1}{1 - (V_R/V_B)^N} \qquad (1)$$

where V_B is the junction breakdown voltage, V_R is the applied reverse voltage and N is a numerical factor which depends upon the type of semiconductor crystal used (N = 3 to 6 for silicon). As V_R approaches V_B, M approaches infinity and the junction breaks down.

Avalanche breakdown occurs when the maximum electric field in the vicinity of the junction reaches a critical value. Since the critical field in silicon is about 3×10^5 V/cm for impurity concentrations less than 10^{16} atoms/cm^3 [1], the breakdown voltage of a planar diffused pn junction is determined to a first order by the doping concentration on the lightly doped side [2]. This result follows from the derivation of the breakdown voltage for the one-sided step-junction (OSSJ) model. In this model, the acceptor doping concentration on the p^- type side is much greater than the donor doping concentration on the n^- type side. To a good approximation, the breakdown voltage of a power MOSFET is determined by the impurity concentration in the lightly doped epitaxial layer.

A 100 volt power MOSFET requires a background (epitaxial layer) concentration of 3.5 to 6 x 10^{15} atoms/cm^3, while a 500 volt MOSFET requires a lower concentration in the region of 1 x 10^{14} atoms/cm^3[3]. Since the resistivity of the epitaxial layer is inversely proportional to the dopant concentration, a high voltage MOSFET has a higher on-resistance than a low voltage counterpart with the same die size. In almost all available power MOSFETs, the on-resistance is approximately proportional to the breakdown voltage raised to the 2.5 power [4]. This relationship is shown in Figure 4.

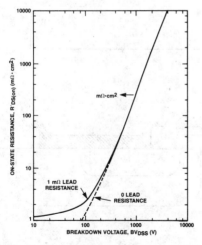

The Observed Relationship Between R'$_{DS(on)}$ Multiplied by Chip Size (R'$_{DS(on)}$) and Breakdown Voltage for DMOS Power FETs, Shows That BV$_{DSS}$ is Proportional to R'$_{DS(on)}$ Raised to the 2.5 Power
Figure 4

It is important to ensure that the epitaxial layer is no thicker than necessary, if the MOSFET is to have minimum on-resistance. The epitaxial layer thickness is a major factor in the design of an optimized power MOSFET. If the epitaxial layer is too thick, the device will have an excessively high on-resistance for its breakdown voltage specification. Conversely, if the epitaxial layer is too thin, reach-through breakdown occurs below the desired breakdown voltage rating.

2. Reach-through breakdown

This mode of breakdown occurs when the expanding depletion region in the n^- type epitaxial layer reaches the heavily doped n^+ substrate. The spreading of the junction depletion region is inversely proportional to the square root of the impurity doping concentration [1]. As the depletion region enters the substrate, it begins to spread much more slowly, and the electric field rises rapidly towards the critical value necessary to initiate avalanching. When it reaches this critical value, localized avalanching occurs, and the device breaks down.

Commercially available power MOSFETs are designed by first choosing the background concentration to give the required avalanche breakdown voltage. The epitaxial layer thickness is then chosen so that at the avalanche breakdown voltage, the depletion region in the epitaxial layer just reaches the substrate. This epi layer selection process minimizes the device on-resistance for a given breakdown voltage and die size.

3. Punch-through breakdown

This effect occurs when the widening depletion region at the body-drain junction reaches the heavily doped n^+ source diffusion. The field in the body depletion region sweeps electrons from the source across the reverse biased body-drain junction. The leakage current increases rapidly as a consequence.

This breakdown mode is rarely observed in commercial power MOSFETs. They are normally designed to be avalanche breakdown limited.

The Cause of Latchback and Its Avoidance [1]

In power DMOS structures, the source-body contact is made some distance from the channel region. This contact shorts the emitter and base of the parasitic npn transistor (pnp for a p-channel MOSFET), formed by the source, body and drain regions. Figures 5a, b and c show this for the VMOS, LDMOS, and VDMOS structures, respectively.

The source and body diffusions form a distributed series of npn bipolar transistors with their bases separated by the resistance of the body region below the source. To explain the phenomenon of latch-back, a simplified model of the DMOS structure is used. This model, shown in Figure 6, consists of the gate

2

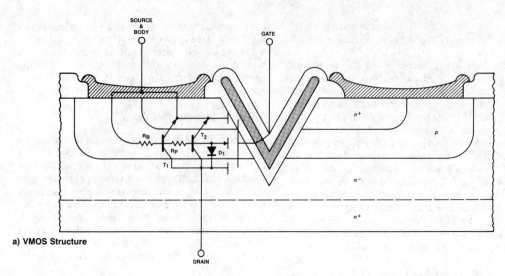

a) VMOS Structure

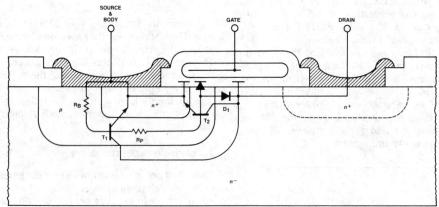

b) LDMOS Structure

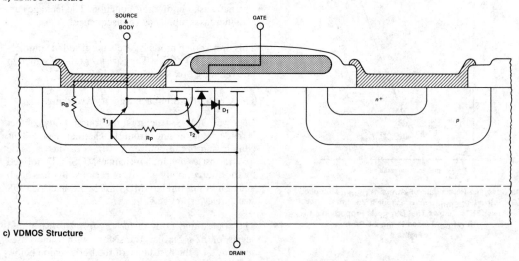

c) VDMOS Structure

Cross-Section and Circuit Models
This Diagram Shows Distributed Parasitic *npn* Bipolar
Transistor Inherent in the Three Common Power DMOS Structures
Figure 5

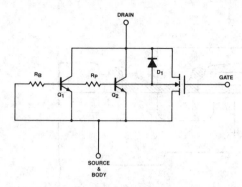

The Simplified *n*-Channel DMOS Model, Used to Explain the
Phenomenon of Latchback, Includes Only the Parasitic
Bipolar Transistors at the Extreme Edges
of the Body Diffusion
Figure 6

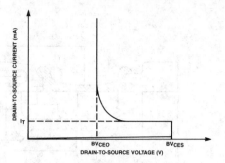

This Curve Shows Idealized Latchback Occurring in a DMOS
Power FET Where BV_{CEO} is 50% of BV_{CES}
Figure 7

controlled channel and only the two *npn* transistors at
the extremes of the body diffusion [1]. The bases of
these two *npn* transistors are separated by a resistance
R_P. The body contact resistance is modeled by a
resistance R_B. The body-drain diode is also included,
and serves as the avalanche current generator in the
model. This model is equally applicable to all three
power MOSFET structures mentioned above, despite
their cross-sectional differences.

With the gate short-circuited to the source, the
leakage current through D_1 increases as the applied
V_{DS} reaches its avalanche breakdown voltage. This
current causes the voltage drop across resistors R_P
and R_B to increase. If this drop is allowed to reach
0.6 volts, Q_2 will begin to conduct current. When this
happens, the sustaining voltage of the MOSFET
suddenly drops from the avalanche breakdown voltage of diode D_1 to a lower value. This lower sustaining
voltage is equal to the collector-emitter breakdown
voltage of Q_2 under forward bias conditions.

Most bipolar transistor manufacturers do not usually
specify the collector-emitter breakdown voltage under
forward bias conditions. They specify two other values of collector-emitter breakdown voltage. These
are the BV_{CEO} (with the base left open circuit), and
the BV_{CES} (with the base short-circuited to the
emitter). The BV_{CES} of transistor Q_2 is equal to the
avalanche breakdown voltage of diode D_1, which sets
the BV_{DSS} limit of the power MOSFET.

The BV_{CEO} of Q_2 approximately equals its collector-
emitter breakdown voltage under forward bias conditions. Most bipolar transistors, including the parasitic device in a power MOSFET, exhibit a BV_{CEO}
that can be as low as 20% of their BV_{CES}. Thus the
latched-back sustaining voltage of a power MOSFET
can be considerably lower than its BV_{DSS}. Figure 7
shows an idealized latched-back breakdown voltage

curve, where the BV_{CEO} is 50% of the BV_{CES} or
BV_{DSS}. The trigger current I_T is the value of avalanche current at which the transition to BV_{CEO}
occurs. Not surprisingly, this current can be shown to
be [1]

$$I_T = \frac{V_{BE(Q_2)}}{R_P + R_B} \qquad (2)$$

where $V_{BE\,(Q_2)}$ is roughly 0.6 volts.

Now that the mechanism which causes latchback in a
power MOSFET has been described, at least two
methods for preventing it, based on the model shown
in Figure 6, can be listed. These are [1]:

1. Reduce the value of $R_P + R_B$ so that any lateral
   body-current flow does not result in the 0.6 volt
   drop necessary to turn on Q_2.

2. Change the shape of the body region such that the
   current flow in avalanche breakdown bypasses R_P
   (R_P is usually $>>$ than R_B).

Most of the modern commercially available DMOS
power FETs use technique #2 to prevent latch-back.
Figure 8 shows how this can be accomplished in both
vertical DMOS and V-groove type devices. The
heavily doped $p+$ regions that extend more deeply
into the epitaxial layer form an integrated bypass
diode that effectively diverts avalanche current away
from the $p-$ type body region (and hence R_P). The
equivalent model for this modified structure is shown
in Figure 9. It has been experimentally verified that if
the avalanche breakdown voltage of D_2 is less than
that of D_1, the circuit will not suffer from latch-back.

The Temperature Dependence of Breakdown Voltage and Leakage Current

As mentioned previously, the breakdown voltage of a
power MOSFET is determined to a first order by the
doping concentration in the epitaxial layer. Since this

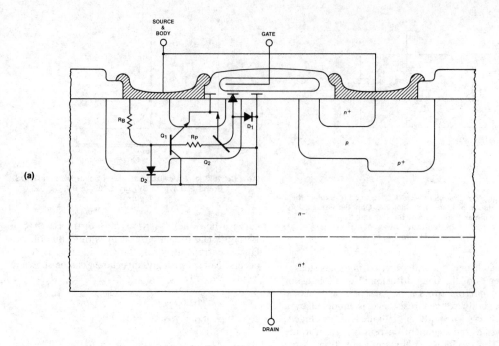

(a)

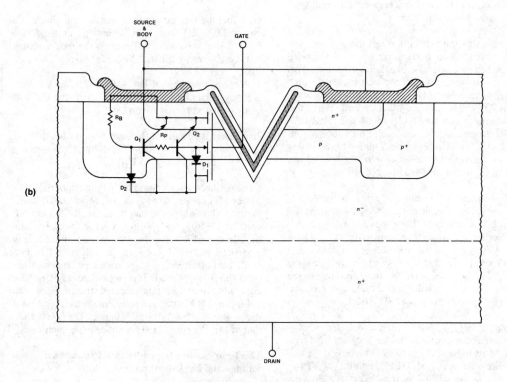

(b)

A Practical Device Modification to Prevent Latchback is the
Inclusion of an Integrated Bypass Diode, D_2, as Shown in
the Modified VDMOS and VMOS Structures Respectively
Figure 8

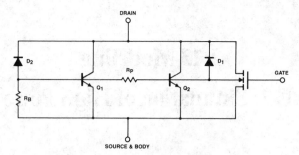

The Circuit Model of CMOS Power FET with an Integrated Bypass Diode Is Very Similar to That in Figure 6, but Now Includes Diode D$_2$
Figure 9

concentration does not vary substantially with junction temperature, one would expect the device breakdown voltage not to vary with temperature either.

In practice however, there is a slight variation in breakdown voltage with temperature (about a 5% increase per 100°C rise in junction temperature). This is due to a second order effect known as "phonon-assisted electron scattering." At elevated junction temperatures, the increased vibrational energy of the silicon lattice atoms causes the mobile charge carriers in the epitaxial layer to move in an increasingly random pattern, relative to the direction of the electric field.

Thus, the mobile carriers do not attain as high an average kinetic energy for a given electric field. Maximum kinetic energy is attained when the carriers move parallel to the field. Consequently, the value of the critical field required to initiate avalanching increases with temperature, and so does the device breakdown voltage.

The drain-to-source leakage through the body-drain junction, on the other hand, is highly temperature sensitive. The leakage current in a reverse biased *pn* junction is mainly proportional to the square of the intrinsic (i.e. undoped) carrier concentration in silicon (N_i). The value of N_i^2 can be written as [2]:

$$N_i^2 = KT^3 e^{-E_g/\kappa T} \tag{3}$$

where K is a constant independent of temperature, E_g is the energy band-gap of silicon (1.12 eV) and κ is Boltzmann's constant. One can see that the T^3 term completely dominates the leakage current over a wide temperature range. The approximate fractional change in the leakage current of a reverse biased silicon *pn* junction is thus given by [2]

$$\frac{1}{I_L}\left(\frac{\partial I_L}{\partial T}\right) = \frac{1}{2}\left(\frac{3}{T} + \frac{E_g}{KT^2}\right)\Bigg|_{V_R} = \text{Constant} \tag{4}$$

which works out to be approximately +8% per °C, or a doubling of leakage current for every 12°C rise in junction temperature.

The body-drain junction leakage in a power MOSFET at a junction temperature of 150°C can therefore be more than three orders of magnitude greater than that at room temperature. This can limit the usefulness of a power MOSFET that exhibits excessive junction leakage current (say several hundred μA) at room temperatures, when operated at temperatures near its maximum data sheet limit (typically 150°C). Frequently, however, transistors with high leakage currents at 25°C do not have the same temperature coefficient as one with low leakage.

There are additional sources of leakage in a power MOSFET originating from the package in which the die is mounted. In a properly packaged device, these leakages are of the order of a few nA and do not constitute a significant part of the total leakage. In a device which has been defectively packaged, however, the additional leakage can reach several hundred nA and may or may not be temperature sensitive. This situation is rarely encountered in properly manufactured and tested devices.

References

[1] R.A. Blanchard, *Optimization of Discrete High Power MOS Transistors*, PhD. Thesis Dissertation, Department of Electrical Engineering at Stanford University, December 1981.

[2] David M. Navon, *Electronic Materials and Devices,* Houghton Mifflin Company, 1975.

[3] S.M. Sze and G. Gibbons, "Solid State Electronics," September 1966.

[4] Chenming Hu, "A Parametric Study of Power MOSFETs," PESC 1979 Conference Record.

2.13 Modeling

2.13.1 SPICE 2: Simulation of High Power MOSFET

Introduction

The power MOSFET offers superior performance due to its high switching speed, low power voltage-driven gate requirement, ease of paralleling, and absence of secondary breakdown phenomenon that troubles the Bipolar Junction Transistors. Herren Jr., Nienhaus and Bowers developed a computer model for high power MOSFET[1]. This model takes into account the deviation of the device transfer characteristic (Figure 4c) from the ideal square law characteristic by the inclusion of a lumped source resistance. One of the chief attractive features of the model is that its parameter determination procedure requires only the manufacturer's published device data.

In the following article, the computer model is briefly described, and the model parameter determination procedure for the SPICE 2 program is explained. A computer model of the device type IRF330 was generated based on this procedure. Simulation results of both resistive and inductive switching circuits using this device are presented.

SPICE 2[2] is a network analysis program using nodal analysis techniques and has the advantages of programming ease and relatively fast execution time. Version 2E2 was used in the following programs.

High Power MOSFET Model[1]

Figure 1 shows the high power MOSFET model for an n-channel MOSFET. A similar model exists for the p-channel MOSFET device. The substrate and the source are assumed to be shorted internally. J_D is a non-linear current source depending on VGS, VGD, V_T (the threshold voltage), β (the device conductance

constant), and KD (the channel length modulation constant). In the SCEPTRE version of the model, C_{GD} and C_{DS} are non-linear capacitors. However, C_{GD} is modeled as a constant capacitor in the SPICE 2 version of the model[1]. C_{DS} is given by:

$$C_{DS} = \frac{C_{DO}}{\left(1 + \dfrac{V_{DS}}{\phi}\right)^n}, \text{ where}$$

C_{DO} = capacitance value at zero V_{DS}
ϕ = junction contact potential = PB
n = an exponent between $\frac{1}{3}$ and $\frac{1}{2}$

In the SPICE 2 version of the model values of $\phi = 1.0$ and $n = \frac{1}{2}$ have been assumed. Thus, the model parameters to be determined are 1) V_T, 2) β, 3) KD, 4) R_S, 5) R_D, 6) R_G, 7) C_{DO}, 8) C_{DG}, 9) C_{GS}.

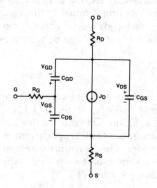

High Power MOSFET Model (Ref. 1)
Figure 1

Figure 2 shows the SPICE 2 built-in level-1 *n*-channel MOSFET model[2] based on the model proposed by Schichman and Hodges. This model is adapted to the requirements of the high power MOSFET model as indicated in the next section. *n*-channel and *p*-channel devices are taken care of by using different built in SPICE 2 models, viz., NMOS and PMOS. For *n*-channel MOSFETs VTO, the gate threshold voltage is considered positive for enhancement mode devices and negative for depletion mode devices. For *p*-channel MOSFETs, VTO is negative and positive for enhancement mode devices and depletion mode devices, respectively.

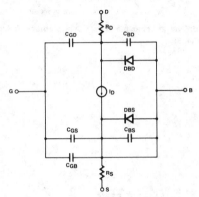

SPICE 2 Built-In MOSFET Model (Ref. 2)
Figure 2

Determination of Model Parameters[1]

Following is a step by step procedure to determine the model parameters based on Reference 1.

STEP 1: Figure 3 shows a set of output (dc) characteristics drawn in an exaggerated manner. Choose three characteristic curves L1, L2, and L3 corresponding to gate-to-source voltages V_{GS1}, V_{GS2}, and V_{GS3}, respectively, according to the following guidelines.
  a) L1-curve with the highest drain current
  b) L3-curve close to $I_D = 0$ line
  c) L2-curve approximately half-way between the above two curves.

STEP 2: Choose points P1 (I_{D1}, V_{DS1}), P2 (I_{D2}, V_{DS2}),...P6 (I_{D6}, V_{DS6}) as per the following guidelines.
  a) P1, any point well into the pinch-off region on line L1.
  b) P2, on line L2 such that $V_{DS2} = V_{DS1} - (V_{GS1} - V_{GS2})$
  c) P3, any point well into the pinch-off region on line L3.
  d) P4 and P5, two points as far apart as possible on the near horizontal straight-line segment of L1 curve.

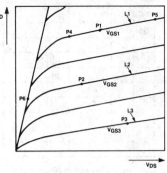

Output Characteristics—To Locate P1 Thru P6
Figure 3

  e) P6, a point on the saturation (ohmic region) line. It may be more convenient to select this point from the saturation (low V_{DS}) characteristics provided by the manufacturer.

STEP 3: Choose V_{T1}, the approximate intersecting point of the transfer characteristic with the $I_D = 0$ axis (Figure 4c), as the starting value of V_T.

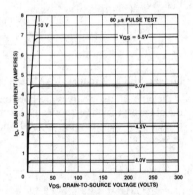

a) Typical Output Characteristics

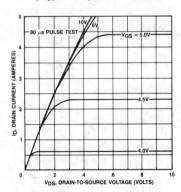

b) Typical Saturation Characteristics
IRF330 Characteristics (Ref. 3)
Figure 4

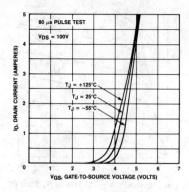

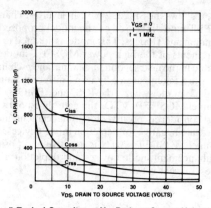

c) Typical Transfer Characteristics   d) Typical Capacitance Vs. Drain-to-Source Voltage

IRF330 Characteristics (Ref. 3)

Figure 4

STEP 4: $R_S = \dfrac{\left(\dfrac{I_{D1}}{I_{D2}}\right)^{\frac{1}{2}} (V_{GS2} - V_T) - (V_{GS1} - V_T)}{(I_{D1} I_{D2})^{\frac{1}{2}} - I_{D1}}$

For the SPICE 2 model, RS = R_S.

STEP 5: Let $\quad x \overset{\Delta}{=} (V_{GS1} - I_{D4} * R_s - V_T)$
$\qquad\qquad y \overset{\Delta}{=} (V_{GS1} - I_{D5} * R_s - V_T)$

$\qquad KD = \dfrac{I_{D4}\, y^2\, (V_{DS5} - V_{GS1} + V_T) - I_{D5}\, x^2\, (V_{DS4} - V_{GS1} + V_T)}{I_{D5}\, x^2 - I_{D4}\, y^2}$

For the SPICE 2 model, LAMBDA = 1/KD.

If the characteristic curves are horizontal in the pinch off region (without any discernible slope), this step may be skipped and KD = ∞.

STEP 6: $\beta = \dfrac{I_{D1}}{(V_{GS1} - I_{D1} R_s - V_T)^2 \left\{ 1 + \dfrac{(V_{DS1} - V_{GS1} + V_T)}{KD} \right\}}$

If KD = ∞, then $\beta = \dfrac{I_{D1}}{(V_{GS1} - I_{D1} R_s - V_T)^2}$

For the SPICE 2 model, KP = 2*β

STEP 7: V_T to be obtained by solving

$I_{D3} = \beta\, (V_{GS3} - I_{D3} R_s - V_T)^2 \left\{ 1 + \dfrac{(V_{DS3} - V_{GS3} + V_T)}{KD} \right\}$

One way to achieve this would be to solve for V_T, iteratively, the rearranged equation.

$V_T = V_{GS3} - I_{D3} R_s - \sqrt{\dfrac{I_{D3}}{\beta \left\{ 1 + \dfrac{(V_{DS3} - V_{GS3} + V_T)}{KD} \right\}}}$

If KD = ∞, then

$$V_T = V_{GS3} - I_{D3} R_s - \sqrt{\frac{I_{D3}}{\beta}}$$

For the SPICE 2 model, VTO = V_T.

STEP 8: Repeat steps 4, 5, 6, and 7 until parameters V_T, R_s, KD, and β converge. It may be necessary to carry out a number of iterations before the parameters converge.

STEP 9: Let b $\triangleq \dfrac{2(V_{GS6} - V_T - V_{DS6})}{I_{D6}}$

$$c \triangleq \frac{1}{\beta I_{D6}} - \frac{(V_{DS6} - I_{D6} R_s)(2V_{GS6} - I_{D6} R_s - 2V_T - V_{DS6})}{I_{D6}^2}$$

$$R_D = \frac{-b + \sqrt{b2 - 4c}}{2}$$

For the SPICE 2 model, RD = R_D.

Note: Since P6 lies on the ohmic region line, determining V_{GS6} is somewhat arbitrary. However, R_D is not found to be critically dependent on this.

STEP 10: Refer to the capacitance vs. V_{DS} curves (Figure 4d).
a) $C_{GD} = C_{rss}$ (at high V_{DS})
For the SPICE 2 model, $C_{GS} = C_{GD}$
b) $C_{GS} = C_{iss} - C_{rss}$ (at high V_{DS})
For the SPICE 2 model, $C_{GD} = C_{GS}$
c) $C_{DO} = C_{oss} - C_{rss}$ (at $V_{DS} = 0$)
For the SPICE 2 model, $C_{BD} = C_{DO}/$ (1x10-10) and PB = 1.0

Note 1: C_{GD} and C_{GS} are approximated by constant value capacitances. Also, the units for C_{GS} and C_{GD} in SPICE 2 are farads/meter. This, however, does not create any error since the default values for channel length and width in SPICE 2 are 1.0 m each.

Note 2: As pointed out in Reference 1, C_{GS} and C_{GD} have been interchanged in SPICE 2 MOSFET model, apparently an error.

Note 3: The unit for C_{BD} in SPICE 2 is farad/meter^2. The default drain area in SPICE 2 is 1 x 10-10 m^2.

STEP 11: Refer to the switching time test circuit (Figure 5). A simplified equivalent circuit during turn-on and turn-off is shown in Figure 6.

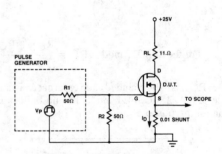

Switching Time Test Circuit (Ref. 3)
Figure 5

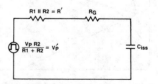

Simple Equivalent Circuit of Gate Input
Figure 6

Let

$$R' \triangleq \frac{R_1 R_2}{R_1 + R_2}$$

$$V'_p \triangleq \frac{V_p R_2}{R_1 + R_2}$$

Considering the turn-on process,

$$R_{G1} = \frac{T_{ON}}{C_{iss} \, ln \dfrac{V'_p}{V'_p - V_T}} - R'$$

where T_{ON} is the turn-on delay.

To consider the turn-off delay, plot the load line on the output characteristics as in Figure 7. This intersects the ohmic region line at point Q. Find the gate-to-source voltage, $V_{GS} = VA$, such that the corresponding pinch-off region characteristic passes through Q when extrapolated as shown.

Then

$$R_{G2} = \frac{T_{OFF}}{C_{iss} \, ln \left(\dfrac{V'_p}{VA}\right)} - R'$$

For the SPICE 2 model, $RG = \dfrac{R_{G1} + R_{G2}}{2}$

Note: RG has to be included as an external resistance, since SPICE 2 MOSFET model has no built-in element for this purpose.

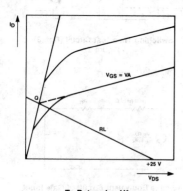

To Determine VA
Figure 7

By adopting the above method, a computer model for the device type IRF330 was generated. The model listings may be seen in Figure 11 which shows a SPICE 2 program of a circuit using IRF330. Here, the NMOS model IRF330 along with resistor RF constitutes the device model.

The capacitor C_{rss} is a highly non-linear function of V_{DS}, especially at low V_{DS} values, whereas it has been approximated as a constant value capacitor. This could be a source of error when using SPICE 2 with this model. The dynamic performance (rise and fall times) was found to be critically dependent upon the value of C_{rss} chosen.

Simulation with the Model

DC Characteristics

The output characteristics, saturation characteristics, and the transfer characteristics were simulated with this device model. Comparing these simulated curves (Figures 8a, 8b, and 8c) with the characteristic curves provided by the manufacturer (Figures 4a, 4b, and 4c), we note that the model simulates the performance of the real device accurately both in the ohmic region and in the pinch-off region. However, in the area in which transition from one region to the other occurs, the model is less accurate.

Switching Characteristics

The real test of a dynamic model is its ability to simulate dynamic behavior of the device in a circuit in a fairly accurate manner. With this in view, the model was tested with resistive loads and with an inductive load under switching conditions.
a) **Resistive Loads:**
i) **Switching Time Test Circuit:**
This circuit, shown in Figure 5, is the test circuit adopted by the manufacturer for measuring the switching times at low voltages (Reference 3). The results of the simulation (Figure 9) indicate that, while the details of the waveforms (Figure 7C, Page 16, Reference 3) are not simulated accurately, yet the 'ON' and 'OFF' delays have the correct values. The rise and fall times are between 20 and 30 nsec while the actual device has typical values of 50 nsec. This discrepancy is probably due to the error introduced in simulating the non-linear capacitor C_{rss} with a fixed value capacitor. In fact, by changing C_{rss} from 20 pF to 40 pF rise and fall times close to the manufacturer's specifications may be obtained. However, since IRF330 is a high voltage device, the value of 20 pF for C_{rss} corresponding to higher V_{DS} values was retained.

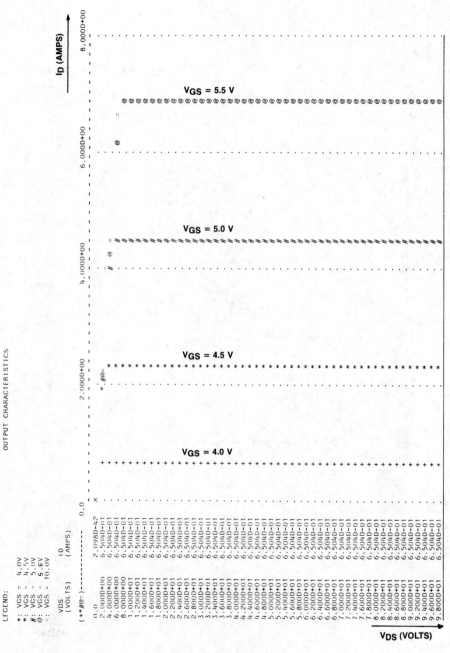

OUTPUT CHARACTERISTICS

LEGEND:

+: VGS = 4.0V
*: VGS = 4.5V
#: VGS = 5.0V
@: VGS = 5.5V
=: VGS = 10.0V

$V_{GS} = 5.5$ V

$V_{GS} = 5.0$ V

$V_{GS} = 4.5$ V

$V_{GS} = 4.0$ V

I_D (AMPS)

V_{DS} (VOLTS)

VDS (VOLTS)	ID (AMPS)
(+*#@=)	
0.0	2.098D-42
2.000D+00	6.504D-01
4.000D+00	6.504D-01
6.000D+00	6.504D-01
8.000D+00	6.504D-01
1.000D+01	6.504D-01
1.200D+01	6.504D-01
1.400D+01	6.504D-01
1.600D+01	6.504D-01
1.800D+01	6.504D-01
2.000D+01	6.504D-01
2.200D+01	6.504D-01
2.400D+01	6.504D-01
2.600D+01	6.504D-01
2.800D+01	6.504D-01
3.000D+01	6.504D-01
3.200D+01	6.504D-01
3.400D+01	6.504D-01
3.600D+01	6.504D-01
3.800D+01	6.504D-01
4.000D+01	6.504D-01
4.200D+01	6.504D-01
4.400D+01	6.504D-01
4.600D+01	6.504D-01
4.800D+01	6.504D-01
5.000D+01	6.504D-01
5.200D+01	6.504D-01
5.400D+01	6.504D-01
5.600D+01	6.504D-01
5.800D+01	6.504D-01
6.000D+01	6.504D-01
6.200D+01	6.504D-01
6.400D+01	6.504D-01
6.600D+01	6.504D-01
6.800D+01	6.504D-01
7.000D+01	6.504D-01
7.200D+01	6.504D-01
7.400D+01	6.504D-01
7.600D+01	6.504D-01
7.800D+01	6.504D-01
8.000D+01	6.504D-01
8.200D+01	6.504D-01
8.400D+01	6.504D-01
8.600D+01	6.504D-01
8.800D+01	6.504D-01
9.000D+01	6.504D-01
9.200D+01	6.504D-01
9.400D+01	6.504D-01
9.600D+01	6.504D-01
9.800D+01	6.504D-01

0.0   2.000D+00   4.000D+00   6.000D+00   8.000D+00

SPICE 2 Simulation Results
Figure 8(a)

2

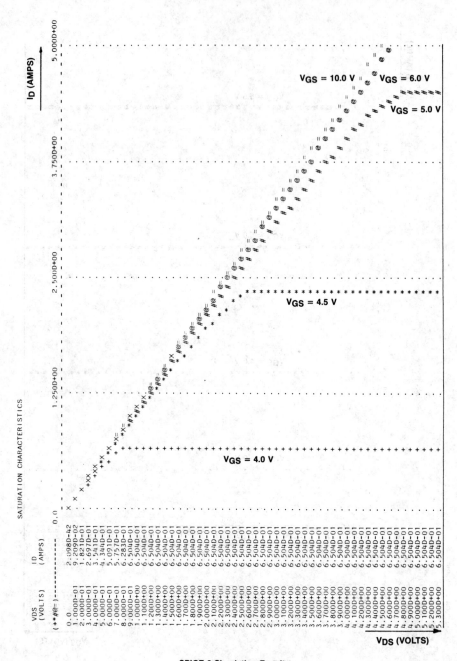

SPICE 2 Simulation Results
Saturation Characteristics
Figure 8(b)

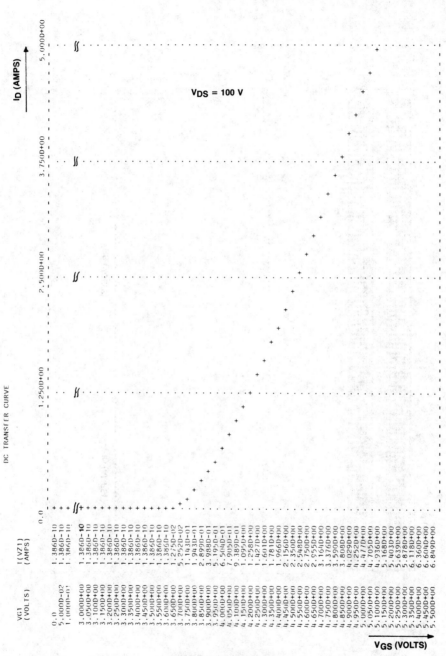

DC TRANSFER CURVE

I_D (AMPS)

$V_{DS} = 100$ V

V_{GS} (VOLTS)

VG1 (VOLTS)	I(VZ1) (AMPS)
0.0	1.386D-10
5.000D-02	1.386D-10
1.000D-01	1.386D-10

VG1 (VOLTS)	I(VZ1) (AMPS)
3.000D+00	1.386D-10
3.050D+00	1.386D-10
3.100D+00	1.386D-10
3.150D+00	1.386D-10
3.200D+00	1.386D-10
3.250D+00	1.386D-10
3.300D+00	1.386D-10
3.350D+00	1.386D-10
3.400D+00	1.386D-10
3.450D+00	1.386D-10
3.500D+00	1.386D-10
3.550D+00	1.386D-10
3.600D+00	1.386D-10
3.650D+00	1.275D-02
3.700D+00	5.252D-02
3.750D+00	1.143D-01
3.800D+00	1.943D-01
3.850D+00	2.899D-01
3.900D+00	3.988D-01
3.950D+00	5.195D-01
4.000D+00	6.504D-01
4.050D+00	7.905D-01
4.100D+00	9.389D-01
4.150D+00	1.095D+00
4.200D+00	1.258D+00
4.250D+00	1.427D+00
4.300D+00	1.601D+00
4.350D+00	1.781D+00
4.400D+00	1.966D+00
4.450D+00	2.156D+00
4.500D+00	2.350D+00
4.550D+00	2.548D+00
4.600D+00	2.755D+00
4.650D+00	2.955D+00
4.700D+00	3.164D+00
4.750D+00	3.376D+00
4.800D+00	3.590D+00
4.850D+00	3.808D+00
4.900D+00	4.029D+00
4.950D+00	4.252D+00
5.000D+00	4.477D+00
5.050D+00	4.705D+00
5.100D+00	4.936D+00
5.150D+00	5.168D+00
5.200D+00	5.403D+00
5.250D+00	5.639D+00
5.300D+00	5.878D+00
5.350D+00	6.118D+00
5.400D+00	6.360D+00
5.450D+00	6.604D+00
5.500D+00	6.849D+00

SPICE 2 Simulation Results
Transfer Characteristics
Figure 8(c)

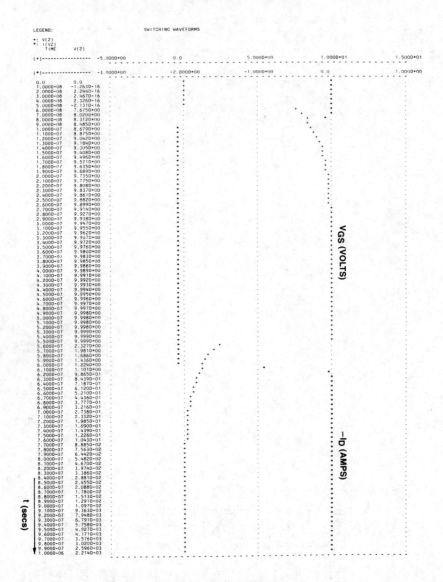

SPICE 2 Simulation Waveforms –
Switching Time Test
Figure 9

ii) Switching Efficiency

One of the important considerations for the operating frequency in switching applications is the efficiency of the switch. Reference 3 (Page 18) describes a test performed to determine the efficiency of the switch at different frequencies with resistive load. Figure 10 shows a SPICE 2 diagram which simulates the test circuit. Figure 10 also shows additional SPICE 2 circuits to measure the device dissipation and the source power delivered. The measurement principle is as follows.

$$G1 = V_{DS}i_D = \text{instantaneous power dissipated in MOSFET}$$

$$G2 = V_Di_D = \text{instantaneous power delivered by source}$$

In one period, T,

$$\Delta V_{C1} = \frac{1}{C_1} \int_{t_1}^{t_1 + T} V_{DS}i_D dt$$

$$\Delta V_{C2} = \frac{1}{C_2} \int_{t_1}^{t_1 + T} V_Di_D dt$$

By selecting $C_1 = C_2 = T$,

$$\Delta V_{C1} = P_1 = \text{average power dissipated in MOSFET}$$

$$\Delta V_{C2} = P_2 = \text{average power delivered by source}$$

$$\text{Switch Efficiency} = \frac{\text{Power input} - \text{MOSFET losses}}{\text{Power input}}$$

$$= \frac{P_2 - P_1}{P_2}$$

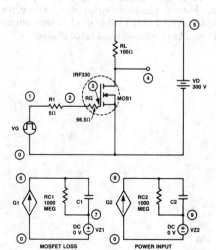

SPICE 2 Simulation Diagram to Test Switching Efficiency
Figure 10

```
*****************
* INPUT LISTING *
*****************
TEST OF MOSFET MODEL
*----------------------------------------------------
*SWITCHING EFFICIENCY DETERMINATION AT 500 KHZ - DEVICE TYPE-IRF330
*----------------------------------------------------
VG 1 0 PULSE(0. 10. 0. .1NS .1NS .9999US 2.US)
R1 1 2 5.
MOS1 4 3 0 0 IRF330
RL 4 5 100.
VD 0 5 DC -300.V
G1 0 6 POLY(2) 4 0 5 4 0 0 0 0 .01
C1 6 7 2.UF
RC1 6 7 1000.MEG
VZ1 7 0 DC 0.V
G2 0 8 POLY(2) 5 0 5 4 0 0 0 0 .01
C2 8 9 2.UF
RC2 8 9 1000.MEG
VZ2 9 0 DC 0.V
*----------------------------------------------------
RG 2 3 66.5
.MODEL IRF330 NMOS(VTO=3.6048 KP=13.44 LAMBDA=0. RS=.1293 CGS=20.PF
+          CGD=660.PF CBD=4.5 PB=1. RD=.7577 LEVEL=1)
*----------------------------------------------------
.TRAN 20.NS 6.2US 1.8US UIC
.PLOT TRAN V(4) V(2) V(1) I(VD) I(VZ1) I(VZ2)
.PRINT TRAN I(VZ1) V(6,7) I(VZ2) V(8,9)
.OPTIONS LIMPTS=1000
*----------------------------------------------------
.END
```

SPICE 2 Listing for the Circuit to Test Switching Efficiency
Figure 11

Figure 12 shows the simulation results, along with experimental data from Reference 3. It is found that the model predicts the efficiencies at different frequencies in a reasonably accurate manner.

Note: The 1000 MΩ resistors across C_1 and C_2 are included to meet SPICE 2 programming requirements.

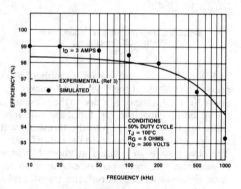

Switching Efficiency vs Frequency
Figure 12

b) Inductive Load Switching:

Figure 13 shows a circuit (Page 18, Reference 3) that was used in order to study the ability of the model to simulate switching circuits with inductive loads. The diode was modeled as a near ideal device, and stray inductances were not included in the simulation model. A capacitive speed-up network was included in the MOSFET gate drive circuit. The inductance value was chosen to be large so that the current remains essentially constant over the interval of interest. The circuit was simulated both under ON and OFF transient conditions. SPICE 2 listings for switching off are given in Figure 14. The ON and OFF simulated waveforms are shown in Figures 15a and 15b, respectively.

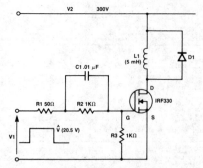

Switching Circuit with Clamped Inductive Load (Ref. 3)
Figure 13

```
******************
* INPUT LISTING *
******************

TEST OF MOSFET MODEL
*--------------------------------------------------------------
*GENERATION OF INDUCTIVE SWITCHING WAVEFORM - DEVICE TYPE-IRF330
*--------------------------------------------------------------
R1 1 2 50.0HMS
R2 2 3 1.K
C1 2 3 .01UF IC=10.V
R3 3 0 1.K
V1 1 0 PULSE(20.5 0. 500.NS .1NS .1NS 1.S 1.1S)
L1 6 5 5.MH IC= 3.0A
D1 5 6 D
V2 0 6 DC -300.V
MOS1 5 4 0 0 IRF330 IC=2.3, 10., 0.
.MODEL D D
*--------------------------------------------------------------
RG 3 4 66.50HMS
.MODEL IRF330 NMOS(VTO=3.6048 KP=13.44 LAMBDA=0. RS=.1293 CGS=20.PF
+           CGD=660.PF CBD=4.5 PB=1. RD=.7577 LEVEL=1)
*--------------------------------------------------------------
.TRAN 5.NS 950.NS 450.NS UIC
.PLOT TRAN V(3) V(5) I(V2)
.OPTIONS RELTOL=.00001
*--------------------------------------------------------------
.END
```

SPICE 2 Listing for Inductive Load Circuit
(Switch-Off Conditions)
Figure 14

In Figure 15a, the gate voltage is applied at 500 nsecs. The voltage at the device gate quickly rises above the final value of 10V due to the presence of the speed-up capacitor, C_1. Until the gate input capacitor, C_{iss}, charges up to the threshold voltage, V_T, the device remains in the cut-off region. The device starts to conduct at about 515 nsecs commutating the free-wheel diode. Since no parasitic elements are considered, this transfer takes place almost instantaneously. V_{DS} remains until

then at approximately 300 volts. V_{DS} now falls at such a rate as to maintain the internal gate-to-source voltage, V_{GS}, (across C_{GS}) of the device, equal to the value required to sustain the load current. In this case, this value of V_{GS} would be 4.65V as may be seen from the output characteristics.

In Figure 15b, the gate voltage is switched off at 500 nsecs. Initially, a negative voltage appears across the gate-to-source, though the final voltage would be zero. This is due to C_1, the speed up capacitor. V_{DS} now rises to maintain the internal gate to source voltage, V_{GS}, once again at approximately 4.65 volts. I_D continues to flow through the MOSFET. When V_{DS} reaches a value a little more than 300V, the diode takes over conduction and the device current drops quickly from full value to zero.

In Figure 15a, the rise times of I_D and V_D are 0 nsec and 50 nsec, respectively. The corresponding values from Reference 3 are 80 nsec and 30 nsec. In Figure 15b, the fall times of I_D and V_D and the time delay (off) are 0 nsec, 50 nsec, and 25 nsec, respectively. The corresponding values from Reference 3 are 40 nsec, 40 nsec, and 60 nsec. A major discrepancy is observed in the rise and fall times of the drain current. This may be attributable to the likely presence of stray inductances in the experimental circuit.

Note: When simulating a switch in series with an inductor using SPICE 2, the voltages across the switch and the inductor are sometimes inaccurately computed under switch off conditions (i.e., switch current and inductor current = 0). The voltages may even, erroneously, appear to be oscillatory. The more stringent relative error tolerance (RELTOL) value used in the program in Figure 14 is to avoid this source of error.

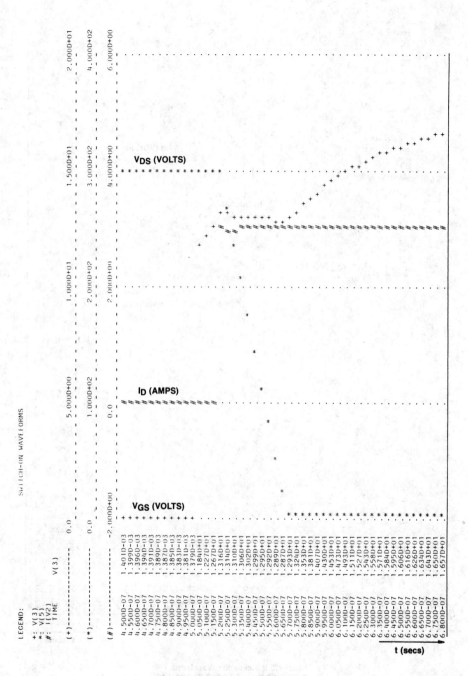

SPICE 2 Simulation Waveforms
Clamped Inductive Circuit (Switch-On)
Figure 15(a)

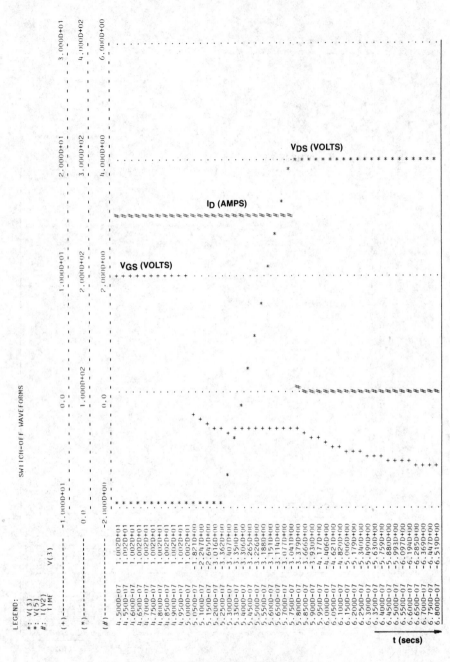

SPICE 2 Simulation Waveforms
Clamped Inductive Circuit (Switch-Off)
Figure 15(b)

t (secs)

Conclusion

The high power MOSFET model due to Herren Jr., Nienhaus and Bowers is seen to be a simple model requiring only the device data published by the manufacturer. The SPICE 2 computer simulation run with the IRF330 device model shows that the model adequately reproduces the static performance of the device. The model performs generally well in the dynamic switching modes. However, the replacement of the non-linear capacitor C_{rss} with a fixed value capacitor in the SPICE 2 version of the model leads to smaller rise and fall times at low V_{DS}. Also, by optimizing the model parameters further, it is possible to improve the performance of the model. As an example, by increasing R_D magnitude, better correlation at low frequencies in the switching efficiency test (Figure 12) may be obtained.

References

[1] P. C. Herren, Jr., H. A. Nienhaus, and J. C. Bowers, "A High Power MOSFET Computer Model," *IEEE PESC Record*, 1980.

[2] L. W. Nagel, "SPICE 2: A Computer Program to Simulate Semiconductor Circuits," Memorandum No. ERL-M520, College of Engineering, University of California, Berkeley, May 1975.

[3] International Rectifier, *HEXFET Databook– Power MOSFET Application and Product Data*, El Segundo, California: International Rectifier Corporation, Semiconductor Division, 1981.

2

2.13.2 The High-Frequency Model

Power MOSFET transistors for high-frequency applications have mushroomed in popularity, partly because of inherent temperature-related advantages unique to the technology. Such advantages as discussed elsewhere include no thermal runaway, no current hogging and, to a lesser extent, a greatly reduced secondary breakdown phenomenon. As a result of increased popularity of the power MOSFET, it is important that the designer has the tools necessary to use CAD (Computer Aided Design) techniques for design optimization. One CAD program that appears to have gained widespread popularity among high-frequency design engineers is known as Compact; consequently, the high-frequency model offered in this handbook was specifically developed for that program. Although models have been offered before, few have achieved a fully satisfactory representation that offers viable results over wide bandwidths [1,2,3].

The principal deficiency of these earlier models is neglecting the parasitic bipolar transistor inherent in all power MOSFET structures. This parasitic *npn* bipolar transistor arises from the interaction of the p^- channel (base), the n^+ source (emitter), and the n^- drain-drift epitaxy (collector); these "layers" can be seen in the cross section of the MOSFET shown in Figure 1.

At DC and frequencies below approximately 15 MHz, it appears that the elements of this parasitic *npn* bipolar transistor have little effect upon amplifier performance. However, at higher frequencies, this parasitic bipolar transistor becomes increasingly significant.

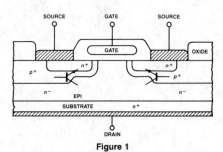

Figure 1

Device Design

The models we have reviewed in this chapter all have one physical commonality: they are *planar* double-diffused MOS structures. However, to maintain the lowest parasitic capacitances, all high-frequency power MOSFETs are fabricated using the truncated etched V-groove technology similar to that shown in Figure 2. Although this structure appears dramatically different from the more familiar planar structure (see Figure 1), in actuality, operation is similar and, as a consequence, the basic planar models previously studied remain essentially unchanged.

The basic construction closely resembles the common four-layer bipolar transistor but with a notable difference—the V-groove is etched anisotropically into the structure to provide access to the p^- channel by means of a metal gate overlay over oxide. As with the more conventional planar DMOS structures, the p^- channel (base) is shorted to the n^+ source (emitter), to effectively mute the *npn* bipolar transistor. For DC

Reprinted from RF Design magazine January/February, 1979 with permission. © Cardiff Publishing Co. 1979

and linear low-frequency models, this parasitic *npn* bipolar transistor is effectively eliminated, but *not so* for high-frequency applications.

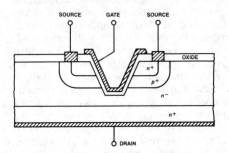

Truncated V-Groove Power MOSFET
Figure 2

The Schematic Model

Using the physical model, the entire electrical path and all the parasitic elements that comprise the schematic model of the high-frequency power MOS-FET can be identified. Figure 3 identifies each element of the schematic model. The element, R_{OS} (1/ G_{OS}), represents the output resistance (conductance) which cannot be physically realized. L_G, L_S, and L_{DP} are not intrinsic (that is, part of the actual semiconductor element) but represent the package parasitic inductances of the VMP4 transistor. R_G and R_{SP} represent resistive losses in both the gate and source metalizations as well as the lead losses. C_{GS} differs from C_{GN} in that the former is the field capacitance, whereas the latter is that parasitic capacitance existing between the gate metal and the $n+$ source diffusion.

The parasitic *npn* bipolar transistor evident in Figure 1 must be considered as contributing parasitic elements (R_B, C_{CG}, and C_{DB}) as well as being a potential parasitic generator, that is, contributing a finite Beta as an active element. The last is possible, and some explanation is necessary. Although the VMOS source is metallically tied to the base to reduce the effects of this parasitic *npn* bipolar transistor, the resistivity of the *p* diffusion, that forms both the DMOS channel and the parasitic transistor's base, has a finite resistance (measured as ohms-per-square). Although at the point of metallic contact the base-to-emitter resistance is effectively zero; nonetheless, as the distance from the short increases, the bulk resistance increases. Thus Figure 1 can be "corrected" as shown in Figure 4. Should a voltage exist across the

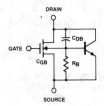

DMOS with a Parasitic *npn* Bipolar
Figure 4

base resistance, R_B, it is conceivable that the parasitic *npn* bipolar transistor can turn on. One obvious means of placing a voltage across this resistance is by coupling the output voltage on the drain through the drain-base capacitor, C_{DB}. However, it was determined that this parasitic *npn* bipolar transistor, acting as an independent parasitic generator, is effectively muted having been found not to be a major contribu-

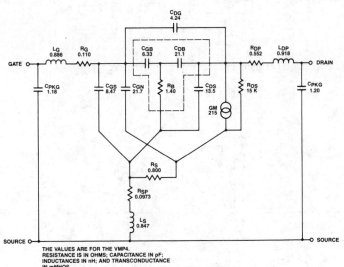

THE VALUES ARE FOR THE VMP4.
RESISTANCE IS IN OHMS; CAPACITANCE IN pF;
INDUCTANCES IN nH; AND TRANSCONDUCTANCE
IN mMHOS.

The Schematic Model (VMP4)
Figure 3

tor to the performance of the VMP4 at HF through VHF (400 MHz). Nevertheless, the contribution of the *npn* bipolar transistor's parasitic elements, R_B, C_{CB}, and C_{BD}, as an RC feedback network are of paramount importance, as illustrated in Figure 5, where the intrinsic gain, $S_{21}(dB)$, of the DMOS model is computed both with and without the contribution of these parasitic elements.

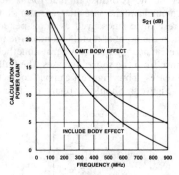

Effect of Including/Excluding the Body Resistance in the Calculation of $S_{21}(dB)$
Figure 5

Experimental Results

Measured scattering parameters were used as the basis for establishing an exact model. Nodal analysis using the Circuit Optimization Program, Compact [6], working with 15 variables, allowed the interconnection of 3 and 4 port networks as illustrated in Figure 6. In the Compact program, variables are identified as negative quantities. The final program used to determine the values of each element of the VMP4 transistor is listed in Figure 7. Please be aware that one does not simply insert measured S-parameters and let the program crank numbers endlessly. No, indeed. First, you must have a good idea of "where you are coming from," and use good judgment as to the initial values. Compact optimizes the initial data to arrive at parameter-fitting model.

R_{IN} (in Figure 6) is used simply as a sense element required to realize a 4-port generator. To reduce its effect, a value of 1E10 ohms was assigned. S-parameters measured at 200 MHz were entered into the program and the computed Y-parameters were then compared with measured values and found to be in close agreement as shown in Figure 8.

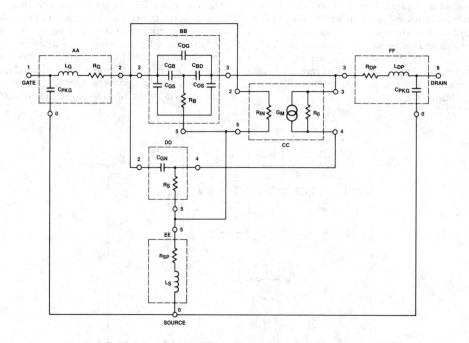

Nodal Analysis Flow Chart for Compact Synthesis
Figure 6

```
CAP    AA    PA        1.18
SRL    BB    SE     −  0.110          −0.886
CAX    AA    BB
CAP    BB    PA     −  8.47
CAP    CC    SE     −  6.33
RES    DD    PA     −  1.40
CAP    EE    SE     −21.10
CAP    FF    PA     −13.50
CAX    BB    FF
CAP    CC    SE     −  4.24
PAR    BB    CC
GEN    CC    VC        .100E+11        0.150E+05      215
CAP    DD    SE     −21.7
RES    EE    PA     −  0.800
CAX    DD    EE
SRL    EE    SE     −  0.973E-01       −0.847
SRL    FF    SE     −  0.552           −0.918
CAP    GG    PA        1.20
CAX    FF    GG
CON    AA    T3        1.00            2.00      0.0
CON    BB    T3        2.00            3.00      5.00
CON    CC    T4        2.00            5.00      4.00      3.00
CON    DD    T3        2.00            4.00      5.00
CON    EE    T2        5.00            0.00
CON    FF    T3        3.00            6.00      0.0
DEF    AA    T2        1.00            6.00
TWO    BB    S1        50.0
SET    AA    BB
PRI    AA    S1        50.0
END
200
END
.77    −146.3   2.14   57.5   .035   −6.0   .759   −146.6
END
.001
1    1    1    .627
END

EOF;
```

**Compact Program Including the Parasitic Elements of
the *npn* Bipolar in a Nodal Analysis
Figure 7**

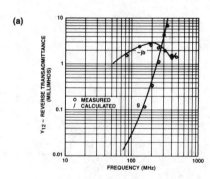

(a)

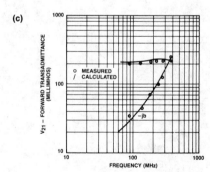

(c)

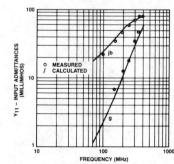

(b)

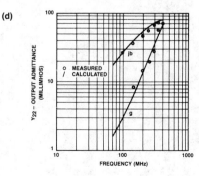

(d)

**VMP4
Figure 8**

Conclusions

This model offers a designer an excellent start in computer aided design of amplifiers.

References

[1] James G. Oakes, et. al., "A Power Silicon Microwave MOS Transistor," *IEEE Trans. Microwave Theory and Techniques,* Vol. MTT-24, June 1976, pp. 305–311.

[2] T. M. S. Heng, et. al., "Vertical Channel Metal-Oxide-Silicon Field Effect Transistor," Final Report, Westinghouse Research and Development Center, ONR Contract N00014-74-C-0012, November 1976.

[3] Hans J. Sigg, et. al., "D-MOS Transistor for Microwave Applications," *IEEE Trans. Electron Devices,* Vol. ED-19, January 1972, pp. 45-53.

[4] Marvin Vander Kooi and Larry Ragle, "MOS Moves into Higher Power Applications," *Electronics,* Vol. 49, June 24, 1976, pp. 98-103.

[5] Arthur Evans, et. al., "High Power Ratings Extend VMOS FETs Domination," *Electronics,* Vol. 51, June 22, 1978, pp. 105–112.

[6] Les Besser, Compact *Reference Manual,* National CSS, Inc., Version 4.50, January 1977.

3.1 Linear Operation

Introduction

Applications for power MOSFETs can basically be divided into two categories: those in which the devices are used as resistive on-off switches and those in which they are not. This latter group of applications is generically described by the term "linear", but it does not necessarily mean that a MOSFET is being used in a linear part of one of its characteristic curves. Rather, the term linear is used whenever a MOSFET is operated in a continuous or non-continuous fashion where the drain current is a function of gate-to-source voltage or vice-versa. Usually this is achieved by operating a MOSFET either as a voltage-controlled current source or as a voltage-controlled resistor with open or closed loop control. What distinguishes "linear" operation from switching operation is that in a switching circuit the drain current is determined by components external to the MOSFET, whereas in a linear circuit this is not necessarily so.

This section provides the potential user of power MOSFETs in linear applications with useful and technically correct information about MOSFET behavior with varying voltage, current and temperature. Each of the topics discussed in this section provides insight into how a MOSFET functions and the underlying physical phenomena that govern its properties.

Output Characteristics

The output characteristics of a MOSFET provide information about how the drain current (I_D) varies with applied drain-to-source voltage (V_{DS}) at various values of gate-to-source voltage (V_{GS}). Figure 1 shows a typical family of output characteristic curves

for a Siliconix IRF630 power MOSFET. Two distinct regions of operation can be observed from this family of curves: the ohmic region where on-resistance is approximately constant and the saturation region where drain current is approximately constant.

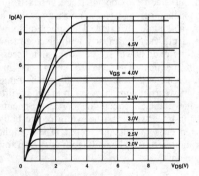

The Output Characteristics of an IRF630 DMOS Power FET at $T_C = 25°C$ Show the Drain Current Increasing Linearly from Zero and then Levelling Off at Constant Values Dependent Only on V_{GS}
Figure 1

With zero applied drain-to-source voltage, the drain current is zero independent of the gate-to-source voltage. As V_{DS} is increased in value, with the gate biased positively at some voltage greater than approximately three volts, the drain current increases linearly at first but then levels off to an essentially constant value. The point at which this occurs is where $V_{DS} = V_{GS} - V_T$, and this voltage is termed $V_{DS(sat)}$. V_T is the device parameter known as threshold voltage, and the section on transfer characteristics discusses this in more detail.

The phenomenon that causes the drain current to level-off or "saturate" with increasing drain-to-source voltage is known as channel pinch-off. Figure 2 shows in cross-section how this occurs in a vertical DMOS FET (like the IRF630) for the conditions $V_{GS} > V_T$ and $V_{DS} > V_{GS}-V_T$.

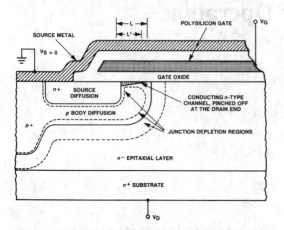

The Pinched-Off Channel in a n-Channel VDMOS Device is Shown for the Conditions $V_{DS} \geq (V_{GS} - V_T)$ and $V_{GS} > V_T$
Figure 2

The depth of the conducting n-type channel below the silicon/silicon dioxide interface varies from zero at the pinched-off end, to a non-zero value at the edge of the source diffusion. The diagram shows the channel depth varying linearly with distance from the pinched-off end; however, this does not necessarily occur in practice. The reason for this is that the channel depth is dependent, among other things, on the impurity carrier concentration in the body diffusion below the channel. Since the carrier concentration varies non-linearly across the width of the body diffusion, so does the channel depth. The straight line approximation to the channel depth is shown for convenience only.

Electrons traveling through the channel "see" no barrier as they approach the depletion region at the pinched-off end [1]. Consequently, the saturated drain current is determined by the rate at which the electrons arrive at the edge of the depletion region. In a first order analysis, this rate is only dependent on V_{GS} and is insensitive to V_{DS}. Thus the drain current would seem to be constant for $V_{DS} > V_{GS} - V_T$.

In reality, however, the saturated drain current does increase somewhat as the applied V_{DS} is increased. Notice that the pinched-off end of the channel is separated from the body-drain pn-junction by a dis-

tance equal to the width of the depletion region in the body diffusion. It is this depletion region and the one in the n- epitaxial layer that sustain the difference between the applied drain-to-source voltage and the value of $V_{DS(sat)}$.

As the applied V_{DS} is increased, the expanding depletion region in the body diffusion causes the pinched-off end of the channel to move towards the source diffusion. Consequently, the effective channel length L′ decreases as V_{DS} increases. It is important that this "channel shortening" effect, as it is known, be minimized since the saturated drain current is proportional to 1/L′. The saturated drain current in the square-law region can be approximated by the equation [1]:

$$I_D(sat) = \frac{\mu \; C_{OX} \; W}{2 \; L'} \; (V_{GS} - V_T)^2 \qquad (1)$$

Where μ is the carrier mobility in the conducting channel, C_{OX} is the gate oxide capacitance per unit area, and W is the total effective channel width.

One can see that if V_{GS} is held constant, the drain current will vary only with L′, assuming negligible drift in junction temperature (i.e., μ and V_T stay approximately constant). As V_{DS} is increased, L′ decreases and I_D increases. This rise in drain current causes the MOSFET to have a finite output conductance in its saturation region. The "horizontal" family of curves shown in Figure 1 thus all have a slight upward slope no matter how slight. It can be shown that the output conductance in the square law region is equal to:

$$G_O = \frac{\partial I_D}{\partial V_{DS}} = - \frac{I_D}{L'} \times \frac{\partial L'}{\partial V_{DS}} \qquad (2)$$

Equation 2 shows that the output conductance is equal to the value of the drain current times the fractional change in effective channel length per volt of V_{DS}. This accounts for the observed rise in output conductance of most devices as the drain current increases.

L′ also varies as a function of the non-linear impurity carrier concentration across the body diffusion. However, almost all of the currently available power MOSFETs show less than 10% channel shortening as V_{DS} is increased from zero to its rated breakdown voltage. This means that few devices will show more than an 11% variation in saturated drain current over the operating drain to source voltage specification.

The family of output characteristic curves for all power MOSFETs show a rather unexpected behavior as far as junction temperature is concerned. Figure 3 shows the family of curves for the same IRF630 used before, except in this case the junction temperature was raised to 125°C. One can see that in comparison to Figure 1, below gate-to-source voltages of 4.0 volts, the family of curves has moved up, while above $V_{GS} = 4.5$ volts, the curves have moved down.

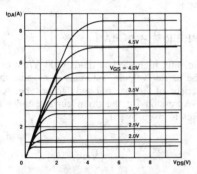

The Output Characteristics of an IRF630 at $T_C = 100°C$ Show That in Comparison to Figure 1, the Top Two Traces Have Moved Down While the Lower Traces Have Moved Up
Figure 3

The device thus has a zero temperature coefficient point between $V_{GS} = 4.0$ volts and 4.5 volts, and this is characteristic of all available power MOSFETs. The physical mechanisms that result in this characteristic behavior will be discussed in detail in the next section.

Transfer Characteristics and Threshold Voltage

The transfer characteristic curve of a power MOSFET shows how the saturated drain current varies with applied gate-to-source voltage for a constant value of drain-to-source voltage. Figure 4 shows the I_D versus V_{GS} transfer curve of an IRF630 MOSFET, with V_{DS} set to 10 volts and a case temperature of 25°C.

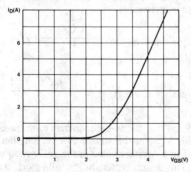

The Saturated Transfer Characteristics of an IRF630 MOSFET at $T_C = 25°C$ Show a Square-Law Behavior up to $I_D = 3A$, Above Which the Curve Becomes Linear
Figure 4

In Figure 5, there are three distinct regions on the device's transfer curve, and each results from differing conduction phenomena.

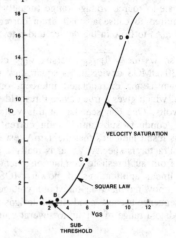

This Diagram Shows the Three Distinct Regions on the Saturated Transfer Characteristic Curve
Figure 5

In region A to B, a small amount of drain current flows, but V_{GS} is below the MOSFET threshold voltage V_T. Since the magnitude of the current that flows in this subthreshold mode of conduction is negligible compared to a device's maximum current handling capacity, very little work has been done to characterize power MOSFETs in this region. A paper by Swanson indicates that there is an exponential relationship between drain current and gate-to-source voltage for subthreshold conduction [2].

In region B to C on the curve in Figure 5, the drain current is proportional to the square of the difference between V_{GS} and V_T. This is characteristic of all MOSFETs, both small signal and power devices, as indicated by equation (1). The actual derivation of equation (1) is outside the scope of this section, but basically it results from applying Gauss's law to find the surface charge density under the gate electrode, and then by performing some simple algebraic and calculus operations to solve for the drain current [3].

Obviously a plot of $\sqrt{I_{D(sat)}}$ versus V_{GS} in the square-law region should be linear. The reason one would plot this data is that the true threshold voltage of a device can be obtained from the extrapolated intercept of the straight line part of the curve onto the V_{GS} axis. It should be noted that the value of threshold voltage found by this method can be considerably different from the threshold voltage range most manufacturers specify on their data sheets.

This is because it is very cumbersome and time consuming to determine the "true" device threshold

3

voltage using the above method in a production environment. Therefore, almost all power MOSFET data sheets specify a pseudo-threshold voltage which is just the gate-to-source voltage range (usually 2 to 4 volts) required to cause a small drain current (frequently 1 mA) to flow in the device under test.

Figure 6 shows the $\sqrt{I_{D(sat)}}$ versus V_{GS} curve of the IRF630 DMOS device in its square-law region. Also shown is the extrapolated intercept onto the V_{GS} axis, which gives a true device threshold voltage of 2.90 volts. One can see that at this voltage the MOSFET conducts substantial drain current (25.6 mA). This result indicates that the 1 mA test current most manufacturers specify V_T at is really a current resulting from subthreshold conduction phenomena. In most linear applications of power MOSFETs, however, knowing the true threshold voltage is of little consequence because good circuit designs usually accommodate a range of V_T up to several volts.

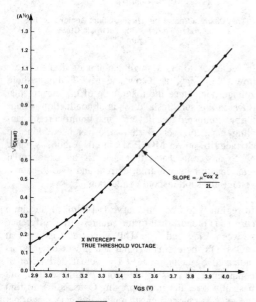

By Plotting $\sqrt{I_{D(SAT)}}$ versus V_{GS}, the True Threshold Voltage of the MOSFET can be Obtained by Extrapolation of the Curve Onto the V_{GS} Axis
Figure 6

Referring to Figure 5 again, it can be seen that as soon as the gate-to-source voltage at point C is reached, the transfer curve becomes linear. This linearization is the result of the effect known as "velocity saturation" occurring. It can be explained as follows:

The lateral electric field accelerating the mobile charge carriers through the pinched-off channel (see Figure 2) is equal to the difference in the voltage at either end of the channel divided by the effective channel

length [4]. The voltage at the surface of the channel at the source end is zero while at the pinched-off end it is equal to $V_{DS(sat)}$ or $V_{GS} - V_T$. Thus:

$$E_{CHAN} = \frac{(V_{GS} - V_T) - 0}{L'} = \frac{V_{GS} - V_T}{L'} \quad (3)$$

As the E field accelerating the carriers increases with gate-to-source voltage, their velocity also increases since [4]

$$V_{CARRIER} \stackrel{\Delta}{=} \mu \, E_{CHAN} \quad (4)$$

At sufficiently high electric fields (10^4 V/cm or 1 V/μm), the carrier velocity saturates at a value that is characteristic of the material and carrier type. A simplified expression for the drain current under these conditions is given by [5]

$$I_{D(sat)} = \frac{C_{ox} \, W \, V_{sat}}{2} (V_{GS} - V_T) \quad (5)$$

where $V_{sat} = 5 \times 10^6$ cm/second for electrons in silicon [6].

The electric fields required to cause the channel carrier velocity to saturate are easily achieved in DMOS power FETs. Since the effective channel length is typically in the 1 to 2 μm range, velocity saturation can occur at gate-to-source voltages just a few volts above threshold. This is evident in Figure 4.

Another important aspect of the transfer characteristic curve depicted in Figure 4 is the transconductance or slope of the curve. Transconductance is defined as

$$g_m = \frac{\partial I_D}{\partial V_{GS}} \bigg|_{V_{DS} = CONSTANT} \quad (6)$$

For the square law and velocity saturated regions, the transconductance is

$$g_m \text{ (square law region)} = \frac{\mu \, C_{ox} \, W}{L'} (V_{GS} - V_T) \quad (7)$$

$$g_m \text{ (velocity sat. region)} = \frac{C_{ox} \, W \, V_{sat}}{2} \quad (8)$$

One can see that the transconductance increases linearly with V_{GS} in the square law region, but then levels off to a constant value in the velocity saturated region. This is characteristic of all short channel DMOS power FETs. No other semiconductor device shows such a linear relationship between input voltage and output current.

In the velocity saturated region, device transconductance is thus a function of two device variables: oxide capacitance per unit area (C_{OX}) and effective channel width (W). For a given value of C_{OX}, it can be seen that the transconductance increases with total W, which itself increases with die size. This is the reason why power MOSFETs having large continuous drain current ratings, and hence a large die size, also have high values of transconductance (in some cases as high as 20 siemens).

Figure 7 shows a typical data sheet transfer curve for the IRF630 device for TC = 25°C, −55°C, and +125°C. It was seen in the previous section that for a specific value of applied V_{GS}, there is zero drift in the drain current with temperature. This is clearly evident in Figure 7 as the point at which all three curves cross through. One can see that below this "zero TC point", as it is known, the temperature coefficient of saturated drain current is positive while above the zero TC point it is negative.

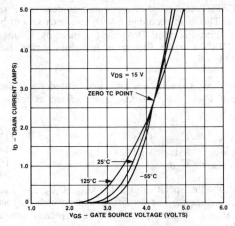

The Transfer Curves of an IRF630 MOSFET at T_C = −55°C, 25°C and 125°C Clearly Show the Occurrence of the Zero Temperature Coefficient or "Zero T_C" Point
Figure 7

Obviously, the zero TC point of a power MOSFET must occur in either the square law region or the velocity saturated region. Using equations (1) and (5), it can be shown that the zero TC point may occur in either of these two regions.

From equation (1), for the square law region,

$$\frac{\partial I_{D}(sat)}{\partial T} = I_D\left(\frac{1}{\mu} \times \frac{\partial \mu}{\partial T} - \frac{2}{V_{GS}-V_T} \times \frac{\partial V_T}{\partial T}\right) \quad (9)$$

and, from equation (5), for the velocity saturated region,

$$\frac{\partial I_{D}(sat)}{\partial T} = \frac{C_{OX}Z}{2}\left((V_{GS}-V_T)\frac{\partial V_{sat}}{\partial T} - V_{sat}\frac{\partial V_T}{\partial T}\right)(10)$$

Experimental data has shown that the derivative of the magnitude of MOSFET threshold voltage with respect to temperature is negative, and reasonably constant at about −3 to −6 mV/°C [7].

The derivatives of carrier mobility and saturation velocity are also both negative over the operating temperature range of a power MOSFET although V_{sat} is less sensitive to temperature than μ is.

By setting equations (9) and (10) equal to zero, the value of V_{GS} at the zero TC point can be determined for both the square law and the velocity saturated regions:

V_{GS} (at zero TC point in SLR) =

$$V_T + \frac{2\mu\, \frac{\partial V_T}{\partial T}}{\frac{\partial \mu}{\partial T}} \quad (11)$$

V_{GS} (at zero TC point in VSR) =

$$V_T + \frac{V_{sat}\, \frac{\partial V_T}{\partial T}}{\frac{\partial V_{sat}}{\partial T}} \quad (12)$$

It is interesting to note the similarity between equations (11) and (12). Both of them indicate that the zero TC point lies at some specific positive voltage above device threshold. This positive offset above threshold is in each case equal to the derivative of threshold voltage with respect to temperature (or two times the threshold voltage in the case of equation (11), divided by the fractional change per degree centigrade in either carrier mobility or saturation velocity). However, it is often difficult to determine in which region the zero TC point lies even from a clear diagram such as that shown in Figure 7.

One should note that the temperature coefficient of drain current at low values of I_D, is dominated by the threshold voltage drift with temperature. This is obvious from equation (9) since its second term becomes quite large as V_{GS} approaches V_T.

Conversely, the temperature coefficient of drain current above the zero TC point is dominated by the mobility drift with temperature in the square law region and by the carrier saturation velocity drift with temperature in the constant transconductance region.

Interelectrode Capacitance

All types of DMOS power FETs, whether lateral, vertical or V-groove, have capacitances between the gate, source and drain terminals that are dependent upon the device geometry and die size. The origin of these capacitances in a closed-cell vertical DMOS device, like the IRF630, is shown in Figure 8.

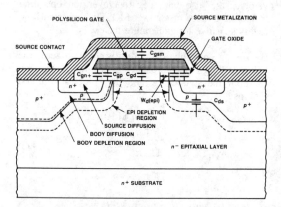

**This Cross Sectional View of a VDMOS Power FET Shows the Origin of the Interelectrode Capacitances
Figure 8**

One can see that the polysilicon gate structure has both capacitance to the source metalization overlaying it, as well as capacitance to the silicon surface below. C_{gsm} is the gate-to-source metal capacitance; C_{gn+} is the gate to n^+ source-diffusion capacitance; C_{gp} is the gate to p^- type body-diffusion capacitance; and C_{gd} is the gate to drain-drift region capacitance. Commonly, the sum of C_{gsm}, C_{gn+} and C_{gp} is referred to as C_{gs}, the total gate-to-source capacitance.

The parasitic *npn* bipolar transistor inherent to the structure of an *n*-channel DMOS FET has a reverse biased *pn*-junction between its base and collector. The depletion capacitance of this junction appears between the drain and source terminals of the MOS-FET and is termed C_{ds}.

Although the capacitances in a power MOSFET are distributed over the whole surface area of the die, for the purpose of simplified circuit analysis they are

usually lumped into single elements as shown in Figure 9. This model is adequate for low-frequency (<5 MHz) small signal analysis but breaks down at higher frequencies. A more elaborate model is required at higher frequencies and must include parasitic package and bond wire inductances (see 2.13.2).

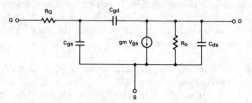

**This Simplified Small-Signal Model of a DMOS Power FET in its Saturation Region is Adequate for Low Frequency (< 5 MHz Analysis)
Figure 9**

Manufacturers data sheets do not generally tabulate C_{gs}, C_{gd} and C_{ds} directly; rather they specify the input, output and reverse transfer capacitances of the MOSFET connected in a common-source configuration. This is in accordance with standards set out by the Joint Electron Devices Engineering Council (JEDEC) and the Electronic Industries Association (EIA).

The three capacitances one finds listed on a data sheet, C_{iss}, C_{oss} and C_{rss}, are defined as follows [8]:

$$C_{iss} = C_{gs} + C_{gd} \qquad (13)$$
$$C_{oss} = C_{ds} + C_{gd} \qquad (14)$$
$$C_{rss} = C_{gd} \qquad (15)$$

Usually the specifications for C_{iss}, C_{oss} and C_{rss} will include the test conditions under which they are measured. These test conditions typically are as follows:

(i)   f   = 1 MHz
(ii) V_{GS} = 0 volts
(iii) V_{DS} = 25 volts

Additionally, many manufacturers include on their data sheets curves showing how the capacitance parameters vary with applied drain to source voltage. Some plot the variation of C_{iss}, C_{oss} and C_{rss} against V_{DS}, but a more useful set of curves would be C_{gs}, C_{ds} and C_{gd} plotted against V_{DS}. Siliconix now supplies this latter set of curves on their data sheets so that the small-signal AC model capacitance parameters can be written down directly. An example of these capacitance versus V_{DS} curves for the Siliconix IRF630 MOSFET is shown in Figure 10. As can be seen, there is only a very slight variation in C_{gs} with applied V_{DS}, but a much more substantial variation in C_{gd} and C_{ds}.

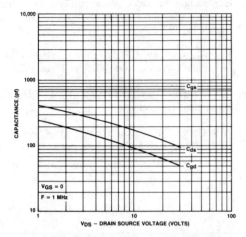

VGS = 0
F = 1 MHz

V_{DS} – DRAIN SOURCE VOLTAGE (VOLTS)

CAPACITANCE (pf)

C_{gs}
C_{ds}
C_{gd}

This Log-Log Plot of Device Capacitance versus Drain to Source Voltage Shows Very Little Variation in C_{gs} but a Much More Substantial Variation in C_{ds} and C_{gd}
Figure 10

One would probably expect C_{gs} not to vary at all with applied V_{DS}, and for all practical purposes, this assumption can be made. Of the three capacitances that constitute C_{gs}, only C_{gp} shows a dependence on V_{DS}. This is due to the widening of the depletion region in the p^- type body diffusion as V_{DS} increases. The further the body depletion-region moves towards the n^+ source diffusion, the smaller C_{gp} becomes. This is because only the undepleted region of mobile charge at the surface of body diffusion, below the polysilicon gate, can form the lower "plate" of C_{gp}. Since most modern DMOS power FETs show typically less than 10% depletion spreading across the width of the body region as V_{DS} is varied from zero to its rated breakdown value, the change in C_{gp} and hence C_{gs} is small.

The substantial decrease in the values of C_{ds} and C_{gd} as V_{DS} is increased is also due to depletion spreading, but in these two cases, it is due to spreading in the epitaxial layer rather than the body diffusion.

Because C_{ds} is a junction depletion-capacitance, an approximate relation for its value per unit area can be derived from the analysis of the one-sided-step-junction (OSSJ) model [9]:

$$C_{ds} \text{ (per unit area)} \sqrt{\frac{qK_s\epsilon_o C_B}{2\,(V_R + \phi_B)}} \quad (16)$$

where q is the elementary electronic charge (1.9×10^{-19}C), K_s is the dielectric constant of silicon (11.7); ϵ_o is the permittivity of free space (8.86×10^{-14}F/cm); C_B is the epitaxial layer background concentration (atoms/cm^3); V_R is the applied reverse voltage to the junction (= V_{DS}); and ϕ_B is the built-in diode potential (= 0.7 V).

For applied values of $V_{DS} >> \phi_B$, equation (16) shows that the value of C_{ds} should vary approximately in proportion to $1/\sqrt{V_{DS}}$. This is in fact the case, although it may not be entirely obvious from Figure 10 which has been plotted on log-log graph paper.

The dependence of C_{gd} on V_{DS} is very similar to that of C_{ds} as seen in Figure 10, even though it is not a junction-depletion capacitance. The reason for this is the lower "plate" of C_{gd} formed by the region of mobile charge in the epitaxial layer below the polysilicon gate (reference Figure 8). This region of mobile charge effectively becomes smaller as the depletion regions between adjacent cells (in a closed cell structure) approach each other with increasing V_{DS}. It should be fairly obvious that the relationship between C_{ds} and V_{DS} is not the same as that of C_{gd} and V_{DS}. In the former, V_{DS} modulates the effective capacitor thickness while in the latter it modulates the effective capacitor plate area.

To determine a relation between C_{gd} and V_{DS}, we must start by carefully examining Figure 8. It can be shown that for a given value of depletion region width in the epitaxial layer, $W_{d(epi)}$, the effective value of C_{gd} per unit area becomes reduced from the value of C_{ox} (the gate-oxide capacitance per unit area) to a lower value given by:

$$C_{gd} \text{ (per unit area)} = C_{ox}\left(1 - \frac{2W_{d(epi)}}{x}\right) \quad (17)$$

where x is the average distance between the metallurgical junctions of adjacent cells.

Again from the analysis of the one-sided-step-junction model, it can be shown that [12]:

$$W_{d \text{ (epi)}} = \sqrt{\frac{2K_s\epsilon_o(V_R + \phi_B)}{qC_B}} \quad (18)$$

Although the curve of C_{gd} versus V_{DS} may look similar to that of C_{ds} versus V_{DS}, one can see that C_{gd} actually varies in proportion to $(1 - K\sqrt{V_{DS}})$ and not $(1/\sqrt{V_{DS}})$.

As far as thermal effects are concerned, the capacitance parameters of a power MOSFET are one of the few that show negligible variation with temperature. In fact, it is a direct consequence of this invariance that the device switching times, which are capacitance dependent, are essentially independent of temperature.

3

References

[1] Richard S. Muller and Theodore I. Kamins, *Device Electronics for Integrated Circuits,* John Wiley & Sons, 1981.

[2] R.M. Swanson and J.D. Meindle, "Ion-Implanted Complementary MOS Transistors in Low-Voltage Circuits," *IEEE J. Solid State Circuits,* Vol. SC-7, April 1972.

[3] David M. Navon, *Electronic Materials and Devices,* Houghton Mifflin Company, 1975.

[4] R.A. Blanchard, *Optimization of Discrete High Power MOS Transistors,* PhD Thesis Dissertation, Department of Electrical Engineering at Stanford University, December 1981.

[5] S.C. Sun and J.D. Plummer, "Electron Mobility in Inversion and Accumulation Layers on Thermally Oxidized Silicon Surfaces," *IEEE Transactions on Electron Devices,* ED-27 No. 8.

[6] F.F. Fang and A.B. Fowler, "Hot Electron Effects and Saturation Velocities in Silicon Inversion Layers," *Journal of Applied Physics,* No. 41, March 15, 1970.

[7] Paul Richman, *MOS Field-Effect Transistors and Integrated Circuits,* John Wiley & Sons, 1973.

[8] EIA or JEDEC Specifications for MOSFET Device Parameter Measurement.

[9] A.S. Grove, *Physics and Technology of Semiconductor Devices,* John Wiley and Sons, 1967.

3.2 Switching Characteristics

Introduction

Power MOSFETs, well-known as majority-carrier devices, are also well-known for their extraordinary switching speeds—far faster than the best power bipolar transistors currently available. When switching speeds are compared, the fall time rather than the rise time is of primary concern. It is here that the MOSFET excels.

When bipolar transistors are used in high-speed applications, a variety of schemes are often used to hasten the fall time, but none of them can equal what the power MOSFET can do without help of any kind! Yet the power MOSFET, like the bipolar transistor, *is a charge-coupled device.* Although MOSFETs differ in design, construction, and intended application, to achieve fast switching times, an understanding of their charge-transfer characteristics is crucial. Consequently, some of the schemes popular with bipolar transistors will also be useful for power MOSFETs.

Improved switching efficiency, which is what the designer looks for in higher-speed switching, is a measure of power loss. In the majority of switching applications, as in switch mode power supplies, efficiency focuses upon switching losses. Aside from saturation losses which, depending upon bias, (see Figure 1) are fixed. Switching losses are a function of switching speed. Consequently, at low switching frequencies it is entirely possible and, indeed, probable that switching speed no longer becomes a primary loss mechanism. Therefore, the important switching characteristics are those involving high-speed switching which is best achieved by using power MOSFETs.

Switching losses (efficiency) are more fully treated in Section 4.1.

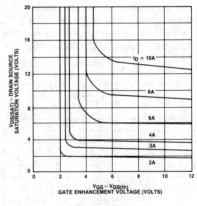

Figure 1

Since early 1981, an increasing number of power MOSFET vendors have added charge-transfer characteristic curves to their data sheets, but they have failed to offer supporting information, leaving the prospective user to question the usefulness of these curves.

Charge-Transfer Characteristics

All power MOSFETs, without the encumbrances of Zener gate protection, whether metal gate or polysilicon, display a DC input resistance of many megohms. When used as a switch, the power required to maintain a quiescent condition (either ON or OFF) is zero whereas with a power bipolar transistor, it is not. In parallel with this high-input resistance, there is an equivalent input capacity consisting of gate-to-source capacitance, C_{gs}, and gate-to-drain capacitance, C_{gd}.

3

For an enhancement-mode MOSFET, the gate voltage must exceed the threshold to begin switching action. For an *n*-channel MOSFET, the direction is positive, and for a *p*-channel, it is negative. Since the input impedance of the power MOSFET is a nearly pure capacitance, the drive must first *charge* this capacitor; likewise, to turn OFF the power FET, it must be *discharged*. For this reason, the power MOSFET is considered to be a charge-coupled transistor. Consequently, for very high-speed switching, a driver is necessary that can both source and sink sufficient current to charge/discharge this input capacitance in a reasonably short time.

If the input capacitance is known, the energy necessary to charge the gate can be determined:

$$W = \frac{1}{2C_{in}dV_{gs}^2} \quad \text{Watt-seconds} \tag{1}$$

Unfortunately, input capacitance is not well-defined because the Miller effect renders Equation 1 nearly useless. Rather than struggling with capacitance, a more viable method is to use the charge-transfer characteristics found on most power MOSFET data sheets. Figure 2 shows the typical charge-transfer characteristics for a Siliconix VN4000A, 400 V, 1.0Ω MOSPOWER FET.

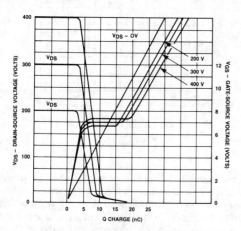

Turn-On Charge Transfer Characteristics
Figure 2

Although Figure 2 may be similar to that supplied in most power MOSFET data sheets, it offers more information than is necessary for this discussion. Focusing attention on a less cluttered Figure 3, where just the 200 V data is reproduced, there are three distinct regions of importance. In Region 1, the gate voltage, V_{gs}, has risen to a level where there is drain current conduction. In Region 2, turn-ON is complete when the drain voltage has switched 90%. In Region

3, the gate voltage rises to its maximum drive potential as the drain voltage slowly settles to V_{SAT}.

Following the gate-charge cycle, first is the initial turn-ON delay $t_{d(on)}$ [Region 1] which continues until conduction begins ($0.9V_{DS}$). During this period, the gate potential is charging the equivalent input capacitance which, in the pre-threshold region, is merely C_{gs}–thus, the fairly constant slope. This pre-threshold capacitance can be calculated from the equation:

$$C_{gs} = \frac{Q_{g1}}{V_{g1}} \tag{2}$$

The most obvious event that distinguishes Region 2 from Region 1 is the abrupt increase in input capacitance identified by the flattening of the gate-charge-transfer curve. As the power MOSFET turns ON, the Miller effect becomes the dominant input capacity. Miller effect capacity can be calculated using the equation:

$$C_{in} = C_{gs} + C_{gd}(1 - A_V) \tag{3}$$

A greatly simplified approach to determine the equivalent input (or Miller) capacitance uses data available from Figure 3.

$$C_{in} = \frac{Q_{g2} - Q_{g1}}{V_{g2} - V_{g1}} \tag{4}$$

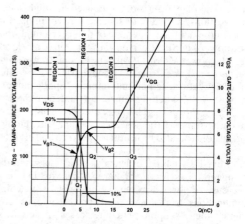

Turn-On Charge Transfer Characteristics
Figure 3

Region 3 reveals two important bits of information, and it poses two questions: Why does the flat charge-transfer, so obvious in either Figure 2 or 3, remain as it does; and why does the drain voltage decay abruptly stop and slowly settle to V_{SAT}? Equation (3) appears to contain *explainable* elements, viz. C_{gs}, the gate-to-source capacity; C_{gd}, the gate-to-drain capacity; and A_V, the voltage gain which equals (dV_{DS}/dV_{gs}).

The Effects of Gate Drive on Feedback Capacitance

The effects of drain-to-source voltage, V_{DS}, upon the interelectrode capacitances are not determined easily from the Capacitance-versus-Voltage plot found on most power MOSFET data sheets (Figure 4) since C_{iss}, C_{oss}, and C_{rss} appear to become asymptotic beyond a few dozen volts. Since C_{gs} and C_{gd} are depletion-dependent, they are voltage dependent. A far better understanding of this dependency is seen when these capacitances are plotted as shown in Figure 5.

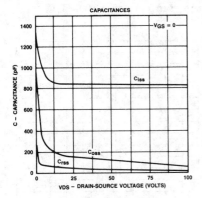

CAPACITANCES

Figure 4

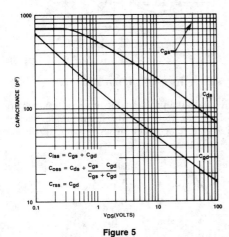

$C_{iss} = C_{gs} + C_{gd}$

$C_{oss} = C_{ds} + \dfrac{C_{gs} \; C_{gd}}{C_{gs} + C_{gd}}$

$C_{rss} = C_{gd}$

Figure 5

Aside from turn-ON delay, $t_{d(on)}$, which is directly affected by the pre-threshold input capacity of Region 1 [see Equation (2)], t_{rise} and t_{fall} are heavily influenced by the Miller effect, which is in turn affected by the gate-to-drain capacity, C_{gd}.

A more careful study of Region 3 reveals that the gate-charge curve is essentially flat until the drain

voltage, V_{DS}, has decayed to V_{SAT}. Then the gate-charge resumes its rise to the level of impressed gate-drive voltage, V_{GG}. The two questions posed are interrelated. By magnifying a portion of Figure 3, a tangential approximately shows that the knee is very close to the gate-to-source voltage. Of particular significance is the resulting potential across C_{gd} during switching, best seen in Figure 6 where a spectacular rise in capacitance occurs.

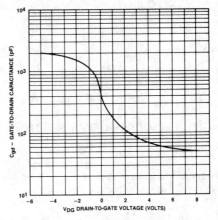

Figure 6

The answers to the questions become obvious. As the drain voltage decays during the switching cycle, the potential across C_{gd} also decays until V_{DS} reaches V_{gs}. Since C_{gd} is depletion dependent, its capacitance rises dramatically as the voltage between drain and gate diminishes *and changes polarity* when V_{DS} drops *below* V_{gs}—as it surely does during the switching cycle. As C_{gd} rises, the Miller capacitance increases even more rapidly and, despite the drop in voltage gain (dV_{DS}/dV_{gs}), the increase in Miller capacity keeps the gate-charge characteristics nearly flat until V_{SAT} is reached.

Switching-Time Calculations
Turn-ON Delay and Turn-ON (Rise) Time

On every power MOSFET data sheet, switching times can be found. The data presented may be measured experimentally rather than by analysis.

The switching times can be computed from the charge-transfer characteristics using the following equations:

$$t_{d(on)} = \frac{Q_{g1}}{V_{g1}} \; R_{gen} \; \ln \frac{V_{GG}}{V_{GG} - V_{g1}} \qquad (5)$$

$$t_r = \frac{Q_{g2} - Q_{g1}}{V_{g2} - V_{g1}} \; R_{gen} \; \ln \frac{V_{GG} - V_{g1}}{V_{GG} - V_{g2}} \qquad (6)$$

where R_{gen} is the source resistance in ohms, and the ratio Q_g/V_g is the equivalent input capacitance.

This, in concert with the source resistance, becomes the familiar RxC time constant. Aside from R_{gen}, all the data necessary to calculate the turn ON delay, $t_{d(on)}$, and rise time, t_r, can be taken from the charge-transfer curves of Figure 2 or 3. Reasonable care is needed since small errors in the data may lead to enormous errors in the calculations. Table 1 provides a comparison between the measured data, taken from an average of 10 devices and calculated data using Equations (5) and (6).

Table 1

ns	Calc.	Meas.	Calc.	Meas.
$t_{d(on)}$	29.9	32	61.7	60
t_r	52.8	52	105	105
$V_{DD}=200\,V$	$R_{gen}=50\,\Omega$		$R_c=100\,\Omega$	

Turn-OFF Delay and Turn-OFF (or Fall) Times

Turn-OFF delay, $t_{d(off)}$, and fall time, t_f, can be readily calculated from the data contained in the charge-transfer characteristics of Figure 2 or 3 in a similar manner using the equations:

$$t_{d(off)} = \frac{Q_{g3} - Q_{g2}}{V_{g3} - V_{g2}}\ R_{gen}\ \ln \frac{V_{GG}}{V_{g2}} \qquad (7)$$

$$t_f = \frac{Q_{g2} - Q_{g1}}{V_{g2} - V_{g1}}\ R_{gen}\ \ln \frac{V_{g2}}{V_{g1}} \qquad (8)$$

A problem arises, however, when t_f data is taken from these charge-transfer curves. A short review of what happens during a typical switching cycle will illuminate the problem and reveal the cure.

Every power MOSFET is bracketed by parasitic capacitances: C_{gs}, C_{gd}, and one not previously mentioned, C_{ds}—the drain-to-source capacity. Before the MOSFET enters into the turn-ON cycle, C_{ds} is fully charged to the rail potential. During the switching cycle, C_{ds} begins its discharge through the ON resistance of the power MOSFET. When the time comes for the MOSFET to turn OFF, the conditions may differ in that rather than exhibiting a time constant consisting of $r_{DS(on)}$ and C_{ds} (MOSFET ON resistance times the drain-to-source capacity), the time constant now becomes R_{load} and C_{ds} (load resistance times the drain-to-source capacity)! If the load resistance/reactance differs from the ON resistance of the MOSFET, the calculation of t_f will be in error *if* the gate charge transfer characteristics depicted in Figure 2 are used. The solution is to define a gate-*discharge* decay characteristic curve.

Gate-Discharge Decay Characteristics

The gate-discharge characteristics are complicated because of the dependency upon the particular appli-

cation since the charge time is dependent upon both the load resistance and *effective* drain-to-source capacitance. The latter is dependent upon the MOSFET's C_{oss} and the particular mounting configuration. A 0.003-inch fiber insulator between a TO3 package and a grounded heatsink, for example, offers 200 pF of additional drain-to-source capacity. Figure 7 identifies a typical gate discharge transfer characteristic for a Siliconix VN4000A MOSPOWER FET that has a resistive load of 100 Ω with the MOSFET mounted to a grounded heatsink using a 3 mil fiber washer. A careful comparison of this figure to Figure 2 shows the V_{DS} rise time of Figure 2 is obviously faster than the corresponding fall time of Figure 7. This confirms that rise time is generally swifter than fall time—even for a power MOSFET.

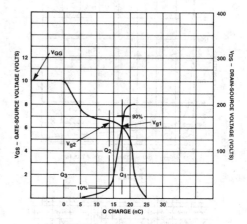

Turn-Off Charge Decay Characteristics
Figure 7

Correlating Turn-OFF (Fall) Time

To illustrate the problem encountered when charge-transfer characteristics are used rather than discharge-transfer characteristics, a comparison is offered in Table 2.

Table 2

ns	Meas.	Calc. Fig. 2	Calc. Fig. 7
t_f	48	32	46
	$V_{DS}=200\,V$	$R_{gen}=50\,\Omega$	

Conclusions

What has been offered in this section suggests that switching times can be accurately determined for a variety of applications and test conditions provided charge-transfer data is offered on the MOSFET data sheet. Furthermore, one is able to recognize the importance of low feedback or gate-to-drain capacitance which suggests that higher speed rise and fall times can be achieved with low C_{gd} FETs.

4.1 Safe Operating Area and Thermal Design for MOSPOWER Transistors (AN83-10)

Introduction

MOSPOWER transistors have evolved into true power devices. Like any power semiconductor, these devices have thermal and electrical limitations which must be observed if acceptable performance and service life are to be achieved.

In general, thermal and electrical characteristics are mutually interrelated so actual limits depend on the particular device application.

To help the user, most data sheets contain information on maximum junction temperature (T_{jmax}), safe operating area (SOA), maximum voltage and current ratings, as well as steady state and transient thermal impedances. Despite the wealth of information presented in a good data sheet, it is not possible to provide graphs and reference tables to cover all possible applications. The designer is still faced with the problem of accurately calculating several quantities such as junction temperature (T_j), total power dissipation (P_T) and correct SOA curve for the application.

The information on how to perform these calculations is scattered. This application note intends to solve the problem by collecting the necessary information in one place and arranging it in a logical order so that the thermal and SOA design for power MOSFETs are no longer a mystery but a relatively simple and direct procedure.

We begin by discussing thermal models for a MOSFET. With this information, we then solve for T_j and P_T and predict the system thermal stability. Finally, we demonstrate the procedure for generating an SOA curve for a particular application.

The MOSFET Thermal Model

The Steady State Thermal Model

Figure 1 gives a simplified thermal system diagram for a MOSFET and the electrical circuit analog for the steady state. By inspecting Figure 1B, we can write an expression for T_j:

$$T_j = T_a + (R_{\theta jc} + R_{\theta cs} + R_{\theta sa}) P_T \qquad (1)$$

For convenience we usually let:

$$R_{\theta ja} = R_{\theta jc} + R_{\theta cs} + R_{\theta sa} \qquad (2)$$

so that:

$$T_j = T_a + R_{\theta ja} P_T \qquad (3)$$

This seems to be a very simple expression, but there is a hidden problem: in a MOSFET, P_T is an exponential function of T_j which leads to some difficulty in calculating T_j and P_T. Before solving that problem, pause to examine the elements of equation (1) and consider to what degree the designer can control these elements:

1. T_a is usually an externally imposed requirement stated in terms of maximum temperature of the heat exhaust medium.

2. T_{jmax} may be imposed externally such as in a military application where T_{jmax} may be limited to $+105°C$ to $+125°C$. In some applications, the overall MTBF de-

4

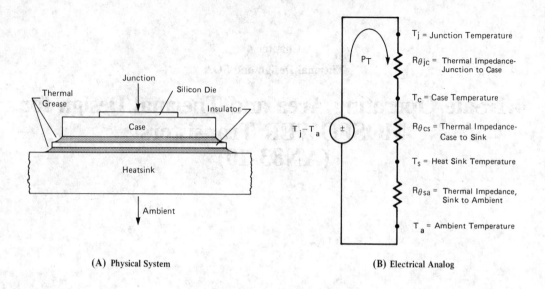

(A) Physical System

(B) Electrical Analog

MOSFET Thermal Model for the Steady State
Figure 1

sired may dictate T_{jmax}. There is, of course, the maximum rating from the manufacturer, to be exceeded only if drastic reductions in operating life are acceptable. In any case, the user makes the descisiion on the acceptable limits.

3. For a given die size, the value of $R_{\theta jc}$ will depend on the package chosen. The same device, for example, will have a smaller thermal resistance in a TO-3 package than in a TO-39 package. Often, however, there is little flexibility in package choice, particularly for larger devices. Another way to lower $R_{\theta jc}$ is to parallel devices. For example, if the choice is between a single IRF440 or two IRF430's in parallel, both possibilities will have the same $R_{DS(on)}$, but the effective $R_{\theta jc} + R_{\theta cs}$ for the two devices will be 1/2 that for the single. Naturally there are disadvantages to paralleling multiple devices, so it is up to the designer to make the tradeoff.

4. $R_{\theta cs}$ is determined by package choice and by the device mounting method on the heatsink. If the device is mounted directly on a smooth, flat surface, without an insulator, with a small amount of thermal grease and the proper screw torque, then $R_{\theta cs}$ will be low. If on the other hand, the mounting uses an ungreased mica washer loosely clamped to a rough surface, the $R_{\theta cs}$ will be large. Reference 1 of the bibliography has a thorough discussion on semiconductor mounting.

5. $R_{\theta sa}$ — The heat sink design is completely under the designer's control within practical and economic limits. Multiple parallel devices may be helpful in reducing $R_{\theta sa}$

because if the heat input to the heatsink is dispersed rather than concentrated (at one point), the effective thermal impedance will be lower for a given heatsink.

The Transient Thermal Model

In many applications, the power dissipated in the MOSFET is pulsed rather than DC. For these applications, the thermal model must be modified to account for thermal capacity introduced by the die, the case, the case insulator, and the heat sink. A thermal model for pulsed operation is shown in Figure 2A where the mass of each component is represented by a capacitor.

As a practical matter, some of the capacitors are much larger than others because of significant mass differences in various parts of the system. For example, we can break down $R_{\theta jc(t)}$ into its components as shown in Figure 2B. The thermal time constants of elements in the packaged device are given in Table 1.

Table 1
Element Thermal Time Constants

Element	Thermal Time Constant
die	50 - 500 μsec
die attach	1 - 5 msec
case	1 - 5 seconds

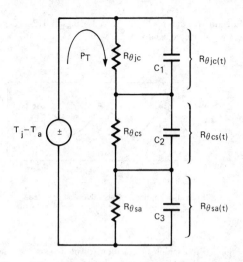

(A) Electrical Analog for the Complete System

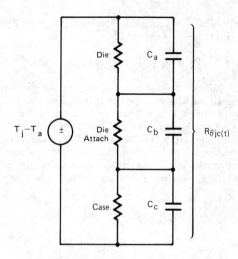

(B) Electrical Analog for the Junction to Case Impedance

Transient Thermal Impedance Model
Figure 2

The dominant time constant depends on the power pulse length. For example, if the pulse width is 100 μsec, then thermal response is determined primarily by the die characteristics. The difference in time constants can be used to detect imperfections in die mounting.

In most systems the heat sink time constant is long compared to the device time constants and does not enter into the calculation except for very long pulses (>10 seconds).

When a pulse of power is applied to this network, the peak value for T_j will depend on peak power and on pulse width (t_p). Figure 3 shows the response of T_j to pulses of various widths but with the same peak value. The shorter the pulse, the smaller the rise in T_j.

The variation in $R_{\theta jc}(t)$ with t_p is shown graphically in Figure 4 where $R_{\theta jc}(t)$ is normalized so that:

$$r(t) = \frac{R_{\theta jc}(t)}{R_{\theta jc}} \qquad (4)$$

For very short pulses, $r(t)$ is quite small, but as t_p is increased $r(t)$ approaches 1, which is the same as saying that for long pulses the transient impedance approaches the steady state impedance.

4

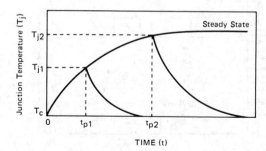

Thermal Response to a Single Power Pulse
Figure 3

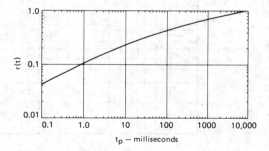

Single Pulse Normalized Transient Thermal Impedance
Figure 4

From this curve we can readily calculate T_j if we know P_T, $R_{\theta jc}$ and t_c, using the expression:

$$T_j = T_c + P_T r(t)\, R_{\theta jc} \qquad (5)$$

Up to this point we have been discussing thermal response to a single pulse; however, most applications have repetitive pulses. Variations in T_j will have the form shown in Figure 5. In this case T_{jmax} may be much higher (for the same t_p) than it would have been in the single pulse case due to the temperature rise resulting from the average power dissipation. The value of $r(t)$ for a repetitive pulse can be approximated from the single pulse curve using the following expression:

$$r(t) = D + (1-D)\, r_1 + r_2 - r_3 \qquad (6)$$

where

$$D = \frac{t_p}{T} \qquad (7)$$

t_p = Pulse width of the power pulse

T = Pulse repetition interval

t_{pw} = Composite pulse widths used to calculate r_1, r_2, and r_3

$r_1 = r(t)$ for $t_{pw} = T + t_p, D = 0$
$r_2 = r(t)$ for $t_{pw} = t_p, D = 0$ taken from the single pulse curve, Figure 4.
$r_3 = r(t)$ for $t_{pw} = T, D = 0$

Equation (6) can readily be solved using the HP-41C program given in appendix A.2.

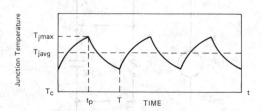

Thermal Response to Repetitive Power Pulses
Figure 5

Figure 6 is a graph of equation (6) plotted on the single pulse graph for several values of D. This graph, usually included in the data sheet, allows the user to determine $r(t)$ by inspection. For duty cycles not plotted, the value for $r(t)$ can be estimated by interpolating between the given curves.

MOSFET Power Losses

There are several possible power loss sources in a MOSFET:

1. P_S — the switching transition loss.

2. P_G — the portion of the drive power dissipated in the gate structure.

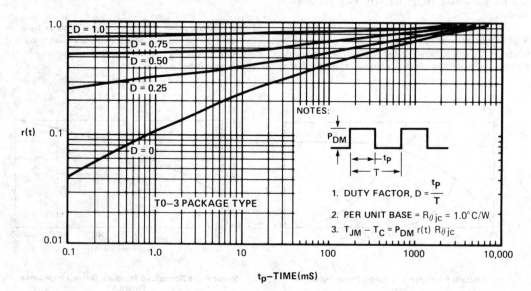

**Normalized Transient Thermal Impedance for Single Uniform
Repetitive Pulses
Figure 6**

3. P_L — the power loss due to drain-source leakage current (I_{DSS}) when the device is off.

4. P_D — the reverse diode conduction and t_{rr} losses.

5. P_C — the conduction loss while the device is on.

Switching Transition Losses

Compared to a bipolar junction transistor (BJT), the switching loss in a MOSFET can be much smaller, but there are still some losses which must normally be accounted for. The switching losses depend on both the switch transition times and the type of load switched. Examples of several typical loads, along with idealized switching waveforms and expressions for P_S, are given in Figures 7 through 10.

Their power losses can be calculated from the general expression:

$$P_S = f_S \left[t_{s1} \int_0^{t_{s1}} V_{DS} I_D \, dt + t_{s2} \int_0^{t_{s2}} V_{DS} I_D \, dt \right] (8)$$

For the idealized waveforms shown in the figures, the integration is quite easy and can be approximated by calculating the area of a triangle or trapezoid.

Table 2 gives a comparison of the switching losses for resistive, clamped inductive and unclamped inductive switching using representative values.

Table 2
Switching Loss Comparison

Figure #	V_{DS}	I_{D1}	I_{b2}	t_{s1}	t_{s2}	f_s	P_S
7	100V	1A	1A	100ns	100ns	100kHz	0.33W
8	100V	1A	1A	100ns	100ns	100kHz	1W
9	150V	0	2A	100ns	400ns	100kHz	6W

Notice how much higher the losses are for inductive switching, particularly the unclamped case. Note also that in Figure 9, t_{s2} is <u>not</u> the turn-off switching time of the FET, rather it is the time the FET remains in avalanche breakdown. Breakdown will persist until the energy in L is dissipated and the current falls to zero.

The capacitive load (Figure 10) has a peculiarity in the turn-off waveform for V_{DS}. If the turn-off time for the MOSFET channel (t_{s2}) is very short and I_{DD} is small (or C is large), then the rise time of V_{DS} is controlled by the load, i.e. the rate of charge of C, not by t_{s2}. On the other

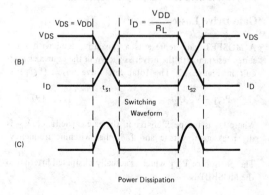

$$(D) \quad P_S = \frac{V_{DS} I_D (t_{s1} + t_{s2}) f_s}{6}, \quad f_s = \text{Switching Frequency}$$

Resistive Load Switching Waveforms
Figure 7

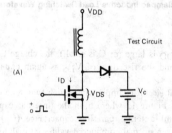

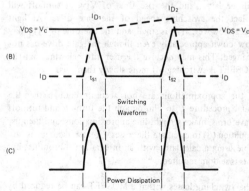

$$(D) \quad P_S = \frac{V_{DS} (I_{D1} t_{s1} + I_{D2} t_{s2}) f_s}{2}$$

Clamped Inductive Load Switching Waveforms
Figure 8

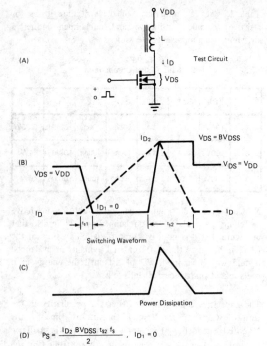

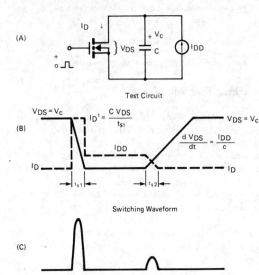

(D) $P_S = \dfrac{I_{D2} \, BV_{DSS} \, t_{s2} \, f_s}{2}$, $I_{D1} = 0$

(D) $P_S = \dfrac{C \, V_c^2}{2} f_s + \dfrac{I_{DD} \, V_c \, t_{s1} \, f_s}{2} = (C \, V_c^2 + I_{DD} \, V_c t_{s1})\left(\dfrac{f_s}{2}\right)$

Unclamped Inductive Load Switching Waveforms
Figure 9

Capacitive Load Switching Waveforms
Figure 10

hand, if I_{DD} is large (or C is small), the charge time of C is short, and the rise time of V_{DS} is agian equal to t_{s2}.

A practical circuit example where such switching occurs is shown in Figure 11 where both the internal capacitance (C_{OSS}) and external parasitic capacitance (C_p) are included. Given normal component values, at high loads I_{DD} will be large, and the rise time of V_{DS} at turn-off will reflect the switching speed of the gate drive. At light loads, however, I_{DD} is small, and the rise time of V_{DS} may slow down appreciably even though the gate drive has not changed. This may deceive the user into believing that the MOSFET is switching off more slowly than it is.

If the approximations are not accurate enough, then the usual procedure is to photograph the turn-on and turn-off waveforms, make straight line approximations, and then use Equation (8) to calculate the losses. Greater accuracy is possible using a calculator with an integration subroutine, but this is seldom required.

The switching losses within a MOSFET can be reduced by using a snubber across the device from drain to source. The usual purpose of a snubber is to reduce voltage and/or current stress on the device. By altering the switching wave-

forms, a snubber can also remove much of the switching loss from the device. This allows it to run cooler, in turn helping reduce conduction losses (P_C). In some cases the use of a snubber can significantly improve overall efficiency; in fact, use of a snubber should be considered even if it is not needed for device protection.

Gate Drive Losses

A MOSFET gate represents a capacitive load with some series resistance to the driver as shown in the equivalent circuit in Figure 12. The total gate drive power (P_{GT}) is

$$P_{GT} = V_{gs} Q_g f_s \qquad (9)$$

Where Q_g is the peak charge in the gate capacitance, V_{gs} is the peak gate voltage and f_s is the switching frequency.

The portion of P_{GT} which is actually dissipated internal to the MOSFET is:

$$P_G = V_{gs} \, Q_g \, f_s \left(\frac{R_G}{R_S + R_G}\right) \qquad (10)$$

Typical values for R_G range from 0.05 to 4Ω depending on device chosen.

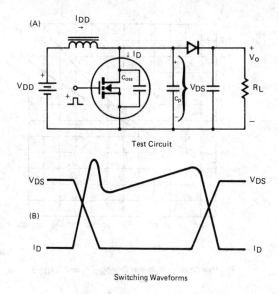

Test Circuit

(B)

Switching Waveforms

**Combined Capacitive and Inductive Load Switching Waveforms
Figure 11**

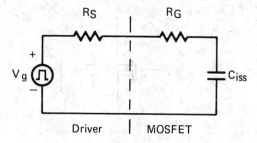

**Gate Drive Equivalent Circuit
Figure 12**

Drain to Source Leakage Current Losses

When a MOSFET is turned off and V_{DS} is still present, a small leakage current (I_{DSS}) flows from drain to source. This will cause some power dissipation (P_L) during the off interval:

$$P_L = I_{DSS} V_{DS} (1-D) \qquad (11)$$

where D is the conduction duty cycle of the switch.

Normally P_L will be small since I_{DSS} is typically only a few microamperes and may be ignored. If, however, T_j is high or if V_{gs}, during the off period, is not well below the threshold voltage of the device, then P_L may become significant. I_{DSS} is not easy to measure in a switching circuit because I_D alternates between amperes and microamperes. If there is some question on the value for P_L, a separate breadboard should be set up which reproduces V_{DS}, V_{gs} and T_j during the off interval on a steady state basis. From this I_{DSS} can be measured directly. As a general rule, I_{DSS} will double for an $11°C$ rise in T_j.

Internal Diode Losses

The MOSFET structure contains an internal diode oriented as shown in Figure 13A. In most respects this is a normal *pn* junction diode, and in some applications it is allowed to conduct during a portion of the operating sequence, eliminating the need for an additional external diode. The diode losses during conduction are proportional to the product of I_{RD} and V_F, which have the usual diode relationship (Figure 13B). The loss in the diode (P_D) can be approximated rather well by:

$$P_D = I_{RD(avg)} \; V_{F(avg)} \qquad (12)$$

Some additional loss will occur during the reverse turn-off interval (t_{rr}). This loss is usually small, but as the switching frequency is raised (> 100kHz), it can become significant. A reasonably accurate loss value can be obtained by photographing the voltage and current waveforms of the diode during the reverse turn-off interval and then applying equation (8) to a piecewise linear approximation of the waveforms.

While the t_{rr} losses are usually small and may frequently be ignored, there is a condition which can greatly increase this loss. The internal diode is actually the base-collector junction of a parasitic BJT. Under some conditions the BJT can be turned on by a rapid dV/dt waveform. When this happens, the current waveform during t_{rr} has a much larger amplitude and lasts longer. This greatly increases the power dissipation and can destroy the device. The best way to detect this condition is by careful observation of the t_{rr} current waveform. An applications note treating this problem in detail will be available from Siliconix.

Conduction Losses

In most applications, the major loss in the MOSFET is due to the non-zero on resistance, $R_{DS(on)}$, through which the drain to source current (I_D) must flow. When conducting as a switch, the device is simply a resistor, and the conduction losses (P_C) are:

$$P_C = \{I_D \, (rms)\}^2 \; R_{DS(on)} \qquad (13)$$

Note the root-mean-square (rms) value for I_D is specified. This is quite different from a BJT in which the average value of the collector current is normally used.

The expression for P_C looks simple, but that is deceiving because $R_{DS(on)}$ is a function of several variables: the junction temperature (T_j), the gate-to-source voltage (V_{gs}), the drain current (I_D), and manufacturing variations.

The dependence of $R_{DS(on)}$ on T_j is shown in Figure 14. This curve has been normalized so that $R_{DS(on)} = 1$ when $T_j = 25°C$. Notice there are two curves: one for a low voltage device and one for a high voltage device. Both curves have a positive temperature coefficient, i.e. as T_j increases, R_{DS} increases. This is characteristic of all power MOSFETs.

4

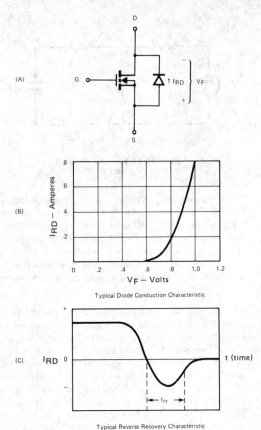

Typical Diode Conduction Characteristic

Typical Reverse Recovery Characteristic

**MOSFET Integral Diode Characteristics
Figure 13**

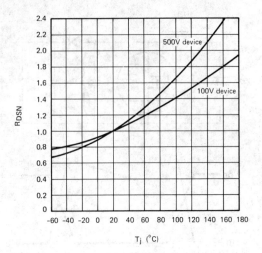

**The Relationship Between $R_{DS(on)}$, T_j and BV_{DSS}
Figure 14**

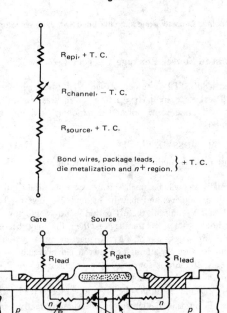

**MOSFET Internal Parasitic Resistances
Figure 15**

Each device type will have a unique temperature curve (among devices of the same type, however, the curves will be very similar) which the manufacturer empirically determines and places on the data sheet. However, the same device type from two different manufacturers may have somewhat different characteristics so the user should exercise care when using this curve.

In general, given the same die size, a high voltage part will have a higher $R_{DS(on)}$ and a higher temperature coefficient than a low voltage device due to the devices internal resistance distribution. Figure 15 shows some of the resistances inherent in a MOSFET. R_{epi} is the resistance of the epitaxial drain region. As the breakdown voltage is increased, both the resistivity and the thickness of the epitaxial region must increase. The temperature coefficient of this region is simply that of a positive temperature coefficient silicon resistor. Both the source resistance and the connections to the die have positive temperature coefficients. The channel resistance is different in that it has a negative temperature coefficient due to the gate. The result of these competing temperature coefficients is that in a low voltage device, the channel resistance is a major portion of the total resistance, and thus the net tempera-

ture coefficient is relatively low. In a high voltage device, however, R_{epi} dominates, and the temperature coefficient is much larger. Variations of 8:1 or more between different device types are not uncommon.

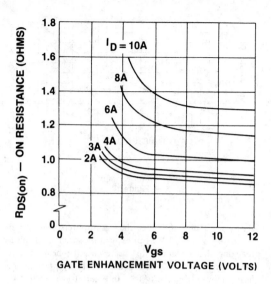

The Effect of Drain Current on $R_{DS(on)}$
Figure 16

The dependence of $R_{DS(on)}$ on V_{gs} and I_D is shown in Figure 16. For each value of I_D, three regions are defined: off, transition, and on. In the off region, the device has a very high resistance (Megohms), and the only current flowing is I_{DSS}. In the transition region, the device becomes a linear variable resistor controlled by V_{gs}. Because the transconductance (g_m) is so large (1 to 30S) in a MOSFET, this region exists only for a very small range of V_{gs}. As V_{gs} is further increased, $R_{DS(on)}$ begins to stabilize; a point is soon reached where a further increase in V_{gs} gives little or no reduction in $R_{DS(on)}$. Normally there is no advantage in raising V_{gs} above 12 to 15V.

As I_D is increased, both the values for V_{gs} in the transition region and the minimum value of $R_{DS(on)}$ will increase. This is due to an inherent effect, much like a JFET, which pinches the current flow in the drain region as I_D increases.

In order to solve Equation (13), the designer must specify T_j, I_D and V_{gs}. V_{gs} is externally imposed, so it is not hard to define. The real "Catch 22" in solving equation (13) is that $R_{DS(on)}$ is a function of T_j which is, in turn, a function of P_T. We can't solve one without the other! To make things even more difficult, the $R_{DS(on)}/T_j$ characteristic is derived empirically; we do not necessarily have a closed form equation to express it. All is not lost, however, as we shall see in the next section.

Determination of P_T and T_j

We can summarize our progress to this point by stating that:

$$P_T = P_1 + I^2_{D(rms)} R_{DS(25°C)} R_{DSN} \quad (15)$$

$$T_j = T_a + R_{\theta ja} P_T \quad (16)$$

where

$$P_1 = P_S + P_G + P_L + P_D \quad (17)$$

As long as P_L is small, P_1 does not vary much with T_j; for this analysis we will presume P_1 is constant.

To find P_T and T_j, we must solve two equations for two unknowns. These are at least two possible solution methods: The first is graphical and uses the data sheet curve for R_{DSN} the second is numerical and uses a calculator. The numerical solution begins by fitting a curve to the empirical data and then proceeds to solve for P_T and T_j by using a converging iterative algorithm. A numerical example and a program for the HP-41 are given in Section 4.2. The following is an example of a graphical solution.

We need to solve equations (15) and (16) simultaneously. This can be done by plotting each equation on the same graph. The desired solution will be the intersection of the two plots. Equation (16) is simply a straight line, but equation (15) is a bit more complex. In equation (15), the only variable is R_{DSN} (Figure 14 is typical). This curve and the desired curve for P_T are the same except that the values of the ordinant must be multiplied by a constant and an offset applied. With a little algebra, we can use this curve directly for a graphical solution. The following shows how this can be done.

The device chosen (an IRF440) is rated for 500V at 8A and comes in a TO-3 package. The case to sink insulator is presumed to be a greased beryllia washer. It is common practice to select a heat sink so that $R_{\theta sa} = R_{\theta jc}$; this will be assumed here.

EXAMPLE 1

Device type	= IRF440	$R_{\theta cs}$	= 0.2°C/W
$R_{DS(on)}$ at 25°C	= 0.8 Ω	$R_{\theta sa}$	= 1°C/W
V_{gs}	= 10V	T_a	= 50°C
$I_{D(rms)}$	= 3A	P_1	= 5W
$R_{\theta jc}$	= 1°C/W		

Solve for P_T and T_j :

STEP 1

Calculate T_j for two values of R_{DSN} using the expression:

$$T_j = T_a + R_{\theta ja} [P_1 + I^2_{D(rms)} R_{DS(25°C)} R_{DSN}] \quad (18)$$

R_{DSN}	T_j
0.4	67.3°C
2.0	92.7°C

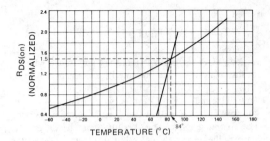

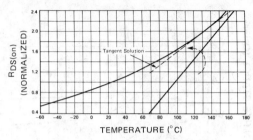

Graph for Example 1
Figure 17

Graph for Example 2
Figure 18

Note the values chosen for R_{DSN} are selected for plotting convenience. Any pair of points that are reasonably well separated will suffice.

STEP 2

Plot the two points on the R_{DSN}/T_j graph for the IRF440 (Figure 17) and draw a straight line which intersects the R_{DSN} curve.

STEP 3

From the intersection of the two curves, read off T_j and R_{DSN}:

$$T_j \quad = 84°C \tag{19}$$

$$R_{DSN} = 1.5 \tag{20}$$

STEP 4

Calculate P_T from the following expression:

$$P_T = P_1 + I^2_{D(rms)} \, R_{DS(25°C)} \, R_{DSN} \tag{21}$$

which for this example gives:

$$P_T = 15.8W \tag{22}$$

Certainly this is a simple means for finding T_j and P_T. To make this procedure even easier, this curve has been made twice normal size in Siliconix's data sheets.

This procedure can be used to give even more information; in particular, it can tell us if the thermal system will be stable, or if a thermal runaway condition exists. The following example will illustrate.

EXAMPLE 2

If I_D in Example 1 is increased to 5A and steps 1 and 2 in the procedure are performed, the graph will appear as

shown in Figure 18. Notice there is no intersection. We do not have a solution! We can explain the meaning of this by realizing that equation (15) represents the heat input to the thermal system, and equation (16) represents the heat removed from the system. In Figure 18 we see that the heat into the system is always greater than the heat being removed, so T_j must rise. There is no equilibrium point. This is a condition of thermal runaway which will destroy the device.

Suppose Example 1 represents a device operating under normal load in a power converter. Further suppose Example 2 represents the same converter but with an output overload. Example 2 shows that the converter will fail under this overload. If the overload must be tolerated then it will be necessary to reduce $R_{\theta ja}$ which will then rotate the line representing equation (16) counterclockwise (as indicated) until an acceptable maximum T_j is obtained. The maximum value for $R_{\theta ja}$ for a given T_j can be calculated by rearranging equation (18):

$$\max R_{\theta ja} = \frac{T_j - T_a}{P_1 + I^2_{D(rms)} \, R_{DS(25°C)} \, R_{DSN}} \tag{23}$$

For example, if the maximum value for T_j in Example 2 is 150°C then:

$$\max R_{\theta ja} = 2.0°C/W \tag{24}$$

This could be achieved by increasing the size of the heat sink so that:

$$R_{\theta sa} = 0.8°C/W \tag{25}$$

In general even a tangent solution (dashed line in Figure 18) is not acceptable because the parameters are seldom known with sufficient accuracy. Supposedly stable tangent solutions on paper may well display thermal runaway in the hardware. A prudent designer will accept only those solutions which have a clear intersection as shown in Figure 17.

The graphical method can also be used to illustrate the effect of improvements in the thermal system. For example suppose we reduce P_1 and $R_{\theta ja}$ from the values used in example 1, as follows:

EXAMPLE 3

IF:

$$P_1 = 3W \qquad (26)$$

$$R_{\theta ja} = 1.8°C/W \qquad (27)$$

THEN:

$$T_j = 74°C \qquad (28)$$

$$P_T = 13.1W \qquad (29)$$

A comparison of Example 1 and 3 is given in Figure 19. This example illustrates how relatively small changes can have a significant effect. This happens because the thermal system is regenerative, i.e. positive feedback. The effects of small changes are exaggerated by positive feedback. Relatively small changes can make the difference between an efficient circuit and one with excessive losses.

The determination of P_T and T_j may also be obtained by using the HP-41 calculator program offered in section 4.2.

To reduce circuit losses, the designer has several variables to work with:

1.   As explained earlier, the thermal impedances are usually designer controlled. From Example 3 we can see that lowering $R_{\theta ja}$ is a very effective tool for reducing P_T and T_j.

2. $R_{DS(on)}$ can be made smaller. Because of the regenerative nature of the system, reducing $R_{DS(on)}$ gives a proportionally larger reduction in P_C. $R_{DS(on)}$ can be reduced in several ways — a larger device, devices in parallel and, to a limited extent, higher V_{gs}.

3.   The value for the drain current is imposed on the device by the circuit. Given the same application, it is sometimes possible to change the current waveform to minimize the value of $I_{D(rms)}$.

For example, in a switchmode power converter, the size of the low pass filter inductor is a variable. If the inductor is small, the switch currents will be triangles, but if it is large,

the switch currents will become trapezoids. This comparison is shown in Figure 20. Given the *same* average current and duty cycle, the rms current will be lower for the trapezoid than the triangle. The relationship between the two is shown in Figure 21. Given that the losses are proportional to $I^2_{D(rms)}$ and considering the thermal regeneration, even a small reduction in $I_{D(rms)}$ will significantly reduce P_C.

Another means for reducing $I_{D(rms)}$ in switchmode converters is to use an alternate circuit topology with a lower rms switch current for the same power processed.

The following example illustrates the reduction in P_C made possible by making L larger and using an alternate topology. For comparison, assume the following operating conditions:

$$\text{input voltage, } V_S \ \ = 20 \text{ to } 30V \qquad (30)$$

$$\text{output voltage, } V_O = 25V \qquad (31)$$

$$\text{output current, } I_O = 3 \text{ Amp} \qquad (32)$$

Two different circuits meeting our requirements are shown in Figure 22. The first (A) is a parallel Quasi-Squarewave Converter. The second (B) is a boost derived circuit in which the switch conduction duty cycle is greater than 50% so that both switches are on simultaneously for a portion of the switching sequence.

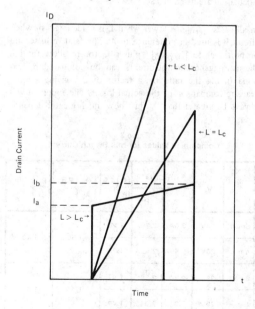

L_c is the value for L which will just maintain continuous inductor current at a given load and operating frequency.

An Illustration of the Effect of Inductor Size on the Drain Current Waveform in a Switchmode Converter
Figure 20

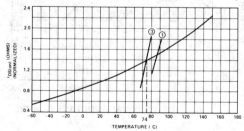

Graph for Example 3
Figure 19

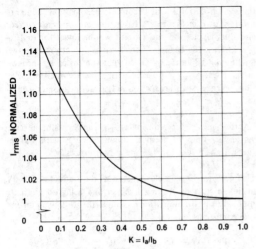

Variation in I_{rms} as a Function of the Trapezoidal Current
Ration (K)
Figure 21

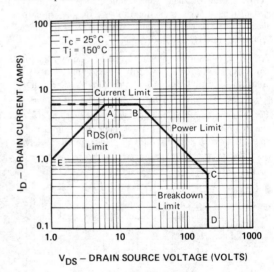

The Converter Circuits Used for the Switch Current Calculations
Figure 22

We can calculate the rms values for the switch current (i_1) in each circuit, first by assuming L is sufficiently large so that the switch current waveforms are rectangles, and second by assuming L is smaller so that the waveforms are triangles, but with the same value of D. The results of this calculation are listed in Table 3.

In all cases i_1(rms) is lower when L is made larger or when circuit B is used in preference to A. The relative losses are proportional to $[i_1(rms)]^2$ plus a factor to allow for the thermal regeneration. Currents can be compared by exponentiating the ratio by a factor of 2.5 giving a more realistic comparison of the actual losses. The exponent will always be greater than 2, just how much greater depends

on $R_{\theta ja}$ and R_{DS}. As a rough guide, L should be at least twice the critical inductance (L_c) required to maintain continuous inductor current at full load over the range of V_s. Making L much larger than $3 \times L_c$ usually isn't worthwhile since most of the benefits have already been achieved.

Operation in discontinuous inductor current mode raises the power losses rapidly to levels even higher than indicated in Table 3. In general, operation at full power in the discontinuous mode is unacceptable from a power loss point of view and is usually avoided at powers above 100W for off-line operation and 25 to 50W for lower voltages.

Table 3
Comparison Values for rms Switch Current

L	V_s	CIRCUIT		$\left(\dfrac{A}{B}\right)^{2.5}$	$\left(\dfrac{A_s}{A_L}\right)^{2.5}$	$\left(\dfrac{B_s}{B_L}\right)^{2.5}$
		A	B			
Small	20V	3.0A	2.79A	1.20	1.36	1.43
Large	20V	2.65A	2.42A	1.25		
Small	25V	2.74	2.35	1.47	1.44	1.44
Large	25V	2.37	2.03	1.47		
Small	30V	2.51A	1.77A	2.39	1.44	1
Large	30V	2.17	1.77A	1.66		

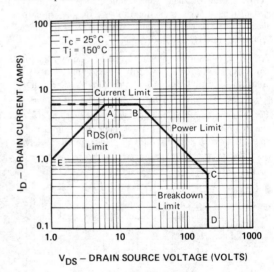

Typical MOSFET SOA Curve
Figure 23

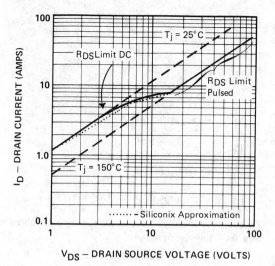

two straight lines (dotted lines Figure 24) in the data sheets since this gives a more accurate definition of this boundary.

The data sheet SOA curve is not unique, even for one particular device. The reason is the SOA boundaries do not generally represent limits which, if exceeded, will result in immediate destruction, but rather these boundaries are limits beyond which the service life of the device becomes unacceptable. What is acceptable in one application may not be in another. To illustrate this point, we have added a third axis to the SOA curve (Figure 25) to represent useful life. As the boundaries are reduced, the useful life of the MTBF increases dramatically at first and then more slowly. This is why in applications where especially long service life or very high MTBF is required, significant derating may be needed beyond manufacturer's specifications.

The SOA curve is also a function of the T_c, T_{jmax}, t_p and D. The data sheet SOA curve will normally be for $T_c = 25°C$, a single pulse, and a few different values of t_p. Few actual applications enjoy these particular conditions. The manufacturers should not be criticized since there are an infinite number of possibilities, and they have merely chosen to standardize on one, to allow a direct comparison between different devices. The MOSFET SOA curve is also patterned after the traditional Bipolar Junction Transistor SOA graph.

For each application, the user must create an SOA curve that reflects his specific requirements. Fortunately this is not a difficult task.

4. Snubbers can be used to remove some or all of P_S. There are many different snubber circuits, and the literature is quite extensive [2-6]. The reduction in P_S will depend on the particular circuit chosen.

5. An HP-41 program for calculating $I_{D(rms)}$ is given in appendix A.1. This will greatly simplify the rms calculation and make it easier to optimize the circuit.

The SOA Curve

A typical example of a MOSFET data sheet SOA curve is given in Figure 23. The curve exhibits four limiting boundaries: maximum current (A-B), maximum power (B-C), maximum voltage (C-D) and the $R_{DS(on)}$ limit (E-A). The current limit is set at a level which limits the current density in the bonding wires and in the die surface metalization to provide reliable operation. The power limit is that power dissipation which will raise T_j from T_c to T_{jmax}. The voltage boundary is determined by the designed breakdown voltage of the device. Actual breakdown voltage of the device is higher than this limit but varies from one device to another and varies with temperature. The SOA boundary is selected to lie inside of all normal variations. The $R_{DS(on)}$ limit is due to the minimum "on" resistance of the device. For example, it is not possible to have 10A flowing through a 2Ω resistor when only 1V is applied.

At the present time, all manufacturers do not draw the $R_{DS(on)}$ boundary the same way. The solid line in Figure 24 shows the actual limit allowing for the heating of the device due to the power being dissipated. The dashed lines represent the boundary for $R_{DS(on)}$ at 25°C and 150°C. Any of the three boundaries may appear on a data sheet. Siliconix has elected to approximate the actual limit with

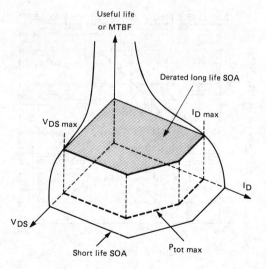

A Three Dimensional Representation of SOA for a MOSFET
Figure 25

How to Create Real-World SOA Curves

Figure 26 is a typical data sheet SOA graph. As pointed out in the previous section, this graph is valid only for one set of conditions which in this case is:

1. $T_j = 150°C$

2. $T_c = 25°C$

3. $D = 0$

4. $t_p = 100 \mu sec$

A typical user might wish to make the following changes:

1. T_c higher

2. T_j lower

3. Repetitive pulses, $D \neq 0$

4. Derating of V_{DS} and I_D

5. A t_p other than that shown on the graph

The new SOA graph can be created with a simple step by step procedure as shown in the following example:

ASSUME:

1. Voltage and current derating factor = 0.8

2. $T_{jmax} = 125°C$

3. $T_c = 85°C$

4. $t_p = 200 \mu sec$

5. $D = 0.25$

6. The device is an IRF440

7. $R_{\theta jc} = 1°C/W$

8. $r(t) = 0.27$ (from Figure 6)

STEP 1: Re-draw the $R_{DS(on)}$ limited boundary.

Points on the boundary for $T_j < 125°C$ can be calculated from the following expressions:

$$I_D = \sqrt{\frac{T_j - T_c}{R_{\theta jc}\ r(t)\ R_{DS(on)}\ R_{DSN}}} \qquad (33)$$

$$V_{DS} = \sqrt{\frac{(T_j - T_c)\ R_{DS(on)}\ R_{DSN}}{R_{\theta jc}\ r(t)}} \qquad (34)$$

For $T_j = 125°C$ the boundary is defined by:

$$V_{DS} = I_D\ R_{DS}\ (25°C)\ R_{DSN}\ (125°C) \qquad (35)$$

A plot of the $R_{DS(on)}$ boundary for this example is shown in Figure 27. The values for R_{DSN} are taken from Fig. 28.

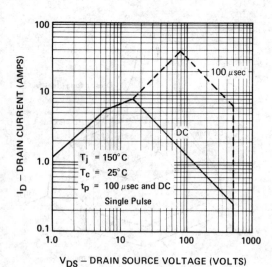

IRF440 SOA
Figure 26

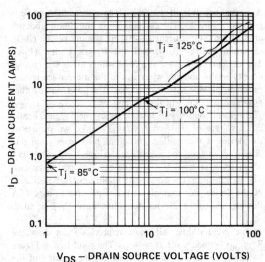

The Modified $R_{DS(on)}$ Boundary
Figure 27

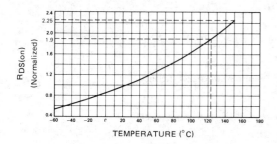

The Effect of Temperature on R_{DS(on)}
Figure 28

STEP 2: Re-draw the I_D and V_{DS} boundaries parallel to the original boundaries but reduce them by a factor of 0.8 as shown in Figure 29. Note in this case the I_D boundary does not exist because the R_{DS} boundary takes precedence.

STEP 3: Re-draw the thermal boundary. The total power for the boundary can be determined from:

$$P_T = \frac{T_j - T_c}{r(t) \; R_{\theta jc}} = 148W \qquad (36)$$

Two points corresponding to this power are:

	V_{DS}	I_D
A	50V	2.97A
B	200V	0.74A

Now plot these two points and draw a straight line between them. This becomes the pulsed power boundary as shown in Figure 29.

The graph in Figure 30 (solid line) is the final SOA curve for the IRF440, given the specified operating conditions. In comparison to the original SOA curve (dashed line), the new curve is very different. Clearly, if you use the un-modified curve, it is highly probable that device failure would occur if the device is operated anywhere near the original limits.

Conclusion

As we have shown, the thermal calculations for a power MOSFET are not particularly complex or time consuming. It has also been shown how important these calculations are to achieve an efficient and reliable design. Most design parameters are under the designer's control, and by juggling the variables around a bit, an efficient design can usually be achieved. If the thermal design is overlooked or given a short shrift, there is small chance for a successful design.

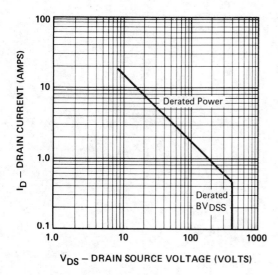

Modified Boundaries for BV_{DSS} and Power
Figure 29

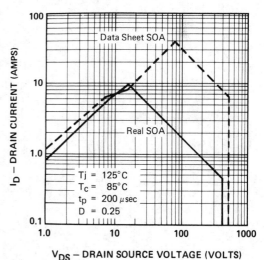

Final SOA Graph for the IRF440
Figure 30

References

[1] W. Roehr, "**Mounting Techniques for Power Semiconductors**," Motorola Application Note AN778, February 1978.

[2] R. Walker, "**Circuit Techniques for Optimizing High Power Transistor Switching Efficiency**," Proceedings of POWERCON 5, May 1978.

[3] T. Undeland, "**Stress Reduction in Power Transistor Converters**," IEEE—IAS Annual Meeting Proceedings, October 1976.

[4] Harada, Ninomiya and Kohno, "**Optimum Design of an RC Snubber for a Switching Regulator by Means of the Root Locus Method**," Proceedings of the IEEE—Power Electronics Specialists Conference, June 1978.

[5] G. Edwing and R. Isbell, "**A New High Efficiency Turn-Off Switching Aid for Power Transistors in Switching Regulators**," IEEE Journal of Solid State Circuits, Volume SC-17, November 6, December 1982, pp. 1210–1213

[6] E. Calkin and B. Hamilton, "**Circuit Techniques for Improving the Switching Loci of Transistor Switches in Switching Regulators**," IEEE Transactions on Industry Applications, Volume IA-12, No. 4, July/August 1976.

4.2 HP-41CV Power MOSFET Thermal Analysis Program

Introduction

MOSPOWER FETs are being used in more and more switching circuits now that high voltage and high current devices are available at reasonable cost. One of the most misunderstood facts about MOSFETs is that they are not immune to thermal runaway when driven with gate-to-source voltages between 10 and 20 volts. Under these conditions, the drain-to-source on-resistance ($R_{DS(on)}$) increases exponentially with junction temperature and, consequently, so does the power dissipated by the MOSFET for a given RMS drain current.

Proper heatsink design, as with bipolar transistors, is essential. Reference 1 gives an excellent account of thermal design for MOSPOWER FETs and presents an accurate graphical approach for selecting a heatsink with the correct thermal resistance.

This application note presents a program for the HP-41CV programmable calculator that computes the total dissipated power (P_T) and the operating junction temperature (T_j) of a MOSPOWER FET. The program features prompting for the ten input variables required and fast convergence to the solution (if one exists).

The ten input variables required are:

1. Regression constant a
2. Regression constant b
3. Fixed MOSFET power loses (P_1)
4. RMS drain current ($I_D(RMS)$)
5. $R_{DS(on)}$ at 25°C
6. Ambient temperature (T_a)
7. Maximum operating junction temperature (T_j(maximum))
8. Thermal resistance junction to case ($R_{\theta jc}$)
9. Thermal resistance case to heatsink ($R_{\theta cs}$)
10. Thermal resistance heatsink to ambient ($R_{\theta sa}$)

Regression constants "a" and "b" are used by the program to determine the value of $R_{DS(on)}$ at different values of T_j. These constants can be obtained from an exponential curve-fit to data points taken from the normalized $R_{DS(on)}$ versus the T_j curve. This curve shows $R_{DS(on)}$ as a function of T_j normalized to its value at 25°C, and is of the form $y = ae^{bx}$. Most manufacturers now include the curve as part of their MOSFET data sheets.

The curve fitting can be done using the exponential curve-fit program EXP which is included in the HP-41 Statistics Pac. If this Pac is not available, then the thermal analysis program can be instructed to calculate constant "b" given the value of constant "a". This method is much faster than a separate curve-fit operation, but the accuracy is reduced slightly. In many cases though, this reduced accuracy is more than adequate for the thermal program to calculate useful results.

Theory

The program uses two iterative algorithms to determine the points on the P_T versus T_j curve at which the tangent line touches it and the $R_{\theta ja}$ line intersects it (refer to Figure 1). $R_{\theta ja}$ is the total thermal resistance from junction to ambient and is equal to $R_{\theta jc}$ + $R_{\theta cs}$ + $R_{\theta sa}$. By finding the point on the P_T curve at which the tangent line touches, the range of T_j for which a solution will occur can be determined. This range will extend from T_a to either T_j(maximum)

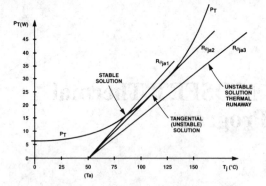

This Diagram Shows the Exponential P_T Versus T_j Curve and Examples of Stable, Tangential and Unstable Solutions
Figure 1

or T_j(tangent), whichever is lower. If T_j(tangent) is lower, the maximum allowed $R_{\theta ja}$ is equal to the reciprocal of the slope of the tangent line. If T_j(maximum) is lower, however, the maximum allowed $R_{\theta ja}$ is then equal to the reciprocal of the slope of the line passing through T_a on the T_j axis and the T_j(maximum) point on the P_T curve. By comparing the actual $R_{\theta ja}$ with the maximum allowed value, the algorithm that finds the intersection point of the P_T curve and the $R_{\theta ja}$ line is only entered if a solution exists. This speeds up program execution when a solution does not exist.

An implicit equation for T_j(tangent) can be derived as follows:

$$P_T = K_1 a e^{bT_j} + P_1$$
(dissipated MOSFET power) $\qquad$ (1)

where: $K_1 = I_{D(RMS)}{}^2 R_{DS(on)}$ @ 25°C

and: $P_1 = P_{switching} + P_{leakage} + P_{gate}$
(fixed losses)

Therefore:

$$\frac{\partial P_T}{\partial T_j} = K_1 a b e^{bT_j} \quad \text{(slope of } P_T \text{ curve at } T_j) \quad (2)$$

also:

$$P_H = \frac{T_j - T_a}{R_{\theta ja}} \quad \text{(dissipated heatsink power)} \quad (3)$$

For the tangent line where $P_T = P_H$ and $T_j = T_j$(tangent), we have:

$$K_1 a e^{bT_j} + P_1 = (T_j - T_a) K_1 a b e^{bT_j}$$

and so:

$$T_j = T_a + 1/b + \frac{P_1}{K_1 a b e^{bT_j}} \qquad (4)$$

By choosing an initial value of $T_j = T_a + 1/b$, the first iterative algorithm evaluates approximations of T_j(tangent) according to:

$$T_{j(i)} = T_a + 1/b + \frac{P_1}{K_1 a b e^{\{bT_j(i-1)\}}} \qquad (5)$$

The convergence condition chosen for this algorithm is that $|T_j(i)-T_j(i-1)|$ be less than 1°C. In tests, the algorithm converged to the required T_j(tangent) in typically three to four iterations.

A second iterative algorithm finds the solution intersection point of the P_T curve and the $R_{\theta ja}$ line. A successive-approximation technique similar to that used in analog-to-digital converter ICs is employed to guarantee rapid convergence to the desired solution point.

The algorithm uses two successive-approximation registers, T_L and T_H. T_L is set initially to T_a while T_H is set to either T_j(maximum) or T_j(tangent), whichever is lower. The flow chart in Figure 2 shows that the approximation to T_j(intersection point) becomes better and better, until the magnitude of the difference between P_T and P_H is less than 1% of P_T. When this condition is reached, the solution intersection point has been found.

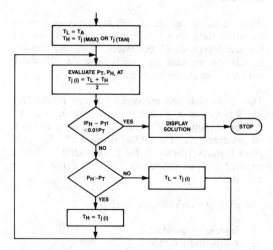

The Algorithm Used to Find the Solution Intersection Point on the P_T Curve Employs a Successive Approximation Technique to Achieve Rapid Convergence
Figure 2

Program Use

This program was written to facilitate easy use. If desired, regression constants a and b can be determined using the HP-41 Statistics Pac curve-fitting routines. Please refer to the Stat Pac Users Handbook for instructions on using the curve-fitting routines.

When the program is executed, input data is prompted for and put into storage registers R_0 through R_9. An editing feature has been incorporated into the input routine that allows one to enter a set of data and then selectively alter only those variables desired. The input routine uses a standard prompt format as follows:

"VARIABLE = CURRENT VALUE"

where "VARIABLE" is the name of the particular variable being prompted for and "CURRENT VALUE" is the current contents of the storage register associated with this variable. If the current value of the variable is acceptable to the user, it can be left unchanged simply by pressing the R/S key. If, however, the variable must be changed before a subsequent run of the program, a new value can be keyed into the x-register and the R/S key then pressed. This action causes the old data in the storage register to be overwritten with the new data. The editing feature also allows one to review a set of data after it has been entered. By restarting the program at its entry point and stepping through the prompts with the R/S key, the previously entered variables can be checked for correctness. The input prompt sequence and register map are as follows:

register	variable	prompt sequence
R_0	a	"A = "
		"CALC B Y/N?"
R_1	b	"B = "
R_2	P_1(fixed losses)	"P1 = "
R_3	I_D(RMS)	"IDRMS = "
R_4	$R_{DS(on)}$ @ 25°C	"RDS25 = "
R_5	T_A	"TA = "
R_6	T_J(max)	"TJMAX = "
R_7	$R_{\theta jc}$	"ROJC = "
R_8	$R_{\theta cs}$	"ROCS = "
R_9	$R_{\theta sa}$	"ROSA = "

After the first variable (a) has been entered, the program will ask if constant b should be calculated from constant a. The prompt "CALC B Y/N?" indicates that the program is ready to do this. Pressing the Y key will cause b to be calculated and put into storage register R_1. Pressing the N key, on the other hand, will cause this step to be bypassed thus leaving the contents of register R_1 unchanged.

The program also uses registers R_{10} through R_{17} as temporary working registers as follows:

register	variable	
R_{10}	K_1	$(= I_D(RMS)^2\, R_{DS(on)}$ @ 25°C)
R_{11}	$T_j(i)$	iteration storage register
R_{12}	$R_{\theta ja}$	(maximum allowed)
R_{13}	$R_{\theta ja}$	(current value)
R_{14}	$T_L(i)$	successive approximation registers
R_{15}	$T_H(i)$	
R_{16}	$P_H(i)$	iteration storage registers
R_{17}	$P_T(i)$	

When the last variable $R_{\theta sa}$ has been entered, the program iteratively calculates an approximation to the junction temperature at the tangent point (T_j(tangent)). It then informs the user of the calculated value with the following display:

"TJTAN = VALUE"

The maximum allowed value of $R_{\theta sa}$ is then determined and compared to the input value of $R_{\theta sa}$. If the input value is too high, the program informs the user with two "beeps" and the following display sequence:

"ROSA TOO HI"

(pause)

"ROSA < MAX VALUE" where "MAX VALUE" is a number equal to the maximum allowed value of $R_{\theta sa}$

(pause)

"TRY AGAIN"

(pause)

"ROSA = CURRENT VALUE"

After the last display (above), the program will stop and wait for a new value of $R_{\theta sa}$ to be entered. If the new value is still too high, the last display sequence will be repeated. When the new value of $R_{\theta sa}$ is acceptable to the program, it will go on and iteratively calculate the solution point. After the solution point is found, the program will stop, and P_T will be displayed. Pressing the R/S key will then cause T_j to be displayed. Pressing the R/S key again will cause the prompt "ROSA = CURRENT VALUE" to be displayed so that new heatsink data can be evaluated.

Example

A test was run to see how well the results from the program matched those obtained from a graphical solution. The data used in example 1 of reference 1 was used as input to the thermal program, after constant "a" had been determined from the normalized $R_{DS(on)}$ curve shown in Figure 3. Determining constant "a" from a graph such as that shown in Figure 3 is a very simple procedure because "a" is just the value of R_{DSN} at $T_j = 0°C$.

The example data was as follows:

Device type = IRF440

Constant "a" = 0.85

Constant "b" (calculated) = 6.50×10^{-3}

Fixed Power losses (P_1) = 5 W

I_D (RMS) = (i) 3 A
$\qquad\quad$ = (ii) 5 A

$R_{DS(on)}$ at 25°C = 0.8Ω

Ambient temperature (T_a) = 50°C

Max. junction temp. ($T_j(max)$) = 150°C

$R_{\theta jc}$ = 1°C/W

$R_{\theta cs}$ = 0.2°C/W

$R_{\theta sa}$ = 1°C/W

The resulting calculator output using the HP82143A thermal printer is shown in Figure 4.

For case (i) with I_D(RMS) = 3A, the program calculated a solution of:

T_j = 84.00°C and P_T = 15.59 W

This agreed very well with the graphical solution of:

T_j = 84°C and P_T = 15.8 W

For case (ii) with I_D(RMS) = 5 A, both the calculator and graphical methods indicated there would be no solution, i.e., thermal runaway would occur. As a bonus, the program also indicated that to have a stable thermal system with T_j below $T_j(max)$, $R_{\theta sa}$ must be less than 0.80°C/W.

Conclusion

It has been shown that a numerical solution to the problem of power MOSFET thermal analysis can be just as accurate as a graphical solution, and, in addition, it eliminates the requirement for any additional paperwork (i.e., graph plotting).

The HP-41CV calculator program presented will find the required solution point (if it exists) within a matter of seconds and will allow selective evaluation of many different MOSFET/heatsink combinations in a short period of time.

This program should prove to be a useful tool to the designer of any power MOSFET switching circuit because it is simple to use, accurate, and fast.

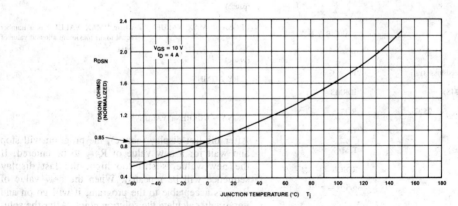

This Curve Shows the Normalized $R_{DS(on)}$ Versus T_j Curve for an IRF440 Power MOSFET, and the Value of Constant a at T_j = 0°C
Figure 3

Prompt	Entry	XEQ "MTAP"
PROGRAM MTAP		
A=0.000		
A=0.000	.850	RUN
CALC B Y/N?		
Y		RUN
B=6.50E-3		
P1=0.00		
	5.00	RUN
IDRMS=0.00		
	3.00	RUN
RDS25=0.00		
	.80	RUN
TA=0.00		
	50.00	RUN
TJMAX=0.00		
	150.00	RUN
ROJC=0.00		
	1.00	RUN
ROCS=0.00		
	.20	RUN
ROSA=0.00		
	1.00	RUN
TJTAN=231.66		
PT=15.59		
		RUN
TJ=84.00		

Prompt	Entry	XEQ "MTAP"
PROGRAM MTAP		
A=0.850		
		RUN
CALC B Y/N?		
N		RUN
B=6.50E-3		
		RUN
P1=5.00		
		RUN
IDRMS=3.00		
	5.00	RUN
RDS25=0.00		
		RUN
TA=50.00		
		RUN
TJMAX=150.00		
		RUN
R0JC=1.00		
		RUN
R0CS=0.20		
		RUN
R0SA=1.00		
		RUN
TJTAN=214.95		
R0SA TOO HI		
R0SA<0.00		
TRY AGAIN		
R0SA=1.00		
	.70	RUN
PT=47.41		
		RUN
TJ=140.00		

**Using the Example Data, the HP-41CV Thermal Program
Produced an Almost Identical Solution to That Produced
Using the Graphical Method Described in Reference 1
Figure 4**

Program Listing

```
01◆LBL "MTAP"        42 PROMPT        83 SF 00          124 PSE          165 RCL 09        206 RCL 11
02 "PROGRAM MTAP"    43 FS?C 22       84 RCL 03         125 CLD          166 +            207 RCL 01
03 AVIEW             44 STO 02        85 X  2           126 RCL 11       167 STO 13       208 *
04 PSE               45 "IDRMS="      86 RCL 04         127 RCL 06       168 X<Y?         209 E  X
05 FIX 3             46 ARCL 03       87 *              128 X>Y?         169 GTO 07       210 RCL 00
06 CF 00             47 PROMPT        88 STO 10         129 GTO 04       170 TONE 7       211 *
07 "A="              48 FS?C 22       89 RCL 05         130 RCL 01       171 TONE 7       212 RCL 10
08 ARCL 00           49 STO 03        90 RCL 01         131 *           172 "ROSA TOO HI" 213 *
09 PROMPT            50 "RDS25="      91 1/X            132 E  X         173 AVIEW        214 RCL 02
10 FS?C 22           51 ARCL 04       92 +              133 RCL 00       174 PSE          215 +
11 STO 00            52 PROMPT        93 STO 11         134 *           175 RCL 12        216 STO 17
12◆LBL 00            53 FS?C 22       94◆LBL 03         135 RCL 10       176 RCL 07       217 100
13 AON               54 STO 04        95 RCL 02         136 *           177 RCL 08       218 /
14 "CALC B Y/N?"     55 "TA="         96 RCL 01         137 RCL 02       178 +            219 RCL 16
15 PROMPT            56 ARCL 05       97 RCL 01         138 +           179 -            220 RCL 17
16 AOFF              57 PROMPT        98 *              139 RCL 06       180 "ROSA < "    221 -
17 ASTO X            58 FS?C 22       99 E  X           140 RCL 05       181 ARCL X       222 ABS
18 "N"               59 STO 05        100 RCL 01        141 -           182 AVIEW        223 X<Y?
19 ASTO Y            60 "TJMAX="      101 *             142 X<>Y         183 PSE          224 GTO 10
20 X=Y?              61 ARCL 06       102 RCL 00        143 /           184 PSE          225 RCL 17
21 GTO 01            62 PROMPT        103 *             144 GTO 05       185 "TRY AGAIN"  226 RCL 16
22 "Y"               63 FS?C 22       104 RCL 10        145◆LBL 04       186 AVIEW        227 X>Y?
23 ASTO Y            64 STO 06        105 *             146 RDN          187 PSE          228 GTO 09
24 X  Y?             65 'ROJC="       106 /             147 STO 06       188 GTO 02       229 RCL 11
25 GTO 00            66 ARCL 07       107 RCL 01        148 RCL 01       189◆LBL 07       230 STO 14
25 RCL 00            67 PROMPT        108 1/X           149 *           190 RCL 05        231 GTO 08
27 1/X               68 FS?C 22       109 +             150 E  X         191 STO 14       232◆LBL 09
28 LN                69 STO 07        110 RCL 05        151 RCL 01       192 RCL 06       233 RCL 11
29 25                70 "ROCS="       111 +             152 *           193 STO 15       234 STO 15
30 /                 71 ARCL 08       112 RCL 11        153 RCL 00       194◆LBL 08       235 GTO 08
31 STO 01            72 PROMPT        113 X<>Y          154 *           195 RCL 14        236◆LBL 10
32◆LBL 01            73 FS?C 22       114 STO 11        155 RCL 10       196 RCL 15       237 "PT="
33 SCI 2             74 STO 08        115 -             156 *           197 +            238 ARCL 17
34 "B="              75◆LBL 02        116 ABS           157 1/X          198 2            239 PROMPT
35 ARCL 01           76 "ROSA="       117 1             158◆LBL 05       199 /            240 "TJ="
36 PROMPT            77 ARCL 09       118 X<=Y?         159 STO 12       200 STO 11       241 RCL 11
37 FS?C 22           78 PROMPT        119 GTO 03        160◆LBL 06       201 RCL 05       242 INT
38 STO 01            79 FS?C 22       120 "TJAN="       161 RCL 12       202 *            243 ARCL X
39 FIX 2             80 STO 09        121 ARCL 11       162 RCL 07       203 RCL 13       244 PROMPT
40 "P1="             81 FS? 00        122 AVIEW         163 RCL 08       204 /            245 GTO 02
41 ARCL 02           82 GTO 06        123 PSE           164 +           205 STO 16        246 END
```

Reference

Rudy Severns, *"Safe Operating Area and Thermal
Design for MOSPOWER Transistors,"* Siliconix
Applications Note AN83-10, November 1983.

Chapter 5

Practical Design Considerations

5.1 High Speed Gate Drive Circuits

Introduction

Many design aids and application notes have been written about driving power MOSFETs directly from CMOS or open-collector TTL logic. These schemes are very attractive because of their simplicity (see "Using Power MOSFET Transistors to Interface from IC logic to High Power Loads" in Section 6.3), but the switching times of MOSFETs driven in this manner are often too slow for many applications. This is not by any means a result of the MOSFET having inherently slow switching times; in fact, any power MOSFET can be switched ON or OFF in less than 10 nsec if desired. When a MOSFET is driven by logic, the loss of its potential high speed switching characteristics is actually due to the limited peak output current available to charge or discharge the effective gate-to-source capacitance.

The larger the die area of the power MOSFET being driven, the larger its effective gate-to-source capacitance ($C_{gs(eff)}$). Consequently, the total gate charge

(Q_g) required to change the gate-to-source voltage from 0 to 10 volts, or vice-versa, also increases. Logic gates, with their very limited peak output current (or $\partial Q/\partial t$) capability, are inevitably the limiting factors if used as gate drivers for modern, large power MOSFETs.

Additionally, a phenomenon known as the "Miller effect" causes the effective gate-to-source capacitance of a common-source connected power MOSFET to assume a much larger value than its static specification during a switching transition. This phenomenon, illustrated in Figure 1, is well known to the designers of analog circuits and is often used to advantage. However, in high-speed gate drive circuits, it necessitates a substantial peak output current capability.

Since $C_{gs(eff)} \triangleq \partial Q_g/\partial V_{gs}$, from Figure 1 it can be seen that this parameter is essentially constant at about 700 pF below $V_{GS} = 6$ volts. $C_{gs(eff)}$

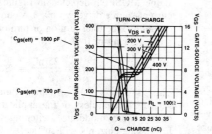
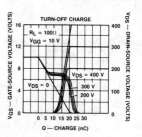

The Q_g Versus V_{GS} Charge Transfer Curves for the Siliconix VNDA40 Geometry Show the Occurrence of Maximum Effective Gate to Source Capacitance at a V_{GS} of Approximately 7 V
Figure 1

increases rapidly when the MOSFET turns on or off at $V_{GS} = 7$ volts, as indicated by the approximately horizontal portions of the turn-on and turn-off V_{GS} versus Q_g charge transfer curves. Above $V_{GS} = 7$ volts, $C_{gs(eff)}$ returns to a lower constant value of approximately 1900 pF. One can see that, due to the large capacitance in the "horizontal" region of the curves, a relatively large amount of charge is required to change V_{GS} by only a fraction of a volt.

It is difficult to quantify the actual value of $C_{gs(eff)}$ in the "horizontal" region of the charge transfer curves because it varies with V_{DS}. In practice, one really doesn't need to know the value of $C_{gs(eff)}$ in the "horizontal" region anyway. The curves shown in Figure 1 provide enough information to accurately determine how fast a MOSFET will switch when driven by a particular circuit.

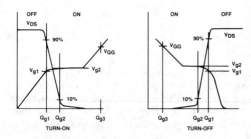

Switching Time Equations

$$t_{d(on)} = \frac{Q_{g1}}{V_{g1}} R_{gen} \ln\left(\frac{V_{GG}}{V_{GG} - V_{g1}}\right) \quad (1)$$

$$t_r = \frac{Q_{g2} - Q_{g1}}{V_{g2} - V_{g1}} R_{gen} \ln\left(\frac{V_{GG} - V_{g1}}{V_{GG} - V_{g2}}\right) \quad (2)$$

$$t_{d(off)} = \frac{Q_{g3} - Q_{g2}}{V_{GG} - V_{g2}} R_{gen} \ln\left(\frac{V_{GG}}{V_{g2}}\right) \quad (3)$$

$$t_f = \frac{Q_{g2} - Q_{g1}}{V_{g2} - V_{g1}} R_{gen} \ln\left(\frac{V_{g2}}{V_{g1}}\right) \quad (4)$$

These Four Equations Enable One to Calculate the Switching Times of a Power MOSFET, Based on Classical Circuit Analysis of the Driver Circuit Model Shown in Figure 3
Figure 2

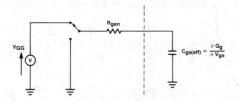

This Simple Circuit Can Be Used to Model a Voltage Source Type Gate Drive Circuit
Figure 3

In Figure 2, the pertinent equations for determining the four switching times $t_{d(on)}$, t_r, $t_{d(off)}$, and t_f are given, based on classical circuit analysis of the driver circuit model shown in Figure 3. In order to minimize the switching times, it can be seen that the output resistance R_{gen} must be made as low as possible. Doing this also increases the immunity to possible mode 1 ($\partial V_{DS}/\partial_t$) breakdown (reference Section 5.4). As an example, if we let $V_{GG} = 10$ V and

$R_{gen} = 2\,\Omega$, the four switching times for the MOSFET depicted in Figure 1 are as follows:

$$t_{d(on)} = 1.7 \text{ nsec}$$
$$t_r = 10.2 \text{ nsec}$$
$$t_{d(off)} = 4.3 \text{ nsec}$$
$$t_f = 4.3 \text{ nsec}$$

For a low value of R_{gen}, we see that the MOSFET will switch very rapidly. However, a low value of R_{gen} also means that the peak MOSFET gate current can be large during the switching transition. Increasing the value of R_{gen} reduces the peak gate current drawn by the MOSFET but degrades the switching speed of the device. It is up to the designer to trade off switching speed against peak MOSFET gate current for a particular application.

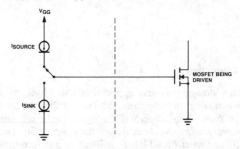

This Circuit May Be Used to Model the Current Source Driver Circuits Such as the Siliconix D469 IC
Figure 4

Certain gate drive circuits will appear to a MOSFET not as a voltage source with a series output resistor R_{gen}, but rather as a current source (or sink). This is the case for both the D469 and DS0026 IC drivers to be discussed later. Figure 4 shows a circuit that can be used to model these two devices. If the peak output capability of a current-source type driver is known, the total turn-on or turn-off time for a power MOSFET driven by it can be very easily determined.

By definition:
$$\Delta t = \frac{\Delta Q}{I} \quad (5)$$

If the MOSFET depicted in Figure 1 is driven by a current-source driver with a 3 A output sink or source capability, the total turn-on or turn-off time, assuming a V_{GS} swing of 0 to 10 volts and vice versa, is:

$$t_{on} \text{ (or } t_{off}) = \frac{25 \text{ nC}}{3 \text{ A}} = 8.33 \text{ nsec}$$

This is a very fast switching time for any power MOSFET regardless of the application, but in many cases, more moderate switching times and hence lower gate currents will suffice.

Let's examine three groups of high-speed gate drive circuits suitable for driving most of the larger power MOSFETs available when the required switching times are less than 50 nsec.

Direct Coupled Drive Circuits Using ICs Alone

Until now, the most popular IC for driving MOSFETs in high speed switching applications has been the DS0026 MOS clock driver made by National Semiconductor. This dual-channel bipolar IC can sink or source a relatively large peak output current (1.5 A) and has inherently small propagation delays within its internal circuitry. It also operates from a wide supply voltage range (22 V).

Unfortunately, the DS0026 has some drawbacks which prevent it from being the "perfect" IC driver. Since the DS0026 requires a large logical "1" input current (10 mA), it is unsuitable for use in applications where the control logic is LS TTL or high-speed CMOS. The large quiescent supply current the device

draws with one output at zero volts (30 mA) means its average power dissipation at low MOSFET duty cycles will be quite large (as high as 0.5 watts). Additionally the DS0026 is configured as an inverting-only driver and requires two external TTL inverters when non-inverting drive is required.

As an alternative, the D469 CMOS quad power MOSFET driver from Siliconix can be used to replace the DS0026 in just about all but the higher speed switching applications. Each driver in the D469 may be configured as being either logically inverting or non-inverting and can sink or source a peak current of 0.5 A. To obtain greater peak output current capability, necessary when driving large capacitive loads, each of the D469s outputs may be connected in parallel if desired.

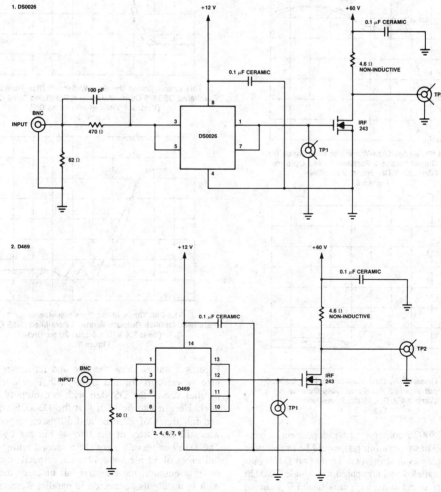

The Schematic Diagram of the Test Fixture Used to Evaluate the Switching Performance of the DS0026 and D469 Shows Identical Load Conditions for Each Device
Figure 5

Both the DS0026 and D469 were incorporated into a test fixture driving identical power MOSFETs with identical loads so that their switching performance could be compared. Figure 5 shows the schematic diagram of this fixture. The power MOSFETs used were Siliconix IRF243s with a maximum static C_{iss} of 1600 pF and a maximum static C_{rss} of 300 pF.

The load resistance for both MOSFETs was approximately 4.6 ohms (non-inductive), and the power supply voltage was 60 V. Both the DS0026 and the D469 were powered from a separate 12-volt power supply and driven by a TTL level pulse train. The pulse train had a pulse width of 150 nsec and a repetition rate of 1 kHz.

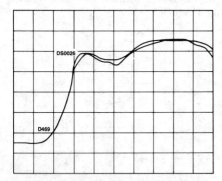

This Graph Shows the Gate Waveform Rise Times for 1 DS0026 Output Versus 2 Paralleled D469 Outputs (Vert: 2.4 V/Div, Horiz: 20 ns/Div)
Figure 6

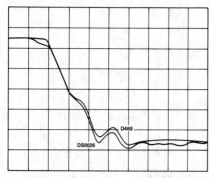

The Fall Times of the Gate Waveforms for 1 DS0026 Output Versus 2 Paralleled D469 Outputs (Vert: 2.4 V/Div, Horiz: 20 ns/Div)
Figure 7

Since the DS0026 contains two independent drivers and the D469 has four, a fair test would be to compare the two devices as dual drivers. Thus, two D469 outputs were paralleled and compared to a single DS0026 output in drive capability. Figures 6 and 7 indicate that on a "per chip" basis, the D469 was essentially the equal of the DS0026. To be precise, the D469 gate waveform's rise and fall times were 36 nsec and

44 nsec, respectively, while the DS0026s were 26 nsec and 42 nsec. It appears that the DS0026 was somewhat faster (by 10 nsec) on the rising edge of its waveform than the D469, but both had very similar fall times.

Some further tests were conducted to see how the D469, with two or more outputs connected in parallel, compared to the DS0026 with both of its outputs connected in parallel.

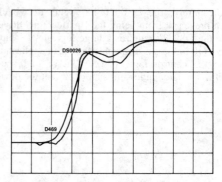

This Graph Shows the Gate Waveform Rise Time for 2 Paralleled DS0026 Outputs Versus 4 Paralleled D469 Outputs (Vert: 2.4 V/Div, Horiz: 20 ns/Div)
Figure 8

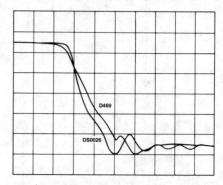

The Fall Times of the Gate Waveforms for 2 Paralleled DS0026 Outputs Versus 4 Paralleled D469 Outputs (Vert: 2.4 V/Div, Horiz: 20 ns/Div)
Figure 9

Figures 8 and 9 show the rise and fall times of the gate waveforms for the D469 with four outputs in parallel versus the DS0026 with two outputs in parallel. Here it can be seen that the DS0026 with rise and fall times of 22 nsec and 30 nsec, respectively, was still the faster of the two — but not by much. The D469 came in a very close second with rise and fall times of 32 nsec and 42 nsec, respectively. Interestingly enough, the rise and fall times of the D469 with four outputs connected in parallel were virtually the same as those with only two outputs in parallel. The rise and fall times for the DS0026, on the other hand, showed some improvement.

The main difference observed between the D469 and the DS0026 was the propagation delay-time each device exhibited. For the D469 this parameter was approximately 50 nsec while that of the DS0026 was in the range of 5 to 10 nsec. Figure 10 clearly shows this difference. It should be noted that Figure 10 is the only one in which the difference in propagation delay is shown. Figures 6 through 9 do not show this because it is easier to compare switching waveforms when the oscilloscope traces are overlaid with one another.

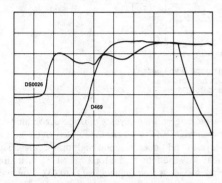

This Graph Clearly Shows the 50 ns Difference in
Propagation Delay Between the DS0026 and the D469
(Vert: 2.4 V/Div, Horiz: 20 ns/Div)
Figure 10

In summarizing the results of this comparative test, we can say that the D469 is really a DS0026 in a "CMOS disguise." Although the DS0026 was faster than the D469 in both cases, its speed advantage was no more than about 10 nsec in any of the rise and fall time measurements. As a two-channel driver the D469 is the obvious choice, especially when one considers its low power consumption compared to the DS0026. In the test fixture shown in Figure 5, the DS0026 drew an average supply current of 30 mA and thus consumed about 360 mW of power while the D469 drew an average current of slightly more than 2 mA and consumed about 25 mW of power. Needless to say, the DS0026 ran quite hot to the touch while the D469 was cold. The 50 nsec propagation delay of each driver in the D469 should not pose any problem at all as long as the MOSFETs in a particular circuit are all driven by D469s.

Direct Coupled Drive Circuits Using Discrete Components

If the IC gate drivers described above begin to become the limiting factor in the performance of MOSFET switching circuits that use them, it is inevitably the result of two factors:

(1) The peak output current capability of the IC is not high enough.
(2) The device has inherent switching times which are too slow.

When it is desired to drive very large single chip or multichip power MOSFETs with switching times in the range of 50 nsec to 20 nsec, a current booster stage may be added to either a D469 or a DS0026 with no loss of switching speed. Figure 11 shows how a complementary MOSFET booster stage can be connected to one output of a D469. The additional output buffer functions in exactly the same manner as a CMOS logic inverter, and its output voltage swings all the way to either power rail. Current is only drawn by the buffer during switching transitions; consequently, the average power it dissipates is proportional to the switching frequency. The buffer may also be used by itself if the TTL input level compatibility of the D469 is not required.

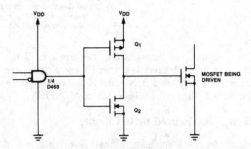

NOTE: Q_1 AND Q_2 COMPRISE 1/2 A VQ7254 POWER MOSFET ARRAY

A Simple CMOS Current Booster Stage Can Be Added
to a D469 Output to Increase Its Output Capability,
or May Be Used on Its Own
Figure 11

Depending on the MOSFETs used in the buffer stage, the available output current can be as high as several amps. For the VQ7254 dual complementary pair shown in Figure 11, the pulsed output current specification is three amps. This is a six-fold improvement over the 0.5 A peak output sink/source capability of each channel of the D469 alone.

For very high speed applications where MOSFET switching times of less than 20 nsec must be achieved, both small internal propagation delays and high peak output current capability are the necessary characteristics of a gate driver circuit. Figure 12 shows a circuit which works very well with large capacitive loads. When the input to the driver is a logical "0," Q_1 is held in conduction by one half of the DS0026, and Q_2 is clamped off by Q_1. When a logical "1" input occurs, Q_1 is turned off and a current pulse is applied to the gate of Q_2 by the other half of the DS0026 through T_1. After about 20 ns, T_1 saturates, and Q_2 is held on by its own C_{gs} and the bootstrap circuit comprised of C_1, D_1 and R_1. For pulses less than 50 μsec wide, the bootstrap circuit may not be needed because the input capacitance of Q_2 discharges very slowly. At the end of the logical "1" input pulse, Q_1 turns on, shutting off Q_2.

5

PERFORMANCE SUMMARY

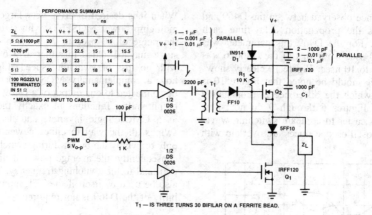

Z_L	V+	V+	t_{on}	t_r	t_{off}	t_f
			ns			
5 Ω & 1000 pF	20	15	22.5	7	15	7
4700 pF	20	15	22.5	15	16	15.5
5 Ω	20	15	23	11	14	4.5
5 Ω	50	20	22	18	14	4
100 RG223/U TERMINATED IN 51 Ω	20	15	20.5*	19	13*	6.5

* MEASURED AT INPUT TO CABLE.

T_1 — IS THREE TURNS 30 BIFILAR ON A FERRITE BEAD.

**A Circuit Capable of Driving Very Large Capacitive Loads Is
Shown in This Diagram Along with Its Tabulated
Performance Under Varying Load Conditions
Figure 12**

One can see that the switching performance of the above mentioned driver is extremely good with large capacitive loads, and it is suitable for driving even the largest currently available power MOSFETs.

High Speed Isolated Drive Circuits

Many switching applications involve driving power MOSFETs from control logic or other circuitry that is ground referenced. The question that inevitably arises is: how does one drive power MOSFETs whose sources are not ground referenced? The solution, of course, is to use some form of DC isolation between either the control logic and the driver or between the driver and the MOSFET(s).

There are basically two viable methods that can be used to provide DC isolation between a MOSFET and its associated control circuitry when necessary. These are:

(I)   Transformer isolation
(II) Opto-coupler isolation

Each method has its own inherent advantages and disadvantages as far as performance/cost trade-offs are concerned. We will examine them in more detail.

(I)   Transformer isolation

Figure 13(a) shows a simple transformer drive circuit that effectively isolates the ground referenced D469 from the *n*-channel MOSFET with a floating source. Unfortunately, this circuit has a major problem: the AC waveform across the primary winding can be asymmetrical. This follows from the fact that the average of the transformer winding volt-seconds product must equal zero as shown in Figure 13(b). Therefore, the gate enhancement voltage of the MOSFET will vary with duty cycle, being greater at low duty cycles and smaller at high duty cycles.

If a wide range of duty cycles must be accommodated then the MOSFET gate will be overdriven at low duty cycles and underdriven at high duty cycles. This can lead to variations in the total MOSFET switching time as is evident from equations (1) to (4). However, for those applications where only a moderate variation in duty-cycle is to be accommodated, this simple circuit can work very well.

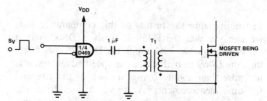

**a) This Simple Transformer Coupled Circuit Provides Isolated
Drive for the Power MOSFET**

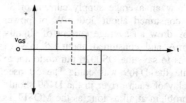

**b) The Resulting Gate Voltage Can Be Seen to Be an
Asymmetrical AC Waveform, Whose Amplitude Varies with
Duty Cycle
Figure 13**

There are transformer isolated drive circuits where V_{GS} does not vary with duty cycle. Figure 14 shows a transformer coupled drive circuit in which the primary winding sees only a symmetrical AC waveform. V_{GS} is fixed by the voltage swing at the output of the D469 drivers and the turns ratio of the transformer. The AC voltage swing across the primary of the transformer can be seen to have a peak to peak value of 2 V_{DD} since one D469 output is at 0 volts while the other is at V_{DD} and vice-versa.

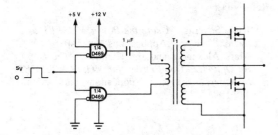

This Transformer Coupled Drive Circuit Generates a
Symmetrical Gate to Source Voltage Waveform for Each
MOSFET, Whose Amplitude is Independent of Duty Cycle
Figure 14

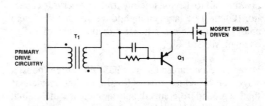

A Bipolar Transistor and Its Associated Components Reduce
the Effective Drive Impedance Presented to the Gate
Terminal of the MOSFET
Figure 15

C_1 is added to the circuit to prevent a DC component from appearing across the primary of T_1 due to a potential lack of symmetry in the switching times of the two drivers used. The value of C_1 is determined by the magnetizing current of T_1. It should be large enough so that the voltage drop across it due to the magnetizing current during one half a cycle is small compared to V_{DD}.

Of course, current booster stages may be added to the D469 outputs in Figure 14 to increase the drive current to the primary winding and hence the gates of the MOSFET(s). Additional isolated windings may also be added to T_1 so that totem pole MOSFETs in bridge circuits may be driven as illustrated in Figure 14.

Often the impedance of the secondary winding(s) that drive the MOSFETs may well be high enough to cause mode 1 $(\partial\ V_{DS}/\partial t)$ triggering problems. By

adding a bipolar transistor Q_1 and its associated components as shown in Figure 15, these problems can be overcome. At turn-off, Q_1 is driven on by the energy in T_1, clamping the gate of Q_2 to its source. Even after all the energy in T_1 is discharged, Q_1 still presents a relatively low impedance to the gate of Q_2.

(II) Opto-Coupler Isolation

Figure 16 shows a different method of providing DC isolation in a totem pole MOSFET "H-bridge" circuit that uses n- and p-channel devices. Here a D469 driver is referenced to the positive power supply rail and an opto-coupler is used to provide DC isolation between it and the low voltage control logic. A -12 volt power supply, referenced to the positive rail, is generated for the D469's ground pin with a zener diode, capacitor, and resistor.

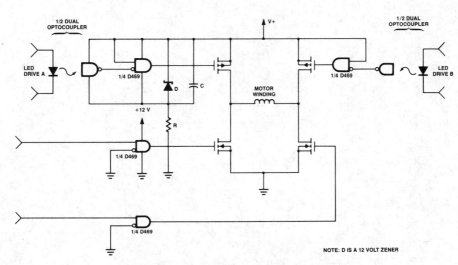

NOTE: D IS A 12 VOLT ZENER

This Schematic Diagram Shows a Different Method of
Providing Isolation — Floating a D469 Driver and Optically
Coupling the Logic Drive to the Upper MOSFETs
Figure 16

5

The D469 in this example provides gate-drive voltages for the *p*-channel MOSFETs that swing from the positive rail to 12 volts below it. The reservoir capacitor (C) must be large enough to ensure that sufficient charge is available for transfer to the gates of the *p*-channel MOSFETs.

Here again a complementary MOSFET current booster stage can be added to the floating D469 as well as to the ground referenced one to increase the output current capability.

References

[1] Rudy Severns, *"Using the Power MOSFET as a Switch,"* Intersil Applications Note A036, 1981.

[2] Mark Alexander and Jim Harnden, *"The D469: An Optimized Quad Driver for MOSPOWER FET Switches,"* Siliconix Applications Note AN83-11, 1983.

5.2 Temperature Compensated Biasing for Power MOSFETs in Linear Amplifiers

Introduction

Many of the highest performance audio amplifiers available today use power MOSFETs in their output stages. The advantages of using power MOSFETs instead of bipolar transistors in an amplifier output stage have been well described in numerous articles[1,2,3]; however, important thermal characteristics of MOSFET output stages are often overlooked. The purpose of this application note is to familiarize the designer with these characteristics and to show how to compensate for them when necessary.

It is a misconception that power MOSFETs have a negative temperature coefficient of saturated drain current at all values of gate-to-source voltage. Figure 1 shows that for large gate-drive voltages (six volts and above) the temperature coefficient of a power MOSFET is negative. In applications where substan-tial drain current (and hence gate-drive) is required, the drain current decreases as the junction temperature increases, causing the device power dissipation to decrease as well. If adequate heatsinking for the power MOSFET is provided then thermal runaway will not occur.

Linear amplifiers, on the other hand, generally require a certain value of quiescent current in the output stage to minimize crossover distortion. This value of quiescent current is normally made less than 100 mA per device to minimize power dissipation with no output signal. The problem here is that the quiescent gate-to-source bias voltage of each MOSFET may only be a few tenths of a volt greater than its threshold voltage (V_T). Under these conditions, the temperature coefficient of drain current is positive, and if not compensated for, can result in thermal runaway in the output stage.

An obvious solution might be to bias the MOSFETs in an output stage at the zero TC point indicated in Figure 1. This would be acceptable if the zero TC point occurred at a drain current value within the required quiescent bias current range of interest. Unfortunately, since most power MOSFETs exhibit a zero TC point at a gate-to-source voltage of between four and seven volts, the corresponding drain current (depending on the device) can be as high as several amps. Such a large value of quiescent bias current is unacceptable because the power dissipation at idle would then be very high. The output devices would inevitably fail due to excessive junction temperature if called upon to deliver any appreciable power to a load.

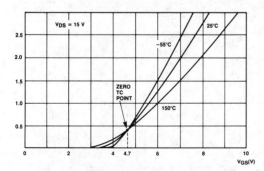

The Saturated I_D Versus V_{GS} Transfer Characteristics of an IRF521 n-channel DMOS Power FET Clearly Show the Occurrence of the Zero TC Point at a V_{GS} of Approximately 4.7 V
Figure 1

Figure 2 shows a typical complementary MOSFET source-follower output-stage and its associated bias and driver circuitry. The biasing network is generally an adjustable DC voltage source connected between the bases of the complementary Darlington bipolar transistor driver stage. If this bias voltage generator is fixed with respect to the junction temperature of the MOSFETs, their bias currents will rise as they heat up. As the bias currents of the MOSFETs rise, however, the quiescent power dissipation increases and causes the MOSFET junction temperatures to rise even further. This is clearly an unstable situation and must be avoided in a viable amplifier design.

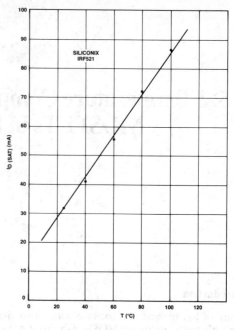

Figure 3

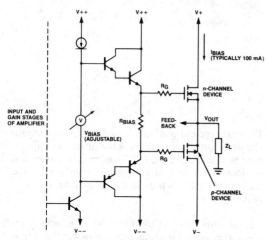

A Commonly Used Driver and Output Stage Configuration in MOSFET Audio Power Amplifiers
Figure 2

This paper investigates the physical phenomena that govern the behavior of power MOSFETs with temperature and presents two simple biasing circuits that effectively compensate for the positive temperature coefficient of drain current at low bias levels.

Theory

The experimentally observed temperature dependence of saturated drain current in a power MOSFET is illustrated in Figure 3. Clearly, under low bias conditions the temperature coefficient of drain current is positive. In order to compensate for such a temperature dependence, we need to know which electrical characteristics of a MOSFET vary with temperature and which ones are dominant at low values of I_D.

Despite their structural differences, the conventional planar MOSFET and a DMOS power FET exhibit similar electrical behavior in the saturated region of operation. Consequently, conventional device models

provide a reasonably accurate description of the I/V characteristics. For $V_{DS} > V_{GS} - V_T$ and $V_{GS} > V_T$, the drain current of a MOSFET is given by[4]

$$I_{D(SAT)} = \frac{\beta}{2} (V_{GS} - V_T)^2 \qquad (1)$$

where

$$\beta = \frac{\mu C_{ox} W}{L'} \text{ (the transistor gain)} \qquad \{A/V^2\}$$

and

μ	=	carrier mobility	$\{cm^2/V.sec\}$
C_{ox}	=	gate capacitance per unit area	$\{F/cm^2\}$
W	=	total gate width	$\{cm\}$
L'	=	electrically effective channel length	$\{cm\}$
V_T	=	threshold voltage	$\{V\}$

It is important to accurately determine the true threshold voltage of a MOSFET in order to obtain agreement between equation (1) and the actual device characteristics.

Most manufacturers specify a convenient, fixed value of drain current (usually 1 mA) at which the "threshold voltage" is measured. True threshold, however, is measured at a current proportional to the transistors W/L ratio[5], and it is often substantially larger than 1 mA. Consequently, the threshold values obtained by measurement of V_{GS} at an I_D of 1 mA and at a value proportional to W/L can differ by as much as 0.5 volts.

The graph of $\sqrt{I_{D(SAT)}}$ versus V_{GS} shown in Figure 4 provides an accurate method for determining V_T as well as β[6]. The straight line portion of the graph corresponds to the "square-law" operating region described by equation (1) and has a slope of $\beta/2$ and an x-intercept equal to the true threshold voltage of the MOSFET. One can see that the equation closely agrees with measured data over a wide range of drain current, including the bias range of most audio power amplifiers.

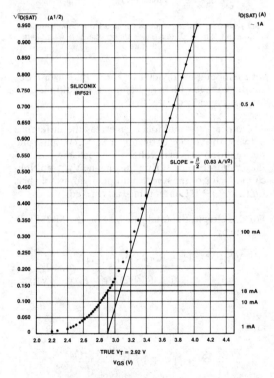

A Method to Determine the True Threshold V_T (and Transistor Gain B/2) from the Saturated Drain Current Transfer Characteristics (@ V_{DS} = 20 V)
Figure 4

By differentiating equation (1) with respect to temperature, some insight can be gained into the temperature dependence of saturated drain current. The two terms in this equation which exhibit a variation with temperature are the carrier mobility and the threshold voltage V_T. Differentiating equation (1), we see

that the temperature coefficient of drain current is as follows[7]:

$$\text{tempco}\,(I_D) = \frac{\partial I_{D(SAT)}}{\partial T}$$

$$= |I_D|\left(\frac{1}{\mu}\,\frac{\partial \mu}{\partial T} - \frac{2}{(V_{GS}-V_T)}\,\frac{\partial V_T}{\partial T} \right)$$

(2)

From device physics, it can be shown that over the temperature range of interest[4,8]

a) for an *n*-channel MOSFET,

- $\dfrac{\partial \mu}{\partial T}$ and $\dfrac{\partial V_T}{\partial T}$ are negative

- $(V_{GS} - V_T)$ is positive

b) for a *p*-channel MOSFET,

- $\dfrac{\partial \mu}{\partial T}$ and $(V_{GS} - V_T)$ are negative

- $\dfrac{\partial V_T}{\partial T}$ is positive

The two terms in equation (2) can be seen to have opposing signs for both *n* or *p*-channel devices.

At low values of the effective gate drive $(V_{GS} - V_T)$ and correspondingly low values of $I_{D(SAT)}$, the second term in equation (2) is dominant, and the tempco is positive.

In conclusion, at low values of $(V_{GS} - V_T)$ the temperature coefficient of drain current is proportional to the temperature coefficient of threshold voltage. Figure 5 shows the experimentally determined threshold variation with temperature of a power MOSFET. Typically this change in V_T is in the range

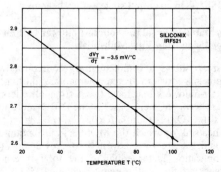

Figure 5

of -2.5 to -6 mV/° C[4,9]. To achieve temperature compensated biasing then, the effective gate drive $(V_{GS} - V_T)$ must remain nearly constant. This can be realized by controlling V_{GS} so that it is varied by an amount equal to the change in threshold voltage with temperature.

We now examine two practical circuits that provide output stage bias stabilization by reducing the MOSFET gate to source voltages as their junction temperatures increase.

Compensated Biasing Circuits

There are basically two means of providing thermal bias compensation in a linear amplifier. The first method is to electrically sense the output stage idle current and use some form of feedback control circuitry to ensure that the bias remains constant. The second method also uses feedback control, but instead of sensing the idle current, the case temperatures (T_C) of the MOSFETs are sensed by a thermal sensor, and the quiescent gate bias is reduced accordingly as T_C increases. The first method works well to keep the idle current in a MOSFET output stage constant, but it often degrades the electrical performance of the amplifier. What one achieves is a very stable amplifier over its operating temperature range; however, the resulting distortion specifications or output voltage swing will very likely be unacceptable[10].

The second method, on the other hand, in no way electrically interferes with the drain-to-source current paths of the MOSFETs in the output stage. As well, it can provide comparable bias stability to the first method, if implemented properly.

Two simple, thermally compensated biasing circuits for MOSFET power amplifiers using either complementary source-follower output stages or complementary common-source output stages will now be described. Both circuits take advantage of the -2.2 mV/° C variation in base-emitter voltage of a bipolar transistor to decrease the gate bias voltage of the MOSFETs in an output stage as they heat up.

Bias Stabilization of the Source-Follower Output Stage Using Dual V_{BE} Multipliers

Figure 6 shows a complementary source-follower output stage driven by complementary darlington emitter-followers that have two V_{BE} multipliers in series between their bases. One VBE multiplier transistor (Q_4) is thermally coupled (i.e., glued) to the heatsink on which the MOSFETs are mounted. R_3 and R_4 are then adjusted to provide a V_{BE} multiplication factor equal to the total threshold voltage variation of the n and p-channel MOSFETs

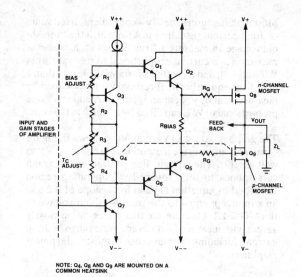

NOTE: Q_4, Q_8 AND Q_9 ARE MOUNTED ON A COMMON HEATSINK

Superior Bias Stability Can be Obtained in a Complementary Source-Follower Output Stage Using Dual V_{BE} Multipliers
Figure 6

with temperature divided by the variation in base-emitter voltage with temperature. This multiplication factor is typically greater than one.

The other V_{BE} multiplier generates the difference between the voltage of the upper V_{BE} multiplier and the required DC bias voltage between the bases of the emitter-followers. By adjusting the V_{BE} multiplication factor of the upper multiplier, adjustable bias for the MOSFET output stage is achieved.

For this circuit, the total gate bias for the MOSFETs is given by

$$V_{BIAS} \cong (1 + R3/R4) \, V_{BE(Q4)}$$
$$+ (1 + R1/R2) \, V_{BE(Q3)} - 4V_{BE} \tag{3}$$

and

$$\frac{\partial V_{BIAS}}{\partial T} = (1 + R3/R4) \, \frac{\partial V_{BE(Q4)}}{\partial T} \tag{4}$$

If $(1 + R3/R4) \cong 3$, then

$$\frac{\partial V_{BIAS}}{\partial T} \cong -6.6 \, \text{mV/° C}$$

This decrease in MOSFET gate bias voltage with temperature effectively compensates for the decrease in V_T of both the n and p-channel devices with temperature.

Bias Stabilization of the Common-Source Output Stage Using a Temperature-Compensated Current Sink

Another means of biasing a complementary power MOSFET output stage is shown in Figure 7. The circuit uses a fixed gate resistor and an adjustable current sink to provide variable bias for the p-channel MOSFET. Transistor Q_2 is thermally coupled to the heatsink on which the MOSFETs are mounted. As Q_2 heats up, its V_{BE} decreases and causes the current drawn through the collector of Q_1 to decrease as well. This reduces the gate drive voltage of the p-channel MOSFET as it and the n-channel device heat up. The op-amp automatically controls the n-channel MOS-FET to mirror the bias current of the p-channel device, which keeps the output centered at zero volts.

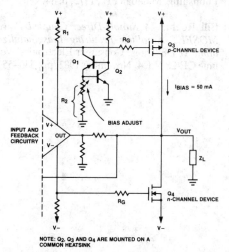

This Complementary Common-Source Output Stage Uses a Temperature Compensated, Adjustable Current Sink for Bias Stability
Figure 7

In this circuit, the gate bias of the p-channel MOSFET is given by

$$V_{GS}(P) = -R_2 I_1 + \frac{V_{BE}(Q2)}{R_1} \tag{5}$$

and

$$\frac{\partial V_{GS}(P)}{\partial T} = -\frac{R_2}{R_1} \times \frac{\partial V_{BE}(Q2)}{\partial T} \tag{6}$$

If $R2/R1 \cong 1.5$, then $\dfrac{\partial V_{GS}(P)}{\partial T} \cong +3.3\,mV/°C$

This effective change in p-channel MOSFET gate to source voltage compensates for the change in its V_T with temperature (which is positive, not negative).

Conclusions

It has been shown that in linear power amplifiers, where the MOSFET bias current levels are low, the temperature dependence of V_T is the dominant factor in the variation of I_D with temperature. By compensating for this effect alone, a high degree of thermal stability over an uncompensated bias arrangement can be achieved.

Figure 8 compares the temperature stability of drain current for the two biasing circuits described with the uncompensated thermal characteristics of the n-channel MOSFET used. Both of the bipolar transistor V_{BE} referenced compensation circuits are extremely effective in reducing bias sensitivity to temperature.

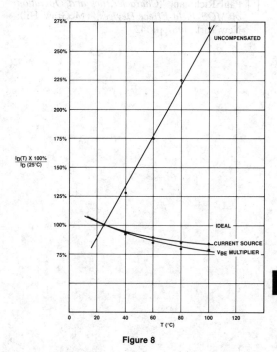

Figure 8

These circuits provide an inexpensive but necessary solution to the problem of bias stabilization in a MOSFET power amplifier without compromising distortion or output voltage swing.

5

References

[1] Robert R. Cordell, *"A MOSFET Power Amplifier With Error Correction,"* presented at the 72nd convention of the Audio Engineering Society, October 1982 (AES preprint).

[2] Erno Borbely, *"High Power High Quality Amplifier Using MOSFETs,"* Wireless World, March 1983.

[3] Tohru Sampei, Shin-ichi Ohashi, *"100 Watt Super Audio Amplifier Using New MOS Devices,"* IEEE Trans. Consumer Electronics, Volume CE-23, August 1977.

[4] Paul Richman, *"MOS Field Effect Transistors and Integrated Circuits,"* Wiley: New York, 1973, pp. 84–86, pp. 124–133.

[5] Paul Richman, *"Characteristics and Operation of MOS Field-Effect Devices,"* McGraw Hill: New York, 1967, p. 142.

[6] Richard S. Muller, Theodore L. Kamins, *"Device Electronics for Integrated Circuits,"* John Wiley; New York, 1977, pp. 360–361.

[7] William N. Carr, Jack P. Mize, *"MOS/LSI Design and Application,"* McGraw-Hill; New York, 1972, pp. 87–88.

[8] Edward Yang, *"Fundamentals of Semiconductor Devices,"* McGraw-Hill; New York, 1978, p. 23.

[9] William Penney, Lillian Lau, editors, Staff of American Microsystems Incorporated, *"MOS Integrated Circuits, Theory, Fabrication, Design and Systems Applications of MOS LSI,"* Krieger Publishing; Malaban FL., 1972, p. 84.

[10] Bill Roehr, *"A Simple-Direct Coupled Power MOSFET Amplifier Featuring Bias Stabilization,"* IEEE Trans. Consumer Electronics, Volume CE-28 No. 4, November 1982, pp. 546–552.

5.3 Parallel Operation
of Power MOSFETs
(TA 84-5)

Introduction

There are many reasons for operating power MOSFETs in parallel, either as multiple die on a common substrate or as individually packaged devices. Compared to paralleled bipolar junction transistors (BJTs), paralleled MOSFETs present fewer problems, require less derating, and provide higher performance. Parallel operation, however, is still not trivial. Problems can and do occur. To parallel devices successfully, the designer must understand both the causes and the cures for the problems. Fortunately, most of the cures are simple and are more a matter of attention to detail than of exotic or complex design.

This discussion attempts to identify all the problems that users potentially could experience and provides an explanation of each problem. Then it will tell how to recognize these problems and will give practical means for eliminating them.

Although the discussion is quite extensive, the final conclusions will show that all of the problems may be easily avoided and that paralleling MOSFETs is reasonable and, in many cases, highly desirable. However, because of the lack of prior, in-depth discussion of paralleling MOSFETs, it is necessary to substantiate this contention.

Throughout the discussion, the object will be to maintain a balance between simple, practical advice and reasonable, theoretical explanations.

Many designers experienced in BJT applications have a well merited aversion to paralleling devices unless it is absolutely necessary. While this is certainly appropriate for BJTs, such thinking is a distinct handicap when using MOSFETs because useful opportunities for improved performance or reduced cost may be missed. It is important to keep an open mind and to read carefully the following arguments.

Motivation for Paralleling

There are many possible reasons for paralleling multiple devices:
- Lower $R_{DS(on)}$
- Lower circuit inductance
- Improved thermal performance
- Compensation for radiation effects
- Lower cost
- Redundancy
- Higher I_D
- Derating

In switching applications, the majority of the power dissipation is due to $R_{DS(on)}$. If $R_{DS(on)}$ is made smaller, the power loss will go down. $R_{DS(on)}$ for a given device is determined by the active die area and the breakdown voltage (BV_{DSS}). $R_{DS(on)}$ decreases more or less linearly with increasing die area and increases exponentially (factor ≈ 2.5) with BV_{DSS}. In theory, $R_{DS(on)}$ may be made as small as desired simply by increasing the die area, but the cost per unit area of the die increases exponentially when the die dimensions are greater than about $0.125'' \times 0.125''$. This increase in cost puts a practical upper limit on die size. Presently this is in the range of $0.25''$ to $0.30''$ square. In addition, as die are made larger, fewer and fewer package choices are available. When $R_{DS(on)}$ must be reduced even further, paralleling is the best alternative.

Because of the rapid increase in cost of the larger die and the restriction on suitable packages, there are often compelling reasons to use smaller die in cheaper packages even when a suitably low $R_{DS(on)}$ device is available in a larger die. For example, a size five die $(0.25'' \times 0.25'')$ will have one half the

$R_{DS(on)}$ of a size four die (0.18″ × 0.18″). However, the size five die will not fit in a TO-220 package. It requires a TO-3. At current prices, the cost of two TO-220s would be twenty to twenty-five percent less than one TO-3. As prices drop further, production increases and competition grows, this price differential should grow wider.

It has been shown (1) that paralleling is an effective means for reducing the junction-to-heatsink thermal impedance ($R_{\theta js}$). To use the previous example, $R_{\theta js}$ for a size five die in a TO-3 package is about 1.2°C/W; for a size four die in a TO-220 package, $R_{\theta js}$ = 1.3°C/W; but two TO-220's in parallel would reduce the effective $R_{\theta js}$ to 0.65°C/W. For the same $I_{D(rms)}$ and heatsink, $R_{DS(on)}$ will be lower because T_j will be lower. This may allow a further reduction in die size or as an alternate, the heatsink may be made smaller thus reducing its cost. The thermal design issues are treated in detail in Reference (1). For high power pulse applications, the limitation may be on $I_{D(peak)}$ rather than $R_{DS(on)}$. Again paralleling provides a means for increasing the allowable peak current. Very often in pulse applications, rapidly rising (high di/dt) current waveforms are required. Sometimes the limiting factor on speed is the package inductance. One means of reducing this inductance is to parallel several smaller devices so that the total effective inductance is decreased in proportion to the number of devices paralleled. Also smaller packages will tend to have lower inductances.

Another problem which can arise in fast pulse applications— if low voltage high current devices are used — is a degradation in the external BV_{DSS} capability caused by voltage spikes generated by the internal connection inductance (2). Figure 1 shows a MOSFET with the internal parasitic inductances. At turn-off, I_D is flowing. As the device is turned off, the voltage polarities will be as indicated. The actual voltage across the junction will be

$$V_{DS} = V_{DD} + (L_D + L_S)\frac{dI_D}{dt}$$

For example, if a 50 ampere, 100 V device with 20 nH of internal parasitic inductance is switched in 20 nsec, the internal voltage spike between the drain and source terminals (which is not visible externally) will be 50 V! The usable value for BV_{DSS} has been reduced to 1/2 its data sheet value. Paralleling several smaller devices, possibly in smaller, lower inductance packages, can reduce the effective value of parasitic inductance and increase the usable operating voltage.

When MOSFETs are subjected to large neutron fluxes (> 10^{13} neutrons/cm²), $R_{DS(on)}$ may rise dramatically. One means to ensure that the post irradiation $R_{DS(on)}$ is not unacceptably high is to start with a low value for $R_{DS(on)}$ (much less than required) in the unirradiated circuit. To achieve this, it is usually necessary to parallel devices.

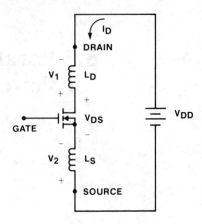

Mechanism Responsible for Internal Voltage Spiking at Turn-off
FIGURE 1

Concerns about Paralleling Devices

When considering a design using parallel FETs, a number of concerns and possible problems arise. Generally these concerns fall into four categories:

1. Steady-state current sharing
2. Thermal stability and maximum T_j
3. Dynamic current sharing during switching
4. Parasitic oscillations

As is so often the case, many problems that exist in theory are not significant in the actual hardware. Nevertheless, in the following discussion, many possible problems will be explained.

Steady-State Current Sharing

The distribution of current among parallel devices is a concern to the designer. With regard to the effects of asymmetrical current sharing, three questions need to be answered:

1. What is the maximum junction temperature among the devices?
2. Does the asymmetry cause a significant increase in total dissipation?
3. Is any device operating outside of its safe operating area (SOA)?

When the current is not distributed equally, some devices may run hotter than others. Because the operating reliability is directly related to T_j, it is important to identify the maximum T_j which can occur. Obviously if any device is operating outside of its rated SOA, the reliability will also be greatly reduced.

Furthermore, it is important to know if the conduction asymmetry is creating a power loss penalty.

The answers to these questions depend on the operating state of the devices. There are two possibilities: linear or switching operation. When the devices are used as switches, V_{GS} will be large (10-15V), and the devices will be fully enhanced. In this mode the device acts like a positive temperature coefficient resistor. Typical $R_{DS(on)}$ versus T_j characteristics are shown in Figure 2. Note that $R_{DS(on)}$ has been normalized (R_{DSN}) to 25°C for comparison purposes.

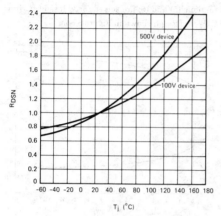

The Relationship Between $R_{DS(on)}$, T_j and BV_{DSS}
FIGURE 2

When the devices are operating in the linear mode, the behavior is quite different. Figure 3 shows a typical transfer characteristics graph (I_D as a function of V_{GS} with T_j as a parameter. The interesting feature of this graph is that above 4.5A, the temperature coefficient (TC) is positive, but below 4.5A, the TC is negative. For this device $R_{\theta js}= 1°C/W$, and if a perfect heatsink is assumed such that $t_c = 25°C$ then the maximum value for V_{DS} at 4.5A is

$$V_{DS} = \frac{T_j - T_c}{R_{\theta jc} \, I_D} = 27.8 \text{ V} \qquad (2)$$

Since this is a 400V device, it is unlikely that the device would be used in the linear mode with such a low value of V_{DS}. Most linear applications would use the device at currents well below 4.5A to exploit the BV_{DSS} capability (400V), and therefore, they would be operating in the negative TC region. Since this is exactly the opposite of the switch mode, the two types of applications will be treated separately.

Current Sharing While Fully Enhanced

When V_{GS} is large (6-8V above V_{th}), a MOSFET is essentially a positive TC resistor, and the current will divide among the paralleled devices in proportion to their individual $R_{DS(on)}$ as illustrated in Figure 4. As was shown in Figure 2, the TC will be positive, so there is a tendency for a device which has a greater

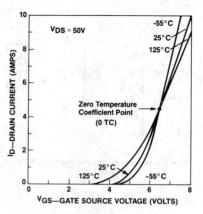

Typical Transfer Characteristics for an IRF330
FIGURE 3

than average current to heat up more than the other devices, increasing $R_{DS(on)}$, which in turn reduces its I_D. The degree of thermally forced current sharing present has been analyzed (3), and an outline of the analysis is presented here.

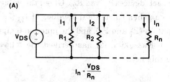

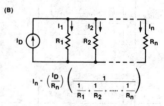

Current Sharing in Parallel MOSFETs
FIGURE 4

A thermal model for two parallel devices is given in Figure 5A. The normal thermal impedances for junction to case ($R_{\theta jc}$), case to heatsink ($R_{\theta cs}$) and heatsink to ambient ($R_{\theta sa}$), are present as well as the thermal coupling between the devices (R_c). Three possibilities are indicated:

1. R_c between nodes T_{c1} and T_{c2} represents die mounted on a common case or header.
2. R_c between nodes T_{s1} and T_{s2} represents the situation in which separate packaged devices are paralleled on a common heatsink.
3. The situation where there is no common coupling between devices, i.e. separate heatsinks, is represented by $R_c = \infty$.

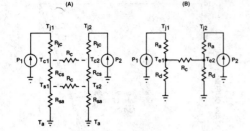

(A) **(B)**

Thermal Models for Two Parallel MOSFETs
FIGURE 5

For analytical purposes, the model in Figure 5A can be simplified as shown in Figure 5B where the values for R_a and R_d depend on the position of R_c. For example, for devices paralleled on a common heatsink,

$$R_a = R_{\Theta jc} + R_{\Theta cs} \qquad (3)$$
$$R_d = R_{\Theta sa} \qquad (4)$$

From this network, it is possible to calculate the values of $R_{DS(on)}$ for each device (R_1 and R_2). Given the values for R_1 and R_2, one can then calculate the values of I_D (I_1 and I_2), power dissipation (P_1 and P_2), and the junction temperatures (T_{j1} and T_{j2}). Kassakian (3) has derived the following expressions for this network.

$$R_1 = R_{10}\left\{ 1 + \frac{I^2 R_1 R_2 A}{(R_1 + R_2)^2}\left[R_2\left(\frac{(R_d + R_a)(\frac{R_c}{R_d} + 1) + R_a}{\frac{R_c}{R_d} + 2}\right) + R_1\left(\frac{R_d}{\frac{R_c}{R_d} + 2}\right)\right]\right\} \qquad (5)$$

$$R_2 = R_{20}\left\{ 1 + \frac{I^2 R_1 R_2 A}{(R_1 + R_2)^2}\left[R_1\left(\frac{(R_d + R_a)\frac{R_c}{R_d} + 1) + R_a}{\frac{R_c}{R_d} + 2}\right) + R_2\left(\frac{R_d}{\frac{R_c}{R_d} + 2}\right)\right]\right\} \qquad (6)$$

Where

R_{10} = 25°C value for $R_{DS(on)}$ in device #1
R_{20} = 25°C value for $R_{DS(on)}$ in device #2
A = Temperature coefficient of resistance. This can vary from 0.5 to 3%/°C depending on the device.
I = $I_1 + I_2$

By itself, this set of equations is not generally useful since the equations are non-linear: i.e. they contain cross products and powers of R_1 and R_2 (the dependent variables). Using numerical methods, this type of equation is usually solved on a computer. However, in most cases this is not necessary as the following example solution demonstrates.

Assume for example, that 10A, 100V devices are being used which have the following characteristics:

$R_{\Theta jc}$	= 1.67° C/W	R_{10}	= 0.12 Ω
$R_{\Theta cs}$	= 1.00° C/W	R_{20}	= 0.16 Ω
$R_{\Theta sa}$	= 1.47° C/W	I	= 20 A
		T_c	= 0.67 %/°C

Assume further that the die are mounted on the same header:

$$R_a = 1.67° C/W$$
$$R_d = 2.47° C/W$$

If Equations (5) and (6) are solved for values of R_c from 0 to 100° C/W, the graph shown in Figure 6 results.

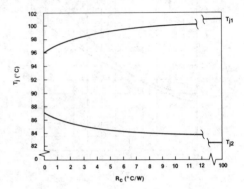

FIGURE 6

Notice that when R_c is small the difference between T_{j1} and T_{j2} is also small (about 9°C) even though R_{20} is one-third larger than R_{10}! The effect on current sharing is almost non-existent (about 0.3A).

As long as R_c is comparable to or smaller than R_a, the difference in junction temperatures should remain small. In practice this should not be difficult to achieve. For multiple die on a common header, this requirement holds very well. For multiple individual packages mounted on a common heatsink, this requirement can be met by having a reasonably heavy common web between devices and by mounting these devices relatively close to each other.

In the past, many people (this author included) have touted thermally forced current sharing as a major feature of parallel device operation in MOSFETs. Yet thermally forced current sharing is not prominent except in the case of very high voltage devices with large TCs and separate heatsinks.

The most common applications for MOSFET switches involve a load impedance that is large compared to $R_{DS(on)}$. In these applications, the total current (I) through all of the devices is determined by the load. In this case, the total power loss is

$$P_T = I^2 R_T \qquad (7)$$

where R_T = the total on-resistance of the parallel devices.

In this case, the distribution of current between the devices is not considered; all that matters is the final value for R_T. The fact that some devices will be slightly warmer than others will increase R_T slightly, but this is usually a second order effect.

From these analyses and other work (5), the following general observations can be made:

1. The current sharing between parallel devices is in proportion to $R_{DS(on)}$.
2. When the devices are well coupled thermally, the differences in T_j are small even when the difference in $R_{DS(on)}$ is substantial.
3. For thermally coupled devices, forced current sharing is insignificant.
4. For high voltage thermally uncoupled devices, some forced sharing can occur, but since this is achieved at the expense of higher T_j and R_T, in most cases thermally uncoupled parallel operation is undesirable.
5. Parallel devices should be mounted on a common heatsink or substrate with a minimum common thermal impedance (R_c).
6. Matching of devices for $R_{DS(on)}$ is usually not necessary unless the range of variation ($\pm 20\%$) would allow too large a value for R_T. Rather than matching or screening devices for $R_{DS(on)}$, it is frequently cheaper and easier to add another device in parallel.
7. If a limit on the maximum value for R_T is desired and matching is acceptable, the matching for $R_{DS(on)}$ should be done at the anticipated average current for each device and the planned value for V_{GS}.
8. To minimize differences in $R_{DS(on)}$ and the final value for R_T, it is important to enhance the devices fully. A V_{GS} of 10 to 15 V is sufficient.
9. The relevant issue in paralleling is not current sharing, per se, but rather the junction temperature differences and any additional power losses. If the ΔT_j and ΔP_T are small, the asymmetry in the current is irrelevant.

Current Sharing During Linear Operation

In the linear mode of operation, the temperature coefficient of ON-resistance is negative. This means that if one device, in a group of parallel devices, is conducting more than its share of current, its temperature will rise. This will further increase this device's share of the total current. This process can lead to thermal runaway and is very similar in nature to the thermal instability present in BJTs. In the case of the MOSFET, the transconductance ($g_m = \Delta I_D / \Delta V_{GS}$) is much lower, and the tendency towards instability is correspondingly less.

Thermal regeneration of this type combined with normal device characterization variations can cause three problems:

1. Large differences in current sharing can occur. The current distribution among parallel devices will vary with temperature and, in some cases, with total current (I).

2. The quiescent operating or "Q" point is ill-defined and varies with temperature.
3. Thermal runaway and subsequent device failure are possible.

The following discussion will examine these problems and demonstrate a simple cure. For the purposes of this discussion, two parallel devices will be used, but the principles exposed apply to multiple parallel devices. Figure 7 shows the reference model for the discussion.

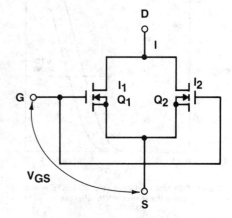

Two MOSFETs in Parallel, Reference Circuit
FIGURE 7

When two non-identical devices are paralleled, a variety of situations can arise depending on the differences between the devices. Figures 8, 9, and 10 show several possibilities as well as the resulting current imbalance. For the moment, the effect of T_j changes will be ignored.

Figure 8 illustrates the effect of differences in g_m. In this case, the current differential increases as I_D is increased.

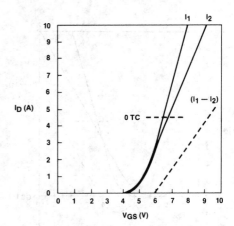

Current Difference between Two MOSFETs where
g_m is the same but Vth is different
FIGURE 8

Figure 9 illustrates the effect of a one volt difference in V_{th}. In this example, all of the current flows through one device until V_{GS} reaches 5V. At that point, the second device begins picking up some of the current. The current asymmetry remains essentially constant for $V_{th} > 6.5V$. The problem here is that if the "Q" point is below 3A, one device will hog nearly all of the current!

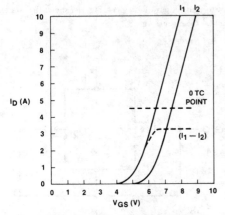

Current Difference between Two MOSFETs where gm is the same but Vth different
FIGURE 9

Most practical applications will represent some combination of the two previous examples. A typical combination is shown in Figure 10. In this particular example, the device carrying the greater portion of the current will depend on the "Q" point.

Now it is necessary to include the effect of the junction temperature on the current imbalance. Looking again at Figure 9, notice that $I_1 > I_2$. It is reasonable to assume that Q_1 will heat up and Q_2 will cool down. An approximation of what will occur is shown in Figure 11. From this figure, it can be seen that the

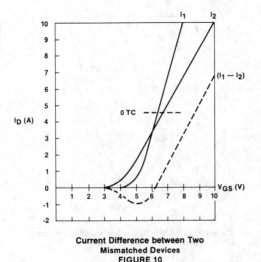

Current Difference between Two Mismatched Devices
FIGURE 10

peak value of the difference between I_1 and I_2 is increased, and the range of V_{GS} over which Q_1 takes all or most of the current is expanded. In most applications, the effect of differential heating in linear operation is to make the problem worse. One way to minimize the effect of asymmetrical currents is to maximize the thermal coupling between the devices (make R_c as small as possible). This is the same conclusion reached for parallel devices operating as switches!

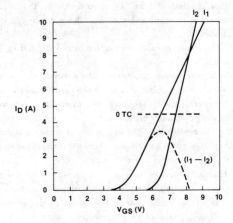

Current Imbalance from Figure 9 when Heating Effects are taken into account
FIGURE 11

Good thermal coupling by itself, however, is not enough to assure good current sharing; it merely reduces the degree of mismatch. For linear applications, some further means to force sharing must be taken. One obvious solution is to match devices. Unfortunately to get really good current sharing, it is necessary to match the entire transfer characteristic. This level of matching would rarely be practical. A useful compromise would be to match the devices at the "Q" point. This is fine in a stable thermal environment, but if the ambient temperature (T_a) varies over a wide range then it is unlikely the devices will remain matched. In any event, matching of devices can be costly and a considerable nuisance in production.

A better solution would be to take randomly selected devices and force them to share. This can be done most easily by using small source resistors to provide negative feedback, as shown in Figure 12A. As an added advantage, this will also stabilize the "Q" point.

The effect of the source resistor can be quantified by examining its effect on the "Q" point of a single device as indicated in Figure 12B. It has been shown (6) that the effective g_m (g_m') will be

$$g_m' = \frac{g_m}{1 + R_s g_m} = \frac{1}{R_s + \frac{1}{g_m}} \qquad (8)$$

To minimize the effect of differences in g_m' it is necessary that

$$R_s \gg 1/g_m \qquad (9)$$

Typical devices will have values for $1/g_m$ in the range of 0.1 to 1.0 ohms.

(A)

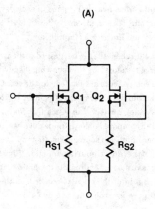

(B)

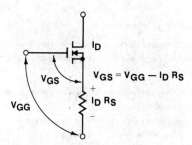

Forced Current Sharing using Source Resistors
FIGURE 12

The effect of R_s on the "Q" point stability (with fixed V_{GG}) is shown in Figure 13. If the operating point is selected as $V_{GG} = V_{GS} = 5.4\,V$ and $R_s = 0$ then $I_D = 1.2\,A$ at T_1, but it will rise to $I_D = 2.7\,A$ at T_2 — an increase of more than 2:1! By making $R_s = 3$ ohms (as indicated by the sloping line), the change in $I_D(\Delta I)$ when the temperature changes from T_1 to T_2 is reduced to $0.25\,A$! This is a very great improvement. In this example, g_m is approximately 2.5S so that $R_s \approx 8/g_m$. Even if R_s is reduced to $1\,\Omega$, ΔI is still only $0.6\,A$.

For those readers not familiar with the graphic technique just used, a few words of explanation may prove helpful. The straight line representing $R_s = 3$ and $V_{GG} = 9\,V$ is shown in a graph representing the values of V_{GS} for given values of I_D. For example, if $I_D = 1.667\,A$, the drop across R_s is 5V, viz. $3 \times$

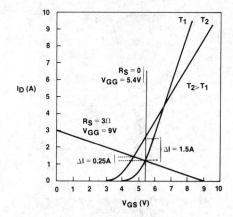

Operating Point Stabilization
FIGURE 13

1.667A. From Figure 12 we know that

$$V_{GS} = V_{GG} - I_D R_s \qquad (10)$$

so that V_{GS} in this case is 4V.

What we are seeking is a simultaneous solution to Equation (10) and the equation represented by the graph of the transfer characteristic. This is done by graphing Equation (10) (the straight line) on the transfer characteristic graph and by noting the intersection of the graphs. In Figure 13, the intersection representing the "Q" point is at $V_{GS} = 5.4\,V$, and $I_D = 1.2\,A$.

This same technique can be applied to parallel devices since stabilization of the "Q" point also stabilizes the current sharing. Using the example of Figure 9, the effect on sharing by adding a 2 ohm resistor in series with each device can be determined. A series of parallel lines are drawn for several different values of V_{GG}, each having a slope corresponding to 2 ohms. Each line establishes a "Q" point with a given value of ΔI. The values for ΔI corresponding to the different values for V_{GG} can now be plotted and compared to the values for the original example (Figure 9). This is shown in Figure 15 where it can be seen that ΔI is reduced from 3.2 to 0.5A! Clearly this is an effective means for equalizing the current distribution as well as stabilizing the operating point.

Source resistors can also improve thermal stability by reducing differences between the transfer characteristics as T_j is varied. This effect can be illustrated by regraphing the example in Figure 13 as a function of V_{GG} rather than V_{GS}. This is done in Figure 16. In effect, the thermal regenerative gain is much lower and the stability greatly enhanced.

Note that Figure 16 is also an alternative method for determining ΔI between non-identical devices; however, it is usually simpler to draw the resistive load lines for R_s, as was done in previous examples (Figures 13 and 14), than to redraw the graph in the form used in Figure 16.

5

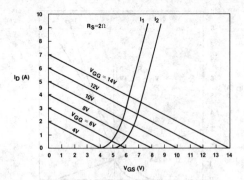

**Rs Load Lines Plotted on the Transfer
Characteristic Graph
FIGURE 14**

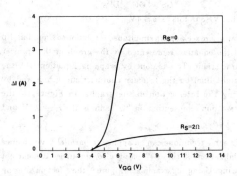

**The Effect of Source Resistors on Current Sharing
FIGURE 15**

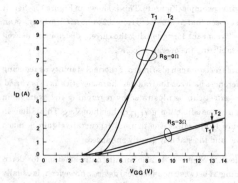

**The Effect of Rs on the Transfer Characteristics
over Temperature
FIGURE 16**

The use of source resistors has several advantages:

1. The operating point is stabilized.
2. Excellent current sharing and equalization of power dissipation is achieved.
3. Transconductance differences are minimized so that the small signal gain is nearly the same in each device.
4. Thermal stability is enhanced.
5. The small signal linearity is improved.

Unfortunately, these benefits are not gained without some cost. The use of source resistors has the following disadvantages:

1. Additional components are needed: i.e. the resistors.
2. The large signal dynamic range is reduced by the voltage drop across the resistors, but since this voltage drop is usually small compared to V_{DS} this is not a severe penalty. On the other hand, the source resistors tend to extend the lower limit of the large signal linear region (Figure 16), and this compensates by extending the dynamic range.
3. The voltage gain of the stage is reduced. In the common source configuration, the voltage gain (A) is

$$A = g_m R_L \qquad (11)$$

where R_L is the load resistance.

From Equation (6), the gain will be reduced to

$$A = \frac{g_m R_L}{1 + R_s g_m} \qquad (12)$$

when source resistors are used.

From the foregoing discussion, the following general observations regarding parallel devices operating in the linear mode can be made:

1. This mode of operation is very different from the switching mode in most respects.
2. To avoid excessive current asymmetry and thermal instability, some positive means must be provided to stabilize the operating point and force current sharing.
3. The simplest means to achieve the above goals is to use small resistors in series with each device's source lead.
4. Good thermal coupling between devices will greatly improve thermal stability and current sharing and help to minimize the size of R_s.
5. Matching of devices, while useful and effective, is usually not necessary. Small source resistors are usually a cheaper and simpler solution. The size of the resistors can be reduced by prescreening devices to eliminate those with larger than average characteristic deviations.
6. Multiple die devices will normally (at least at Siliconix) have the die selected from adjacent positions on the same wafer and will be well matched. In addition, the thermal coupling will be very good. The user cannot, however, add individual source resistors because the die are sealed within the case. If the degree of intrinsic matching in multiple die devices is not adequate, the single die devices can be paralleled using individual source resistors.

Current Sharing During Switching Transitions

When parallel MOSFETs are used as switches, several questions arise concerning device behavior during the switch transitions from OFF to ON and from ON to OFF:

1. What is the current distribution?
2. Can current asymmetries be severe enough to damage a device?
3. Is the switching time affected by current asymmetries?

When a MOSFET makes a transition from one state to another, the device must pass through the linear region, if only momentarily. This means that much of what has been discussed concerning the linear mode of operation applies to dynamic switching. However, there are some differences:

1. The operating time in the linear region is very short (typically 10 to 100 nsec).
2. Because of the rapid transitions, thermal heating effects are usually negligible.
3. The use of source resistors is unacceptable because of the substantial increase in the effective value of $R_{DS(on)}$.
4. The external parasitic and intentional resistances, inductances, and capacitances must be taken into account.

An equivalent circuit representing two parallel FETs is given in Figure 17. Differences in current during switching can be caused by:

1. Differences in the external circuit elements.
2. Differences in capacitive or inductive device characteristics.
3. Differences in device characteristics during linear operation.

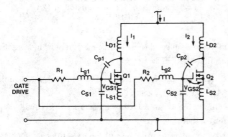

Equivalent Circuit for Two Parallel MOSFETs During Switching Transitions
FIGURE 17

If for example, Q_1 and Q_2 are identical, but the drive circuit impedances are different, Q_1 and Q_2 will not turn on simultaneously. This problem, however, is under the control of the designer. The following steps can be taken to minimize this problem:

1. Minimize the values for R, L_g, L_s, L_D, C_p and C_s.
2. There will be practical limits on how small the external impedance can be. When that limit is reached, every effort should be made to equalize the values for the remaining

impedances. Using symmetrical layouts is one way to do it. An example of a symmetrical layout for a parallel switch connection (push-pull) with three paralleled devices is given in Figure 18. Many other practical symmetrical layouts are possible.

3. Good layout techniques are vital. The effect of L_s on the switching waveform is shown in Figure 19. Obviously L_s should be made as small as possible. Further examples of good layout technique are given in Figures 20 and 21.

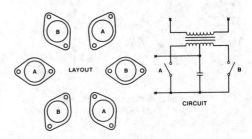

Symmetrical Layout Example
FIGURE 18

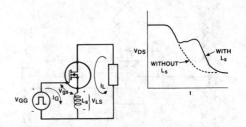

The Effect of Common-Source Inductance on Switching
FIGURE 19

Differences in package inductance and inter-terminal capacitance are amazingly small ($\approx \pm 5\%$) and usually can be ignored. This is fortunate since without a complex matching procedure, there is little that can be done anyway.

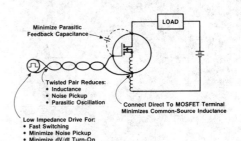

Suggested Drive Circuit Layout
FIGURE 20

Differences in device characteristics, on the other hand, can cause differences in current distribution during switching transitions. To illustrate what occurs, the techniques previously used for linear operation can be applied to the example in Figure 9. If a clamped inductive load is assumed, Figure 22 shows the switching waveforms. If $I_D = I_1 + I_2 = 12A$ then the current during switching will assume the value shown in

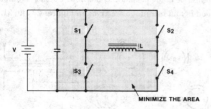

High di/dt Paths in an H-Bridge
FIGURE 21A

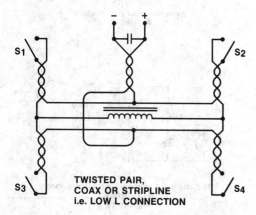

TWISTED PAIR,
COAX OR STRIPLINE
i.e. LOW L CONNECTION

Low Inductance Connections for High di/dt Paths
FIGURE 21B

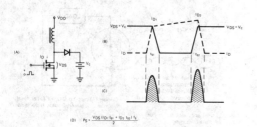

Clamped Inductive Load Switching Waveforms
FIGURE 22

Figure 23. If V_{GS} is a linear function of time (a ramp) then the graphs for I_1 and I_2 become their current switching waveforms as a function of time! The salient feature of these waveforms is that even though the two devices have very different threshold voltages, the current spike for Q_1 is small (about 1.7A). In most cases, when unmatched devices are paralleled, only relatively small current differentials are observed. A good series of actual oscillographs is given in Reference (5) to illustrate this point. Applications do exist where from 20 to 50 devices are paralleled. In these applications, it is theoretically possible (but relatively unlikely) to generate destructive current spikes. Such spikes can be avoided by prescreening devices to eliminate those devices that are radically different from the median.

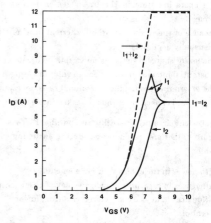

Values for Drain Current during Changed Inductive Switching when $I_1 + I_2$ is limited to 12A
FIGURE 23

The following observations can be made regarding switching in parallel MOSFETs:

1. Careful circuit layout is needed to minimize and equalize parasitic impedances.

2. Minimizing differential gate impedances will equalize V_{GS} for each device.

3. Except for large numbers of devices, the current spikes during switching transitions will usually be well within the limits of the device capability.

4. If many devices are paralleled, simple prescreening which measures I_D at a given value of V_{GS} in the transition region (i.e. not fully enhanced) should be sufficient. This will eliminate the unusually different devices.

Parasitic Oscillations in MOSFETs

Most power MOSFETs presently available are very fast devices with appreciable gain at frequencies up to 300 MHz. This high frequency gain, coupled with the internal and circuit parasitic inductances and capacitances normally present, make it possible for unwanted parasitic oscillations to occur. The frequency of oscillation can range from 1 to 300 MHz.

The oscillations occur while the device is in the active mode where the transconductance is large. When the device is off or when V_{GS} is large and the device is fully on, oscillations do not occur. This means that in a switching application, the oscillation will occur on a transient basis during the turn-on and turn-off transitions. In an application where the device is biased on to some fixed point in the linear region, the oscillations can be continuous.

An example of a parasitic oscillation during a turn-on transition is given in Figure 24. Figure 24A shows the normal drain-to-source (V_{DS}) and gate-to-source (V_{GS}) voltage waveforms when no oscillation is present; Figure 24B shows the same waveforms when an oscillation is present. An expanded portion of the oscillation envelope is given in Figure 25. In this case, the oscillation frequency is approximately 85 MHz. Although the oscillation amplitude shown is not very high, voltages of 100V or more are possible at the peak of the envelope.

The existence of parasitic oscillations can have the following consequences:

1. Gate rupture due to overvoltage.
2. Gate rupture due to overheating of the gate structure.
3. Increased power dissipation in the device.
4. Increased voltage and current stress in associated circuit components.
5. HF to VHF electromagnetic interference (EMI).

None of these are normally acceptable, and, in nearly all cases, the parasitic oscillations must be eliminated. Many so called "mysterious" failures in MOSFETs are due to parasitic oscillations.

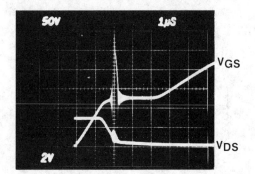

FIGURE 24B

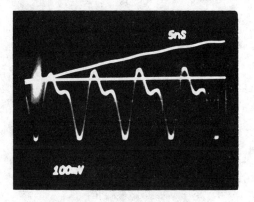

Expanded Time Base Showing Free Running Parasitic Oscillation
FIGURE 25

A single device can oscillate by itself. Parallel devices can also oscillate as if they were a single device; this is referred to as a "common mode" oscillation. In addition, paralleled devices can oscillate in a "differential mode." The analysis for each mode is very similar, but the parasitic circuit elements controlling the oscillations are different.

Common Mode Oscillation

Figure 26 gives a model of the internal and external parasitic capacitances and inductances in a typical application. To analyze this circuit for oscillations, a simplified incremental model (Figure 27) can be used. Figure 27 is not exact but will still give useful answers. The following assumptions have been made and are normally valid:

1. $L_{Ge} \gg L_{Gi}$
2. $L_{De} \gg L_{Di}$
3. $L_{Se} \gg L_{Si}$
4. $R_{Ge} \gg R_{Gi}$
5. $C_1 = C_{gse} + C_{gsi}$
6. $C_2 = C_{gde} + C_{gdi}$
7. $C_3 = C_{dse} + C_{dsi}$

A more complex model could, of course, be used at the cost of greatly increased computational difficulty.

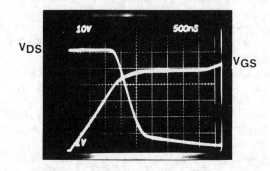

Non-Oscillating MOSFET
FIGURE 24A

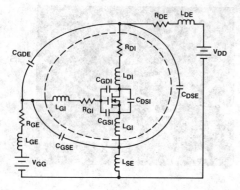

Equivalent circuit
FIGURE 26

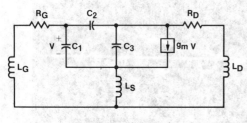

Incremental Model
FIGURE 27

The characteristic equation for the incremental model is (7)

$$a_1S^4 + a_2S^3 + a_3S^2 + a_4S + 1 = 0 \qquad (13)$$

Where

$$a_1 = C_e^2 L_e^2 \qquad (14)$$

$$a_2 = C_e^2[R_G(L_D + L_S) + R_D(L_G + L_S)] + g_m L_e^2 C_2 \qquad (15)$$

$$a3 = R_D R_G C_e^2 + g_m C_2[R_G(L_D + L_S) + R_D(L_G + L_S)] + L_G(C_1 + C_2) + L_D(C_2 + C_3) + L(C_1 + C_3) \qquad (16)$$

$$a_4 = g_m R_D R_G C_2 + R_G(C_1 + C_2) + R_D(C_2 + C_3) + g_m L_S \qquad (17)$$

$$L_e^2 = L_D L_G + L_D L_S + L_G L_S \qquad (18)$$

$$C_e^2 = C_1 C_2 + C_1 C_3 + C_2 C_3 \qquad (19)$$

g_m = transconductance of the MOSFET

Note that equation (13) is fourth order even though there are six reactive elements in the model. The reason for this is that C_1, C_2, and C_3 form a ring of capacitors which means that the voltages on any two determine the voltage across the third. This reduces the equation from sixth to fifth order. L_D, L_G, and L_S form a cut set of inductors where the current in any two inductors determines the current in the third. This further reduces the order from fifth to fourth.

In some applications, the model may be further simplified if either L_D or L_S is very small. The coefficients of the characteristic equations for these cases are

$$L_D = 0$$

$$L_e^2 = L_G L_S \qquad (20)$$

$$a_1 = L_G L_S C_e^2 \qquad (21)$$

$$a_2 = C_e^2(L_S R_G + L_G R_D + L_S R_D) + g_m L_G L_S C_2 \qquad (22)$$

$$a3 = R_D R_G C_e^2 + g_m C_2(L_S R_G + L_G R_D + L_S R_D) + \qquad (23)$$
$$L_G(C_1 + C_2) + L_S(C_1 + C_3)$$

$$a_4 = g_m R_D R_G C_2 + R_G(C_1 + C_2) + R_D(C_2 + C_3) + \qquad (24)$$
$$g_m L_S$$

$$L_S = 0$$

$$L_e^2 = L_D L_G \qquad (25)$$

$$a_1 = L_D L_G C_e^2 \qquad (26)$$

$$a_2 = C_e^2(L_D R_G + L_G R_D) + g_m L_D L_G C_2 \qquad (27)$$

$$a3 = R_D R_G C_e^2 + g_m(L_D R_G + L_G R_D)C_2 + \qquad (28)$$
$$L_G(C_1 + C_2) + L_D(C_2 + C_3)$$

$$a_4 = g_m R_D R_G C_2 + R_G(C_1 + C_2) + R_D(C_2 + C_3) \qquad (29)$$

The practical question which must be answered is: "Does the circuit oscillate or not?" The characteristic equation can answer this question. If any of the roots of the polynomial are negative (i.e. lie in the right half-plane or have the term s-α) then the circuit will oscillate. There are several ways this can be determined.

1. Examine the polynomial for negative coefficients. In this particular case, all of the known quantities have positive signs, and therefore, all of the coefficients will be positive.

2. Use a calculator or computer, with a root finding program, to see if any negative roots exist. For example, the HP41-CV with a math pack can find the roots for up to fifth order polynomial and would be very useful.

3. The Routh-Hurwitz procedure (7) can be used to determine if negative roots exist. This procedure can conveniently be done on a programmable calculator.

Giandomenico (8) has suggested an alternate means for determining stability by treating the incremental model as a feedback system and then modeling the circuit using SPICE. If the equivalent open loop gain and phase shift at 0° and a gain of 1 coincide, the circuit will oscillate at that frequency. The SPICE model suggested by Giandomenico is shown in Figure 28. This approach would be particularly useful if more complex models were used such as the complete high frequency RF model (9).

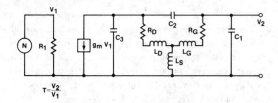

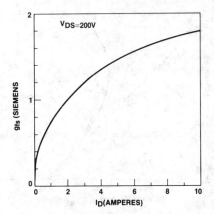

Transcendance as a Function of I_D
FIGURE 30

Whatever approach is used, appropriate values for the model elements must be known. The values for the internal and external parasitic inductances, the external parasitic capacitances, and the parasitic resistances can be measured or estimated from simple calculations (10). These elements do not change value during the operating cycle, but g_m, C_{gd}, and C_{ds} do vary.

Typical variations for intra-terminal capacitances are shown in Figure 29. Clearly C_{ds} and C_{dg} will change dramatically as the device is switched. The variations in g_m during switching transitions are shown in Figures 30 and 31. g_m is not very sensitive to V_{DS} except for low values, but it does vary as I_D changes. From these graphs, for the particular device being used, the simultaneous values for C_{gd}, C_{ds}, and g_m can be determined from the I_D/V_{DS} load line. The stability analysis can then be performed at several points on the load line to determine if oscillations are possible.

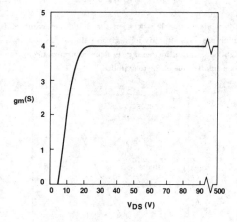

Transcendance Variation as a Function V_{DS}
FIGURE 31

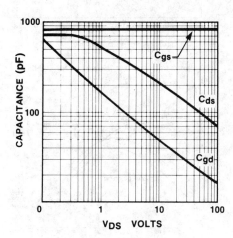

Intraterminal Capacitance Variation
FIGURE 29

This analysis procedure is relatively complex, and unless the whole business is pre-programmed into a computer, it is unlikely that most designers would use this for solving day to day design problems. The equations can be solved for a few examples, however, and from these examples, general trends

can be identified. From these, appropriate means for suppressing oscillations and estimates of the proper values for damping elements can be found.

A typical example of the effect of g_m, L_G, L_S, and R_G on stability is shown in Figure 32. Examining this graph, several trends can be clearly seen:

1. The larger the value of g_m, the smaller the region of stability.
2. The smaller the value of L_G, the larger the region of stability.
3. Even a very small value for L_S will greatly improve the region of stability.
4. Small increases in R_G greatly increase the region of stability.
5. As the I_D rating is increased and the BV_{DSS} rating decreased, g_m will increase. This implies that large, low voltage devices or multiple devices in parallel will be more prone to oscillation.

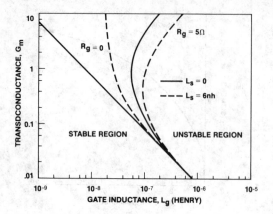

Transcondutance as a Function of Gate Inductance
FIGURE 32

A typical relationship between the maximum value of L_G which provides stability for a given value of R_G is shown in Figure 33. Again the message is clear, minimize L_G and then use a small value of R_G to stabilize.

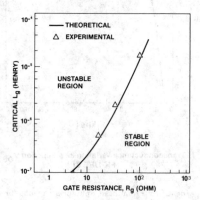

Critical Gate Inductance as a Function of Gate Resistance
FIGURE 33

Differential Mode Oscillations

Oxner (∗) has demonstrated that two or more parallel MOSFETs can oscillate in a differential or "push-pull" mode. Kassakian (4) has shown how to analyze this case. The following discussion uses his method.

When two devices are paralleled, the equivalent circuit, including parasitics, becomes double that shown in Figure 26. The circuit is quite complex and more than a little intimidating. The circuit can, however, be greatly simplified by recognizing that in the *differential* mode, the circuit can be simplified by assuming that all of the nodes on the plane of symmetry are at incremental ground. This means that all elements common to *both* devices can be eliminated. In addition, the method of

∗ See Appendix

half-circuits can be used to further reduce the equivalent circuit complexity. The result of these simplifications is topologically identical to Figure 27. This means that we can use equations (13) through (29) without modification. We *cannot*, however, make the same assumptions regarding the relative values of the parasitic elements. The following assumptions apply:

1. All elements *common* to both devices are ignored.
2. The internal inductances (L_{Gi}, L_{Di}, L_{Si}) cannot be ignored. In fact, in devices using multiple internal chips, the only differential inductances present are the internal bond wires!

With representative values for the circuit elements, we can again generate some informative examples as shown in Figures 34 and 35.

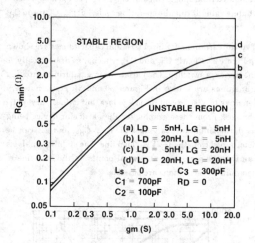

Minimum Value of Gate Resistance for Stability
FIGURE 34

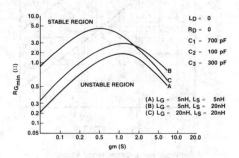

The Effect of Changing L_S and L_G on $R_{G(min)}$
FIGURE 35

When $L_S = 0$ (Figure 34), the following points can be made:
1. The higher the value for g_m, the larger R_G must be.
2. The larger L_D, the larger R_G required.
3. Increasing L_G increases R_G, but this is not a strong function.

When $L_D = 0$ (Figure 35), which is more often the case than $L_S = 0$, the relationships change a bit:

1. R_G no longer increases monotonically with g_m. This is in agreement with Figure 32.
2. For values of g_m above 1.5S, increasing L_G actually improves stability!
3. Typical values for R_G needed to assume stability are small (in the range of 0.5 to 5 ohms) even in large devices (high g_m).

How to Recognize Oscillation Problems

It is important that the designer recognize when oscillation problems might appear. Normally the first indication of this problem is the failure of devices with gate-to-source shorts or with high leakage ($I_{GSS} > 100nA$). The best means for detecting oscillations is to place a scope probe between the gate and source terminals, directly on the device. The scope should have a bandwidth of at least 200 MHz since it is very difficult to detect VHF oscillations with an instrument that responds to only a few MHz. A low capacitance probe should be used, and as little stray inductance as possible should be introduced. It is possible for the measurement process itself to alter the circuit operation and to suppress or induce the oscillation!

Prevention of Oscillation in MOSFETs

From the theoretical and experimental work on this problem, it is clear that preventing parasitic oscillations is not a major problem and can be accomplished by observing the following guidelines:

1. Minimize the parasitic inductances and capacitances. In particular L_G, L_S, and C_{gd} should be made as small as possible. Making the parasitic elements smaller raises the resonant frequency. As the frequency is increased, the gain of the device will decrease, and the resistive damping present will become more effective. The net result is a reduced likelihood of oscillation.
2. Use small (1-5 ohm) differential resistors in the gate lead of each device. Because of the silicon gate structure of Siliconix devices, most of the needed resistance will already be present. R_G should, of course, be non-inductive. Carbon composition resistors are particularly good.
3. Any resonant circuit has a non-zero value of Q. The higher Q is, the slower the oscillation will build up. The time constant (τ) will be

$$\tau = \frac{2Q}{\omega} \qquad (30)$$

If the switch transition time is short compared to τ, then the oscillation will not appear even though the circuit is potentially unstable.
4. Minimize the differential values of L_S and L_D.
5. For the differential mode of oscillation, ferrite beads provide both R_G and increased L_G and can be very effective in suppressing oscillations.

Appendix
Controlling Oscillation in Parallel Power MOSFETs

Introduction

Perhaps the most easily controlled environment for the study of parasitic oscillation is to examine the charge-transfer characteristics of parallel-wired power MOSFETs. The circuit is shown in Figure 1.

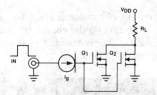

Circuit for Monitoring Charge Transfer
Figure 1

Since we are looking at high-frequency parasitic oscillation, it is vital that the circuit use direct point-to-point wiring—a few misplaced nanohenries may upset the results.

Examining the Conditions for Oscillation

In Figure 2, the resulting charge-transfer characteristics are divided into three well-defined regions. In region 1, the gate voltage has charged the input capacitor of the power MOSFET to where turn-ON just begins (V_T); in region 2, we witness not only the completion of the turn-ON cycle but also the effects of Miller capacitance upon the charge characteristics. In region 3, we see the drain-to-source voltage settling slowly to VSAT followed by a resumption of the gate charging cycle.

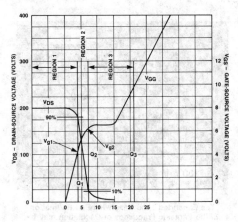

A Typical Charge-Transfer Characteristic
Figure 2

Here we must point out that if we were to parallel power MOSFETs in such a manner as to ensure parasitic oscillation, we would discover a well-defined envelope of oscillation *only in the immediate vicinity of the transition between regions 2 and 3.* This well-defined envelope is shown in Figure 3 where we should carefully note its position on the charge curve. The gate voltage has passed threshold (V_T), and the drain-to-source voltage has decayed to a point approaching *but not having reached* V_{SAT}. Oscillation found anywhere else is caused by other phenomena, perhaps excessive wiring inductance in the test circuit.

To understand the circumstances resulting in oscillation where it appears on the charge-transfer characteristic, we need to closely examine what changes take place in the power MOSFET.

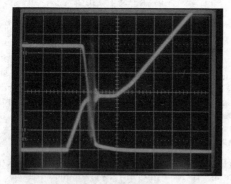

Envelope of Parasitic Oscillation Situated within Region 2 of the Charge Characteristics
Figure 3

By expanding our oscilloscope time base, we can arrive at a close approximation of the frequency of oscillation. The result is given in Figure 4 for a pair of parallel-coupled Siliconix VN4000A MOSPOWER FETs.

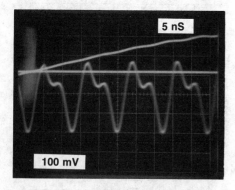

5 nS

100 mV

The Expanded Time Base where an Estimation of the Parasitic Frequency of Oscillation may be Roughly Determined
Figure 4

Causes of Oscillation

Oscillation results from a resonance induced by the combination of parasitic capacitive and inductive elements including, in particular, the loop inductance between discrete power transistors coupled with the characteristically high gain of the power MOSFETs.

Since the interelectrode capacitances of a power MOSFET are voltage dependent, behaving as we see in Figure 5, this parasitic oscillation becomes conditional upon the applied drain voltage.

In Section 3.2 we resolved that the prime cause of the extended Miller effect on the gate charge as well as the slow settling of the drain-to-source voltage to V_{SAT} after turn-ON, was the result of a many-order-of-magnitude change in the gate-to-drain capacitance, C_{gd} as shown in § 3.2-Figure 6 (and again in Figure 6). The meteoric rise in gate-to-drain capacitance only begins as the power MOSFET gate charge *enters into the transitional area between regions 2 and 3.*

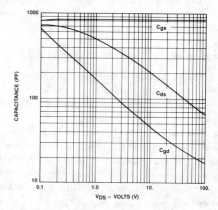

The Interelectrode Capacitance of a VN4000A MOSPOWER FET as a Function of Drain Voltage
Figure 5

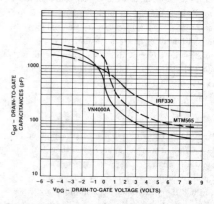

Gate-Drain Capacitance versus Drain-Gate Voltage
Figure 6

Identifying the Oscillator

A modification of the original charge-transfer circuit allowed for the injection of an external rf signal. Additionally, provision was made to monitor the phase difference between the drains using a Hewlett-Packard HP8405A Vector voltmeter. (See Figure 7 for the modified circuit.)

Using this modified test fixture, the parasitic oscillation can be effectively "locked" to the frequency of

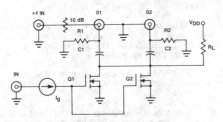

Modified Charge Transfer Circuit
Figure 7

the injected rf signal. By this, the true frequency of oscillation can be easily determined. The result of such a phase-locked parasitic oscillation is illustrated in Figure 8.

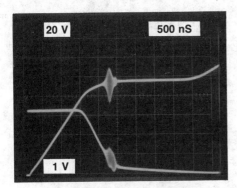

Phase-Locked 'Parasitic' Oscillation
Figure 8

If this parasitic oscillation results from interaction between parasitic capacitance and inductance, we will see the oscillating frequency rise with increasing voltage, and, indeed, such is the case. While tracking this phase-locked parasitic oscillation, we may measure the phase relationship from drain-to-source discovering a *180-degree phase shift!* Consequently, we are able to identify this parasitic oscillation as stemming from a differential-mode or push-pull mode.

Inducing Parasitic Oscillation

If the parallel-connected power MOSFETs are susceptible to parasitic oscillation, yet perhaps not oscil-

lating because of the circumstances, we often are able to induce oscillation by injecting an rf signal.

The charge-transfer characteristic curve given in Figure 9 shows no visible sign of oscillation. In Figure 10 we have the same power MOSFETs in the same circuit, but it shows *induced* oscillation. Had we adjusted this circuit to match the conditions for parasitic oscillation (without inducing the same), the MOSFETs would have broken into parasitic oscillation in the *exact same location on the charge transfer curve!* Compare Figure 10 with Figure 8.

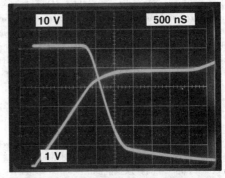

Non-Oscillating Parallel-Coupled MOSFETs
Figure 9

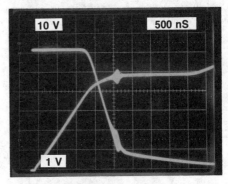

Induced Parasitic Oscillation
Figure 10

Why Some Will Not Oscillate

Particular combinations of parallel-coupled power MOSFETs, with a history of instability can, in some circuits, be found to be "rock stable." Aside from the possibility of simply having insufficient operating voltages, the possibility exists where our layout may effectively prevent unwanted parasitic oscillation. Table I offers one example where the gate length played a determining role in preventing oscillation. However, we should be careful to note that this table also identifies the principal cause for failure. Few power MOSFETs can withstand peak-to-peak, gate-to-source voltages exceeding 40 V.

Table 1
Oscillation Amplitudes for Various Gate Lead Lengths

Length of gate lead (cm)	$V_{DD} = 80V$ (V_{pk-pk})	$V_{DD} = 160V$ (V_{pk-pk})
3	20	100
4	17	86
5	3	44
5.5	0	30
6	0	8
6.5	0	0

Conclusions

In this discussion, a wide variety of possible problems has been considered. In each case, the potential problems have been shown to be either no problem at all or curable through some simple circuit means. In particular, it has been shown that matching of devices is rarely needed. From this discussion, it is quite clear that paralleling MOSFETs is relatively easy if a few simple rules are followed:

1. Provide good thermal coupling between devices.
2. Use good circuit layout practices.
3. Use small series gate resistors to suppress oscillation.
4. Other than the gate resistors needed for oscillation suppression, minimize the differential gate impedances.
5. In linear applications, use small differential source resistors to stabilize the operating point and force current sharing.
6. Carefully examine the circuit waveforms for any signs of parasitic oscillation or current spiking.

For those readers desiring more information, the articles referenced below should prove useful.

References

(1) R. Severns, "Safe Operating Area and Thermal Design for MOSPOWER Transistors," Siliconix Application Note AN83-10, November 1983.

(2) W. Schultz, "High Current FETs, A New Level of Performance," Power Conversion International, March 1984, pp. 43-46.

(3) J. Kassakian, "Some Issues Related to the Behavior of Multiple Paralleled Power MOSFETs."

(4) J. Kassakian and D. Lau, "An Analysis and Experimental Verification of Parasitic Oscillation in Paralleled Power MOSFETs," IEEE. Tran. Electron Devices, Vol. ED-31, NO 7, July 1984, pp. 959-963.

(5) K. Gauen, "Match Power — MOSFET Parameters for Optimum Parallel Operation," EDN, February 23, 1984, pp. 249-267.

(6) K. Gauen, "Paralleling Power MOSFETs in Linear Applications," Proceedings of PCI 1984, April 1984, Atlantic City, pp. 144-151.

(7) R. Saucedo and E. Schiring, "Introduction to Continuous and Digital Control Systems," MacMillan, 1968, p. 96.

(8) D. Giandomenico, "Anomalous Oscillations and Turn-off Behavior in a Vertical Power MOSFET," University of California Research Project, December 1983; later published in the proceedings of POWERCON 11, April 1984.

(9) E. Oxner, "Meet the Power MOSFET Model," Siliconix Application Note AN80-2, 1980.

(10) F. Terman, "Radio Engineers Handbook," McGraw-Hill, 1943, pp. 47-64.

5.3.1 Anomalous Oscillations and Turn-Off Behavior in a Vertical Power MOSFET

Abstract

When power MOSFETs are connected in a resistively loaded inverter circuit, the devices often oscillate. Experimental results indicate that the oscillation is a small-signal instability. Using a small-signal model and the Routh-Hurwitz criterion, analytic expressions were derived to determine the conditions for oscillation. The resulting expression indicates that the maximum allowable gate inductance for which the MOSFET will not oscillate scales roughly as the inverse of the device area. Computer simulations of the loop gain using SPICE demonstrate that the addition of resistance at the gate of the MOSFET or the addition of common source inductance decreases the tendency to oscillate.

An anomalous behavior is observed when the power MOSFET is turned off, in which the MOSFET turns off and then turns back on for several microseconds. This behavior is caused by the parasitic clamping diode of the power MOSFET and is not related to the cause of the oscillation.

Introduction

Power MOSFETs have been found to exhibit oscillations in the range of 1 MHz when they are biased in a simple inverting configuration as shown in Figure 1. This oscillation is unacceptable for linear applications and may destroy the device by exceeding the maximum gate voltage. For switching applications, the power MOSFET is biased in the saturation region only during the transition from ''on'' to ''off'' and vice versa. Thus, this oscillation can only occur for relatively slow switching applications.

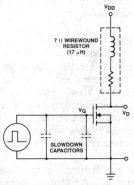

The power MOSFET is connected in an inverting configuration. Devices by several manufacturers were used all with ~400 V, 7 A voltage and current ratings. Slowdown capacitors were added at different distances from the MOSFET gate to effectively AC ground the input of the circuit.

Figure 1

The origin of this oscillation is not due to any unexpected phenomena inside the power MOSFET, but is due to a small signal instability of the parasitic elements of the power MOSFET and of the external circuit. Using a small signal equivalent circuit, the oscillation is studied here by applying the Routh-Hurwitz criterion to determine the conditions for oscillation. Computer simulations of the equivalent circuit were also performed to demonstrate the effects of varying critical circuit values.

An anomalous turn-off behavior was also observed during the study of the oscillation. It was found that when a power MOSFET is turned off into a resistive load with some parasitic inductance, the MOSFET apparently turns itself on a second time for several microseconds. Study of turn-off was pursued because it was believed the feedback required for oscillation might be related to this strange turn-off behavior. While it appears that the turn-off behavior is not related to the oscillation mechanism, its study is still of interest to the general understanding of the power MOSFET. There are several unexpected complications of the turn-off behavior due to the reverse recovery characteristics of the parasitic diode.

Theory and Observations of Oscillation in the Power MOSFET

Experimental Observations

To determine the tendency of the power MOSFET to oscillate, the device was connected in the inverter circuit shown in Figure 1. A pulse generator was used as an input source so that the power dissipation of the power MOSFET could be easily regulated by controlling the duty cycle. The use of the pulse generator also allowed easy analysis of the start-up transient of the oscillation.

The sequence of oscillographs shown in Figure 2 demonstrates the tendency of the power MOSFET to oscillate in a simple inverter circuit as the gate voltage is increased from values just below the threshold, at threshold, and well above threshold. Note that the oscillation is sustained only prior to when the device is about to conduct. When the gate voltage is increased or decreased about this point, the oscillation becomes a damped ringing.

A number of explanations for this oscillation were examined which include: 1) the possibility of a negative output conductance that has been observed by several researchers [1,2]; 2) the active operation of the parasitic bipolar transistor which is known to be a cause of early dV/dt breakdown in the power MOSFET[2]; and 3) the possibility of a small-signal instability caused by greater-than-unity loop gain.

The possibility of negative output conductance as a cause of oscillation does not seem to be a likely explanation because prior research has determined that the negative output conductance is of a thermal origin. Calculations show that the thermal response time of the power MOSFET is much larger than the period of the observed 1 MHz oscillations. This leads to the conclusion that the output conductance is not negative in the frequency range of the oscillations.

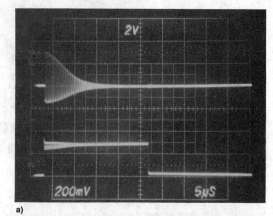

a)

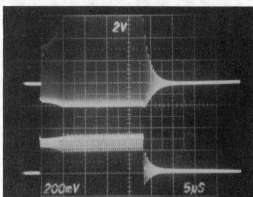

b)

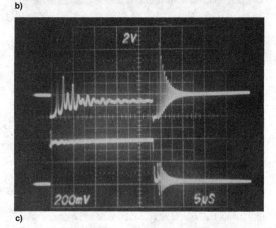

c)

Typical Observed Oscillations for the Inverter Circuit for
a) $V_{gs} \approx V_T$; b) $V_{gs} > V_T$; and c) $V_{gs} \gg V_T$.
Upper Traces: V_{supply} = 10 V/div;
Middle Traces: V_{drain} = 10 V/div; Lower Traces: V_{gs} = 2V/div
Figure 2

The interaction of the parasitic bipolar transistor by dV/dt breakdown initially seemed to be a likely cause of the oscillation. This suspicion was strengthened by the anomalous turn-off behavior in which the MOSFET turns back on for an extended period. Interaction of the parasitic bipolar transistor requires that a source of current be injected across the base-emitter junction to turn it on. This situation can occur when a fast rising voltage is applied to the drain/collector of the transistor that causes a rapid expansion of the base-collector depletion region. The charge displaced from the base depletion region moves towards the source contact causing a resistive voltage drop in the base region. If this voltage drop exceeds approximately 0.6 volts, the emitter base junction becomes forward biased and the bipolar transistor turns on, shunting the cut-off MOSFET.

For the above type of mechanism to be responsible, the oscillation can only exist in a large signal state since a high dV/dt at the drain is required to activate the parasitic bipolar transistor. An experiment to test this hypothesis demonstrated that the parasitic transistor is not involved but that the oscillation is due to a small-signal instability. The experiment involved removing the exciting transient introduced by the rising edge of the input pulse by shunting the gate of the MOSFET with a 0.1 μF capacitor. This reduced the tendency of the MOSFET to oscillate as might be expected by AC grounding the MOSFET gate. However, when the capacitor is decoupled from the gate by moving it 3 inches towards the pulse generator, the circuit oscillated readily. The slowly increasing oscillation amplitude seen in Figure 3 started at near zero amplitude, and this indicated that the oscillation is indeed due to a small-signal instability. Thus, a small-signal analysis should reveal the conditions for oscillation for the power MOSFET.

Routh-Hurwitz Analysis

Having established that the oscillation is due to a small-signal instability, the simple inverter circuit, shown in Figure 1, is modeled using the standard small-signal AC model as seen in Figure 4. A set of state equations can be written to describe this circuit, from which the characteristic polynomial can be found. The location of the roots of the characteristic polynomial in the complex frequency plane then will determine whether the circuit is stable.

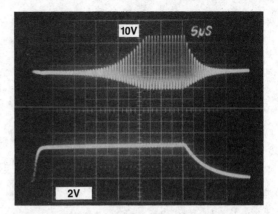

Oscillation of the Power MOSFET when a 0.1 μF Capacitor is Connected ~3 Inches from the Gate of the MOSFET
Figure 3

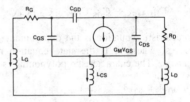

Small Signal AC Model of the Power MOSFET Inverter Circuit Shown in Figure 1
Figure 4

Writing KVL around the three loops and KCL at the three nodes for the circuit shown in Figure 4 yields

$$\dot{I}_g L_g - \dot{I}_{cs} L_{cs} = -I_g R_g + V_{gs} \tag{1}$$

$$\dot{I}_d - \dot{I}_{cs} L_{cs} = -I_d R_d + V_{ds} \tag{2}$$

$$V_{gd} + V_{ds} = V_{gs} \tag{3}$$

$$C_{gs} \dot{V}_{gs} + C_{gd} \dot{V}_{gd} = -I_g \tag{4}$$

$$C_{ds} \dot{V}_{ds} - C_{gd} \dot{V}_{gd} = -I_g - V_{gs}g_m \tag{5}$$

$$I_g + I_d + I_{cs} = 0 \tag{6}$$

These equations can be cast into a matrix state equation:

5

$$
\begin{bmatrix} \dot{i}_d \\ \dot{i}_g \\ \dot{v}_{gs} \\ \dot{v}_{ds} \end{bmatrix}
=
\begin{bmatrix}
-\dfrac{(L_g + L_{cs})\,R_d}{L_{eff}^2} & \dfrac{L_{cs}\,R_g}{L_{eff}^2} & -\dfrac{L_{cs}}{L_{eff}^2} & \dfrac{L_g + L_{cs}}{L_{eff}^2} \\[2mm]
\dfrac{L_{cs}\,R_d}{L_{eff}^2} & -\dfrac{(L_d + L_{cs})\,R_g}{L_{eff}^2} & \dfrac{L_d + L_{cs}}{L_{eff}^2} & -\dfrac{L_{cs}}{L_{eff}^2} \\[2mm]
-\dfrac{C_{gd}}{C_{eff}^2} & -\dfrac{C_{gd} + C_{ds}}{C_{eff}^2} & -\dfrac{C_{gd}g_m}{C_{eff}^2} & 0 \\[2mm]
-\dfrac{C_{gs} + C_{gd}}{C_{eff}^2} & -\dfrac{C_{gd}}{C_{eff}^2} & -\dfrac{(C_{gs} + C_{gd})\,g_m}{C_{eff}^2} & 0
\end{bmatrix}
\begin{bmatrix} I_d \\ I_g \\ V_{gs} \\ V_{ds} \end{bmatrix}
\tag{7}
$$

where L_{eff}^2 and C_{eff}^2 are defined as

$$L_{eff}^2 \equiv L_d L_g + L_{cs} L_d + L_{cs} L_g \tag{8}$$

$$C_{eff}^2 \equiv C_{ds} C_{gs} + C_{ds} C_{gd} + C_{gd} C_{gs} \tag{9}$$

The characteristic equation is the determinant of the matrix $A - sI$ where A is the state equation matrix written above. The characteristic polynomial is then

$$P(s) = a_0 s^4 + a_1 s^3 + a_2 s^2 + a_3 s + a_4 \tag{10}$$

where

$$a_0 = C_{eff}^2 L_{eff}^2 \tag{11}$$

$$a_1 = C_{eff}^2 \left[(L_d + L_{cs}) R_g + (L_g + L_{cs}) R_d \right] + C_{gd} g_m L_{eff}^2 \tag{12}$$

$$a_2 = C_{eff}^2 R_d R_g + C_{gd} g_m \left[(L_d + L_{cs}) R_g + (L_g + L_{cs}) R_d \right] + (C_{gs} + C_{gd}) L_g + (C_{gd} + C_{ds}) L_d + (C_{gs} + C_{ds}) L_{cs} \tag{13}$$

$$a_3 = C_{gd} g_m R_d R_g + (C_{gs} + C_{gd}) R_g + (C_{gd} + C_{ds}) R_d + g_m L_{cs} \tag{14}$$

$$a_4 = 1 \tag{15}$$

The order of the characteristic polynomial is determined by the number of energy storage elements which contribute a state variable. Note that although there are a total of six energy storage elements, the characteristic polynomial is only fourth order. This is because the three capacitors, C_{gs}, C_{ds}, and C_{gd},

form a loop, and the inductors form a cutset. Kirchhoff's Voltage Law constrains one of the capacitor voltages to be the sum of the other two; hence, the three capacitors contribute only two state variables. Likewise, although there are three inductances, there are only two independent state variables because Kirchhoff's Current Law constrains the current in one of the inductors to be the sum of the other two. Thus, the six energy storage elements have only four independent state variables resulting in a fourth order characteristic polynomial.

Although there is no analytic solution for the roots of a fourth order polynomial, stability only requires that all the poles lie in the left half s-plane. The Routh-Hurwitz criterion states that a necessary and sufficient condition for all of the roots of an nth order equation to lie in the left half of the s-plane is that all of the Hurwitz determinants, D_k $(k = 1, 2, \ldots n)$, must be positive. The Hurwitz determinants for a fourth order polynomial are given by

$$D_1 = a_1 \tag{16}$$

$$D_2 = \begin{vmatrix} a_1 & a_3 \\ a_0 & a_2 \end{vmatrix} \tag{17}$$

$$D_3 = \begin{vmatrix} a_1 & a_3 & 0 \\ a_0 & a_2 & a_4 \\ 0 & a_1 & a_3 \end{vmatrix} \tag{18}$$

$$D_3 = \begin{vmatrix} a_1 & a_3 & 0 & 0 \\ a_0 & a_2 & a_4 & 0 \\ 0 & a_1 & a_3 & 0 \\ 0 & a_0 & a_2 & a_4 \end{vmatrix} = a_4 D_3 \tag{19}$$

To simplify the evaluation of the roots of the characteristic polynomial, it would be desirable to reduce the order of the polynomial by setting some of the capacitors or inductors to zero. Unfortunately, the removal of one inductor and one capacitor will not reduce the order of the polynomial because the number of independent state variables is not decreased. However, the algebra in solving these equations is still simplified enormously if some of the circuit elements can be set to zero, and thus, R_g and L_{cs} are set to zero. Setting R_g and L_{cs} to zero represents a typical situation since these quantities are usually very small ($R_g \ll \omega L_g$ and $L_{cs} \ll L_d$). By inspection, D_1 is always a positive quantity since all of the component values are positive. An algebraic evaluation of D_2 reveals that when L_{cs} is zero, D_2 is always positive. Further, since $D_4 = a_4 D_3$, and $a_4 = 1$, the circuit's stability is indicated by the sign of the third Hurwitz determinant, D_3.

The algebraic expression for D_3 when $L_{cs} = R_g = 0$ is

$$D_3 = \{ (C_{gd} + C_{ds}) R_d [C_{gd} g_m L_g R_d + (C_{gs} + C_{gd}) L_g + (C_{gd} + C_{ds}) L_d]$$
$$- C_{eff}^2 L_g R_d - C_{gd} g_m L_d L_g \} (C_{eff}^2 L_g R_d + C_{gd} g_m L_d L_g) \tag{20}$$
$$- C_{eff}^2 (C_{gd} + C_{ds})^2 L_d L_g R_d^2$$

This equation is second order in all variables except L_g which fortunately is one of the variables which can control the oscillation. Solving the above equation for L_g with the constraint that $D_3 > 0$ yields

$$L_g < \frac{(C_{gd} + C_{ds})^2 g_m L_d^2 R_d}{(C_{eff}^2 R_d + C_{gd} g_m L_d) (g_m L_d - C_{gd} g_m R_d^2 - C_{ds} g_m R_d^2 - C_{gd} R_d)} \tag{21}$$

for

$$0 < g_m L_d - C_{gd} g_m R_d^2 - C_{ds} g_m R_d^2 - C_{gd} R_d \tag{22}$$

Note that when the equation on the bottom is negative, the sign of the inequality is changed. Then the condition for stability is always satisfied because L_g is a positive quantity.

For a typical circuit, the load inductance terms dominate in equations (21) and (22). The condition for stability as given by equation (21) can then be reduced to

$$L_g < \frac{(C_{gd} + C_{ds})^2 R_d}{C_{gd} g_m} \tag{23}$$

For the inverter circuit studied here, this condition is a relatively severe constraint. Typical numerical evaluations of equations (21) and (22) for a conventional MOSFET show that these constraints are both easily satisfied.

An examination of equation (23) reveals that the critical value of L_g is roughly inversely proportional to the gate area. Note that the capacitances and transconductance scale proportional to the area while both the on resistance and the load resistance is inversely related to the area. Substituting these dependencies into equation (23) shows that the maximum gate inductance for which the circuit will be stable decreases with increasing device size. Thus, as MOSFETs grow larger in size, they have a greater oscillation tendency.

Computer Simulation Using SPICE

The analytic solution in the previous section has limitations because of the assumptions that $R_g=0$ and $L_{cs}=0$ and because of the general complexity of D_3. More significantly, the analysis does not show how close the circuit is to oscillation but only if the circuit is or is not in oscillation.

Using the circuit simulation program SPICE, the closed loop gain and phase margin are easily found for any particular circuit configuration. To set up the circuit of Figure 1 for a loop gain analysis, the circuit connection to the gate of the MOSFET is broken and an AC test voltage generator is applied to the gate (see Figure 5). By setting the voltage of the AC supply to unity, the voltage at the disconnected circuit node is then the loop gain. If when the phase is zero and the loop gain is less than unity, the circuit will be stable.

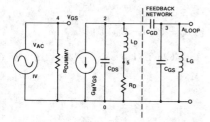

Loop gain analysis of the power MOSFET inverter and its small signal AC model shown in Figure 4. The gate feedback is disconnected and replaced by an AC voltage source. The loop gain is measured at the disconnected node of the gate feedback circuit with the AC voltage source set to unity.
Figure 5

As an example, the validity of the previously derived analytic expressions was tested by using typical circuit values and values of L_g calculated so that the circuit is just below, at, and above the threshold of oscillation. The results are shown in Figures 6, 7, and 8. To see the effects of the addition of R_g or L_{cs}, the circuit description in the SPICE input file was modified to include these elements. The results shown in Figures 9 and 10 indicate that the addition of R_g decreases the loop gain at the frequency where the phase is zero, and thus decreases the tendency of the MOSFET to oscillate. Similarly, the addition of common source inductance tends to decrease the loop gain and decreases the oscillation tendency.

```
1••••••••12/08/83 •••••••• SPICE 2G.5A (23NOV82) •••••••••11:32:06•••••
0LG LESS THAN CRITICAL
0••••  INPUT LISTING          TEMPERATURE =  27.000 DEG C
0•••••••••••••••••••••••••••••••••••••••••••••••••••••••••••••••••••••
LG    0 3   10N
CGS   3 0   200P
CGD   3 2   200P
CDS   2 0   1000P
LD    2 5   20U
RD    5 0   7
GM    2 0 4 0 3
RDUMMY 4 0   100
VG    4 0   AC 1
.PLOT AC V(3) VP(3)
.AC OCT 25 500K 1.6MEG
.WIDTH OUT=80
.END
0•••••••••••••••••••••••••••••••••••••••••••••••••••••••••••••••••••••
```

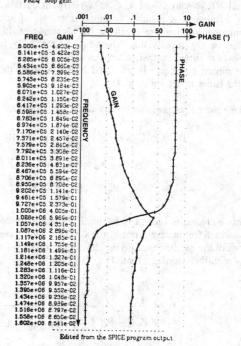

```
0LEGEND  Loop gain: •  Phase +
X
   FREQ   loop gain

              .001    .01     .1      1      10
                                              ► GAIN
             -100   -50      0      50     100
   FREQ    GAIN                                ► PHASE (°)
5.000e+05  4.903e-03
5.141e+05  5.422e-03
5.285e+05  6.005e-03
5.434e+05  6.660e-03
5.586e+05  7.399e-03
5.743e+05  8.235e-03
5.905e+05  9.184e-03
6.071e+05  1.027e-02
6.242e+05  1.150e-02
6.417e+05  1.293e-02
6.596e+05  1.458e-02
6.783e+05  1.649e-02
6.974e+05  1.874e-02
7.170e+05  2.140e-02
7.371e+05  2.457e-02
7.579e+05  2.840e-02
7.792e+05  3.308e-02
8.011e+05  3.891e-02
8.236e+05  4.631e-02
8.467e+05  5.594e-02
8.706e+05  6.890e-02
8.950e+05  8.706e-02
9.202e+05  1.141e-01
9.461e+05  1.579e-01
9.727e+05  2.373e-01
1.000e+06  4.005e-01
1.028e+06  5.966e-01
1.057e+06  4.351e-01
1.087e+06  2.896e-01
1.117e+06  2.165e-01
1.149e+06  1.755e-01
1.181e+06  1.499e-01
1.214e+06  1.327e-01
1.248e+06  1.205e-01
1.283e+06  1.116e-01
1.320e+06  1.048e-01
1.357e+06  9.957e-02
1.395e+06  9.552e-02
1.434e+06  9.236e-02
1.474e+06  8.989e-02
1.516e+06  8.797e-02
1.558e+06  8.650e-02
1.602e+06  8.541e-02
```

Edited from the SPICE program output

SPICE Simulation of the Inverter's Loop Gain
with the Gate Inductance Less than
the Critical Value of $L_g=16.85\mu H$
(Note That the Maximum Loop Gain (0.597) is Less than Unity)
Figure 6

CRITICAL
0••••   INPUT LISTING   TEMPERATURE = 27.000 DEG C
0•••
```
LG   0 3  16.85N
CGS  3 0   200P
CGD  3 2   200P
CDS  2 0  1000P
LD   2 5   20U
RD   5 0    7
GM   2 0 4 0 3
RDUMMY 4 0   100
VG   4 0   AC 1
.PLOT AC V(3) VP(3)
.AC OCT 25 500K 1.6MEG
.WIDTH OUT=80
.END
```
1•••

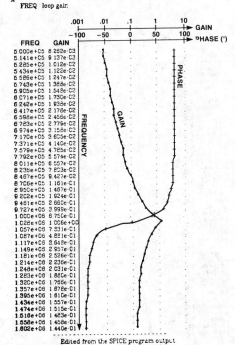

LEGEND: Loop gain: * Phase: +

SPICE Simulation of the Inverter's Loop Gain
with the Gate Inductance Selected Such That
the Circuit is on the Verge of Oscillation
(Note That the Maximum Loop Gain is Unity)
Figure 7

LG GREATER
0••••   INPUT LISTING   TEMPERATURE = 27.000 DEG C
0•••
```
LG   0 3   20N
CGS  3 0   200P
CGD  3 2   200P
CDS  2 0  1000P
LD   2 5   20U
RD   5 0    7
GM   2 0 4 0 3
RDUMMY 4 0   100
VG   4 0   AC 1
.PLOT AC V(3) VP(3)
.AC OCT 25 500K 1.6MEG
.WIDTH OUT=80
.END
```
1••

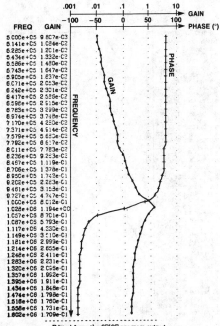

LEGEND: Loop gain: * Phase: +

SPICE Simulation of the Inverter's Loop Gain
with the Gate Inductance More than
the Critical Value of L_g=16.85μH (Note That
the Maximum Loop Gain (1.19) is Greater than Unity)
Figure 8

5

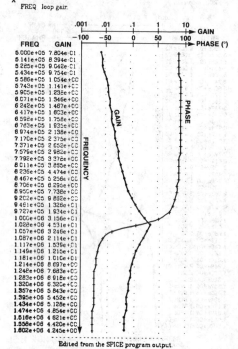

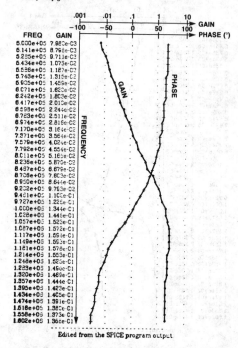

SPICE Simulation of the Inverter Circuit's Loop Gain
to Show the Effects of Gate Resistance
(Note That the Gain is Less than Unity at the Frequency
Where the Phase is Zero, and Hence the Circuit is Stable)
Figure 9

SPICE Simulation of the Inverter Circuit to Show the Effects
of Common Source Inductance or the Loop Gain
(Note That the Maximum Loop Gain is Decreased (0.159)
Which Decreases the Tendency to Oscillate)
Figure 10

Turn-off Behavior

The turn-off behavior of the power MOSFET is shown in Figure 11. Initially, the gate voltage is high and the MOSFET is fully turned on. When the gate voltage falls, the drain voltage rises quickly and then abruptly drops to the "on-voltage," where it remains for several microseconds before ringing and settling to the supply voltage.

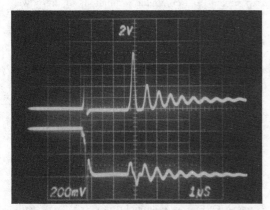

a)

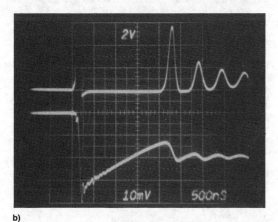

b)

Oscillograph of the Turn-off Behavior
of the Power MOSFET Inverter;
a) Upper Trace: V_D = 10 V/div; Lower Trace: V_{gs} = 5 V/div;
b) Upper Trace: V_D = 10 V/div; Lower Trace: I_D = 0.5 A/div
Figure 11

The sequence of events is shown in Figure 12. Before the MOSFET is turned off, energy has been stored in the inductor. When the device is turned off, the output capacitance of the power MOSFET and the inductor cycle for slightly more than one half cycle in which the inductive energy is dumped into the

output capacitance and then returned to the inductor with the opposite current polarity. Neglecting the nonlinearity of the output capacitance and the damping of the output load resistance, the time for 1/2 cycle is

$$t \approx \pi \sqrt{LC} \qquad (24)$$

After this time, the parasitic diode is forward biased and the voltage across the inductor is

$$V_L = V_{DD} + V_{D-fwd} - (-I_d R_L) \qquad (25)$$

If the damping losses are neglected, then

$$I_d \approx V_{DD}/R_L \qquad (26)$$

and

$$V_L \approx 2V_{DD} + V_{D-fwd} \qquad (27)$$

The drain voltage remains near zero during the time the diode is conducting. The current through the diode is dictated by the L-R circuit and is

$$I_d = \frac{2V_{DD} + V_{D-fwd}}{R}\left(1 - e^{-tR/L}\right) - \frac{V_{DD}}{R} \qquad (28)$$

The time at which the diode current is zero is

$$t_{on} = (L/R) \ln\left[\frac{2V_{DD} + V_{D-fwd}}{V_{DD} + V_{D-fwd}}\right] \qquad (29)$$

For a reasonably large supply voltage, $V_{DD} \gg 0.7$ volts, the above equation can be expanded in a Taylor series to give

$$t_{on} \approx \frac{L}{R}\left[1 - \frac{V_{D-fwd}}{2V_{DD}}\right] \ln 2 \qquad (30)$$

Neglecting the reverse recovery of the diode for the moment, the drain voltage rises at this time and then the drain voltage rings and settles to the supply voltage.

Second Order Effects

A comparison of experimental data with the calculated results of equation (30) reveals a large discrepancy in the "on-time." Also, the oscillograph of Figure 11 shows that the power MOSFET appears to turn on a third time. These discrepancies were found to be due to the effect of the reverse recovery of the diode (see Figure 12). Instead of turning off when the forward diode current crosses zero, the diode continues to conduct in the reverse direction before

5

the drain voltage begins to rise. Consequently, the total turn-on time expressed by equation (30) must be modified to include the reverse recovery time t_{rr}:

$$t_{on} \approx \frac{L}{R} \left[1 - \frac{V_{D-fwd}}{2V_{DD}} \right] \ln 2 + t_{rr} \qquad (31)$$

Since the inductor once again has a downward current direction after the diode's reverse recovery, the situation is the same as when the MOSFET first turned off but with a lower initial current level, and thus the cycle repeats again.

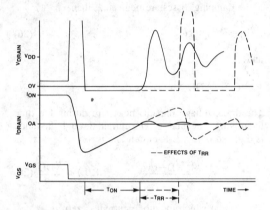

**Diagram of the Power MOSFET's Turn-off Behavior
Showing the Effects of t_{rr}
Figure 12**

Another discrepancy is the dependency of the turn-off time on the supply voltage. From equation (31) the time is independent of the supply voltage when the supply voltage is much greater than 0.7 volts. However, when the peak voltage during turn-off is near the breakdown voltage, the MOSFET will conduct and dissipate some of the stored energy, resulting in a decrease in the turn-on time. In our experiment, even with a low supply voltage of ten volts and a load inductance of $17 \, \mu H$, the peak voltage sometimes exceeded 450 volts and resulted in a decreased turn-on time with increasing supply voltage.

Conclusions

It has been found experimentally that the power MOSFET oscillates due to a small signal instability. The small signal instability was confirmed by analyzing a small signal model of a power MOSFET inverter circuit both analytically and with computer simulations. Using the Routh-Hurwitz criterion to de-

termine if the circuit was stable, a set of expressions was derived that places a constraint on the maximum value of gate inductance for which the circuit will be stable. From this expression it may be seen that as the area of a MOSFET increases, the maximum allowable gate inductance decreases, which may explain why conventional MOSFETs do not have as great a tendency to oscillate as power MOSFETs.

Experiments and computer simulations of the power MOSFET inverter circuit reveal that the addition of small amounts of gate resistance ($R_g > \omega_{osc} L_g$) or common source inductance effectively stops the oscillation tendency while not severely degrading the performance of the MOSFET. Unfortunately, there appears to be no way of reducing the tendency of the power MOSFET to oscillate without degrading its performance.

The anomalous turn-off behavior was found not to be caused by dV/dt breakdown of the parasitic bipolar transistor. Instead, it is caused by the intrinsic diode, clamping the output voltage to ground. The reverse recovery of the diode both extends the time during which the MOSFET is apparently on and causes the power MOSFET to seemingly turn "on" repeatedly as the output voltage settled to its final value.

The turn-off behavior will not be observed in a typical power MOSFET circuit because when the MOSFET is used to drive an inductive load, the drain is usually clamped to the positive supply voltage. However, when the power MOSFET is used to drive an unclamped inductive load, the parasitic diode can extend the time required for the drain voltage to settle at its final value.

References

[1] D.K. Sharma and K.V. Ramamathan, "Modeling Thermal Effects on MOS IV Characteristics," IEEE Electron Dev. Let., Vol. EDL-4, p. 362, Oct. 1983.

[2] Yasuhisa Omura, Eiichi Sano, and Kuniki Ohwada, "A Negative Drain Conductance Property in a Super-Thin Film Buried Channel MOSFET on a Buried Insulator," IEEE Tran. Elec. Dev., Vol. ED-30, Jan. 1983.

[3] D.S. Kuo, C. Hu, and M.H. Chi, "dV/dt Breakdown in Power MOSFETs," IEEE Electron Dev. Let., Vol. EDL-4, p. 1, Jan. 1983.

5.3.2 Thermally Forced Current Sharing in Paralleled Power MOSFETs

Abstract

A thermal model is proposed and analyzed to determine the extent of thermal forcing of current sharing in paralleled power MOSFETs. Both package paralleled and chip paralleled devices are examined. It is shown that the effects of thermal forcing on the current distribution among paralleled devices are relatively insignificant. The most reliable operation of paralleled devices is achieved if they are tightly coupled thermally.

Introduction

The popularity of the power MOSFET arises from its low drive requirement, fast switching speed, and the relative ease with which it may be paralleled to achieve higher power ratings. Since manufacturing yields are presently such that single devices with ratings above approximately 5 kVA are not economically justifiable, it is this latter advantage which permits employing the device in applications above a few kW. Paralleling can be done by the manufacturer at the chip level, or by the user at the packaged device level.

This paper considers the influence of the temperature dependence of the drain-source resistance on static current sharing among paralleled devices. The greatest effect of thermal forcing is achieved when the paralleled devices are thermally isolated from one another, thus permitting them to equilibrate at different temperatures. In practical applications complete isolation can only be approximated, in part due to the physical constraints on equipment size and component arrangement, and in part due to the conflicting requirements imposed by the need to eliminate electrical parasitics.

A static thermal model for a pair of paralleled devices is proposed and analyzed. The model includes the parametric variability necessary to represent both chip and package paralleled devices, and to represent various degrees of thermal coupling between the two devices. Analysis of the model produces a set of coupled non-linear equations which are solved numerically using an iterative algorithm. The drain currents and "junction" temperatures of both devices are then determined for a variety of conditions.

The drain-source resistance of a power MOSFET is dominated by the bulk spreading resistance of the drain region, and not the inverted channel resistance. The temperature dependence of this resistance is caused principally by the variation of the drift mobility. As a consequence, the resistance exhibits a positive temperature coefficient whose value is dependent upon conductivity type, but relatively independent of device construction. The TCR itself does exhibit some temperature dependence, which is more pronounced for high voltage devices. For n-channel, 400 V devices this coefficient varies from approximately 0.6%/°C at 25°C, to a value of approximately 1.2%/°C at 150°C. For purposes of this analysis, the non-linear behavior of the TCR is not considered, and the high temperature value is assumed for all devices. The drain-source resistance as a function of temperature, $R_{ds}(T)$, is thus

$$R_{ds}(T) = R_{dso}(1 + A(T_j - T_a)) \qquad (1)$$

where R_{dso} is the drain-source resistance measured at T_a, and A is the TCR.

Because the current distribution among paralleled devices is governed by their relative drain-source resistances, it is often believed that the positive temperature coefficient of resistance forces current sharing among paralleled devices. [1,2,3] In order to determine the extent to which such forced sharing occurs, it is necessary to assume a thermal system and solve the resulting set of non-linear equations for the individual drain-source resistances.

The Thermal System

Figure 1 shows the static thermal model for two devices in parallel. The resistances R_{jc}, R_{cs}, and R_{sa} represent the "junction" to case, case to sink, and sink to ambient thermal resistances, respectively. (Although there is no heat generating junction in a MOSFET, the term is used here generically to represent the heat source.) The voltages T_j, T_c, T_s, and T_a, represent the junction, case, sink, and ambient temperatures, respectively. The resistance R_c represents the thermal coupling between devices. For multiple chips on a common header, the coupling occurs between the case nodes in the model. For two packaged devices on a single heat sink, the coupling occurs between the sink nodes in the model. The current sources P_1 and P_2 represent the rate at which thermal energy is being dissipated in devices 1 and 2, respectively.

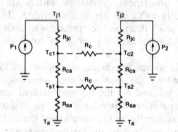

Static Thermal Model for Two Paralleled Devices
Figure 1

The electrical connection of the devices is shown in Fig. 2, where V_{ds} is the on-state drain-source voltage, and I is the net drain current. The non-linearity in this system arises because the distribution of the net current, I, between the devices depends upon the relative values of R_{ds1} and R_{ds2}, as do the relative dissipa-

tions. Therefore the sources P_1 and P_2 in Fig. 1 each depend on both T_{j1} and T_{j2}.

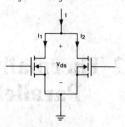

Electrical Connection of Paralleled Devices
Figure 2

The input parameters for this problem are the net drain current, I, and the reference drain-source resistances, R_{ds01} and R_{ds02}. The variables T_{j1}, T_{j2}, I_1, I_2, P_1, and P_2 are then determined. For symbolic simplicity, the subscripts "ds" are not carried forward in the following analysis and discussion. The operating values of the drain-source resistance are referred to as R_1 and R_2, while the initial, or reference, values are R_{10} and R_{20}. Also, literal subscripts are used for thermal resistances, while numerical subscripts are reserved for electrical resistances.

The model of Fig. 1 can be generalized slightly by recognizing that the point of thermal coupling can be moved by varying the relative values of R_{jc}, R_{cs}, and R_{sa}. Therefore, in the equations that follow, R_{jc}, R_{cs}, and R_{sa} have been combined into an equivalent resistance R_d which represents the thermal circuit between the point of coupling and ambient, and an equivalent resistance R_a which represents the thermal resistance between the heat source and the point of coupling. This more general thermal model, in which chip paralleled or package paralleled devices may be represented by simply changing the relative values of R_a and R_d, is shown in Fig. 3.

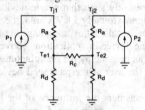

Generalized Static Thermal Model
Figure 3

The model of Fig. 3 may now be analyzed. In terms of the temperatures of the coupled nodes, T_{e1} and T_{e2}, the junction temperatures are

$$T_{j1} = P_1 R_a + T_{e1} \tag{2a}$$

$$T_{j2} = P_2 R_a + T_{e2} \tag{2b}$$

The temperatures of the coupled nodes, relative to $0°C$, may now be determined in terms of P_1 and P_2,

$$T_{e1} = \frac{[P_1(R_c + R_d) + P_2 R_d]}{[(R_c/R_d) + 2]} + T_a \tag{3a}$$

$$T_{e2} = \frac{[P_2(R_c + R_d) + P_1 R_d]}{[(R_c/R_d) + 2]} + T_a \tag{3b}$$

The electrical dissipation in each device, as a function of net drain current is

$$P_1 = [I(R_1 \| R_2)]^2 / R_1 \tag{4a}$$

$$P_2 = [I(R_1 \| R_2)]^2 / R_2 \tag{4b}$$

As indicated earlier, the on-state drain-source resistances are

$$R_1 = R_{1o}(1 + A T_{j1}) \tag{5a}$$

$$R_2 = R_{2o}(1 + A T_{j2}) \tag{5b}$$

Finally, (2), (3), (4), and (5) can be combined to yield R_1 and R_2 as functions of the net drain current, I, and the thermal environment,

$$R_1 = R_{1o} \left\{ 1 + \frac{I^2 R_1 R_2 A}{(R_1 + R_2)^2} \left[R_2 \left\{ \frac{(R_d + R_a)\left(\frac{R_c}{R_d} + 1\right) + R_a}{\frac{R_c}{R_d} + 2} \right\} + R_1 \frac{R_d}{\frac{R_c}{R_d} + 2} \right] \right\} \tag{6a}$$

$$R_2 = R_{2o} \left\{ 1 + \frac{I^2 R_1 R_2 A}{(R_1 + R_2)^2} \left[R_1 \left\{ \frac{(R_d + R_a)\left(\frac{R_c}{R_d} + 1\right) + R_a}{\frac{R_c}{R_d} + 2} \right\} + R_2 \frac{R_d}{\frac{R_c}{R_d} + 2} \right] \right\} \tag{6b}$$

Results

For a given set of input parameters, (6a) and (6b) are solved numerically using an iterative algorithm. Both TO220 and TO3 packages were considered for individually packaged devices in parallel, and for individual chips in parallel a R_{cs} value of one-half that for the package was assumed. This latter is a convenient assumption since it is satisfied automatically as R_c is "moved up" toward the "junction" in the model of Fig. 3.

Although a wide variety of different conditions were studied, the essential conclusions of this work can be demonstrated by the results of a few specific cases. In each case it is assumed that the devices are paralleled at the chip level, i.e., $R_a = R_{jc}$. The parameters for these cases are:

Case 1: (100 V, 10 A, TO220)
$R_{jc} = 1.67°C/W$
$R_{cs} = 1.00°C/W$
$R_{sa} = 1.47°C/W$
$R_{1o} = 0.12$ ohm
$R_{2o} = 0.16$ ohm
$I = 20$ A
$A = .0067$

Case 2: (400 V, 3.5 A, TO220)
$R_{jc} = 1.67°C/W$
$R_{cs} = 1.00°C/W$
$R_{sa} = 1.47°C/W$
$R_{1o} = 0.75$ ohm
$R_{2o} = 1.00$ ohm
$I = 7$ A
$A = .011$

Case 3: (400 V, 11 A, TO3)
$R_{jc} = .83°C/W$
$R_{cs} = .20°C/W$
$R_{sa} = .60°C/W$
$R_{1o} = 0.2$ ohm
$R_{2o} = 0.3$ ohm
$I = 22$ A
$A = .012$

Equations (2)–(6) were solved for R_c varying from 0 to $100°C/W$. The effect of forced current sharing is maximized for large R_c, which permits the necessary temperature difference between devices. This result is, of course, independent of the relative values of R_a and R_d.

5

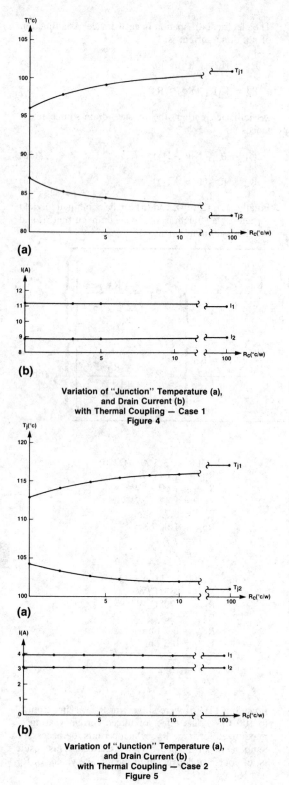

**Variation of "Junction" Temperature (a),
and Drain Current (b)
with Thermal Coupling — Case 1
Figure 4**

**Variation of "Junction" Temperature (a),
and Drain Current (b)
with Thermal Coupling — Case 2
Figure 5**

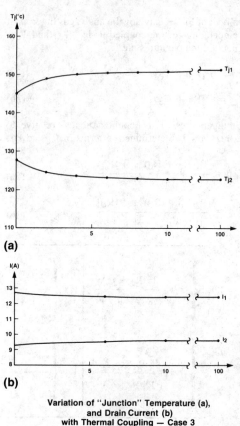

**Variation of "Junction" Temperature (a),
and Drain Current (b)
with Thermal Coupling — Case 3
Figure 6**

Figures 4, 5, and 6 show the variations of "junction" temperature and drain current with thermal coupling. The results show the effect of forced current sharing to be practically non-existent. In fact, if the T_j's are constrained to be equal (by setting $R_a = 0$, $R_d = R_{jc} + R_{cs} + R_{sa}$, and $R_c = 0$), the drain currents differ by less than 5% from their values for the case of total thermal isolation (approximated by $R_c = 100°C/W$). Even though current sharing is not enhanced by thermal isolation, the figures show that the "junction" temperatures can be substantially different under such conditions. Reliability of the hot device is thus compromised with no apparent benefit.

Conclusions

Since the continuous drain current specification for a power MOSFET is a reflection of the maximum junction temperature limit, even though I_1 in Figs. 4(b), 5(b), and 6(b) exceed this specification the device is quite happy as long as $T_j < T_{jmax}$. One is therefore led to the conclusion that forced current sharing is an illusion, and furthermore, that devices to be paralleled need not even be carefully matched for R_{dso}. Also, it is clear from the analysis that paralleling can be used to its best advantage if chips are paralleled by the manufacturer using techniques which result in the minimum value of R_c.

References

[1] *HEXFET Databook*, International Rectifier Corporation, 1981, p. 20.

[2] Oxner, E.S., *Power FETs and Their Applications*, Prentice-Hall Inc., N.J., 1982, p. 8.

[3] *MOSPOWER FET Design Catalog*, Siliconix, 1982, p. 6–4.

5

5.3.3 An Analysis and Experimental Verification of Parasitic Oscillations in Paralleled Power MOSFETs

Abstract

An analysis of the small signal dynamic model of the power MOSFET is presented which predicts the existence of high frequency parasitic oscillations when these devices are electrically paralleled. It is shown that the existence of these oscillations is a strong function of the small signal transfer admittance, g_m, and the differential mode drain, gate and source resistances. The sensitivity of the oscillations to these parameters is determined. Experimental data verifying the qualitative aspects of the analytical results is presented. It is concluded that the problem is potentially most severe for devices which are paralleled by the manufacturer at the chip level. A practical solution to the problem is the introduction of differential mode gate resistance, either as lumped components, or by the use of polysilicon overlays.

Introduction

Anecdotal user reports have indicated "mysterious" failures of paralleled power MOSFETs. At the same time, the appearance of oscillations during the switching transitions of such paralleled devices has been reported in the literature.[1,2,3] The understanding of this behavior is of importance not only for reliably paralleling packaged devices, but also for successfully fabricating large devices using paralleled chips. In this latter case, since the user has fewer remedial options available (because of the lack of multiple gate access), the responsibility rests with the manufacturer to assure that the device is stable under all operating conditions.

Phenomenologically, the oscillations occur at frequencies between 50 and 250 MHz, are observable at both the gate and drain terminals of either device, and have had observed amplitudes exceeding the gate-source breakdown voltage. The existence of the oscillation is a strong function of the gate drive source resistance, but for reasons to be shown, this is believed to be due to the extended duration of the saturation region transition for larger drive resistance. It will also be shown that, given a saturation region residence of sufficient duration to support an oscillation, the inception of this oscillation is a function of only the *differential* components of the gate and drain resistance, and the parasitic drain and source inductances.

An earlier paper[4] has shown the theoretical effect of gate and drain impedances on the parasitic phenomenon. This paper extends the previous analysis to include the effect of source impedance, and verifies the theoretical results experimentally.

The Dynamic Model

Figure 1 shows two electrically paralleled power MOSFETs, Q_1 and Q_2. It is assumed that these devices are identical, and that the layout of their interconnection is symmetrical. Since oscillations occur only when the devices are in their active (saturation) region, it is the incremental model of the circuit of Fig. 1 operating in this region which is important. This model is shown in Fig. 2, which also includes parasitic elements necessary to the creation of an oscillatory system. The inductance, L_G, and resistance, R_G, represent the parasitic impedance between the common drive source connection and the gates of the two devices. The inductances L_D and L_S represent the parasitic drain and source impedances, respectively. The resistor R_D represents the incremental drain-source resistance, which is primarily the resistance of the drift region of the drain. Because of the assumed symmetry of the problem, the parasitic elements are disposed symmetrically between the two devices. The absence of parasitic elements in series with R_S and R_L is justified below.

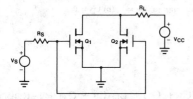

A Pair of Electrically Paralleled MOSFETs
Figure 1

Because of the symmetry and linearity of Fig. 2, the problem may be transformed into differential and common mode coordinates and the method of half circuits used to obtain the relevant characteristic equations. Establishing the plane of symmetry and opening all circuit branches cutting this plane quickly establishes the fact that, to the extent that a single device operates stably, the common mode response of the model cannot include oscillatory behavior. Essentially, the two devices are decoupled in the common mode. Thus the oscillations must be a result of the differential mode circuit.

The differential mode circuit may be derived by recognizing that all nodes on the plane of symmetry are at incremental ground potential. The resulting model is shown in Fig. 3. It is now clear that since R_S and R_L are common mode elements that do not appear in the differential mode circuit, they will not have any effect on the circuit's unstable behavior.

The same is true of any parasitic elements which might have been included in these branches in the model of Fig. 2.

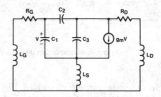

Incremental Model, Including Parasitic Elements for Saturation Region Operation of the Paralleled Devices of Figure 1
Figure 2

The differential mode circuit model of Fig. 3 may now be analyzed to determine its characteristic polynomial. However, rather than confront this rather complicated analysis, two separate situations are considered. The first, shown in Fig. 4(a), assumes no parasitic source inductance, and the second, shown in Fig. 4(b), assumes no parasitic drain inductance. In both cases the effects of gate resistance and parasitic gate inductance are determined. In this way it is possible to gain some insight to the oscillatory behavior as a function of the parasitic elements. The results of the detailed analysis of the effects of drain resistance is presented only for the case $L_D = 0$. It is believed that this situation will be more commonly encountered in practice because the physical location of the drain contact (chip substrate) results in very low differential drain inductance (compared to the corresponding differential source inductance) when devices are paralleled. For example, the heat sink will frequently be the common drain connection for T03 or T0220 packages. This produces an electrostatic geometry exhibiting a low inductance.

5

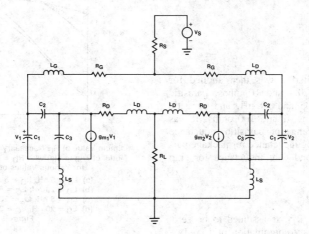

Differential Mode Circuit for the Model of Figure 2
Figure 3

(a)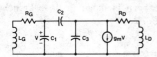

(b)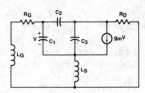

Differential Mode Circuits for the Model of Figure 2 Assuming: (a) $L_S = 0$, (b) $L_D = 0$
Figure 4

The differential mode circuit model of Fig. 4(a) is analyzed to determine its characteristic polynomial, which is

(1a) $\quad S^4(L_D L_G C_e^2) + S^3(R_G L_D C_e^2 + g_m L_D L_G C_2) + S^2(L_G(C_1 + C_2) + L_D(C_2 + C_3) + g_m R_G L_D C_2)$

$\quad\quad + S(R_G(C_1 + C_2)) + 1 = 0$

where $\quad C_e^2 = C_1 C_2 + C_2 C_3 + C_3 C_1$

A similar analysis of Fig. 4(b) yields

(1b) $\quad S^4(L_G L_S C_e^2) + S^3((RL)_e C_2 + g_m L_S L_G C_2) + S^2(L_G(C_1 + C_2) + L_S(C_1 + C_3) + g_m(RL)_e C_2$

$\quad\quad + R_G R_D C_e^2) + S(R_G(C_1 + C_2) + R_D(C_1 + C_3) + g_m(L_S + R_G R_D C_2)) + 1 = 0$

where $(RL)_e = R_D L_S + R_G L_S + R_D L_G$

These polynomials may now be evaluated for the existence of right half plane zeros by use of the Routh-Hurwitz criterion.

Theoretical Results

Characteristics typical of a 100 V, 10 A power MOS-FET were selected for use in evaluating (1). These characteristics are:

$$C_1 = 700 \text{ pF}$$
$$C_2 = 100 \text{ pF}$$
$$C_3 = 300 \text{ pF}$$

The Routh-Hurwitz criterion was then applied for values of g_m between 0.1 and 10 mhos, parasitic inductances between .05 and 50 nH, and parasitic gate resistances between .01 and 100 ohms. It was felt that the selected range of parasitic inductance encompassed both wiring inductance for package paralleled devices, and bond wire inductance for chip paralleled devices.

Behavior with $R_D = 0$

If the drain resistance, R_D, is assumed to be zero, it was found that for any combination of L_D, L_G, and g_m, some value of R_G could be found below which (1a) possesses a pair of rhp zeros. This minimum value of R_G necessary to assure no rhp zeros

of (1a) is plotted in Fig. 5 as a function of g_m for several combinations of L_D and L_G.

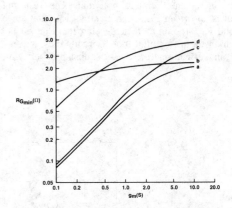

Minimum Value of R_G Necessary to Assure Stability of the Model of Figure 4(a) with $R_D = 0$, as a Function of g_m, for Various Values of L_D and L_G:
(a) $L_D = 5$ nH, $L_G = 5$ nH
(b) $L_D = 20$ nH, $L_G = 5$ nH
(c) $L_D = 5$ nH, $L_G = 20$ nH
(d) $L_D = 20$ nH, $L_G = 20$ nH
Figure 5

Figure 6 shows the results of solving (1b) (non-zero source inductance), again assuming $R_D = 0$. The interesting conclusion of this calculation is that for a ratio L_G/L_S greater than approximately 2.6, (1b) possesses no rhp roots.

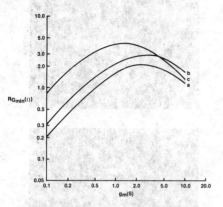

Minimum Value of R_G Necessary to Assure Stability of the Model of Figure 4(b) with $R_D = 0$, as a Function of g_m, for Various Values of L_G and L_S:

(a) $L_G = 5$ nH, $L_S = 5$ nH
(b) $L_G = 5$ nH, $L_S = 20$ nH
(c) $L_G = 20$ nH, $L_S = 20$ nH
Figure 6

It is clear that under the assumption $R_D = 0$, the model of Fig. 2 predicts unstable behavior for a wide range of practical values of the parasitic parameters.

Behavior with $R_D \neq 0$

Figure 7 shows the results of applying the Routh-Hurwitz criterion to (1b) for $L_G = L_S = 20$ nH. The minimum value of R_G necessary to assure no rhp roots is plotted as a function of g_m for various values of R_D. These curves show that as R_D increases, the region of g_m over which oscillations are possible decreases, until at some critical value of R_D no rhp roots will exist even if $R_G = 0$. For this specific example, this critical value is 1.9 ohms. Figure 8 shows the behavior of the range of g_m over

which rhp poles can exist as a function of R_D for $R_G = 0$. Since the assumed device capacitance parameters are typical for devices with widely differing $R_{DS(on)}$ specifications, the results of this analysis imply an aggravation of the oscillation problem for low voltage devices, where $R_{DS(on)}$ is typically fractions of an ohm.

Effect of R_S

In a system exhibiting a pair of rhp poles, the resulting oscillation will grow exponentially with a time constant proportional to its Q. The higher the Q, the slower will be the growth of the oscillation amplitude. Therefore, the duration of validity of the model of Fig. 2 is critical in determining whether these oscillations will be observed. In this respect, the gate drive source resistance, R_S, of Fig. 1 is important because it, together with the device input and reverse transfer capacitances, C_{iss} and C_{rss}, will determine the rise and fall times of the drain variables, and thus the duration of time spent in the saturation region. This is consistent with experimental results which have shown oscillations to appear in the presence of relatively large gate drive source resistance, but to disappear when the drive resistance was reduced.

Experimental Results

A series of experiments was performed to verify the qualitative aspects of the analytical results presented above. It is first shown that the oscillations appear primarily in the differential mode circuit. A critical value of the ratio L_G/L_S is observed, and the effect of R_D to restrict the range of g_m over which oscillations are possible is verified.

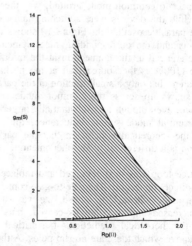

Theoretical Range of g_m Over Which Oscillations Persist, as a Function of R_D for $R_G = 0$ and $L_G = L_S = 20$ nH
Figure 8

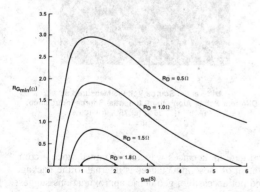

Minimum Value of R_G Necessary to
Assure Stability of the Model of Figure 4(b),
as a Function of g_m for Various Values of R_D
Figure 7

Figure 9 shows the schematic of the experimental circuit. A pair of IRF530's were placed adjacent to each other on a common heat sink. The pulsed gate source consisted of a DS0026 clock driver being driven by a 74LS123 monostable. The load resistor was a 2 ohm non-inductive Dale NH250 and the clamping diode was a fast recovery UES1403. The gate drive source resistor, R_S, was adjusted to ensure that the saturation region residence time was long enough to permit the oscillations to develop. In addition to switching tests, steady-state oscillatory behavior was observed by placing the devices in their saturation regions using a DC gate drive.

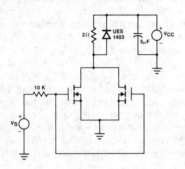

Circuit Used to Experimentally Verify Theoretical Conclusions
Figure 9

To verify that the oscillations are differential mode in nature, an explicit differential source inductance was created by connecting the source terminals together with a small loop of wire (r = 0.35″). Connection was then made to common mode ground from the middle of this loop with a relatively long (3″) piece of wire. The voltage waveform along the loop, with respect to common mode ground, was then observed while the devices were switching and exhibiting the parasitic oscillation. Figure 10 shows these waveforms and the corresponding measurement location referenced to the connection at the middle of the loop (100% is the source lead at the package). It is apparent that only a small fraction of the parasitic voltage shows up across the common mode source inductance, even though it is substantially larger than the differential mode source inductance, thus confirming the theoretical conclusion that the parasitic oscillation is a differential mode phenomenon.

The source loop was next removed and replaced by a short, direct connection. The gate-to-gate connection was then made with a loop whose size was successively increased until oscillations ceased, thus confirming the theoretical prediction of a critical ratio L_G/L_S above which there are no rhp poles. Although a quantitative measurement of the differential source

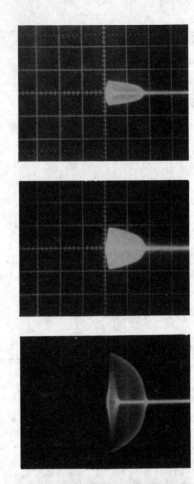

(a)

(b)

(c)

(d)

Differential Source Voltage Measured at Different Points Along the Differential Source Inductance; (a) 10%, (b) 25%, (c) 75%, (d) 100%
Figure 10

and gate inductances was not attempted since a component of these inductances is internal to the package and not accessible, geometric approximations suggest that the observed critical ratio was approximately 2.

The source and gate connections were next both made with short straight wires and the pulsed gate drive was replaced by a DC source. The gate voltage was then increased from zero until oscillations were observed and the corresponding drain current recorded. Increasing the gate voltage further eventually caused the oscillations to cease. The drain current at this point was also recorded. The transconductance, g_m, at these points was then obtained from measurements or the manufacturer's data sheets, depending on the current level. It was observed that the resulting range of g_m was sensitive to the applied drain voltage, V_{cc}. The range of g_m supporting oscillations was thus determined as a function of V_{cc} and is shown in Fig. 11.

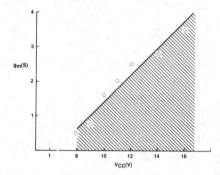

Experimental Range of g_m Over Which Oscillations Persist, as a Function of V_{cc}
Figure 11

Discussion

The experimental results reported above are essentially consistent with the theoretical analysis. The major discrepancy is between the theoretically predicted and experimentally observed ranges of g_m over which oscillations can occur. For the IRF530 the maximum specified value of $R_{DS(on)}$ is 0.18 ohms, which, from Fig. 8, implies a very large range of g_m based on the theoretical analysis. However, Fig. 11, which is based on measurements made for $R_G = 0$, shows that the observed range is quite restricted, especially for low values of drain voltage. The range also increases rapidly as the drain voltage is increased. There are two plausible explanations of this latter behavior. The first is that the lightly doped drain region of the device is partly depleted when $V_{DS} > 0$. This depleted region increases as V_{DS} increases, resulting in a reduction in the contribution of the n-drain region to the bulk drain resistance. The second is that the assumed values of the capacitances in the model of Fig. 2 were those for the devices operating deep in the saturation region where the capacitances are ap-

proximately constant. The measurements, however, were made at relatively low values of V_D where the manufacturer's data shows C_{iss}, C_{rss}, and C_{oss} changing quite rapidly as a function of V_{DS}. Analysis has shown that both of these phenomena will give rise to an increase in the range of g_m over which oscillations can occur as V_D is increased, a conclusion consistent with the qualitative aspects of Figs. 8 and 11.

The quantitative discrepancy between the analytical and experimental results can be attributed to several causes. The first is that the parameters for the model of Fig. 2 were measured at a frequency of 1 MHz, whereas the observed frequency of oscillation was 65 MHz. Although the distribution and value of the small signal energy storage elements are not likely to vary much between 1 and 65 MHz (because they are dominated by the MOS and junction structures, neither of which has strong frequency dependencies in this range), parasitic loss mechanisms within the device will likely exhibit a strong dependence upon frequency, with most of them producing losses which increase with frequency. For instance, the channel body resistance was not taken into account in the model, although Oxner has shown this to be an important component of R_G at high frequencies. [5] These additional parasitic losses will decrease the analytically predicted range of g_m over which oscillations may persist, as can be inferred from Fig. 7 for the case of an increasing R_G. An additional reason for the discrepancy is that measurements showed L_G and L_S to be approximately 50 nH, whereas the analysis resulting in Fig. 8 assumed values of 20 nH.

Conclusions

An incremental, differential mode circuit model has been used to analytically predict experimentally observed oscillations in paralleled power MOSFETs. The oscillations depend upon the differential mode parasitic gate, drain, and source impedances, and not upon the common mode elements most accessible to the circuit designer. The region of potential oscillation increases as R_D decreases, suggesting that the parasitic oscillation problem may be most severe for low voltage, high current devices. It is shown that the introduction of differential mode gate resistance can eliminate the potential for instability, and that increasing switching times can aggravate the tendency to oscillate.

These results are particularly significant for chip paralleled devices, for in this case the small parasitic inductance could give rise to very high frequency oscillations which could build up quickly relative to even "fast" switching times. In addition, these oscillations may not be observable because all the elements of the differential mode model of Fig. 3 are

5

inside the package. This suggests that devices to be paralleled at the chip level be fabricated using silicon gates of an appropriate sheet resistivity, or that small, discrete resistors be inserted in series with the gates of chips fabricated using metal gate technology.

For package paralleled devices, these results support the use of differential gate resistors [2,3] and gate drives which achieve fast switching times. In addition, the existence of a critical L_G/L_S ratio supports the use of ferrite beads on gate leads to increase L_G.

Acknowledgement

The authors are grateful to Kurt Ware and Tom Lee, who as undergraduate students contributed to initial aspects of this work. This research was supported in part by unrestricted grants from Lutron Electronics Co., and the Gould Foundation.

References

[1] HEXFET Databook, International Rectifier Corporation, 1981, p. 20.

[2] Sloane, T.H., H.A. Owen, Jr., and T.G. Wilson, "Switching Transients in High Frequency High Power Converters Using Power MOSFETs," IEEE PESC Record, 1979.

[3] Oxner, E.S., Power FETs and Their Applications, Prentice-Hall Inc., N.J., 1982, p. 129.

[4] Kassakian, J.G., "Some Issues Related to the Behavior of Multiply Paralleled Power MOSFETs," Proceedings of IPEC-Tokyo, March 1983.

[5] Oxner, E.S., "Meet the V-MOSFET Model," RF Design, Jan/Feb 1979, pp. 16–22.

5.3.4 Power FET Paralleling

Introduction

Paralleling of power MOSFETs is not always as easy as it may seem. When all the devices are fully enhanced (fully *ON*), the problem is minimal both because the tolerance of $R_{DS(on)}$ is small and because the positive temperature coefficient of FET conduction resistance tends to enforce proper current sharing through junction temperature differentials. During the switching period, however, current sharing can be a problem. If this problem is not attended to, the surge current in one of the devices during switching may exceed the ratings of the device. This is especially true for high-frequency high-pulse-current applications. In this paper, the causes of dynamic current imbalance will be presented and recommendations for alleviating dynamic current imbalance will be suggested.

Dynamic Current Sharing

The causes of dynamic current imbalance can be classified into two categories: device related and circuit related. Device parameters that directly affect dynamic current imbalance include gate threshold voltage (V_T), transconductance (g_{fs}), and input capacitance (C_g). Circuit parameters critical to dynamic current imbalance include the parasitic inductances of both the gate-drive circuit (L_G) and the power circuit (L_p) as well as the gate-drive circuit source resistance (R_G). The reason both V_T and g_{fs} are involved in current imbalance during switching should be obvious. Devices with lower V_T and/or larger g_{fs} will carry more current at any V_{GS} in their linear region. Gate parameters C_{iss}, R_G, and L_G affect the time required to charge the gate capacitance from a gate-voltage source with a fixed impedance. A FET with a gate-drive circuit that has a smaller time

constant will take more than its share of current during switching.

Device Parameter Values

Typical gate-to-source threshold voltages $V_{GS(th)}$ for commercially available devices range from 2 to 4 V. Transconductance, g_{fs}, depends on the device current rating. In general, the larger the current rating, the larger the g_{fs}. For 5-A devices, the values range from 1 to 3 S. For 10-A devices, they range from 5 to 10 S. The magnitude of gate capacitance C_{iss}, which is the sum of gate-source and gate-drain capacitances, depends on the chip size. The higher the voltage and current ratings of the device, the larger the chip size, and, consequently, the larger the gate capacitance. Typical values of C_{iss} for commercially available devices range from approximately 5000 pF for a 500 V, 20 A device to 150 pF for a 100 V, 3 A device. The tolerance of C_{iss} for a given device is normally within $\pm 30\%$.

Temperature Effects

Device gate-source threshold voltage decreases with temperature. A typical device threshold decreases aproximately 5 mV per °C. Transconductance, g_{fs}, also decreases with temperature. The temperature coefficient is -2% per °C.

It can be seen from the above characteristics that if one of the paralleled devices heats up due to current imbalance, the negative temperature coefficient of V_T tends to aggravate while the negative temperature coefficient of g_{fs} tends to alleviate current sharing problems during switching. This is different from conduction state current sharing in which both the positive temperature coefficient of channel resistance $R_{DS(on)}$ and the negative temperature coefficient of g_{fs} contribute to reduction of current imbalance.

Recommendations for Controlling Current Sharing Problems

There are basically two approaches to control dynamic current imbalance: by means of device parameter matching and by means of circuit techniques.

Device Parameters Matching

Dynamic current imbalance can be minimized by matching the three device parameters described in the previous section, i.e. V_T, g_{fs}, and C_{iss}. The sensitivity of current imbalance to each parameter is different, however. Of the three parameters mentioned, matching of g_{fs} and C_{iss} is less critical. For transconductance mismatch, the worst case current imbalance is limited to the percentage of tranconductance mismatch. Furthermore, if a device heats up due to mismatch, the negative temperature coefficient of transconductance tends to reduce the imbalance. The effect of input capacitance mismatch depends on the external gate-drive impedance as described in a following paragraph.

Of the three device parameters to be matched, the most critical one is the gate threshold voltage V_T. The amount of current imbalance due to V_T mismatch is essentially unlimited. The negative V_T temperature coefficient further worsens dynamic current imbalance.

Circuit Means for Controlling Imbalance

There are several circuit techniques recommended for minimizing dynamic current imbalance. All, however, result in some additional alterations to circuit performance. Thus, trade-off decisions must be made by the circuit designer.

Gate Drive Circuit

The optimum circuit technique for improving current sharing during switching is merely to increase switching speed. By increasing the gate–drive circuit speed, one can reduce the time required to charge and discharge the gate capacitance and thus reduce the

duration of current imbalance during switching. To increase the drive circuit speed, one should reduce the gate-drive impedance (see Figure 1). It should be noted, however, that extremely small values of R_G may not be enough to damp undesirable oscillation in the paralleled gate circuits. Parasitic inductances around the gate circuit must also be minimized.

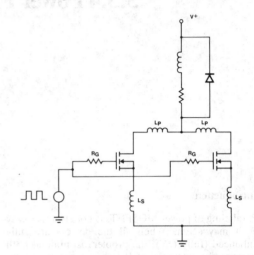

FET Switching Circuit with Two Devices in Parallel
Figure 1

Conclusions

The causes of current imbalance in paralleled MOSFET switching power circuits have been examined, and some techniques have been recommended for improving imbalance. It has been shown that some parameters are more critical than others in affecting the imbalance. Threshold voltage V_T mismatch, for example, can be very critical to dynamic current sharing. The amount of current imbalance depends both on device parameters and circuit parameters. Reference 1 gives several numerical examples of the degree of current imbalance for some specific power MOSFETs.

Reference

Forsythe, James, "Techniques for Controlling Dynamic Current Balance in Parallel Power MOSFET Configuration," 1983 Powercon 8.

5.4 dV$_{DS}$/dt Turn-On in MOSFETs (TA84-4)

Introduction

Under certain conditions, when a rapidly rising voltage waveform is applied from drain to source of a supposed "off" MOSFET, the device will turn-on. dV/dt turn-on is well known in SCRs and bipolar junction transistors (BJTs). This phenomenon in MOSFETs is similar in many respects except that dV/dt turn-on occurs at much higher rates than it does in either SCRs or BJTs. Because spurious turn-on can be destructive, it is of serious concern in those applications where it might occur.

Several different modes of dV$_{DS}$/dt turn-on have been identified, and in most cases, turn-on may be eliminated or the effect of turn-on may be reduced to an acceptable level.

To eliminate this problem, the mechanisms of the turn-on modes, the consequences of spurious turn-on, the means for reducing or eliminating turn-on, and the applications where turn-on is likely to be a problem, must be understood. The following discussion intends to provide the basic information on dV$_{DS}$/dt turn-on needed by the design engineer.

Equivalent Circuit Model

Figure 1 is an equivalent circuit model for a power MOSFET which includes the parasitic elements within the device and within the drive circuit (R$_D$, L$_D$, V$_d$).

The parasitic NPN BJT, an integral part of the device structure, is present in all VMOS and DMOS devices. To minimize the effect of the BJT, R$_b$ is made as small as possible, and H$_{FE}$ is low. Nonetheless, in some circumstances the BJT can be turned on.

The impedance of the drive circuit (Zgs) may take many forms, but the series R, L, battery is representative. R$_D$ comes from several sources: resistance deliberately placed in series to limit the peak current or reduce the rise time of V$_{gs}$, bypass capacitor ESR, wiring resistance, the on resistance of drive switches, etc. L$_D$ is due to the physical size and layout of the drive circuit. If a transformer is used for drive isolation, the primary-to-secondary leakage inductance will add to L$_D$. When series diodes or BJTs are used in the

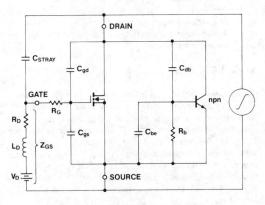

N Channel MOSFET Equivalent Circuit
Figure 1

drive circuit, an equivalent series battery (V$_D$) is present. In some cases, a DC bias may be present, so V$_D$ could be either positive or negative. C$_{stray}$ represents the parasitic capacitance from drain to gate due to the circuit layout. This capacitance may be as large or larger than the intrinsic C$_{gd}$ if careful layout is not used.

R$_G$ is the internal gate resistance. The value of this resistance varies from 0.05 to 5.0Ω depending on the manufacturer and the size of the device. Siliconix devices typically run from 0.5 to 4.0Ω.

The intraterminal capacitances of a MOSFET are functions of V$_{DS}$ as shown in Figure 2*. Because Cgd and Cds are functions of V$_{DS}$, spurious turn-on will depend on both dV$_{DS}$/dt and V$_{DS}$, which complicates the analysis. In the following discussion, the body-drain diode will be mentioned. This diode is the base-collector junction of the parasitic BJT.

* NOTE: Because C$_{be}$ >> C$_{db}$, C$_{db}$ ~ C$_{ds}$

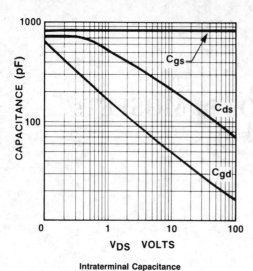

Intraterminal Capacitance
Figure 2

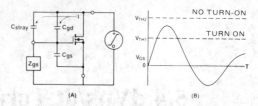

Mode 1 Equivalent Circuit
Figure 3

The following modes of dV_{DS}/dt turn-on have been observed:

1. Device quiescent: Z_{gs} high, MOSFET turns on.
2. Device quiescent: Z_{gs} low, BJT turns on.
3. Body-drain diode conducting: Z_{gs} low, BJT turns on.
4. Body-drain diode conducting: Z_{gs} low, BJT goes into avalanche breakdown.

Mode 1 Turn-On

The equivalent circuit for mode 1 turn-on is given in Figure 3 along with a generalized waveform for V_{gs}. The exact nature of the V_{gs} waveform will be a function of Z_{gs} and C_{stray}, but in general, the lower the impedance presented by Z_{gs} and the lower the value for C_{stray}, the lower the peak value of V_{gs}. If V_{gs} exceeds V_{th}, the device will turn on. In most applications, the amplitude of I_D, during turn-on will be limited by $R_{DS(on)}$. This mode of turn-on is not usually destructive, although it will increase the power losses. Figures 4 and 5 show actual turn-on waveforms when Z_{gs} is resistive.

Even when the FET is not turned on, a pulse of drain current will be present due to the intraterminal capacitances. An idealized waveform for this case is shown in Figure 6A, and a more realistic waveform is shown in Figure 6B. There will always be some series inductance which, when combined with the nonlinear capacitances, produces a distorted, damped sinusoid.

This current pulse will be reflected into the switch which is generating the positive dV_{DS}/dt.

One additional complication to mode 1 turn-on has been suggested but not observed. If the FET is turned on, it is possible for parasitic oscillations to occur. This would raise V_{gs} and increase the power dissipation. It is possible for the gate to be ruptured if V_{gs} becomes large; however, if oscillation is not present during normal switching, it is unlikely that it will occur during mode 1 turn-on.

The best way to detect potential mode 1 turn-on is to observe the gate-to-source voltage waveform with an oscilloscope. If a significant voltage spike appears during the drain voltage transition then some positive action will have to be taken to reduce the pulse amplitude.

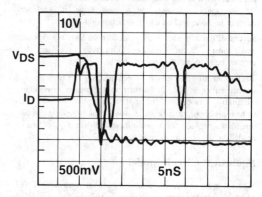

Mode 1 Spurious MOSFET Turn-On. Rg = 10 ohms
Figure 4

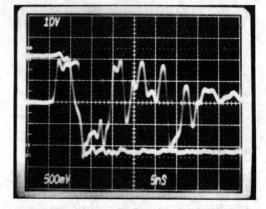

Mode 1 Spurious MOSFET Turn-On. Rg = 100 ohms
Figure 5

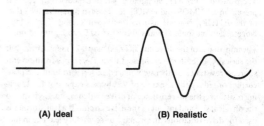

(A) Ideal   (B) Realistic

Pulse Drain Current Waveforms
Figure 6

Mode 2 Turn-On

The equivalent circuit for mode 2 turn-on is given in Figure 7, along with V_{DS} and I_D waveforms. The mechanism for this mode of turn-on is quite simple. As V_{DS} slews, current will flow through C_{db}. If enough current flows so that the voltage across R_b exceeds ~0.65 V, some of the current will flow into the base, turning on the BJT. In contrast to mode 1, the acceptable dV_{DS}/dt rate is only a function of the device design, not of the external circuit elements.

The result of mode 2 turn-on can be harmless or catastrophic depending on V_{DS} and the magnitude of the base current.

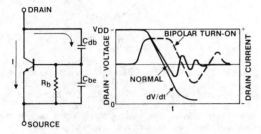

Mode 2 Turn-On Equivalent Circuit and Typical Waveform
Figure 7

In a BJT, the collector-emitter breakdown voltage is a function of base-emitter impedance and of the current flowing in the device. This point is illustrated in Figure 8. For low values of R_b and zero base current injection, BV_{DSS} is equal to BV_{CEX} for the BJT. This is the data sheet voltage rating for the device. Once base current is injected, however, the breakdown voltage drops dramatically to a level $< BV_{CEO}$, which may be as low as $\frac{1}{2} BV_{CEX}$. If V_{DS} is $> BV_{CEO}$ and base current is injected, the device will go into avalanche breakdown. Furthermore, if I_D is not limited by the external circuit, the device can go into secondary breakdown and will be destroyed.

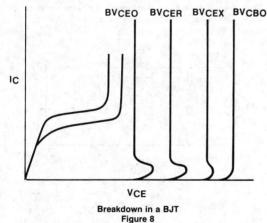

Breakdown in a BJT
Figure 8

So much for the bad news. Now for the good news. In modern MOSFETs, the dV_{DS}/dt rate at which mode 2 turn-on can be initiated is generally very much higher than most applications will ever see. The dV_{DS}/dt limits for Siliconix devices are shown in Figure 9. Realize that the rates shown correspond to transition times of one to three nsec. Very few applications require transitions this rapid. Even testing at these rates is a formidable problem.

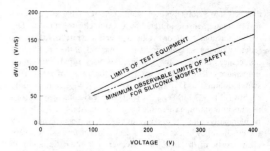

dV_{DS}/dt Limits in Power MOSFET
Figure 9

Mode 3 and Mode 4 Turn-On

Frequently the body-drain diode is used as a catch diode for an inductive load. In a few applications, such as sinewave synthesis or some motor control schemes, the positive dV_{DS}/dt is applied while the diode is conducting. Figure 10 shows the equivalent circuit for this condition. Just prior to the application the voltage ramp, I_R is flowing through the BJT in a reverse direction. Much of the current will flow through R_b and the collector-base diode. Even in the inverted mode, however, the BJT will have some current gain, and a portion of I_R will flow in reverse through the emitter, as indicated. As a result of this current flow, the device is saturated with charge, i.e., it is hard on at the moment the dV_{DS}/dt is applied.

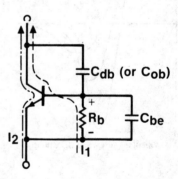

Mode 3 Equivalent Circuit
Figure 10

Typical voltage and current waveforms for mode 3 operation, with and without turn-on, are shown in Figure 11. Figure 11A shows the waveforms for the case where the BJT does *not* turn-on and

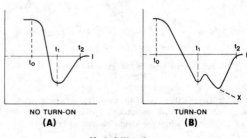

NO TURN-ON
(A)

TURN-ON
(B)

Mode 3 Waveforms
Figure 11

thus acts like a normal *p-n* junction diode. Looking at the waveforms, the sequence is initiated by a switch closure at t = t_0. The current through the diode begins to reverse at a rate (dI/dt) which is determined by the switch transition time and/or the circuit series inductance. If the switch transition is very rapid, the circuit inductance will dominate; conversely, if the transition is slow, the switch dominates. During the interval t_0-t_1, V_{DS} changes only slightly (–1 to +1 volt). V_{DS} does not change significantly until I_D reaches its maximum value and begins to fall. At t = t_1, V_{DS} begins rising rapidly. The rate of dV_{DS}/dt at this point is determined by the circuit series inductance, the peak current *and* the recovery characteristic of the diode. The recovery characteristic is in turn a function of the initial dI/dt. This relationship is illustrated by Figure 12. The more rapid dI/dt, the greater will be the peak amplitude of I_D, Q_{rr} will be larger and trr shorter. More important, the dI_D/dt during the interval t_1-t_2 becomes greater as the initial dI/dt is increased. In some diodes, dI_D/dt can increase dramatically and can form a sharp step or snap as indicated in Figure 13. At lower levels of initial dI_D/dt, the step may not be present, but as the initial dI_D/dt is increased, the step appears. This current step during recovery can generate extremely high rates of dV/dt in the circuit.

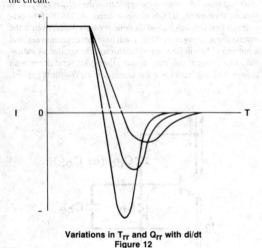

Variations in T_{rr} and Q_{rr} with di/dt
Figure 12

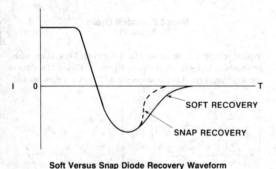

Soft Versus Snap Diode Recovery Waveform
Figure 13

A further complication exists: the diode recovery characteristics are also influenced by the magnitude of I_D during reverse conduction. This relationship is shown qualitatively in Figure 14.

Figure 11B shows the waveforms when turn-on is present. Two cases are shown: recovery without damage and catastrophic

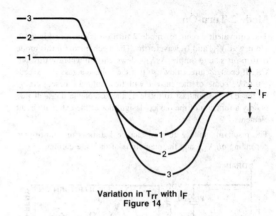

Variation in T_{rr} with I_F
Figure 14

failure. Notice that the BJT turn-on occurs during the interval t_1-t_2 and is triggered by the dV_{DS}/dt resulting from the recovery characteristic of the diode. The dV_{DS}/dt during the interval t_1-t_2 is related to the switch transition time, but it is *not* a direct function. This is very different from modes 1 and 2 where the current waveforms are coincident with and directly proportional to the transition time.

Mode 3 turn-on is an indirect function of transition time and is more directly controlled by the circuit and the recovery characteristic of the device. In terms of rating the device for dV_{DS}/dt capability, the appropriate voltage transition is not the switch transition but rather the transition which appears across the diode during the second portion of the recovery interval.

This complex interaction deserves a more detailed examination. The following discussion uses a successive approximation to explain what happens during mode 3 operation.

An equivalent circuit for mode 3, where turn-on does not occur and the device can be viewed as a diode, is given in Figure 15. Note that the inductive load current is assumed to be constant during the recovery interval and is represented by a current source. L_1 and L_2 are the circuit parasitic inductances.

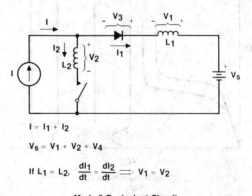

$$I = I_1 + I_2$$
$$V_S = V_1 + V_2 + V_4$$
$$\text{If } L_1 = L_2, \quad \frac{dI_1}{dt} = \frac{dI_2}{dt} \implies V_1 = V_2$$

Mode 3 Equivalent Circuit
Figure 15

The reverse recovery current waveform (I_1) can be approximated by triangular waveforms as shown in Figure 16. From this approximation, the voltage (V_3) across the diode is easily determined, as shown. The critical point in the waveform is the rapid and large positive transition which occurs when the waveform

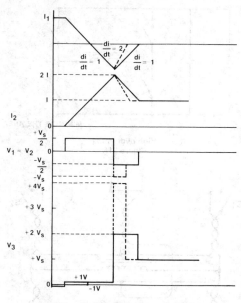

Triangular Waveform Approximation
Figure 16

di/dt reverses. The amplitude of the transition is a function of di/dt during recovery—a diode characteristic. A diode with a snap characteristic would generate a very rapid, high amplitude voltage spike.

Figure 16 indicates an infinite dV_{DS}/dt, which is not what is observed. A better approximation would be to use a trapezoidal waveform as indicated in Figure 17. Figure 17 also shows the actual waveforms. From this figure, we see that the large voltage transient still exists. Its amplitude is essentially the same, but the dV_{DS}/dt is finite. We can see that the reverse recovery characteristic of the diode determines both dV_{DS}/dt and the amplitude of the transients.

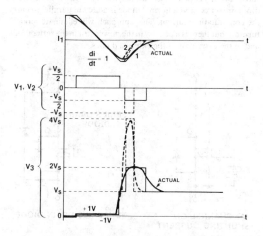

Trapezoidal Waveform Approximation
Figure 17

The device could respond to the voltage transient in several ways:
1. If the amplitude is small and dV_{DS}/dt moderate, the device recovers normally.
2. Mode 3 turn-on could occur without avalanche breakdown.
3. Mode 4 turn-on could occur; i.e. the device avalanches before the BJT turns on.
4. Mode 3 turn-on could occur, lowering BV_{DSS}, leading to avalanche breakdown and perhaps second breakdown.

Each of these turn-on scenarios is potentially destructive.

The complexity of the interaction between the parameters makes it difficult to characterize FETs for mode 3 and 4. Figure 18 displays the results of preliminary testing of Siliconix devices. Competitive devices were also tested and found to be no better at best, and much worse in some cases. While this curve is hardly definitive, it does show that modes 3 and 4 turn-on levels are an order of magnitude lower than mode 2. The dI/dt levels used in the testing were in the range of 150 A/usec. This is representative for motor control or low frequency AC synthesizer; however, LF transmitters and sonar transducer drivers would be expected to have much higher dI/dt rates which would further reduce the acceptable dV/dt.

Fortunately, mode 3 operation is restricted to a few special applications such as:
1. Motor drives and sinewave inverters where multiple repetitive pulsing of the switches occurs.
2. Highly reactive resonant loads, such as found in LF transmitters, sonar transducer drivers, and induction heating equipment.
3. Synchronous rectifiers where the body-drain diode is allowed to conduct.

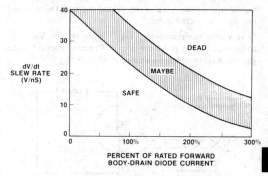

Mode 3 Performance
Figure 18

The Effect of Radiation and Tj on dV_{DS}/dt

The effects of Tj and radiation on dV_{DS}/dt behavior have not been discussed. While little testing has been done, trends can be identified. Vth has a negative temperature coefficient of four to eight mV/°C, so that at high temperatures Vth is lower. This will lower the rate at which mode 1 turn-on occurs. At higher temperatures, V_{be} will decrease in the BJT, and the gain will increase. Both of these effects will reduce the acceptable dV_{DS}/dt rate for modes 2 and 3.

Ionizing radiation has two effects for total doses above 10KRad(Si). First Vth is reduced, and second the gain and diode recovery time of the BJT are reduced. This implies that mode 1 performance is degraded, but modes 2 and 3 should be improved.

5

How to Avoid dV_DS/dt Problems

In most circuits, dV_{DS}/dt turn-on may be avoided. There are many ways to accomplish this:

1. Reduce dV_{DS}/dt
2. Low impedance drive circuits
3. Good circuit layout practice
4. Negative gate bias
5. Series drain diode
6. Snubbers
7. Turn-on of the FET during commutation
8. Use MOSFETs with good dV/dt performance
9. Current fed topologies

The most obvious way to reduce dV/dt problems is to slow the switching speed. This can be accomplished easily by adding resistance in the gate as shown in Figure 19. The diode is added to provide a low impedance when the device is off so that mode 1 problems do not arise.

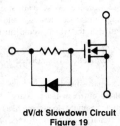

dV/dt Slowdown Circuit
Figure 19

The best means to avoid mode 1 turn-on is to use a low impedance driver and good circuit layout. Figure 20 illustrates how to design a good drive circuit. It is also very helpful to reduce the noise and ringing in the overall circuit.

In very noisy environments, it may be necessary for Vgs to be negative during the switch off interval. The noise coupled to the gate will have to overcome a potential of $V_{th} + V_{bias}$.

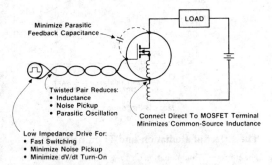

Circuit Layout to Reduce Mode 1 Turn-On
Figure 20

Figure 21 shows how to lay out an H-Bridge switch-mode power converter primary to minimize noise, ringing, and coupling from one point of the circuit to another. The principle is quite simple; first identify those paths in the circuit where the switched currents are flowing and then minimize the inductance and radiating area of these current loops. These techniques will not only reduce dV_{DS}/dt turn-on but will be beneficial in reducing EMI, eliminating voltage spikes and generally cleaning up the circuit waveforms.

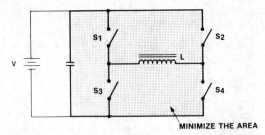

High di/dt Paths
Figure 21A

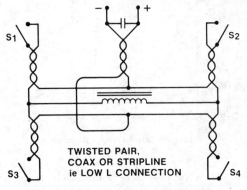

TWISTED PAIR, COAX OR STRIPLINE ie LOW L CONNECTION

Low Inductance Connections
Figure 21B

Mode 3 performance may be improved by reducing or eliminating the current in the body-drain diode. Two ways to do this are shown in Figure 22. In (A) the FET channel is turned on to divert current away from the body diode. This can be very effective as shown in Figure 23. Figure 23 is a superposition of the FET and diode conduction characteristics in a VNE003A. In this particular device, the diode would not start conducting until $I_D > 30$ Amps.

Another method which eliminates diode conduction entirely is shown in (B). A low voltage diode is placed in series with either the drain or source, and a normal diode is added in parallel to carry the commutation current. This approach will eliminate mode 3 turn-on, but there is a penalty in the power cost in the series diode and in the extra cost of the external diodes.

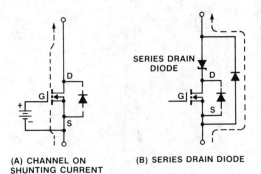

(A) CHANNEL ON SHUNTING CURRENT

(B) SERIES DRAIN DIODE

Schemes for Reducing Body-Drain Diode Conduction
Figure 22

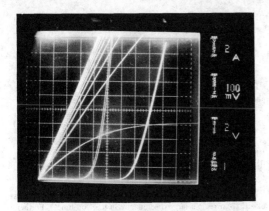

Body Diode/Forward Conduction Comparison
Figure 23

In extreme cases, snubbers may be used to reduce the dV_{DS}/dt, the transient amplitude and to limit current pulses if turn-on should occur.

The effect of dV/dt turn-on will depend on the impedance of the circuit in which the device is being used. For example, the circuit shown in Figure 24A is a voltage fed switch-mode power converter. When S1 is closed, a positive dV_{DS}/dt will appear across S2. If S2 is turned on then a short circuit will appear across the transformer winding. There is little to limit the current because the source impedance is very low. Given the same circumstances in Figure 24B, the series inductor will limit the rate of current rise. The probability of damaging the switches will be reduced.

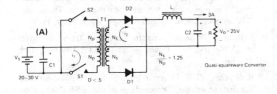

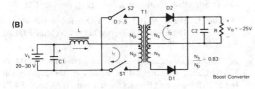

Voltage and Current Fed Converters
Figure 24

Conclusions

dV_{DS}/dt turn-on does exist in power MOSFETs, so the following conclusions may be drawn:

1. Mode 1 is primarily a circuit problem which can best be eliminated by using low impedance drives and good circuit layout techniques.
2. Mode 2 occurs only for very high values of dV_{DS}/dt; therefore, this is seldom a problem.
3. Modes 3 and 4 are potentially the most troublesome but occur only in specialized applications. Modes 3 and 4 may be reduced or eliminated with circuit modifications.

Appendix

Another scenario for mode 2 turn-on has been postulated but not observed. Figure 25 shows a simplified boost converter where the MOSFET is represented by a perfect switch in parallel with the intrinsic BJT. When the switch is opened, some of the inductor current (I_L) will flow into C_{db}. If I_L is large enough, it is conceivable that the BJT could be turned on. The effect on the turn-off waveform would be similar to storage time in a BJT, i.e. extended turn-off time.

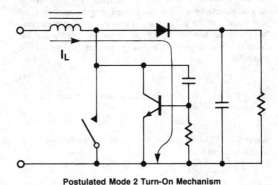

Postulated Mode 2 Turn-On Mechanism
Figure 25

Conceivably, several turn-off pulses could occur as the BJT turns on stealing base current. Like all mode 2 turn-on, this effect will only be possible for very rapid transitions.

A mode of anomalous turn-off ringing has been identified by Giandomenico [6]. This ringing could easily be mistaken for the mode turn-on just discussed or for parasitic oscillation. The ringing occurs when there is excessive unclamped drain inductive and is described by Giandomenico, et al, as follows:

"When the power MOSFET inverter circuit shown in Figure 26 is suddenly turned off, the drain voltage rises quickly and then abruptly drops to the "on-voltage" where it remains for several

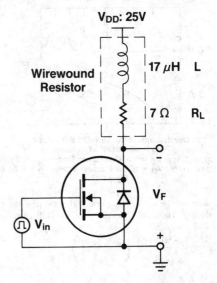

The Power MOSFET Connected in an Inverting Configuration
Figure 26

microseconds before ringing and settling to the supply voltage (see Figure 27A). This behavior was initially thought to be due to dV/dt breakdown and was thought to be related to the "on-state" oscillations. However, this turn-off behavior is in fact due to the clamping diode between the drain and source which is intrinsically incorporated into the vertical MOSFET structure. A typical display of the drain voltage and current during turn-off is shown in Figure 27B from which the turn-off phenomenon can be explained.

"When the device is turned off, the magnetic energy stored in the load inductance is transferred into the output capacitance and then returned to the inductor with the opposite current polarity. Neglecting the non-linearity of the output capacitance and the damping of the output load resistance, the time for the current reversal is:

$$t = \pi \sqrt{LC} \tag{1}$$

"After this time, the parasitic diode is forward biased, and the voltage across the inductor is:

$$V_L = V_{DD} + V_F - I_D R_L \tag{2}$$

where V_F is the forward diode voltage. If the damping losses are neglected, then:

$$I_D \approx -V_{DD}/R_L \tag{3}$$

and

$$V_L \approx 2V_{DD} + V_F \tag{4}$$

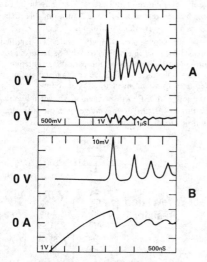

Oscillograph of the Turn-Off Behavior of the Power MOSFET Inverter
a) Upper Trace: V_D = 10V/Div; Lower Trace: V_{gs} = 5V/Div.
b) Upper Trace: V_D = 10V/Div; Lower Trace: 1_D = 0.2A/Div.
Figure 27

"The drain voltage remains near zero during the time the diode is conducting. The current through the diode is dictated by the L-R circuit and is:

$$I_D = \frac{2V_{DD} + V_F}{R_L} \left(1 - e^{-t\,R_L/L} \right) - \frac{V_{DD}}{R_L} \tag{5}$$

"Neglecting reverse recovery, the time at which the diode current is zero is:

$$t_{on} = \left(\frac{L}{R_L} \right) \ln \left[\frac{2\,V_{DD} + V_F}{V_{DD} + V_F} \right] \tag{6}$$

"For a reasonably large supply voltage, $V_{dd} >> 0.7$ volts, the above equation can be expanded in a Taylor series to give:

$$t_{on} = \left(\frac{L}{R_L} \right) \left[1 - \frac{V_F}{2V_{DD}} \right] \ln 2 \tag{7}$$

"Neglecting the reverse recovery of the diode for the moment, the drain voltage rises at this time and then rings and settles to the supply voltage.

"A comparison of experimental data with the calculated results of eq. (7) reveals a large discrepancy in the "on-time." Further, the oscillographs of Figure 27 show that the power MOSFET appears to turn on a third time. These discrepancies are due to the reverse recovery of the diode (see Figure 27B). Consequently, the total turn-on time expressed by equation (7) must be modified to include the reverse recovery time, trr:

$$t_{on} \approx \frac{L}{R_L} \left[1 - \frac{V_F}{2V_{DD}} \right] \ln 2 + trr \tag{8}$$

"Since, after the diode's reverse recovery, the inductor once again has a downward current direction, the situation is the same as when the MOSFET first turned off but with a lower initial current level, and thus the cycle repeats again."

Bibliography

1. J. Redonfey, "The Dv/DT Phenomenon in Power Transistors," The power transistor in its environment, Thompson-CSF, Semiconductor Division, Applications book, 1978.

2. J. Rockot, "Reverse Transistor Action In Transistor Inverters," Westinghouse Power Semiconductor Division, Tech Tip, pp. 1-13.

3. R. Severns, "dV/dt Effects in MOSFET and Bipolar Junction Transistor Switches," Proceedings of the IEEE Power Electronics Specialists Conference, June 1981, Boulder CO.

4. D. Kuo, C. Hu & M. Chi, "dV/dt Breakdown in Power MOSFETs," IEEE Electron Device Letters, Volume EDL-4, No. 1, January 1983.

5. W. Slusark et al., "Catastrophic Burnout in Power VDMOS Field-Effect Transistors," Proceedings of the 21st annual IEEE, Reliability and Physics Symposium, 1983, pp. 173-177.

6. D. Giandomenico, et al., "Analysis and Prevention of Anomalous Oscillations and Turn-Off Behavior in a Vertical Power MOSFET," private report done for Siliconix Inc., later published in the proceedings of POWERCON 11, April 1984, Dallas.

7. E. Oxner, "Static & Dynamic dV/dt Characteristics of Power MOSFETs," PCI Proceedings April, 1984, pp. 132-43.

5.5 Inverse Diodes of Power MOSFETs

Introduction

A parasitic anti-parallel diode is inherently built in the process of fabricating any power MOSFET transistor. This diode has voltage and current ratings equal to the FET, but has, in general, a slow reverse recovery. Depending upon circuit application, the parasitic diode could be a bonus or a problem, for the user. The slow recovery speed of the diode may result in a large recovery current spike and cause high power dissipation in an opposing FET switch. In this paper, first the formation of the parasitic diode will be described, then the problems associated with the diode in power circuits will be explained. Remedies for the problems will also be suggested.

Formation of Parasitic Diode

A DMOS device structure is used to explain the formation of a parasitic diode in the power MOSFET. As shown in Figure 1, source metalization overlaps both the p and the n^+ regions. The reasons for this are, first, to stabilize the voltage of the p region so it does not drift when the device switches and, second, to suppress the action of the parasitic bipolar junction transistor in the structure thereby ensuring proper operation of the MOSFET. This metallization, however, connects an anti-parallel pn diode from source to drain. All MOSFET device structures, such as VMOS and HEXFET, have similar diode formation and, therefore, power MOSFETs will not block reverse voltage.

Because the parasitic diode uses the same Epi and p regions as the MOSFET, the voltage rating of the diode is the same as that of the companion MOSFET. The diode current rating is also the same as that of the MOSFET. Therefore, the parasitic diode is often

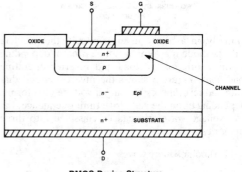

DMOS Device Structure
Figure 1

considered to be a bonus for the user. However, the reverse recovery speed of the diode is normally between 200 and 400 nanoseconds. This can present problems in some circuit applications.

Problems of Using Parasitic Diodes in the Circuit

In a power circuit with a single FET, the parasitic diode is seldom used and seldom presents any problems. However, in a power circuit with a loop consisting of two serially connected FETs and a voltage source, the diode may present a problem. Commonly used half bridge and bridge inverters are examples. Under certain operating conditions with these converters, the recovery of the diode causes problems as will be explained below.

Figure 2 shows a half bridge in which two FETs are used. In first quadrant operation, inductive current free-wheels through D_2 and the load when Q_1 turns

OFF. When Q_1 turns on again, there is a temporary short circuit formed by the loop of V_s, Q_1, and D_2 during the reverse recovery time of D_2. This will cause a large power spike in Q_1 and D_2 and may damage Q_1. The reverse recovery time of D_2 depends, among other parameters, on the amount of forward current flowing at the instant Q_1 turns ON. If a discontinuous mode of operation is chosen, there will not be current flowing in D_2 when Q_1 turns ON; therefore, the problem described will not exist. A similar situation, of course, applies to Q_2 and D_1.

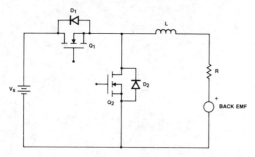

Two-Quadrant DC Chopper Circuit
Figure 2

Similarly, in a bridge inverter circuit, the problem with the diode may or may not be present depending on the nature of the load and inverter operating conditions. Figure 3 shows the inverter circuit diagram and several possible output voltage waveforms. Depending upon the gate drive timing sequence, the load voltage waveform can be classified into three categories:

1. Bipolar square wave

2. Unipolar square wave

3. Quasi-square wave

An analysis of whether the diode will present problems will be given for each of the above three cases. Generally speaking, however, if the inverter circuit requires commutation of current from a diode to the opposing FET in the same totem pole then a severe current spike will occur. If the current is commutated from a diode to its companion FET then the problem will not occur. This is because the diode in the first case is subjected to reverse bus voltage, but in the second case, it is subjected to essentially zero voltage when the diode current is terminated. The next paragraph will discuss the load conditions under which the diode problem will exist for the three basic load voltage waveforms mentioned.

Figures 4, 5, and 6 show the load current waveforms and associated switch timing for different loads. Figure 4 shows the waveforms for the bipolar square wave output voltage. As can be seen, for R-L loads,

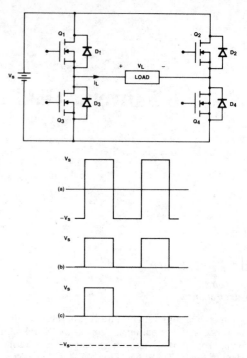

Bridge Inverter and Three Basic Output Voltage Waveforms
(a) Bipolar Square Wave
(b) Unipolar Square Wave
(c) Quasi-Square Wave
Figure 3

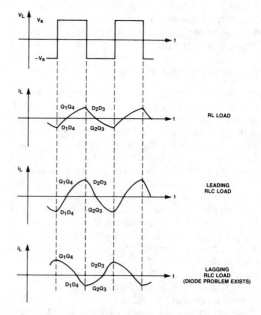

Bridge Inverter Load Current Waveforms for the Case of
Bipolar Square Wave Load Voltage
Figure 4

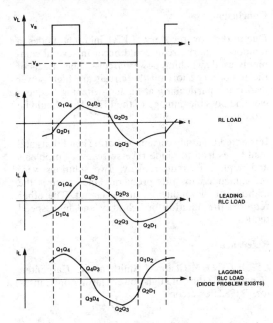

**Bridge Inverter Load Current Waveforms for the Case of
Quasi-Square Wave Load Voltage
Figure 5**

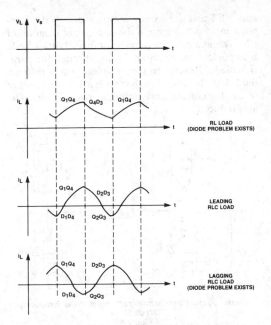

**Bridge Inverter Load Current Waveforms for the Case of
Unipolar Square Wave Load Voltage
Figure 6**

diode current is always commutated by the companion FETs and, therefore, a diode recovery problem would not occur. For an R-L-C load, the current waveform could either lead or lag the voltage waveform. If the natural frequency of the RLC is higher than the inverter switching frequency then the current waveform will lead the voltage waveform. Figure 4, shows that the diode problem will not exist in this case because diode current is commutated by the companion FET. If, however, the current lags the voltage, which means the natural frequency of the RLC load is less than the inverter switching frequency, then the diode current will be commutated by the FET on the opposing part of the totem pole, and the diode recovery problem will occur. This has direct ramifications on the diode requirements for any resonant bridge converter.

Figure 5 shows the waveforms for the case of unipolar square-wave output voltage. For an R-L load, the load current may be continuous and, when it is, the diode current will be commutated by the opposing FET, and the diode recovery problem will occur. If the load current is discontinuous then the diode problem will not occur. Again, for the R-L-C load, only the lagging case presents problems. This is similar to the situation in the case of the bipolar voltage waveform. Figure 6 shows the waveforms for a quasi-square wave output voltage. The situation is similar to Figure 4. Among the three load conditions discussed, only the RLC "lagging" condition presents problems.

Circuit Remedy

Since the diode is inherent in the MOSFET and its recovery time is much longer than the FET's switching time, a circuit remedy must be found in order to use the FET effectively. Figure 7 shows a scheme using a center-tapped inductor in each totem pole to limit the current surge during the diode reverse recovery time. This arrangement requires additional free wheeling diodes around this inductor to release inductive energy when the FET is cut OFF. Another circuit remedy is to use a Schottky diode, DS, in series with

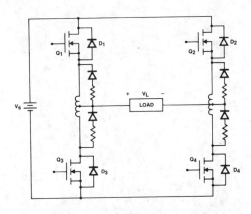

**Inverter Circuit Using Center Tapped Inductors to Reduce
Temporary Current Spike
Figure 7**

each FET and use a fast recovery diode as the free-wheeling diode. Figure 8 shows a circuit diagram of this arrangement. The purpose of the Schottky diode is to prevent current from flowing through the parasitic diode. Reverse current now has to flow through diode D$_F$. Because of the fast reverse recovery nature of DF, the current spike during recovery is considerably reduced.

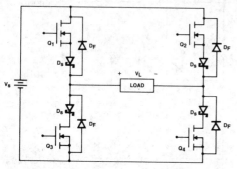

Inverter Circuit Using Schottkys to Prevent the Temporary Circuit Short
Figure 8

Conclusion

Due to the slow recovery of FET inherent diodes, a temporary short circuit could occur in several commonly used switching power circuits. The effects of the temporary short on the circuit are high power dissipation particularly at higher switching frequencies and possible damage to the FET. This should be avoided.

It is noted that only certain operating conditions and load types lead to diode reverse recovery problems and require a circuit remedy. Two solutions were recommended for this problem. One requires the addition of two inductors in the circuit. The other requires the addition of Schottky and high speed diodes.

Reference

Lee, F. C., D. Y. Chen, M. Smith, and G. Carpenter, "Characterization of High Power Darlington Transistors," 1982 Powercon.

5.6 MOSFETs Move In On Low Voltage Rectification (TA84-2)

The efficiency penalty imposed by the offset voltage inherent in Schottky and *pn* junction diodes has been a perennial thorn-in-the-paw for designers of low voltage, high current power supplies. Recent power MOSFET design advances allow designers to economically replace Schottky diodes and provide higher conversion efficiency in many applications. Using power MOSFETs will remove that thorn.

To exploit the opportunity provided by the MOSFET, several questions must be answered:

1. How is a MOSFET optimized for synchronous rectifier service?

2. What are the performance limits?

3. How is an optimized device used in a circuit to take advantage of its characteristics?

4. What is the basis for choosing between a MOSFET or a Schottky diode in a given application?

The following discussion addresses each of these questions.

When we have answered those questions, it will be quite clear that MOSFET synchronous rectifiers are a practical alternative to Schottky rectifiers, particularly for output voltages below 5V.

MOSFET Structures

The present trend of power MOS manufacturers is to first provide transistor selections that perform across a broad range, and then to optimize devices for certain applications such as low voltage rectification. MOSPOWER transistors may be used as synchronous rectifiers for any voltage, but they are receiving the most attention in the 10–50 volt breakdown range.

Many designs for power MOSFETs are presently in use. In most cases these designs were originally optimized for the 100V to 400V range and, as a result, do not represent the practical limits of ON resistance for low voltages.

Figure 1 shows the performance of contemporary transistors normalized by multiplying total chip area (including all inactive chip area) by its on-resistance.

This figure allows comparison of different size devices on the basis of the normalized resistance. The figure includes both theoretical and practical limits. From these limits, it is clear that below 100V, significant improvement remains to be made.

Figure 2a shows the cross section of a contemporary MOS transistor. Device on-resistance can be minimized in several ways. First, the thinnest and lowest resistivity epitaxial layer consistent with the breakdown voltage is employed. Second, gate width must be optimized. If it is too wide, surface utilization is poor, while if it is too narrow, the JFET formed by body regions results in an unwanted increase in total resistance. The surface geometry also requires optimization to squeeze maximum performance from the devices. Figure 2b compares layout efficiency of various surface geometries. While the geometries vary only a few percent, one geometry (square-on-hex) is more efficient than the rest. Keep in mind that

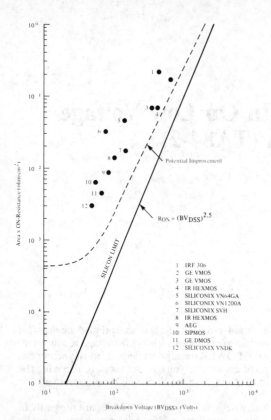

Normalized $R_{DS(on)}$ Comparison
Figure 1

Area × ON-Resistance (ohm-cm²)

Breakdown Voltage (BV$_{DSS}$), (Volts)

SILICON LIMIT

Potential Improvement

$R_{ON} \propto (BV_{DSS})^{2.5}$

1	IRF 306
2	GE VMOS
3	GE VMOS
4	IR HEXMOS
5	SILICONIX VN64GA
6	SILICONIX VN1200A
7	SILICONIX SVH
8	IR HEXMOS
9	AEG
10	SIPMOS
11	GE DMOS
12	SILICONIX VNDK

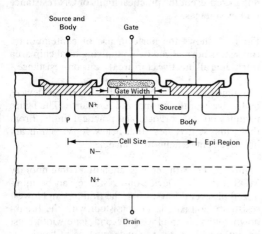

Cross Section of a MOSPOWER Transistor
Figure 2a

the optimum design depends on breakdown voltage and drain current (in high current devices). Another means for reducing resistance is to increase chip area. In practice this means paralleling several die.

	Square on square grid	Circle on square grid	Hexagon on square grid	Square on hexagonal grid	Circle on hexagonal grid	Hexagon on hexagonal grid
Source geometry and grid						
Unit cell	□	○	○	□	○	○
Coefficient of cellular geometries	1.0	0.8862	0.9306	1.0746	0.9523	1.0

The Efficiency of Various MOSFET Surface Geometries
Figure 2b

Low voltage, high current devices using some of these optimizing techniques are beginning to appear. Two such devices presently available are listed in Table 1. These are practical synchronous rectifiers, but they are by no means fully optimized. A factor of two or more improvement should be possible.

Table 1

The ratings of the lowest on-resistance commercially available MOSPOWER transistors.

DEVICE:	SILICONIX VNC003A	MOTOROLA POWER MODULE
PACKAGE:	TO-3	Epoxy Module
ON-RESISTANCE:	.035	.018
BREAKDOWN:	60V	60V
CURRENT:	60A	100A

Electrical Characteristics of Power MOS Transistors as Synchronous Rectifiers

The performance of a MOSPOWER transistor as a synchronous rectifier differs significantly from its use in either linear or switching applications. As shown in Figure 3, the MOS transistor will conduct current in an opposite direction to normal. The transistor should be operated so the intrinsic diode is not turned on, or the performance of the circuit may be altered, as shown in Figure 4 where current continues to flow for some time after the voltage is reversed. When the diode conducts, minority carriers exist. The presence of these minority carriers may have a second effect on circuit performance. The limit of a power MOS transistor is reduced when carriers are present in the body-drain junction region. When a MOSPOWER transistor is rapidly switched off, displacement current flows to discharge the body-drain junction capacitance. Figure 5a shows this behavior schematically. Figure 5b shows the dV/dt behavior of Siliconix devices without minority carriers present at the moment of transition. The dV/dt turn-on threshold is greatly lowered when minority carriers are present. Reduction can be one to two orders of magnitude below that shown in Figure 5b.

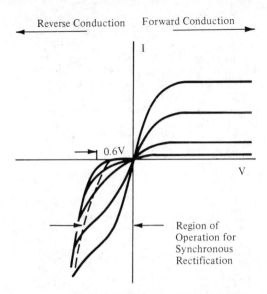

The Current versus Voltage Behavior of a MOSPOWER Transistor in Both the Forward and Reverse Directions
Figure 3

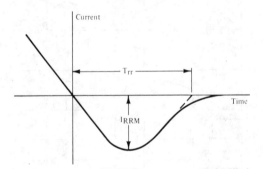

The Reverse Recovery Characteristics of the Integral Diode of a MOSPOWER Transistor
Figure 4

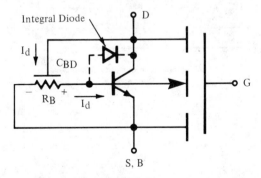

A Schematic Showing the Effect of Displacement Current (I_d) on MOSPOWER Transistor Performance
Figure 5a

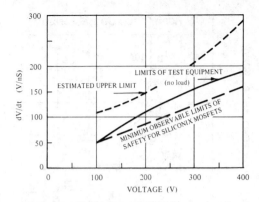

The dV/dt Limits of Siliconix MOSPOWER Transistors When the Diode is Not Conducting
Figure 5b

Several other considerations must also be made when using power MOS transistors for synchronous rectification. Gate-to-source voltage must be limited to manufacturer's maximum data sheet specifications. This requirement sounds straight forward, but the dynamic behavior of circuits, particularly switching circuits such as those found in switch mode power supplies, often results in unexpected transients. Careful design and analysis will prevent any unexpected voltage excursions.

The resistor-like characteristics of power MOS transistors are not maintained as the current increases for two distinct reasons. First, increase in current forces more carriers through the channel region, causing a transverse voltage drop. The channel region narrows, and the current that flows through the device saturates. This increase in ON-resistance as a function of gate voltage and channel current is shown in Figure 6. An increase in current also produces an increase in channel resistance due to device heating. Figure 7 shows the positive temperature coefficient of power MOS transistor resistance.

Designing in MOSFET Synchronous Rectifiers

Gate Drive Timing

Gate drive timing is critical to proper circuit operation. What constitutes proper timing varies with the type of converter circuit used. The popular quasi-square wave converter circuit (Figure 8) is a good example. Voltage and current waveforms are given in Figure 8b, and the proper timing for I_1 and I_2 conduction is a 50% duty cycle drive which is in phase with V_S. While this is the most obvious timing scheme, it is not the only workable one. Figure 9 shows the waveform for I_1 and I_2 which exists in this circuit if a normal diode is utilized instead of a MOSFET.

ON Resistance Characteristics

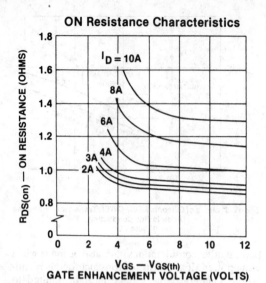

The Device Resistance as a Function of V_{GS} and I_{SD}
for the IRF 730
Figure 6

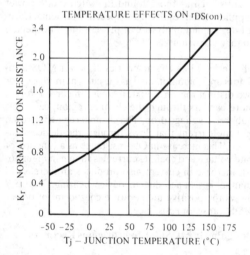

The Normalized Resistance of a MOSPOWER Transistor as a
Function of Temperature for the IRF 730
Figure 7

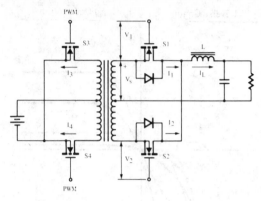

Quasi-Square Wave Converter
Figure 8a

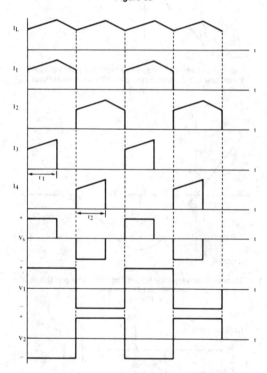

Quasi-Square Wave Converter Waveforms
Figure 8b

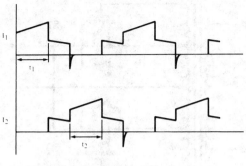

**Diode Rectifier Current Waveform
in a Quasi-Square Wave Converter
Figure 9**

Notice that during the intervals when both S_3 and S_4 are off, I_L divides more or less equally between the two rectifiers. In a diode this change in waveform isn't significant from a loss point of view. If we change the MOSFET gate drive timing as shown in Figure 10, the same current waveforms (without trr spikes) in the synchronous rectifiers will reduce power losses. The improvement occurs because conduction losses in a MOSFET are proportional to the rms current, and the extended conduction interval provides a lower rms waveform for the same average current.

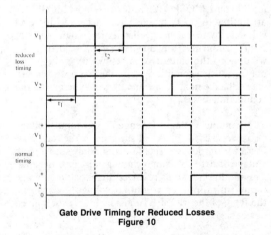

**Gate Drive Timing for Reduced Losses
Figure 10**

Consider what happens if the gate timing does not correspond to the ideal waveforms. There are at least two possibilities worth analyzing—too short and too long conduction intervals. If the conduction interval of S_1 is terminated early, I_1 must continue to flow (the switch is in series with an inductor!), and it will, through the integral diode within the MOSFET. This is not catastrophic, but it will increase the losses because of the higher forward drop of the diode. It will also introduce reverse recovery current spikes and may aggravate the dV/dt problem. If the conduction interval is too long so that S_1, S_2, and either S_3 or S_4 are on simultaneously, then for some period of

time the secondary will be short circuited, and a large current spike will appear in the primary switches. This event is clearly undesirable. Fortunately, MOSFETs can switch rapidly and are easy to drive. These problems can be avoided with a little care when designing the drive circuit.

In other circuits, timing waveforms and consequences of mistiming may be different. Two other converter circuits and their idealized waveforms are shown in Figures 11 and 12. In the buck-derived converter, there is no advantage to extending the conduction interval beyond 50% as there was in the quasi-square wave converter. On the other hand, if the conduction interval is too long, it does not cause current spikes in the primary switch due to the current limiting action of the input inductor. This performance is definitely an improvement over the quasi-square wave circuit. This circuit also accommodates too short a conduction interval by allowing the integral diode to conduct.

(a) Schematic

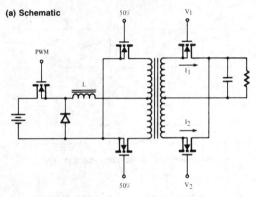

(b) Waveforms

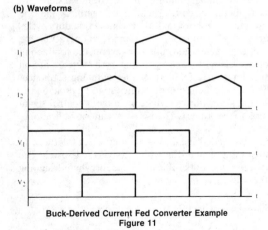

**Buck-Derived Current Fed Converter Example
Figure 11**

The boost-driven circuit is even more tolerant of long conduction intervals. The conduction interval may be increased up to 50% with no effect on circuit operation. For drive circuit simplicity, 50% drive is usually used. For duty cycles beyond 50%, the primary

5

(a) Schematic

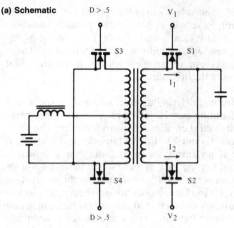

(b) Waveforms

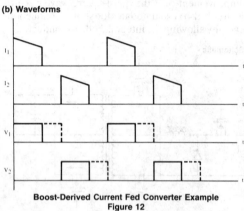

Boost-Derived Current Fed Converter Example
Figure 12

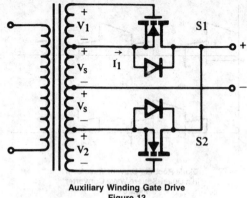

Auxiliary Winding Gate Drive
Figure 13

inductor will again limit switch current transients. In fact, in this circuit, synchronous rectifiers may be used to control output if their conduction duty cycles are greater than 50%. In this mode of operation, the output voltage is controlled by varying the conduction duty cycles of S_1 and S_2 so that both switches are on simultaneously for part of the switching sequence. This allows the regulation function to be done on the secondary without having to couple a control signal back to the primary. This is a significant simplification.

Clearly, gate drive timing and degree of tolerance to mistiming depend on which converter is used. The designer must take this into account when designing the drive circuits.

Gate Drive Circuits

The simplest way to provide gate drive is to use an auxiliary winding on the transformer secondary (Figure 13). The advantage of this method is its simplicity, but there are some disadvantages. For example, if this scheme is used in the quasi-square wave converter, when S_3 and S_4 are both off (a condition which occurs

twice each switching cycle), $V_1 = 0$, and there is no gate drive! I_1 continues to flow, but now it goes through the integral diode—increasing the power losses. There is another drawback to using auxiliary winding drives in the quasi-square wave family of converters. Transformer winding voltages are directly proportional to input voltages and will vary with the input voltage. For example, if the gate drive voltage is 12V when the input voltage is low, then when the input voltage is doubled (not an uncommon requirement), gate voltage will also double to 24V. Except for Siliconix MOSPOWER devices, most MOSFETs are limited to ±20V on the gate. Exceeding this level will destroy the device (Siliconix rates their gates at ±40V with 100% testing to 50V). Adjusting the winding so the gate drive is below 20V at high line may produce a condition where there is not sufficient gate enhancement at low line, thereby increasing the conduction losses.

The winding voltage dependence on input voltage also increases the peak voltage seen by the rectifiers. For a 5V output, the rectifier commonly sees a peak voltage of 20 to 25V, not including noise or transients. This requirement for a relatively high rectifier voltage rating (in proportion to output voltage) means that the R_{DS} of the MOSFET must be higher for a given device size.

All of these problems may be overcome by using another circuit topology. The circuits in Figures 11 and 12 both have correct timing and relatively constant gate voltage when used with an auxiliary winding. They also subject the rectifier to a reverse voltage of twice the output voltage (10V in 5V supply) regardless of the input voltage.

Even the best designed converter has some noise, ringing, and/or transients appearing across the transformer windings. When auxiliary winding drive is used, these transients are coupled directly into the gate where they may cause improper or mistimed

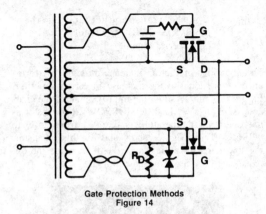

Gate Protection Methods
Figure 14

turn-on and turn-off or even gate destruction. Frequently it is necessary to provide some gate protection (Figure 14). It is vital that gate protection be right at the devices with a minimum of series inductance and be capable of nsec response time. Not all transient suppressors and zeners will respond quickly enough. Notice also in Figure 14 that the source connection for the gate drive is brought out separately. This technique reduces pick-up from the high current in the source lead. Further protection can be provided by bringing the drive winding to the device as a twisted pair.

Figure 14 also shows damping resistors (R_D) across the gates. These resistors are added because drive circuit and winding leakage inductance can resonate with gate capacitance to generate a ringing voltage on the gate. Some form of damping should be used since this ringing may damage the gate or cause spurious turn-on or turn-off. The R-C network on the gate performs this function. Either series or shunt damping resistances may be used. The series resistance, however, reduces switching speed. If, by careful layout and transformer design, the ringing frequency is 1 MHz or above, a ferrite bead may be used for damping. This technique also helps to prevent high frequency parasitic oscillation.

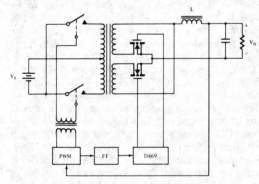

Independent Gate Drive Circuit Example
Figure 15

An alternative to using an auxiliary winding is to use a separate drive circuit derived from the primary switch drive circuits as shown in Figure 15. The example shown is for a quasi-square wave circuit. This drive scheme allows for proper timing of the gate drive.

The advantages of using independent gate drive are several:

1. The amplitude of V_{GS} may be optimized at a predetermined value.

2. The negative part of the drive may be eliminated, reducing drive power and gate stress.

3. Higher switching speeds are usually possible.

4. Gate protection is much easier.

5. Precise drive timing is easily achieved.

6. The rectifiers may perform control functions in some circuits (Figure 12).

7. In some applications, bi-directional power flow is desired. Use of a synchronous rectifier with independent drive allows bi-directional power flow [3,4,5].

The most obvious disadvantage is increased circuit complexity. This additional complexity may not be great, so it is worthwhile to consider the advantages.

Avoiding dV/dt Problems

Whatever drive scheme used, the integral diode should not be allowed to conduct. This will reduce conduction loss and eliminate diode reverse recovery current spikes. Even more important is the preservation of static dV/dt characteristics. Most MOSFETs remain quiescent with an applied dV/dt of 50V–100V/nsec as long as there is no reverse diode current. It is extremely unlikely for a low voltage power supply to even approach this limit. However, if diode current is present, the dV/dt capability may fall to a value as low as 0.5V/nsec. Such rates are possible even in low voltage supplies. If a device is triggered on when the voltage is reversed, the transformer secondary is effectively short circuited and some damage could result.

Thermal Design Considerations

As shown in the previous section, R_{DS} is a positive function of junction temperature (T_j). When I_D is controlled by a current source (the inductor), positive thermal feedback is present. As the device becomes hotter, R_{DS} increases which in turn increases power dissipation, and the junction temperature increases even further. For this reason, it is imperative to observe good thermal design practices. The achievable efficiency is very much dependent on the thermal design. A discussion of this subject is available in a Siliconix Applications Note [6].

Equipment using cryogenic cooling has become relatively common. For the designer of power supplies to be used with such equipment, low temperature cooling provides a unique opportunity to drastically lower conduction losses. A MOSFET will work just fine at liquid nitrogen temperatures ($-196°C$) and even at liquid helium temperatures ($-265°C$). For $T_j = -196°C$, R_{DS} is 20–25% of its value at $T_j = +125°C$. As an added bonus, heat transfer is much better because of the higher thermal conductivity of silicon at these temperatures. The switching properties are not significantly affected. If cryogenic cooling is available, it should be used.

The operation of a MOSPOWER transistor does not depend on the injection of carriers across a junction, nor is the minority carrier lifetime a factor. As long as sufficient majority carriers are present in the source region, drain current will flow when a bias is present on the drain and voltage is applied to the gate. Carrier freeze-out in the source and drain regions does not impact device performance until the conduction electron (or hole) concentration is reduced to a small percentage of its original value.

The main factor affecting MOSPOWER transistor operation is the mobility of the carriers as they flow through the body and drain regions. This mobility increases, so the device on-resistance should decrease. The drain current of a vertical DMOS transistor is given by

$$I_{DS} = C_0 \, Z \, V_{SAT} \, (V_{GS} - V_{GS\,(TH)}) \qquad (1)$$

while its on-resistance is

$$R_{DS(on)} = - \frac{1}{\beta \, (V_{GS} - V_{GS(TH)})} \approx \frac{1}{g_m} \qquad (2)$$

where

$$\beta = \frac{Z}{L} \, \mu_e \, C_0$$

In these equations,

L	= the channel length
I_{DS}	= the drain to source current
C_0	= the capacitance per unit area of the gate oxide
Z	= the amount of active source perimeter
V_{SAT}	= the saturation velocity of electrons in silicon
V_{GS}	= the gate-to-source bias voltage
$V_{GS(TH)}$	= the threshold voltage of the MOSPOWER transistor

g_m	= the transconductance of the MOSPOWER transistor
μ_e	= carrier mobility

The equations predict an increase in device performance proportional to the increase in carrier mobility. Figures 16 and 17 show the improvement in the characteristics of a VNE003A MOSPOWER transistor as the operating temperature is reduced from 25°C to $-196°C$. These figures show that for this transistor, the g_m increases almost a factor of 2, while the on-resistance decreases the same amount.

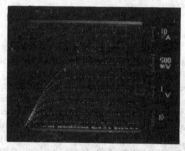

The Characteristics of a VNE003A MOSPOWER Transistor at 25°C (g_m = 6.5S, $R_{DS(on)}$ = 56 mΩ)
Figure 16

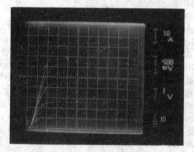

The Characteristics of a VNE003A MOSPOWER Transistor at $-196°C$ (g_m = 12S, $R_{DS(on)}$ = 31 mΩ)
Figure 17

MOSPOWER transistors are also beginning to be used in their third quadrant of operation. In this quadrant, the carriers flow from the drain to the source, opposite to normal current flow. The intrinsic body-to-drain diode is in parallel with the MOSPOWER transistor operated in this direction, as shown by Figure 18. A MOSPOWER transistor operated in this fashion has high current capability with a low forward voltage drop in one direction, and the normal breakdown voltage characteristic in the other direction. This device performance makes MOSPOWER transistors attractive for use as synchronous rectifiers. The characteristics of the VNE003A MOSPOWER transistor in the third quadrant are shown in Figures 19 and 20. These photographs are of particular interest, because of the offset voltage present when the

MOSPOWER transistor is operated at $-196°C$. This offset voltage is present in Figure 17, but is more easily measured here.

To demonstrate the possibilities of low voltage operation an actual power converter was built and tested:

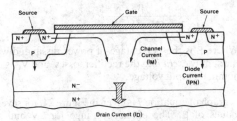

The Carrier Paths in a MOSPOWER Transistor Used as a
Synchronous Rectifier
Figure 18

The Characteristics of a VNE003A MOSPOWER Transistor in
the Third Quadrant at 25°C (V_{offset} = 180 mV,
R bulk = 29 mΩ)
Figure 19

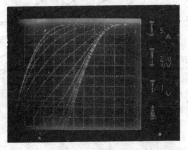

The Characteristics of a VNE003A MOSPOWER Transistor
in the Third Quadrant at $-196°C$ (77°K)
(V_{offset} = 460 mV, R bulk = 18.8 mΩ)
Figure 20

The 5 V, 100 watt power supply shown schematically in Figure 21 was constructed for operation at liquid nitrogen temperatures. For ease of testing, only the MOSPOWER transistors were cooled to this temperature. All other components were kept at room temperature above the nitrogen container. (The PWM 25 was used as the control I.C., but it was found to stop functioning properly at liquid nitrogen temperatures.) The MOSPOWER transistor operated as expected at LN2 temperatures.

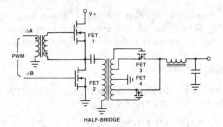

A 5 V, 100 W Half Bridge SMPS Using MOSPOWER Transistors
as the Switching Elements and a Synchronous Rectifiers
Figure 21

Assuming that the majority of the power dissipation occurs in the rectifiers, the power saved by using MOSPOWER transistors in place of junction devices is given by

$$P_{SAVED} = I^2_{LOAD} \left[R_{bulk} - R_{DS(on)} \right] + \quad (4)$$

$$LOAD \ V_{offset}$$

For a 6045 Schottky diode and the VNE003A the percentage of the output power saved as a function of I_L has been calculated at 25°C and $-196°C$. The results of these calculations are shown in Figure 22. This figure shows that at room temperature, a MOSPOWER transistor is a slightly more efficient rectifier at low currents, but is less efficient as the load current increases. However, at liquid nitrogen temperatures, the MOSPOWER transistor is more efficient until the load current approaches 25 amperes. As MOSPOWER transistors optimized for low voltage operation are introduced, the percentage of power saved will increase significantly. For SMPSs operating below 5 volts or greater than 100 volts, MOSPOWER transistors are particularly attractive as the output rectifiers. Below 5 volts the offset of a Schottky diode at very low temperatures results in a significant power loss. For outputs above 100 volts, fast switching *pn* diodes have a large offset voltage and a high bulk resistance.

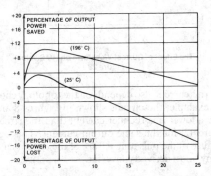

The Percentage of the Output Power Lost or Saved by Using
a VNE003A MOSPOWER Transistor to Replace a 6045
Schottky Diode at $-196°C$ and 25°C.
Figure 22

It is possible to operate the entire power converter in LN_2. The effect of low temperature operation on the other components is discussed in Appendix 1.

$I_{D(rms)}$Reduction

In a MOSFET, conduction loss (P_C) is due to R_{DS}. It can be expressed simply as

$$P_C = I_D^2 \text{ (rms) } R_{DS} \qquad (5)$$

Notice that it is the rms value of I_D which counts. Since R_{DS} is also a function of I_D.

$$P_C \propto I_D^\alpha \qquad (6)$$

Where $\alpha > 2$. Obviously high efficiency requires that I_D (rms) be as small as possible. There are practical ways to do this. One method is to make the averaging inductor larger which changes the shape of I_1 and I_2. Figure 23 gives examples of typical waveforms for large and small inductors. In both examples, average current is the same, but rms current is lower when a larger inductor is used. If we take the ratio of the two values and raise this ratio to the 2.5 power, the power loss difference is almost 40%! Clearly the value of the averaging inductor is critical.

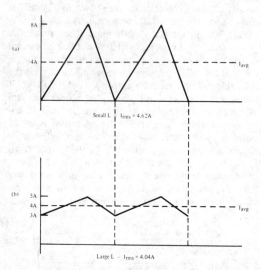

The Effect of Current Waveform on RMS Value
Figure 23

The rms current also depends on the topology chosen. For example, the circuits shown in Figures 8 and 11 have a lower rms rectifier current than the circuit in Figure 12, for the same output current. This occurs because of the pulsating nature of the output current in any boost derived converter.

As previously indicated, in some cases gate drive timing may also be used to reduce $I_{D(rms)}$.

MOSFET–Schottky Comparison

In low voltage output power supplies, overall efficiency (η) is dominated by losses in the rectifiers. The relationship [7] between efficiency and the rectifier forward drop (V_r) can be expressed by

$$\eta = \frac{1 - \beta}{1 + (V_r)/(V_O)} \qquad (7)$$

Where β is the percentage of power dissipated with the supply *excluding* the rectifier losses, and V_O is the nominal output voltage.

A graph of equation 7 is given in Figure 24, for several values of β. The efficiency range for 2V and 5V outputs and for $V_r = 0.4V$ to $0.6V$ is also shown. From Figure 24, it is clear that with a low output voltage, the rectifier forward drop is a major factor in overall efficiency regardless of other circuit losses.

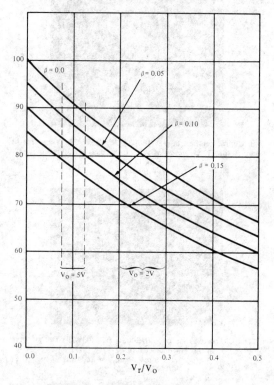

Efficiency as a Function of Forward Drop (V_r)
Figure 24

In a Schottky diode, there are other losses in addition to forward drop. The most prominent of these losses is reverse leakage current which can be substantial when the junction temperature is in the range of 100°C to 150°C (the normal device operating range). Another problem arises from the large capacitance of these diodes. Losses due to charging and discharging this capacitance are relatively small, but in high frequency (> 100kHz) inverters where the primary switches may have transition times of 20nsec or less, large current spikes may be generated. These spikes may increase dissipation and peak power stress in the primary switches.

In a MOSFET, reverse leakage current is significant and capacitance may be as much as an order of magnitude lower. This reduction is a significant improvement particularly for high frequency conversion. Given the very fast transition times possible with a MOSFET, they should be usable as synchronous rectifiers at frequencies up to 1 MHz in switch mode converters, and at even higher frequencies in resonant converters.

The power lost (P_S) in a Schottky diode is expressed as

$$P_S = I^2_{L(rms)}R_B + I_L V_{OS} + P_1 \qquad (8)$$

where: I_L = average output current
$I^2_{L(rms)}$ = RMS diode current
R_B = diode bulk resistance
V_{OS} = diode offset voltage
P_1 = loses due to leakage

The power lost in the MOSFET (P_M) is expressed as

$$P_M = I^2_{L(rms)}R_{DS(on)} \qquad (9)$$

where the switching loss is assumed negligible. To determine which device is superior in a particular application, take the difference between equations 8 and 9. The result in terms of power saved when a MOSFET is used may be expressed as

$$P_{saved} = I^2_{L(rms)} (R_B - R_{DS})$$
$$+ I_L V_{OS} + P_1 \qquad (10)$$

A graph of equation 10 can be used to determine which device is more efficient. If P_{saved} is positive, then the MOSFET has the edge; if however, P_{saved} is negative, then the Schottky is better. A graph of equation 10, using typical values for R_B and V_{OS} and a range of values for R_{DS}, is shown in Figure 25. This graph makes clear where each device is superior and shows the critical role played by R_{DS}.

There are additional factors not taken into account by the graph. As rectifier efficiency is changed, the RMS currents in the transformer windings, the primary switches and the filter capacitors will change. Most of these losses are $I^2_{(rms)}$ dependent, and as rectifier efficiency is improved, the value for β will also improve significantly. Relating this to Figure 25, the zero crossing points will not change, but the degree of improvement in the areas where each device is superior will be enhanced.

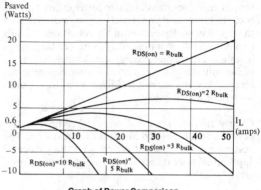

Graph of Power Comparison
Figure 25

The temperature effect on R_{DS}, R_D, and V_{OS} and P_1 is also not included. R_{DS} and R_B increase with temperature as does P_1, and V_{OS} increases as T_j is reduced. Consequently as T_j is reduced, P_{saved} increasingly favors the MOSFET. Consider this behavior when comparing devices.

Putting a Price on Efficiency

To this point, we have been emphasizing efficiency. For many applications, the decision on the device selection is based on the following economic considerations.

1. Direct device cost.

2. Power converter selling price.

3. Life cycle cost of the system converter.

On a device-to-device basis, MOSFETs have historically not looked very promising due to high prices and high R_{DS}. However, prices have drastically decreased, and further reductions are expected. Manufacturers now know how to build very low R_{DS} devices, so the cost gap between MOSFETs and Schottkies will be much narrower in the future.

A more useful comparison is the manufacturing cost of a power supply using each type of device. In this context, the improved efficiency possible with MOSFETs becomes important. With higher rectification efficiency, less cooling is required. This efficiency translates into either smaller heat sinks or (in higher power units) the elimination of a cooling fan. The lower circuit currents due to the higher efficiency also allows cost and size reduction in switches, filter capacitors, and magnetics. The net result is an overall cost reduction.

For telephone companies and large computer manufacturers, a third level of comparison (life cycle cost) becomes important. For these types of users, improved efficiency translates into lower utility bills for both the direct load power *and* the power used to cool the equipment. Reduction in air conditioning loads reduces overall operating costs as well as the capital invested in the cooling equipment. In such systems, a higher initial cost for power conversion equipment is acceptable when improved efficiency reduces operating cost proportionately.

Clearly there are several ways to compare costs between Schottkies and MOSFETs as rectifiers. The best comparison method depends on the particular application, but a direct device-to-device comparison is rarely useful.

Conclusion

Changes in pricing and lower R_{DS} make the MOSFET competitive with Schottky diodes in low voltage rectification applications. To gain the most advantage from using MOSFETs as synchronous rectifiers, the user should carefully consider circuit implementation and overall economic issues. A simple substitution of one device for another is definitely not the key to success.

References

[1] M.S. Adler and R.R. Westbrook, "Power Semiconductor Switching Devices — A Comparison Based on Inductive Switching," *IEEE Transactions on Electron Devices,* ED–29, Number 6, June 1982, pp. 947–952.

[2] M. Chi and C. Hu, "Some Issues on Power MOSFETs," *PESC Record,* June 1982 (392).

[3] G. Cardwell and W. Neel, "Bilateral Power Conditioner," IEEE Electronics Specialist Conference Record, June 1973, pp. 214–221.

[4] H. Matsuo and K. Harada, "New DC-DC Converters With an Energy Storage Reactor," IEEE Transactions on Magnetics, Volume Mag–13, Number 5, September 10/7, pp. 1211–1213.

[5] R.D. Middlebrook, S. Cuk and W. Behen, "A New Battery Charger/Discharger Converter," IEEE Power Electronics Specialists Conference Record, June 1978.

[6] R. Severns, "Safe Operating Area and Thermal Design for MOSPOWER Transistors," Siliconix, inc. Applications Note AN83–10.

[7] R. Kagan, M. Chi and C. Hu, "Improving Power Supply Efficiency with MOSFET Synchronous Rectifiers," Proceedings of POWERCON 9, 1982, p. D4.

Appendix 1. Operation of Other Power Converter Components at Low Temperature

The Electrical Behavior of Semiconductor Devices

The behavior of semiconductor devices such as diodes and transistors and of monolithic integrated circuits made up of these and other components varies greatly with temperature. A basic understanding of the effects of temperature may be gained by examining the variations of three key semiconductor parameters with temperature. These three parameters are:

(1) Carrier mobility: The ease with which conduction electrons and holes move through the semiconductor crystal varies greatly with temperature. The mobility depends on the total number of impurities and the effect of crystal lattice vibrations on carrier movement. As the temperature of silicon is lowered from room temperature, carrier mobility increases as shown in Figure 26. The carrier mobility peaks between $100°K$ and $200°K$, depending on doping concentration, but is typically a factor of 2 or 3 above room temperature mobility.

(2) The number of dopant atoms ionized: As the temperature is reduced, the percentage of dopant atoms that remain ionized (and hence contribute to the conduction process) begins to decrease when KT approaches the ionization energy of the dopant atoms. For common dopant atoms in silicon, the temperature at which dopant "freeze-out" begins is approximately $125°K$, as seen in Figure 27.

(3) Minority carrier lifetime: The relative position of the Fermi level and centers that control the recombination rate of carriers varies significantly with temperature. As the temperature is reduced, the probability that a trapping center will capture both a hole and a conduction electron increases, and minority lifetime therefore decreases.

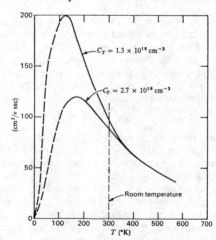

The Effect of Temperature on the Mobility of Holes in Silicon [6]
Figure 26

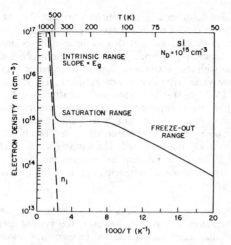

Conduction Electron Concentration in a Silicon Substrate Doped with 10^{15} Donor Atoms per cm^3 [7]
Figure 27

pn Junction Diodes

The reduction in temperature increases the forward voltage drop of a *pn* junction diode as seen in Figures 28 and 29. Figure 28 shows the forward voltage drop of the body-to-drain diode of a VNE003A MOS-

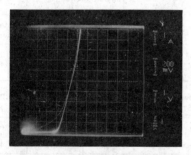

The Current vs Voltage Characteristics of the Body-to-Drain Diode of a VNE003A MOSPOWER Transistor at 25°C
(V_{offset} = .66 V, R bulk = 43.5 mΩ)
Figure 28

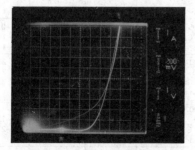

The Current vs. Voltage Characteristics of the Body-to-Drain Diode of a VNE003A MOSPOWER Transistor at −196°C (77°K)
(V_{offset} = 1.33 V, R bulk = 45.5 mΩ)
Figure 29

POWER transistor at 25°C, while Figure 29 shows the voltage drop of the same diode at −196°C. The photographs show that the bulk resistance (R bulk) of this diode is approximately the same at both temperatures, but that the offset voltage (V_{offset}) increases by a factor of two from .66 volts to 1.33 volts. This behavior results from the freeze-out of donor and acceptor atoms. As the temperature is further decreased, this offset will increase until the diode stops functioning altogether.

Schottky Barrier Diodes

Reducing the operating temperature of a Schottky barrier diode from 25°C to −196°C increases its forward voltage drop as shown in Figure 30 and 31. It can be seen from the photographs that the offset voltage increases by about a factor of two while the bulk resistance decreases. The decrease in bulk resistance is the result of an increase in carrier mobility at the lower temperature. However, the forward voltage drop at any current level increases as the temperature is reduced.

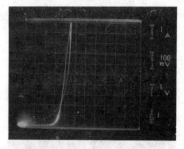

The Current vs. Voltage Characteristics of a 6045 Schottky Barrier Diode at 25°C (V_{offset} = .34 V, R bulk = 10 mΩ)
Figure 30

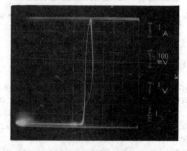

The Current vs. Voltage Characteristics of a 6045 Schottky Barrier Diode at −196°C (77°K) (V_{offset} = .61 V, R bulk = 7 mΩ)
Figure 31

Bipolar Transistors

Bipolar transistors operate through the injection of minority carriers from the emitter region through the base region to the collector region. The current gain

5

of a bipolar transistor depends upon both the injection efficiency of the emitter-base junction and the lifetime of the injected carriers as the flow across the base region. Both of these phenomena grow less efficient as the temperature is lowered, so the performance of a bipolar transistor should decrease as it cools.

This predicted performance is verified by Figure 32 and 33. The two figures show the electrical performance of a 2N6579 bipolar transistor at 25°C and at −196°C. The significant decrease in h_{FE} and the accompanying increase in $V_{CE(SAT)}$ are evident.

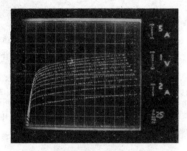

The Current vs. Voltage Characteristics of a 2N6597 Bipolar Transistor at 25°C (h_{FE} = 7)
Figure 32

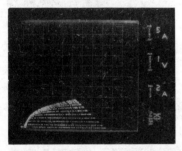

The Current vs. Voltage Characteristics of a 2N6597 Bipolar Transistor at −196°C (77°K) (h_{FE} = .9)
Figure 33

Other Electrical Characteristics

Two parameters, device leakage and breakdown voltage, vary in a similar fashion for all of the devices discussed above. Device leakage decreases as the temperature is reduced, since this current is thermally generated. This current is an exponential function of the temperature. The breakdown voltage also decreases as the temperature decreases though not nearly as rapidly. The breakdown voltage decreases only a few percent per 100°C.

Integrated Circuits

Both bipolar and MOS monolithic integrated circuits are made up of transistors, resistors, diodes, and ca-

pacitors fabricated in the silicon substrate. The behavior of diodes and transistors over temperature has been discussed, but not that of integrated resistors and capacitors. Diffused resistors decrease in absolute value with temperature, though the temperature coefficient varies with dopant concentration. The change of diode, resistor and transistor characteristics does not track with temperature; integrated circuits specified for the commercial or military temperature range operate poorly as the temperature decreases, finally failing to function as designed. The temperature at which an I.C. ceases to function and whether the failure is recoverable upon reheating or not, is circuit dependent.

Design of integrated circuits to operate at very low temperatures is possible; however, the performance of bipolar transistors at reduced temperatures makes them less attractive than MOS transistors for I.C. design. Table 2 shows that n-channel and p-channel MOS transistors operate to liquid helium temperatures (4.2°K). The relative stability of their threshold voltage and predictable increase in g_m allow optimization of the device geometries for low temperature operation.

Table 2
The Characteristics of Various MOS Transistors Over Temperature [5]

TYPE	TK	β	Vth (Volts)
SD1011	293	3408	3.18
Lateral	77	7776	4.33
DMOS n-channel	4.2	—	—
2N6661	293	3362	1.15
V-groove	77	7442	2.54
DMOS n-channel	4.2	2783	1.82
IRD1101	293	3678	2.65
Vertical	77	7938	3.91
DMOS n-channel	4.2	8712	4.17
IRD9120	293	3698	−3.16
Vertical	77	6498	−3.83
DMOS p-channel	4.2	5408	−4.1
MO405	293	42	1.1
Metal Gate	77	128	1.7
NMOS	4.2	149	1.4
MO405	293	35	−0.7
Metal Gate	77	169	−1.1
PMOS	4.2	149	−1.8
CD4007	293	752	2.2
CMOS	77	576	2.7
Metal Gate	4.2	171	2.2
n-channel			
CD4007	293	907	−1.3
CMOS	77	1290	−1.4
Metal Gate	4.2	984	−2.0
p-channel			
VN30A	293	1048	0.4
V-groove	77	8712	2.4
n-channel	4.2	9248	1.3

Thermal Characteristics and Safe Operating Area of Components

The ability of electronic components to dissipate heat at very low temperatures is considerably enhanced. The power dissipation of a component is determined by the thermal characteristics of the device multiplied by the temperature difference between the power dissipating region of the component, and its ambient temperature. Operation of circuits or systems at liquid nitrogen or liquid helium temperatures multiplies the power dissipation capability of a typical device by a factor of 3 to 6. An effect that considerably enhances the power dissipation of silicon devices is its significant increase in thermal conductivity at low temperature. As seen in Figure 34, the thermal conductivity of silicon increases a factor of 10 between room temperature and liquid nitrogen temperature. Other materials including copper show an increase in thermal conductivity over this temperature range as well.

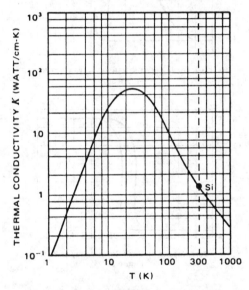

The Thermal Conductivity of Silicon Over Temperature [8]
Figure 34

The safe operating area of a typical bipolar or MOS-POWER transistor must be modified when the transistor is operated at very low temperatures. A comparison of the SOA of a bipolar and a MOS-POWER transistor is shown in Figure 35. The secondary breakdown characteristics of a bipolar transistor do not allow its operation at full current and voltage regardless of the power handling capability of the package. A MOSPOWER transistor with its absence of secondary breakdown may operate at full current and voltage with adequate power dissipation. The limit set by the breakdown voltage is still present, but as discussed earlier, is reduced with decreasing temperature. The current limit set by the

bond wire size increases because the conductivity of aluminum—the common lead wire material—increases with decreasing temperature as shown by Figure 43.

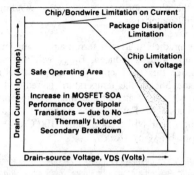

The Safe Operating Area of a Typical Bipolar and MOSPOWER Transistor
Figure 35

Component Reliability

The mechanisms that lead to the failure of semiconductor devices are characterized as having a rate of change determined by the Arrhenius equation:

$$R = A \, e^{-E_A/KT} \tag{11}$$

where

R = specific rate of change
A = constant determined by the component
E_A = the activation energy (eV)
K = Boltzmann's constant
T = temperature in degrees Kelvin

Activation energies for failure mechanisms in semiconductors have been found to range between .3 and 1.1 eV. Reduction of the operating temperature to liquid nitrogen or liquid helium levels significantly increases device lifetime when it is limited by thermally activated mechanisms.

Nature is unlikely to be too kind to the designer of a power conversion system for operation at very low temperatures. Reducing the operating temperature is likely to increase mechanical stress caused by differential expansion rates with temperature, causing a new class of failures.

The Behavior of Passive Components at Very Low Temperatures

The performance of each of the passive component types used in the design of an SMPS changes in its own fashion as the temperature is lowered. In this section, the behavior of passive devices is discussed and related to the underlying physical mechanism.

Magnetic Components

A typical power converter uses magnetic components such as inductors and transformers. To successfully design a low temperature converter the designer must know the characteristics of the core, winding, and structural materials.

Theory predicts that in a ferromagnetic material, as the temperature is reduced, B_{max} will increase and in practice this is exactly what happens. As shown in Figure 36, as the temperature is reduced B_{max} increases, rapidly at first and then leveling off as T approaches 0°K. For example, reducing the temperature of a 50% Ni-Fe core from $+100^{\circ}$C to -200°C increases B_{max} by 18%.

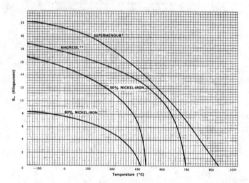

Typical Variation of Flux Density (B_m) with Temperature [3]
Figure 36

Theory also predicts that the core losses will increase as T is reduced. This is due to increased conductivity of the core material and increased hysteresis losses. Again, theory and practice agree. However, the amount of increase in loss varies greatly from one material to another, as shown in Figures 37 through 42. In a ferrite material such as TDK H5B2 (Figures 37 and 38) the core loss increases dramatically. On the other hand, in the 80% Ni-Fe material (Figures 41 and 42) the increase in core loss is barely discernible.

Copper or aluminum is usually used for the winding material. As shown in Figure 43, both materials have a large positive temperature coefficient, resulting in greatly reduced resistance at low temperatures. The resistance of copper, for example, drops by a factor of 13 between $+100^{\circ}$C and -200°C. This increase in conductivity can be used to reduce the winding resistance and subsequent losses. One point should be kept in mind, however, for AC currents the skin depth varies as the square root of the resistivity. In a high frequency converter, as the resistivity of the windings is lowered, less of the material is conducting. Thus, the effective winding resistance does not drop as rapidly as the resistivity of the winding material.

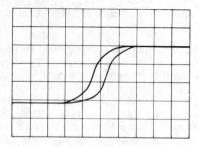

Core Losses of TDK H5B2 at 25°C
Figure 37

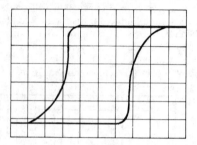

Core Losses of TDK H5B2 at -196°C (77°K)
Figure 38

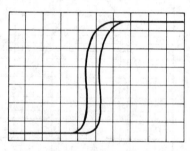

Core Losses of .5 mil Supermalloy at 25°C
Figure 39

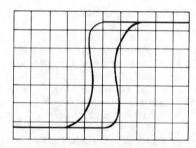

Core Losses of .5 mil Supermalloy at -196°C (77°K)
Figure 40

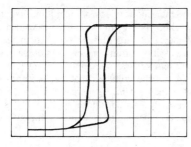

Core Losses of .5 mil 80% Ni-Fe at 25°C
Figure 41

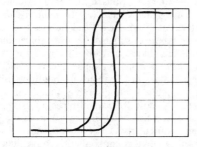

Core Losses of .5 mil 80% Ni-Fe at −196°C (77°K)
Figure 42

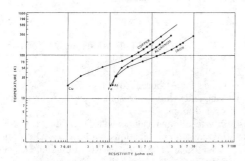

Resistivity of Copper, Aluminum, and Iron as a Function of Temperature [4]
Figure 43

it appears that by lowering the operating temperature from +100°C to −200°C, the insulation voltage stress may be increased by a factor of 2 to 5 for the same life in many film insulators.

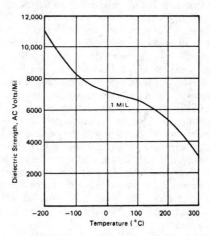

Temperature Effect on the AC Dielectric Strength of Type H Kapton Film [2]
Figure 44

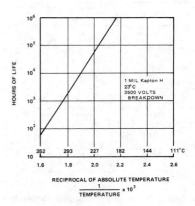

The Effect of Temperature on the Life of 1 mil Kapton H [2]
Figure 45

For high voltage applications, the properties of the insulating material become critical. The net effect of low temperatures is to improve the electrical properties of the insulation, at the expense of the mechanical properties. The effect of temperature on the dielectric strength of .001″ type H Kapton film is shown in Figure 44. We see that by reducing the temperature from +100°C to −200°C the dielectric strength almost doubles, a most desirable result. In a practical application, the usable voltage rating of an insulator is as much a function of its resistance to corona-induced deterioration as it is of dielectric strength. An indication of how strong a function of temperature the life of an insulating film is, can be seen from Figure 45. While the available data is certainly incomplete,

The principal problem in magnetic components caused by lowering the operating temperature is increased brittleness in the materials. In particular the plastic bobbins and some insulating films can become very brittle at low temperatures. A large temperature range can also bring about problems due to the different coefficients of expansion of the various materials in the magnetic structure. These mechanical problems can, however, be overcome by careful design, handling, and mounting.

Given appropriate materials, low temperature magnetic operation is advantageous, since it allows lower losses and small size to be achieved.

5

Capacitors

Many types of capacitors are used in power converters; some of these are usable at low temperatures and others are not. The temperature characteristics of several types of capacitors are shown in Table 3. The film and mica capacitors are very good, showing only small drops in capacitance. More surprising is the relatively good behavior of the solid tantalum types, some of which hold up very well even at 4.2°K! The wet tantalums are not usable at low temperatures due to large capacitance changes and very high ESR values. Very little information is available concerning the ESR of film capacitors at low temperatures, but we do know that the dissipation factor goes up somewhat, and that the resistivity of the foil or metalization goes down. For high current filtering applications the ESR is dominated by the metalization resistance, so that lower temperatures should result in a lower ESR. The lower ESR, coupled with the increased thermal conductivity of the capacitor body, should mean that the current rating of a film capacitor is much greater at lower temperatures. Unfortunately this is only an educated guess at the time.

An important benefit of low temperature operation in film capacitors is the improvement in voltage capability of the insulating film, as outlined in the magnetics discussion. From an energy storage point of view, low temperature operation is a big plus, since $U = CV^2$ and the voltage capability is improved. A factor of 5 increase in V, increases U by 25 times. This is very attractive in those applications, such as pulsed loads, when large amounts of energy must be stored.

Resistors

While resistors are not used in the power conversion process, they are needed for the control functions. Many types of resistors exist and are readily available. The only type *not* suitable for low temperature operation is the ubiquitous carbon composition resistor. Besides having a large temperature coefficient, the resistance values display considerable hysteresis during temperature cycling. Table 4 summarizes the measured properties of several types of resistors. For most purposes, the metal film resistors are more than adequate, particularly at LN_2 temperatures. For precision voltage dividers or similar uses, the bulk metal resistors are remarkably good, and should be adequate even at LHe temperatures.

TYPE	MANF	Value (μF)			% Change fr. 300°K	
		300°K	77°K	4.2°K	77°K	4.2°K
Ceramic	Corning	1.036	0.457	0.387	−55.9	−63.5
		1.001	0.451	0.374	−54.9	−62.6
Mica	Elmenco	1.002	0.997	0.990	− 0.5	− 1.2
		0.997	0.993	0.990	− 0.4	− 0.7
Film (polysulfone)	S & EI	9.923	9.452	0.309	− 4.8	− 6.2
		9.955	9.552	9.444	− 4.1	− 5.1
Film (polysulfone)	S & EI	1.195	1.126	1.115	− 5.8	− 6.7
		1.246	1.176	1.161	− 5.8	− 6.8
Film (polysulfone)	S & EI	1.017	1.012	1.009	− 0.5	− 0.8
		1.099	1.035	1.027	− 5.8	− 6.5
Film (polycarbonate)	S & EI	1.221	1.166	1.155	− 4.5	− 5.4
		1.233	1.170	1.158	− 5.1	− 6.1
Tantalum	Kemet	1.045	0.973	0.936	− 6.9	−10.4
		0.998	0.909	0.854	− 8.9	−14.4

Table 3
The Change in Capacitor Values Over Temperature [5]

TYPE	MANF	Value (Ohms)			% Change fr. 300°K	
		300°K	77°K	4.2°K	77°K	4.2°K
Thick film	Corning LC81	0.9995	1.0231	1.295	+2.3	+29.6
		1.004	1.0459	1.713	+4.6	+71.2
		0.999	1.0403	1.927	+4.0	+92.7
Bulk metal	Vishay	9.996	9.974	9.972	−0.2	− 0.2
		9.9987	9.9768	9.9871	−0.2	− 0.1
Metal film	Corning NC55	1.0007	1.0008	1.0248	+0.1	+ 2.4
		1.0015	1.0131	1.0361	+1.1	+ 3.5
		1.0003	1.0098	1.0283	+1.0	+ 2.8
Metal film	Flt. spares	0.9973	1.0031	1.0161	+0.6	+ 1.9
		2.6077	2.6173	2.6489	+0.4	+ 1.6
		0.9989	1.0039	1.0217	+0.5	+ 2.3

Table 4
The Change in Resistor Values Over Temperature

References

[1] J.K. Watson, "Applications of Magnetism," Wiley - Interscience, 1980, p. 163.

[2] Boeing Aerospace Co., "High Voltage Design Guide for Airborne Equipment."

[3] Magnetics Inc., "Design manual featuring tape wound loops."

[4] Private communications with Dick Gilbert, LSMC.

[5] Private communications with Mike Dix, Staff Scientist at NASA Ames Research Center, Sunnyvale, CA.

[6] G.L. Pearson and J. Bardeen, "Electrical Properties of Pure Silicon and Silicon Alloys Containing Boron and Phosphorus," *Physics Review*, 75, 865 (1949).

[7] S.M. Sze, *Physics of Semiconductor Devices*, 2nd edition John Wiley & Sons, 1981, p. 26.

[8] C.Y. Ho, R.W. Powell, and P.E. Liley, *Thermal Conductivity of the Elements—A Comprehensive Review*, American Chemical Society and American Institute of Physics, New York, 1975.

[9] R.S. Kagan, M. Chi, and C. Hu, "Improving Power Supply Efficiency with MOSFET Synchronous Rectifiers," *Proceedings of Powercon 9*, p. D-4 (1982).

5.6.1 Using Power MOSFETs as High-Efficiency Synchronous and Bridge Rectifiers in Switch-Mode Power Supplies (TA83-1)

Introduction

Switch-mode power supplies (SMPS) are well-known for their many features: high efficiency, light weight (and accompanying small size), and outstanding regulation. These features are principally as a result of their high switching frequency.

With the recent introduction of improved pulse-width modulator (PWM) integrated circuits capable of clock frequencies to 500 kHz and beyond, and power MOSFETs having saturation voltages superior to comparable power bipolar transistors, we must now turn our attention to the rectifiers.

Historically, when high-speed rectifiers were required, if the voltage was low, say less than 30 V, the Schottky-barrier diode rectifier was the optimum choice. However, because of their inherently low peak reverse voltage rating, if our voltage was much higher we had no alternate choice but to use a fast-recovery p-n diode whose PIV ratings could be chosen to be more than adequate for most applications.

We have now reached a barrier where the Schottky-barrier diode and the fast-recovery p-n diode limit performance. This barrier prevents high voltage and high clock frequency SMPS design for two reasons: Schottky-barrier diodes are limited to a PIV generally below 80V; and, fast recovery diodes have excessive reverse recovery time.

Low ON-resistance, high voltage power MOSFETs break through this barrier and offer the potential for higher efficiency, high-voltage switch-mode power supplies.

The Problem Defined

Before we address the problem, we must first acknowledge that for low voltage, high-frequency SMPS applications the Schottky-barrier diode offers satisfactory performance as a rectifier. It is when we attempt to achieve efficient service at high voltages and high clock frequencies that the problems appear.

Since the Schottky-barrier diode fails at high voltages, we are forced to consider alternate rectifiers. Heretofore, our only choice was to use fast recovery rectifiers in either the bridge or full-wave configuration, as shown in Figure 1.

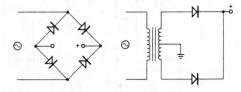

Bridge & Full-Wave Rectifiers
Figure 1

There are, unfortunately, two fundamental problems when using, or trying to use fast-recovery diodes in either of these configurations (Figure 1).

The first and major problem is that for high clock frequency SMPS applications these fast-recovery diodes are simply not fast enough! To allay any argument to the contrary, let us acknowledge that there are high voltage and high current fast diodes with reverse recovery times in the low nanosecond region. But, let us not forget that the new, high-efficiency switch-mode power supplies now appearing in the market as well as those under development are PWM switchers not high-frequency sinusoidal switchers. In other words, the diodes must rectify square waves, or at least something very close — perhaps trapezoidal. If we were to overlay the optimized SMPS square-wave output over a "typical" fast-recovery diode's conduction characteristic (Figure 2), our problem becomes apparent: we discover that we have conduction when we should have none!

An issue might be raised that PWM switch-mode power supplies do not deliver square waves as we find in Figure 2, but, in fact, taking into account the minority-carrier storage time of the typical switching bipolar transistor(s), the SMPS generally exhibit a "dead" time. That is, as we see in Figure 3, we have a period where both the bipolar transistor's storage time as well as the reverse recovery time of the rectifying diode may recover without fear of the crowbar effect.

Our answer to this argument is that we are addressing high-efficiency switch-mode power supply design. Dead time seriously affects efficiency as we see in Figure 4.

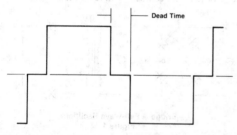

A Square Wave Superimposed Over the Diode Conduction Waveform Showing the Crowbar Effect of Reverse Recovery Time (t_{rr})
Figure 2

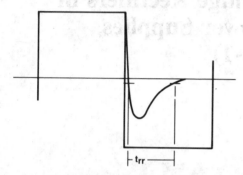

The Typical Output Waveform of a Switch-Mode Power Supply Showing Dead Time
Figure 3

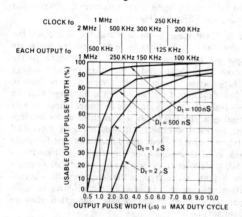

Dead Time Affects Usable Pulse Width
Figure 4

Since we are discussing 'high-efficiency' SMPS design, we can anticipate an output square-wave without dead time, or at least a very short dead time. Here we will discover that the wave is directly affected by a diode's reverse recovery time. In a bridge rectifier, for example, when one diode in the totem pole begins conduction, its counterpart (in the same totem pole) enters into reverse recovery (Figure 2) and for a brief period we have a crowbar. This effect is clearly seen in Figure 5.

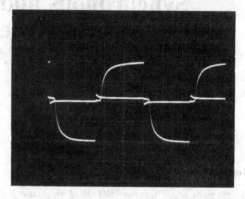

Fast Recovery Diode Bridge Rectifier at 400 kHz Showing Effect of Reverse Recovery Time on Rectification Waveform
Figure 5

Our second problem involves the voltage drop across the rectifier. A bridge results in two (2) diode drops, as shown in Figure 6, whereas a full-wave rectifier has but one diode drop. Again, our efficiency is affected to our detriment.

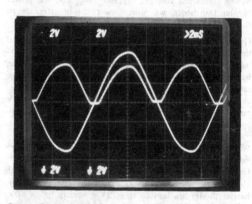

Output (Positive) Waveform Superimposed Over Sinusoidal Input Showing a 1.2 V Two-Diode Drop (Bridge Rectifier)
Figure 6

Enter the Power MOSFET

Resolving Problem #1

Because of the market penetration of power MOSFETs, many of us are familiar with their outstanding features which we need not review. All we need to remember is that power MOSFETs exhibit no minority-carrier storage time.

This statement does, however, require qualification. When we categorically state, "no minority-carrier storage time," we are, in fact, referring to the basic MOSFET, assuming no parasitic elements. Intrinsic to every power MOSFET is a base-drain p-n diode which is, indeed, a parasitic element often considered in many applications as a "beneficial" diode (Figure 7). As we continue our study of synchronous rectification, we perhaps need to be reminded that any p-n diode will exhibit minority-carrier storage time which, in turn, will result in a finite reverse-recovery time.

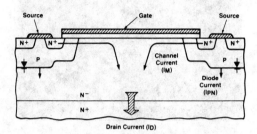

**The Carrier Paths in a MOSPOWER Transistor
Used as a Synchronous Rectifier
Figure 7**

Consequently, if our commission is to use power MOS-FETs as *rectifiers* we must somehow avoid activating this parasitic body-drain diode.

Now that we have "set the stage" so to speak, let's review some perhaps not-so-familiar power MOSFET characteristics.

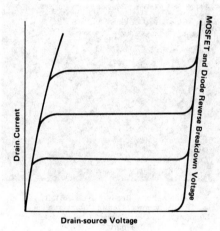

**First Quadrant Output Characteristics
Figure 8**

MOSFET Characteristics

The classic first-quadrant saturation characteristics of a power MOSFET, shown in Figure 8, are generally well-known. However, as long as we have a gate bias enhancing current flow; the first quadrant characteristics are also duplicated in the *third quadrant!* In the familiar first quadrant (Figure 8) the ultimate operational limit is break-down voltage. Third quadrant operation differs in that here we discover our operational limit is when the "parasitic" body-drain diode enters conduction and 'captures'

the channel current heretofore excited by the gate bias potential. A clearer understanding of this phenomenon is shown in Figure 9.

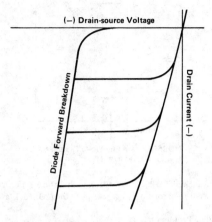

**Simplified Third Quadrant "Output" Characteristics
Figure 9**

Before moving on we need to pause to reflect on why we need to understand third-quadrant operation. We are, in effect, planning to replace the rectifier diode with a power MOSFET. We need to review the electrical schematic of a full-wave rectifier, shown in Figure 10.

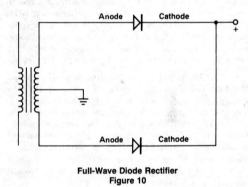

**Full-Wave Diode Rectifier
Figure 10**

Here we have a pair of rectifier diodes in common-cathode connection. Note the polarity of the diode pair: anodes tied to the a-c source, cathodes tied in common to output a positive potential. If we envision the electrical symbol of the power MOSFET and superimpose this symbol in lieu of the rectifier diodes of Figure 10, we discover that our power MOSFET is installed "backwards," that is, we have the source tied to the a-c source and the drain outputting the d-c, as shown in Figure 11.

If our intent is to have the MOSFET act in lieu of the diode, we now must synchronize the MOSFET's turn-ON with the positive-going a-c cycle and turn-OFF when the a-c swings negative. Furthermore, we must never allow the parasitic p-n body-drain diode to participate. If we do, its reverse-recovery time will crowbar the other (MOSFET) rectifier. Remember, we're rectifying the output of a

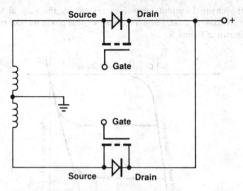

MOSFETs as a Full-Wave Rectifier
Figure 11

high-efficiency SMPS square-wave with little or no dead time. We'll focus on how to synchronize the MOSFET later.

Resolving Problem #2

Interestingly enough, the solution of our first problem is also the solution to our second problem! Once we remove the diode we no longer have a diode drop. We do, however, have an IR drop, and it is here where we must now focus our attention.

Selecting the Power MOSFET

If, indeed the IR voltage drop is a problem needing resolution, it most certainly is if we expect to use power MOSFETs to resolve Problem #1. There is only one way for us to prevent turn-ON of the parasitic body-drain diode: we must maintain an IR drop across the power MOSFET below V_f — the forward voltage of the diode.

Not only do we need a low IR drop across our power MOSFET (which simply means we need an ultra-low $r_{DS(on)}$), we also need to select a MOSFET whose breakdown voltage ratings suffice to withstand the peak inverse voltage (PIV) that we will expect from our SMPS output. We are now approaching a paradoxical problem: as we seek ultra-low $r_{DS(on)}$ power MOSFETs while simultaneously seeking a reasonably high breakdown voltage, we may discover that our costs get in our way! The relationship between $r_{DS(on)}$ and breakdown voltage is:

$$r_{DS(on)} = k \ BV^{2.5}$$

Consequently, the only way to achieve low $r_{DS(on)}$ is to increase the area of the semiconductor chip which, in turn, forces the price upward exponentially!

Since, as we learned earlier, the third quadrant output characteristics duplicate those of the first quadrant, we do have a reasonably simple means to select suitable power MOSFETs from the plethora of power MOSFETs available. All we need is a curve tracer, such as the Tektronix 576. By superimposing the first quadrant output characteristics over the reverse diode's forward characteristics, we, in effect, can duplicate the critical parameters from which we then can make a suitable selection based on both breakdown voltage and maximum permissible

current before we reach V_f — forward voltage turn-ON of the parasitic body-drain diode. Figures 12, 13 and 14 provide typical characteristics taken from three popular power MOSFETs.

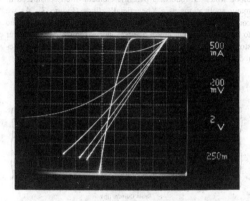

Siliconix inc. Type VN64GA
"Third Quadrant" Output Characteristics
Superimposed Over Forward Diode Breakdown
Figure 12

The power MOSFET shown in Figure 12 offers a 60V BV_{DSS} and a nominal $r_{DS(on)}$ of 0.4 Ohms at a drain current of 10A. The topology is V-groove double-diffused MOS.

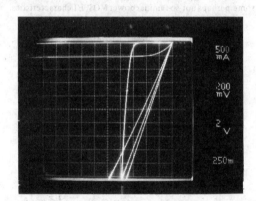

International Rectifier Type IRF130
"Third Quadrant" Output Characteristics
Superimposed Over Forward Diode Breakdown
Figure 13

The IRF 130, whose characteristics are displayed in Figure 13, is rated by the manufacturer at 100V with a typical $r_{DS(on)}$ of 0.14 Ohm at a drain current of 6A. The structure is a vertical double-diffused MOS (VDMOS).

The characteristics shown in Figure 14 are for a VNE003A VDMOS structure rated at 100 V with a typical $r_{DS(on)}$ of 0.035 Ohm at a drain current of 60 A.

We are able to interpret these data by remembering that we wish to pass as much current as possible without exceeding V_f — the turn-ON voltage of the intrinsic parasitic body-drain diode. For all three illustrations, the turn-ON

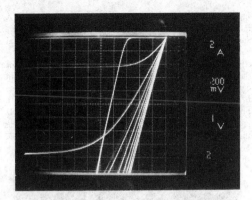

Siliconix inc. VNE003A
"Third Quadrant" Output Characteristics
Superimposed Over Forward Diode Breakdown
Figure 14

voltage, V_f, lies close to $-0.6V$. Consequently, we can make some reasonable approximations of the maximum current that each of these MOSFETs is capable of handling (at 25°C). For the VN64GA (Figure 12) 2A; for the IRF-130 (Figure 13) 4.2A; and for the VNE003A (Figure 14) > 20 VNE003A (Figure 14) > 20.

In a practical sense we can dismiss the VN64GA as Schottky-barrier diode rectifiers are easily capable of handling considerably more than 2A and withstand 60V. Were it not for the high standoff voltage of the IRF-130 (Figure 13), we might dismiss it as well. The VNE003A (Figure 14) offers us a little more, but, again, if it were not for the 100V rating, it, too, might be dismissed.

We should by now recognize the fundamental limitation of today's power MOSFETs: too high $r_{DS(on)}$. But, on the other hand, we must not lose sight of the limitation of the Schottky-barrier rectifier diodes, as well as the limitations of the so-called fast recovery p-n diode.

Synchronous Bridge Rectifiers Using Power MOSFETs

Setting aside our quest for low $r_{DS(on)}$ n-channel power MOSFETs, let's turn our attention to the basic bridge rectifier (shown in Figure 1) and recall that the diode bridge rectifier suffers from a 2-diode drop (see Figure 6). If this bridge consisted of Schottky-barrier diodes across we might attain twice the normal voltage rating seeing that we use a totem pole of diodes across the rails. On the other hand, using conventional fast-recovery diodes, voltage is secondary but we're again troubled when rectifying high efficiency SMPS square waves with little or no dead time.

If we could implement a quad array of MOSFETs using low $r_{DS(on)}$ n- and p-channel devices, we could escape the limitations of breakdown voltage, as well as offer high efficiency, because of the lack of minority-carrier storage time which is inherent in all MOSFETs.

Again, of course, we're strapped with the $r_{DS(on)}$ problem inherent in today's MOSFETs and especially with p-channel devices. Because of the reduced mobility of

p-type material, we discover that for an equivalent $r_{DS(on)}$ p-channel MOSFET our costs are considerably higher than for its n-channel equivalent (with the same voltage rating).

The Circuit

The basic synchronous bridge rectifier circuit using both n- and p-channel MOSFETs follows that of the diode bridge much like we did for the full-wave rectifier shown in Figure 11. The circuit is shown in Figure 15.

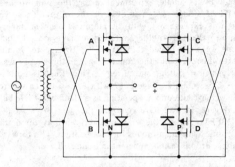

Synchronous Bridge Rectifier Using n- & p-Channel MOSFETs
Figure 15

The operation of this MOSFET bridge rectifier is quite straight-forward. At the half-cycle when the upper rail swings positive the gate bias on FETs B and D is also positive and the gate bias on FETs A and C is negative. Since FETs A and B are n-channel, only FET B turns ON. However, since FETs C and D are p-channel, only C turns ON. With FETs B and C, ON current flows with the right terminal positive. Likewise, when the lower rail swings positive the gate bias on FETs A and C is positive; the bias on FETs B and D is negative and FETs A and D turn ON. Current continues to flow in the same direction as before, and we have gone through one cycle.

If we begin our examination of this bridge rectifier by first applying a sinusoidal input and observing the output from both the positive and negative terminals with respect to one rail we obtain a waveform as shown in Figure 16.

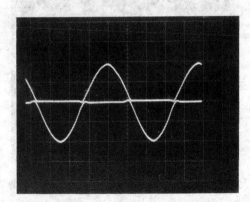

Output Waveforms of Synchronous Bridge Rectifier
Figure 16

We originally faulted the diode bridge rectifier for having a 2-diode drop, which regardless of current generally amounted to at least 0.6V per diode and perhaps more, depending upon the rectifier used. For MOSFETs, our voltage drop is dependent upon $r_{DS(on)}$ and current flow; in other words, the classic IR drop. Consequently, it is definitely to our advantage to use MOSFETs with the lowest ON resistance.

We can obtain a better appreciation for the diode drop effect by a simple experiment using the synchronous MOSFET bridge. We first observe the output (positive for this experiment) with the input waveform superimposed. Because of the limitations of laboratory equipment we were forced to go to a 60 Hz line to obtain worthwhile current (the function generator simply was incapable of outputting current). The effect is noticeable if we compare Figure 17 with Figure 18. In Figure 17, we have the bridge operating as a quad MOSFET; but in Figure 18 we have removed the gate drive to the MOSFETs causing the bridge to rely entirely upon the intrinsic body-drain diodes for rectification. The forward voltage drop is readily apparent in Figure 18.

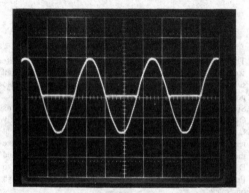

Output Waveform (Positive) MOSFET Bridge Rectifier
Figure 17

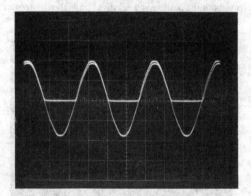

Output Waveform (Positive) Body-Drain Diode Bridge Rectifier
Figure 18

In both Figures 17 and 18, the output current was 1.2 A into a pure resistance load. The total $r_{DS(on)}$ of the n- and p-channel MOSFET pair was 0.6 Ohm for a total voltage drop (at 1.2 A) of 0.72 V which accounts for the slight 'thickening' of the trace in Figure 17. Because of the present limitation of p-channel $r_{DS(on)}$, as the load current increases we can expect a corresponding increase in the voltage drop. The output waveform, showing both positive and negative swing for a square-wave of 250 kHz is shown in Figure 19. The slope of the positive wave was the fault of the pulse generator used for driving the MOSFET bridge. (Compare Figure 19 with Figure 5.)

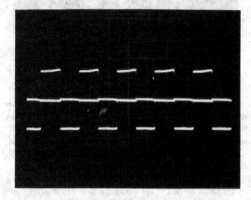

Output Waveform Showing Both Polarities with 250 kHz
Square-Wave Drive
Figure 19

Synchronous Full-Wave Rectifier Using Power MOSFETs

The synchronous full-wave rectifier, shown in Figure 11, depends upon low $r_{DS(on)}$ n-channel power MOSFETs as the rectifier elements. For this illustration we will implement the VNE003A whose characteristics are shown in Figure 14.

To synchronize the MOSFETs we must turn-ON the gates with a gate-to-source voltage of at least +10V to ensure optimum conductance. In a SMPS we are able to add "tickler" windings to the secondary windings of the output transformer to accomplish this, as shown in Figure 20.

The Parts List for the SMPS shown in Figure 20 is as follows:

R1	5.6k	R4	10k	C1	0.01
R2	75k	R5	7.5k	C2	0.001
R3	56k	R6	3.6k	C3	5uF
T1	see text			C4	470uF

Q1, Q2 VN0600A
Q3, Q4 VNE003A

PWM25 — Siliconix Pulse Width Modulator I/C

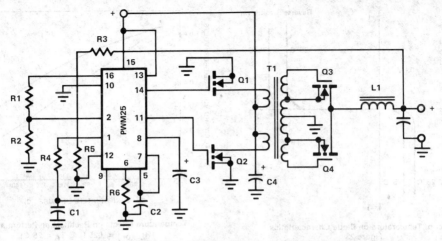

Switch-Mode Power Supply Using Synchronous MOSFET Rectifiers
Figure 20

Transformer Design, T1

Undoubtedly the transformer output stage represents the major design problem for most SMPS designs. Although this lies outside the general scope of this article, a few words should suffice to give us a more comfortable feeling. Aside from the secondary winding we have a pair of "tickler" windings that provide the necessary gate voltage to 'synchronize' the gate turn-ON.

The transformer has an Indiana General E-core #8031-1. Over a nylon bobbin we first wind a 16-turn primary consisting of 8 strands of #30 AWG enamel (Litz wire) over which is wrapped a copper foil acting as a Faraday shield. The secondary has a total of 28 turns of #24 AWG enamel with both a center tap as well as additional taps at ±12 turns which provide tickler windings of 8 turns each. With a primary of 10V/turn and a secondary of 3V/turn, we have a transformer turns ratio of 1:3.33.

In Figure 21 we are able to view the output waveform of one tickler and one-half of the secondary (from center-tap). We see that the tickler winding supplies 12.5V of gate potential — adequate for complete turn-ON of the VDMOS synchronous rectifier. We are now ready to attach the VNE003A (Figure 14) to this secondary.

Another view of interest, shown in Figure 22, is the output from both secondaries of the transformer where we see the 180° phase reversal common to all center-tapped secondaries.

Feeding both VNE003As, we exit through suitable filtering at a fixed 5V at 10A. In our experimental SMPS we were power limited by our laboratory supplies used for primary power. Nonetheless, from Figure 14, we can anticipate a reasonable current before the intrinsic body-drain diodes turn-ON.

We have intentionally left the effects of temperature upon performance until last. As our operating temperature in-

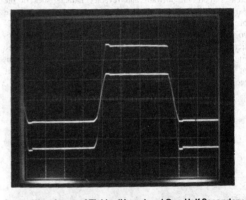

Output Waveforms of Tickler (Upper) and One-Half Secondary (5 VDIV) with Direct Drive from SMPS (Figure 19)
Figure 21

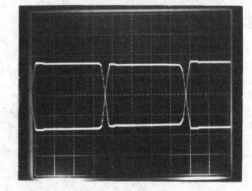

Waveforms Across Secondary
Figure 22

5

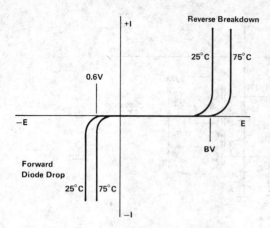

Effect of Temperature on Diode Characteristics
Figure 23

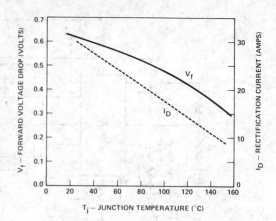

Temperature Effects on Rectification Performance
$[R_{DS(on)} = .025 \; \Omega \; @ \; T_A = 25°C]$
Figure 24

creases, both the forward and reverse characteristics of the parasitic diode follow those of any silicon diode: reverse breakdown improves and V_f (forward diode turn-ON voltage) decreases (Figure 23). Since synchronous rectification depends on third quadrant current flow (see Figure 9) with a voltage drop less than V_f, any reduction of V_f will force a corresponding drop in third quadrant rectification current.

An additional complication arises from the positive resistance coefficient of our power MOSFET. As the temperature increases, so does the ON resistance. Consequently, the I-R (rectification current times $r_{DS(on)}$) drop increases with temperature, thus further restricting our third quadrant rectification current. Figure 24 offers an interesting relationship between V_f, current and temperature for the Siliconix VNE003A MOSPOWER FET.

5.6.2 Use MOSPOWER Transistors as Synchronous Rectifiers in Switched-Mode Power Supplies

Introduction

The demand for low cost and high conversion efficiency is fueling the overwhelming acceptance of switched-mode power supplies (SMPS). To ensure cost-effective design of these inherently high-efficiency supplies, switching frequencies must rise.

When MOSPOWER transistors are used as the switching elements in a power supply, the 50 kHz frequency barrier imposed by bipolar transistors is eliminated. Removal of this barrier has allowed SMPS with operating frequencies up to 500 kHz to be constructed. At these higher frequencies, the size of the magnetic and capacitive components is decreased and the switching losses are lower.

Yet the efficiency of the higher-frequency "switchers" is now limited by the characteristics of the rectifier diodes used in the output circuitry.

Schottky-barrier diodes, with their rapid switching speed, are adequate rectifiers for applications where the output is between 5 and 100 volts. Below 5 volts, the diode offset voltage may result in excessive power dissipation. Above 100 volts, the breakdown voltage limit of the Schottky diode is exceeded and alternatives must be found.

Some MOSPOWER transistors are more efficient rectifiers than Schottky diodes, in the problem areas described above. The purely resistive characteristics of MOSPOWER devices allow efficient rectification of extremely low voltages. Also, their high breakdown voltage ratings allow rectification of voltages as high as 1000 V.

5.6.2 Reprinted by courtesy of PCI, April 1983.

This article examines the characteristics of MOSPOWER transistors as synchronous rectifiers and compares their behavior with that of Schottky diodes. The performance of MOSPOWER transistors versus Schottky diodes is then compared in a 5 V, 100 W SMPS and a detailed circuit description is given.

MOSPOWER Transistors and Schottky Diodes: How They Work

A cross-section of the typical Schottky diode is shown in Figure 1. The Schottky barrier diode operates through the flow of majority-carrier electrons from n-type silicon into the barrier-metal contact. Rectification occurs at the interface between the lightly doped n-type silicon layer and the Schottky barrier-metal. The p-type guard ring increases the breakdown

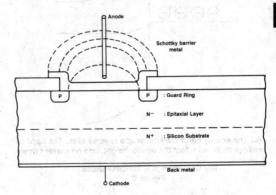

Cross-Section of a Typical Power Schottky Diode
Figure 1

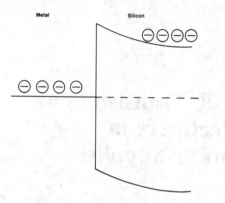

(a) The energy band diagram with no bias applied. The electrons on both sides of the barrier are in equilibrium.

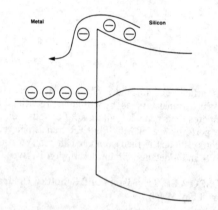

(b) The energy band diagram with a forward bias. The applied voltage reduces the barrier height, and electrons flow.

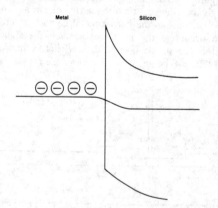

(c) The energy band diagram with a reverse bias. The applied voltage does not affect the barrier height, and no current flows.

The Operation of a Schottky Diode
Figure 2

voltage of the Schottky diode, by reducing the field strength at the perimeter of the junction. It also serves to protect the Schottky diode portion of the junction, in case of excessive reverse voltage.

The specific performance of a Schottky diode depends on its size, the barrier metal used, and the doping profile of the silicon wafer. Figure 2 shows the general operation. Figure 2a illustrates the energy band of an unbiased Schottky diode. Applying a forward bias to the diode lowers the barrier height as seen from the silicon side of the junction. Conduction electrons flow from the silicon into the metal comprising the other side of the diode. The electrons that flow over the barrier and into the metal, disappear into the sea of electrons within the metal.

When a reverse bias is applied to the diode, the barrier between the metal and the silicon (as seen from the metal side of the junction) is not lowered; no carriers flow. The transition from forward to reverse bias occurs rapidly, with no significant reverse recovery time. Electrons that flow into the metal disappear rapidly, because electrons are also the majority carriers in the metal. Carriers are not free to flow back across the barrier as the polarity of the voltage changes.

The MOSPOWER transistor whose cross-section and symbol are shown in Figure 3 is also a majority carrier device. The conduction mechanism in this device, is controlled by an applied voltage between the gate and the source, with an applied drain-to-source voltage. The operation of a MOSPOWER transistor showing the energy band diagrams and the majority carrier flow is illustrated in Figure 4.

With the drain biased positively with respect to the source, but with no bias from gate to source, the energy barrier of the body-region separates the source and drain regions. As the gate voltage is biased positively with respect to the source, the body-region barrier is decreased until at the threshold voltage, V_{TH}, the n-type source and drain-regions are joined. Electrons are then able to flow freely from the source-region to the drain-region. Under this condition the surface of the body-region is inverted; this conductive region is called the channel-region. Electrons are the majority carriers throughout the structure, as in the Schottky diode.

When a negative voltage is applied between the gate and the source, the body barrier height is increased. The electrons in the source and drain regions are now separated even further from each other in terms of

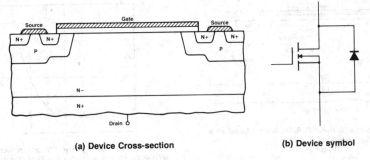

(a) Device Cross-section   (b) Device symbol

Figure 3

energy. This behavior is called accumulation at the surface of the body region.

When the gate-to-source voltage is switched from positive to negative, the conducting *n*-type region connecting the source and drain-regions is converted to *p*-type. The carrier flow then stops abruptly.

When using a MOSPOWER transistor as a synchronous rectifier, the convention of biasing the source and drain terminals is reversed. Now, the current flows in the direction that forward-biases the body-to-drain diode, shown in Figure 3b. Figure 5 shows the current flow of the body-to-drain diode in parallel with the channel current of the MOSPOWER transistor.

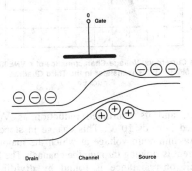

(a) The energy band diagram with no gate bias but a positive drain-to-source bias. The electrons in the source and the drain regions are separated by the "body region" barrier.

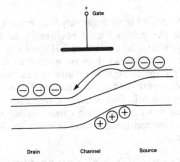

(b) The band diagram with a positive gate bias and a positive drain-to-source bias. The electrons in the source flow freely to the drain, because the barrier height is lowered.

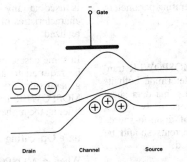

(c) The band diagram with a negative bias and a positive drain-to-source bias. The electrons in the source and the drain region are separated by the body region barrier enhanced by the gate bias.

The Operation of an N-Channel MOSPOWER Transistor
Figure 4

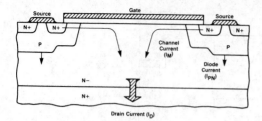

The Current Paths in a MOSPOWER Transistor Used as a
Synchronous Rectifier
Figure 5

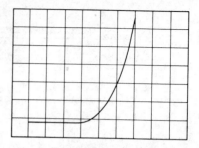

The Current vs. Voltage Characteristics of a VNE003A
MOSPOWER Transistor in the First Quadrant with $V_{GS} = 0$
Figure 6a

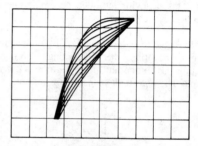

The Current vs. Voltage Characteristics of a VNE003A
MOSPOWER Transistor in the Third Quadrant
Figure 6b

Figure 6a shows the current versus voltage curve of a MOSPOWER transistor operating in the first quadrant. As seen in the same figure, with zero or negative gate-to-source voltage and positive drain-to-source voltage, very little current flows, and the device is essentially off until the breakdown voltage of the device is exceeded. Its peak-inverse-voltage (PIV) rating is equal to its drain-to-gate breakdown specification (V_{DGR}).

When the drain is biased negatively with respect to the source and the gate is biased positively, the MOSPOWER transistor is in quadrant 3 (Figure 6b). If the voltage across the device does not exceed its forward voltage V_F (typically 0.6 V for silicon); the body-to-drain diode is off, and all the current flows in the conducting channel. This is ensured by satisfying the requirement that the peak drain current multiplied by the channel on-resistance is less than the diode forward voltage.

Reverse Characteristics of a MOSPOWER Transistor

To use a MOSPOWER transistor as a synchronous rectifier, one must understand its operating parameters in the reverse mode.

On-Resistance

The on-resistance ($R_{DS(on)}$) of a MOSPOWER transistor and the RMS drain current determine both its voltage drop and power dissipation. This data sheet gives the device on-resistance for one or two gate-to-source voltages; but the curve of drain-to-source voltage versus V_{GS} for various currents should be consulted before a design is attempted.

Figure 7 illustrates this curve for a Siliconix Inc. VNE003A 100 V/0.035 Ω/250 W @ T_C = 25°C, device. At values of V_{GS} below about five volts, the drain current is flowing in the body-to-drain diode. With gate-to-source voltages greater than about 5 V, the channel begins to conduct, and the diode begins to turn off. Notice the change in slope as the conduction path changes from the body-to-drain diode to the channel, and its continued reduction as V_{GS} is increased towards 10 V. This change in slope indicates the minimum voltage at which the majority of the current is flowing through the channel. The MOSPOWER on-resistance is found by dividing the drain-to-source voltage of Figure 7 by the drain current. For the VNE003A MOSPOWER transistor, this graph is shown in Figure 8. If a smaller value of V_{GS} is used at any current level, the rapid switching characteristics of the MOSPOWER transistor are not realized.

In some cases, paralleled MOSPOWER devices may be required to achieve the required $R_{DS(on)}$. As a general rule, the higher the PIV (V_{DGR}) rating of a MOSPOWER transistor used as a synchronous rectifier is, the higher the minimum channel on-resistance.

Safe Operating Area

When a MOSPOWER transistor is used as a synchronous rectifier, it is operated well within its rated SOA. This is due to the low voltage drop across the device.

Switching Times

The switching characteristics of a MOSPOWER transistor used as a synchronous rectifier are essentially

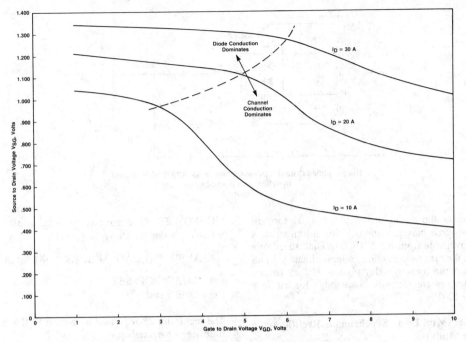

The Source-to-Drain Voltage vs. Gate-to-Drain Voltage for a
100 V, 0.035 Ω VNE003A in the Third Quadrant for Various
Values of I_D [1]
Figure 7

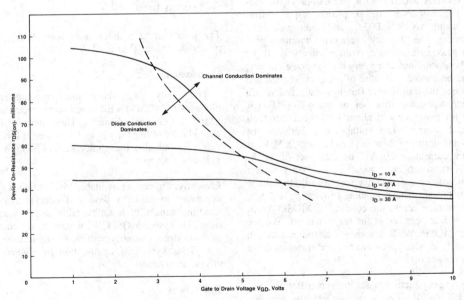

The Device On-Resistance vs. Gate-to-Drain Voltage for a 100 V,
0.035 Ω VNE003A in the Third Quadrant for Various Values of I_D [1]
Figure 8

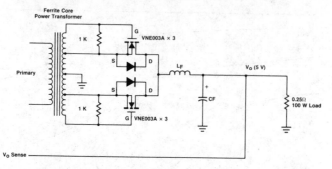

The Simplified Output Configuration of a 5 V, 100 W SMPS Using Synchronous Rectification
Figure 9

the same as those measured under normal forward drain-to-source bias conditions. This information is readily available from the VNE003A data sheet. We find that the turn-on delay time is approximately 30 nS (max) and the turn-off delay time is 150 nS (max). These devices are certainly fast enough for use in a 500 kHz SMPS.

A 100 W SMPS Using Synchronous Rectifiers (A Case Study)

A 5V, 100 W power supply operating at 500 kHz was built and tested with both 50HQ-020 Schottky diodes and VNE003A MOSPOWER transistors as the rectifiers. Figure 9 shows the simplified output configuration using six VNE003As and their gate-drive arrangement. The "tertiary" gate drive windings generate a peak gate-to-source voltage of about 10 V. Since these windings carry very little average current, they can be wound with wire of small cross-sectional area. When the transformer flux has stabilized after a switching transistor, one set of the MOSPOWER devices is conducting with about +10 V between their gates and sources. The voltage drop between the sources and drains of these devices is about 0.24 V, at the drain current of 20 A. The other set of MOSPOWER transistors have about −10 V between their gates and sources. A negative source-to-drain voltage equal to the secondary winding-to-winding voltage, less the voltage across the conducting MOSPOWER device, will appear across the set that is off. We see that the MOSPOWER transistors alternate between quadrant 3, in which they are on, and quadrant 1, in which they are off.

In analyzing conduction losses in the rectifier devices, the MOSPOWER devices are modeled as a resistance equal to $R_{DS(on)}$. Diode rectifiers such as Schottky-barrier and *pn*-junction devices, however, must be modeled as an offset voltage source, V_{OFFSET}, in series with an incremental or bulk resistance R_{BULK}. The power saved by using

MOSPOWER synchronous rectifiers instead of Schottky diodes is simply:

$$P_{SAVED} = I_{LOAD}^2 [R_{BULK} - R_{DS(on)}]$$
$$+ I_{LOAD} V_{OFFSET} \qquad (2)$$

from Ohm's law.

The rectifying devices used in the test SMPS had the following characteristics:

VNE003A (3 in parallel): $R_{DS(on)} - 0.012\ \Omega$

50HQ020: $R_{BULK} = 0.006\ \Omega$,

$V_{OFFSET} = 0.360\ V$

The load current in both cases was 20 A, and the power saved was thus 2.40 W.

Conclusion

A brief analysis has shown that replacing Schottky diodes with MOSPOWER synchronous rectifiers in an SMPS has desirable benefits. These benefits include not only the savings in dissipated power at low output voltages, but also the ability to rectify high output voltages.

Presently, the cost of available MOSPOWER devices compared to Schottky diodes of equivalent current-handling capability is unfavorable in many applications. However, MOSPOWER devices optimized for third-quadrant operation will be available soon. Look out, Schottkys! Stiff competition in SMPS is just around the corner.

References

[1] VNE003 Data Sheet, Siliconix Incorporated 1983.
[2] R.S. Kagan, M. Chi and C. Hu, *"Improving Power Supply Efficiency with MOSFET Synchronous Rectifiers,"* Proceedings of Powercon 9, D-4, 1982.

5.7 Power MOSFETs and Radiation Environments (TA84-3)

Many applications for power MOSFETs require operation in high radiation environments. These applications are usually for space or military equipment although some civilian applications at particle accelerators, nuclear power plants, medical facilities, and industrial equipment also exist.

Like any semiconductor device, the electrical characteristics of a MOSFET will be altered if it is exposed to sufficiently high levels of radiation. The nature of the changes will depend on the type, intensity, and duration of the radiation.

With proper circuit design, MOSFETs can operate successfully at very high levels of irradiation. This paper will provide the basic information required for successful power MOSFET designs.

Types of Radiation

There are two general categories of radiation: ionizing and nonionizing. The response of a MOSFET is quite different for each category. Ionizing radiation can be either electrons, protons, x-rays, or gamma radiation. The name "ionizing radiation" is derived from the effect these types of radiation have in the semiconductor: i.e. they produce ionized electron-hole pairs. Neutrons are non-ionizing and tend to disrupt the atomic or crystalline structure of the semiconductor rather than produce electron-hole pairs. The result of this difference in interaction with the semiconductor material is a difference in the change of electrical characteristics for each type of radiation.

Irradiation Scenarios

Many different irradiation scenarios are encountered in practice. A few typical scenarios are:

1) **Geo-synchronous Orbit** Equipment intended to operate in geo-synchronous orbit is exposed to a large flux of ionizing radiation including high energy electrons and protons. The radiation damage in this environment is cumulative over the life of the mission with total doses of 50 to 200K rad(Si) for seven to ten year missions. The dose rate is relatively low.

2) **Nuclear Weapons Burst** The typical radiation environment 1300 meters away for a 40 KT, near surface, nuclear blast is given in Table 1.

5

TABLE 1

PULSES PRODUCED AT 1300 M FROM 40 kT WEAPON (NEAR SURFACE)

Type of Pulse	Time of arrival of Pulse Maximum(s)	Pulse Width	Remarks
Prompt gamma	5×10^{-6}	3×10^{-8}	High/low dose rate
Electromagnetic (EMP)	5×10^{-6}	10^{-3}	
Neutron & delayed gamma radiation	5×10^{-5}	10^{-3} (50%/dose)	Low dose rate
		10 (90%/dose)	High dose

For other blast scenarios, particularly exoatmospheric, the timing and the intensity of the radiation pulses will differ but will have the same general characteristics. The prompt gamma rate can be in the range of 10^7 to 10^{11} rad/s, and the total neutron fluence may range from 10^{11} to 10^{14} neutrons/cm². The electromagnetic radiation does not directly affect the device, but transient voltages induced in the associated circuitry by the EMP can damage the MOSFET.

3) **Nuclear Plants** A variety of radiation levels is possible in a nuclear plant. Most of the environments are quite benign, but there are two significant radiation environments for the safety instrumentation. In some areas, the instrumentation may see gamma radiations of 0.3 rad/hour, which over a 40 year life is a total dose of 10^5 rad(Si). In the event of a nuclear accident, the total dose is postulated to reach 20 to 40 Mrad(Si). During such an accident, the equipment is expected to continue operating!

The Effect of Ionizing Radiation

Figure 1 shows a simplified representation of a MOSFET structure. When ionizing radiation is absorbed within this structure, electron-hole pairs are created. Because of the radical difference in charge mobility between the gate oxide and the body semiconductor, the effect of ionization in each of these regions is different.

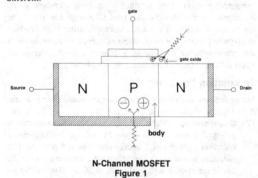

N-Channel MOSFET
Figure 1

In the body region, the mobility is relatively high, and the ions re-combine or are swept out quickly. In the gate oxide where the mobility is very low and the difference in mobility between holes and electrons is quite large, positive charge can be trapped in the oxide for extended periods of time.

In a low dose rate environment, the formation of ions is slow, and very little radiation induced drain-source current (I_{DSS}) will exist. In a prompt gamma or flash x-ray environment, however, the degree of ionization will be many orders of magnitude larger. I_{DSS} can become destructively large due to the induced photo-current. This problem is aggravated by the presence of a parasitic BJT in the MOSFET structure (Reference Figure 1). Photocurrents generated in the p region act like injected base current and, at the high levels of radiation present in a prompt gamma exposure, may turn on the parasitic BJT even though the base is shorted to the emitter.

If the circuit in which the device is operating has a low impedance and significant stored energy, large currents may flow and damage the device. Additionally, second breakdown may occur in the BJT which could destroy the device.

Figure 2 shows a typical I_{DSS} response for a prompt gamma pulse. Notice that the current pulse in the MOSFET is longer than the radiation pulse. This is due to storage time in the BJT.

Graphs of typical pulse amplitudes and storage times from Volmerange and Witteles[1] are given in Figures 3 and 4. From

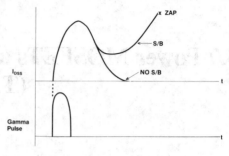

Prompt Gamma Pulse Response
Figure 2   © 1982 IEEE ref. #1

Figure 3, it can be seen that for dose rates above 10^9 rad/s, I_{DSS} becomes very large.

Because of the very short pulse length and the low total dose (10-1000 rad(Si)), very little charge is left in the gate oxide. If the device is not destroyed, it will continue to function normally after the prompt dose with minor characteristic changes.

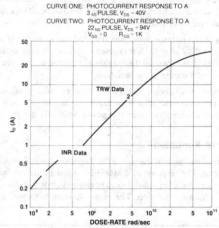

Photocurrent Due to Prompt Gamma
Figure 3   © 1982 IEEE ref. #1

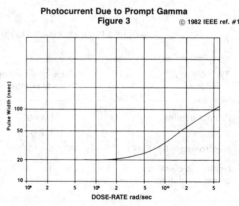

Pulse Width Response to Gamma Pulse
Figure 4   © 1982 IEEE ref. #1

For long term, low rate irradiation, the trapping of positive charge in the oxide is the dominant effect. When an ion pair is formed in the gate oxide, as shown in Figure 5, the electron has much higher mobility than the hole and thus tends to escape from the gate oxide structure leaving a net positive charge trapped in the gate. In effect, the trapped charge is equivalent to a battery in series with the gate as indicated in Figure 5. The apparent threshold voltage (V_{th}) is decreased in n-channel devices and increased in p-channel devices as the trapped charge is increased. The rate at which electrons escape is a strong function of the gate bias (V_{gs}). If V_{gs} is positive, the amount of charge trapped, for a given irradiation, will be greater than if $V_{gs} = 0$. If V_{gs} is negative, the amount of trapped charge is further reduced.

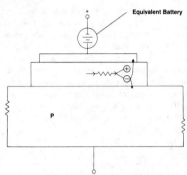

Ion Trapping in the Gate Diode
Figure 5

Test irradiations of MOSFETs show that the trapped charge is proportional to the radiation at levels up to 50 KRad and then begins to level off. Figures 6 and 7 are typical examples of the shift in V_{gs} as a function of total dose. These are curves for individual devices. When a number of devices of the same type are irradiated, there will be differences in the response of individual devices even from the same lot code. Figures 8 and 9 show typical variations between devices of the same type.

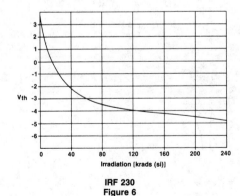

IRF 230
Figure 6

A further complication which needs to be taken into account is that there is more than one value for V_{th}. The data sheet V_{th} is the value of V_{gs} at which a particular current is flowing, usually one to ten mA, and represents the point at which the device is just turning on. From the viewpoint of a switching application, the device may not be on at all at one mA. The threshold of interest is the point at

which five or ten amperes are flowing. This effect is shown in Figure 10 where V_{gs} versus irradiation is plotted with I_D as a parameter.

In a p-channel device, the shift in threshold voltage is also negative, but in this case that means V_{th} is increasing and *more* drive voltage is required to turn on the device. An example for a p-channel device is given in Figure 11.

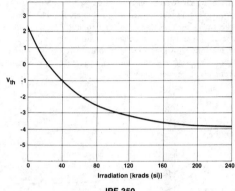

IRF 350
Figure 7

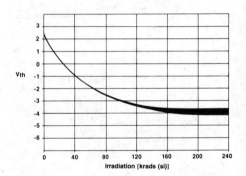

IRF 230
Figure 8

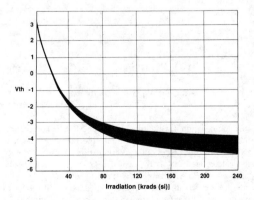

IRF 350
Figure 9

5

The value of V_{gs} during irradiation plays a major role for determining the shift in V_{th}. Figure 12 indicates the shift in V_{th} for different values of V_{gs}. Because more charge is trapped, the more positive V_{gs}, the greater the shift in V_{th}.

The total trapped charge or, equivalently, the shift in V_{th} for a given total dose is not stable with time. After irradiation V_{th} will gradually return to a value near the pre-irradiation value; this is called annealing. Inside the gate oxide, there are a few free electrons which over a period of time will combine with the trapped holes thus reducing the net positive charge. The time scale of the annealing process is a strong function of the ambient temperature-years at room temperature or minutes at 250°C. For long missions with low dose rates, annealing can significantly reduce the V_{th} shift associated with a given total dose, especially if operation is at elevated temperatures.

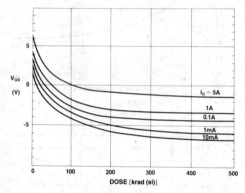

V_{th} Shift for Different Levels of I_d
Figure 10   © 1982 IEEE ref. #1

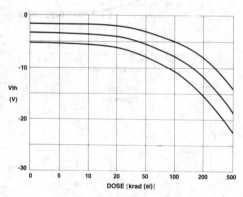

P-Channel V_{th} Shift with Radiation
Figure 11   © 1982 IEEE ref. #1

Ionizing radiation has a third effect on the MOSFET: it reduces the breakdown voltage. To maximize the breakdown voltage (BV_{DSS}), various types of fixed termination are used to reduce the voltage gradients as the depletion layers approach the die surface. The SiO_2, which covers the field termination and the interface between the oxide and semiconductor, will trap both positive and negative charge. At moderate irradiation (1-50 krad(Si)) levels, the trapped charge is primarily positive. At higher levels, negative trapping begins to occur which tends to cancel the positive trapped charge.

The effect of trapped positive charge is to locally increase the field gradients which in turn reduces BV_{DSS}.

The change in BV_{DSS} is a function of the device voltage rating, the specific design details, and the voltage across the device (V_{DS}) during irradiation. Figures 13 through 15 show the behavior of typical devices. Notice in particular the role of V_{DS} during irradiation. The higher V_{DS} is, the greater the reduction in BV_{DSS}. Also notice the drop in BV_{DSS} at 10-50 KRad(Si) levels followed by substantial recovery for two of the devices. It is reasonable to expect that annealing will reduce the trapped charge in low dose rate environments.

The Effect of Neutron Irradiation

Neutrons do not create ions, but they do disrupt the crystal lattice. The effect of neutron damage is to increase the resistance of the drain region. In a power MOSFET, as the breakdown voltage is increased, both the thickness and the resistivity of the drain region are increased. The result is that the relative increase of $R_{DS(on)}$ is much greater in higher voltage devices. A comparison of three devices with different voltage ratings is given in Figure 16. Up to a fluence of 10^{13} neutrons/cm^2 little effect is seen. When the fluence is raised to 10^{14} neut/cm^2, the low voltage device is unaffected, but the high voltage device has a drastically higher $R_{DS(on)}$.

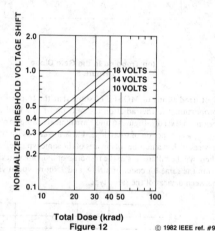

Total Dose (krad)
Figure 12   © 1982 IEEE ref. #9

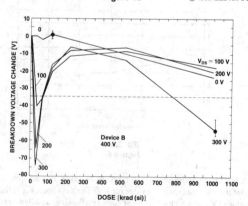

BV_{DSS} Shift with Radiation
Figure 13   © 1983 IEEE ref. #10

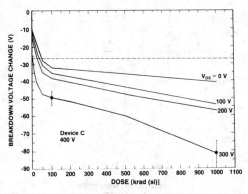

BV_DSS Shift with Radiation
Figure 14
© 1983 IEEE ref. #10

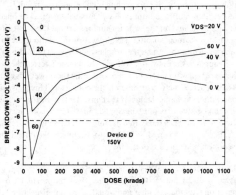

Figure 15
© 1983 IEEE ref. #10

It should be kept in mind that the graph shown in Figure 16 is representative of the change in $R_{DS(on)}$. However, there can still be large variations in the actual shift seen in a particular device. The trend is clear: The higher the voltage rating, the greater the susceptibility to neutron damage.

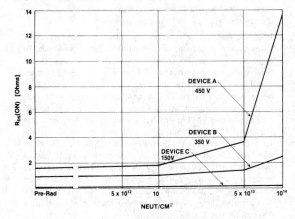

Plot of $R_{DS(on)}$ vs Neutron Fluence
Figure 16
© 1981 IEEE ref. #2

Neutron irradiation testing often shows a shift in V_{th} as well as an increase in $R_{DS(on)}$. The shift in V_{th} is not due to the neutrons but to the gamma radiation normally present in neutron test sources. Some care has to be taken with the data, to separate the two effects. If V_{th} is changed with a fixed V_{gs}, $R_{DS(on)}$ may appear to change. This is particularly a problem in testing P-channel devices where V_{th} may rise sufficiently so that the device is not fully enhanced and $R_{DS(on)}$ is correspondingly increased.

Other Radiation Damage

Both ionizing and neutron irradiation will cause other parameter shifts such as lower gm. Since in most cases, the shifts are small and represent second order effects, we have chosen to ignore them in this discussion. At very high levels of irradiation (>1 MRad(Si) or 10^{15} neutrons/cm²), these other effects may not be second order. Unfortunately at this time, very little test data is available at extremely high irradiation levels.

Strategies to Minimize the Effect of Radiation Damage

Several strategies may be employed to minimize the effect of MOSFET radiation damage on circuit operation.

The first and most obvious is to employ shielding. In a terrestrial application, like a nuclear plant, the addition of lead shielding for the instrumentation is quite practical. In a space application, however, large amounts of shielding are out of the question because of the weight yet some degree of shielding is usually possible.

The second strategy is to employ devices specifically hardened for radiation. Many changes are possible in MOSFET design and processing which will improve the radiation resistance. The feasibility of devices with reasonably stable characteristics up to 10 MRad(Si) has been demonstrated (3). It is interesting to note that some of the first generation power MOSFETs, (which are no longer in production), were much more resistant to radiation damage than today's standard products.

Siliconix has begun development of radiation resistant devices which will be available in the future.

Until such time as "rad hard" devices are available, the most practical strategy is to design the circuit so that changes in device characteristics do not degrade circuit operation.

As was shown earlier, there are four distinct characteristic changes which must be accommodated: V_{th} shift, $R_{DS(on)}$ increase, BV_{DSS} reduction and photo-current turn-on. One or more of these may occur in the same application depending on the irradiation scenario.

V_{th} shift is essentially a gate-drive problem. To maintain proper operation, the V_{gs} applied to the gate of an n-channel device must be sufficiently positive (>+10 V) to turn on the device fully before irradiation and yet negative enough (2–5 V below the final V_{th}) to turn off the device after the maximum irradiation, allowing for device-to-device variations. In practical terms, this means a drive waveform such as shown in Figure 17 must be used. One advantage of the negative bias during the off interval is that it reduces the shift in V_{th} at a given total dose. The exact amount of reduction at a given total dose would depend on the bias values and the duty cycle. Figure 18 gives an example of the effect of bias conditions on the shift in V_{th}. In this example it can be seen that even though the gate drive duty cycle is 50% the shift in V_{th} lies much closer to the −10 volts DC bias curve than to the +10 volts bias curve. To obtain an accurate value for the shift, test

5

devices should be irradiated with the actual circuit voltages, currents and duty cycle applied. The use of negative bias, while very effective in maintaining normal circuit operation, does bring with it some problems. In the circuit shown in Figure 19, for example, V_{gs} can reach -30 V. Most MOSFETs are only rated for ± 20 V except for Siliconix parts which are rated for ± 40 V on the gate. The gate to source breakdown voltage can become a limiting factor in the design.

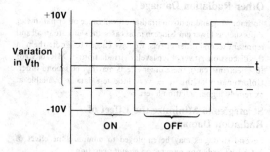

Gate Drive Waveform
Figure 17

Another problem in some circuits is how to couple the asymmetrical (from a volt-second point of view) waveform given in Figure 17, to the gate in a transformer isolated circuit.

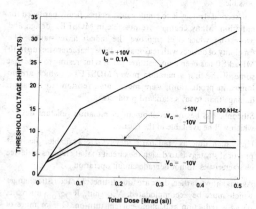

The Effect of Gate Bias on V_{th} Shift
Figure 18

© 1982 IEEE ref. #6

NEGATIVE GATE BIAS EXAMPLES

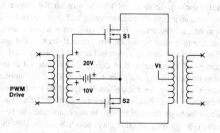

Negative Bias in a Converter
Figure 19A

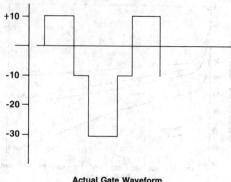

Actual Gate Waveform
Figure 19B

If a power converter circuit is turned on and kept on during irradiation, the negative bias voltage can be self-generated. In the pre-irradiation off-condition, the switches will be off. The bias is not needed to hold off the switches nor for starting. If however, the converter is turned off after a high level of irradiation, the bias voltage will disappear, and the switches will be on. If the bus power is still applied, the converter will be a short circuit across the bus. There are two ways this may be avoided. First the power bus could be disconnected before turning off the converter. This is usually done with a mechanical relay. If the converter is to be turned back on, the negative bias must be applied first. The second method would be to supply a rad hard bias supply which is always present. The bias supply would still have to be interlocked to a bus disconnect relay in case of failure.

The degradation in BV_{DSS} may be circumvented by using a device with an initial BV_{DSS} higher than the normal circuit operation required. Another step would be to minimize the voltage across the device during irradiation. There are a number of topological means for doing this in power converters. For example, instead of using a parallel (or push-pull) converter circuit, a half-bridge circuit could be used, reducing the voltage across the devices by a factor of two.

However, both of these schemes bring a penalty: $R_{DS(on)}$ will be increased unless larger devices are used. If neutron radiation is also present, the use of a higher voltage device will exacerbate the increase in $R_{DS(on)}$ due to neutron damage.

Photo-current turn-on is an easier problem to solve. To prevent damage, the current in the device must be limited to a safe value. In a power converter, the simplest approach is to use a circuit topology which inherently limits the current even when all switches and diodes are in conduction. An example of such a converter circuit is given in Figure 20, and many other circuits of this type exist. In some circumstances however, using this type of topology may not be enough. The problem arises from the turn-on of the parasitic BJT. The circuit shown will limit the current to the normal switch current, but this may be too high for the BJT, which does not normally carry any of the switch current. Second breakdown may then occur. Very short pulses (<100 nsec) may not be damaging, but longer pulses may destroy the FET. One means to eliminate this problem would be to deliberately turn on both FETs during the pulse interval so that much of the current is diverted through the FET channels.

Another protective scheme would be to place a BJT, as shown in Figure 20, to shunt the current. This device is shorted from base to emitter and plays no role in normal circuit operation. It is assumed that the shunt BJT is capable of absorbing the normal switch current without failing.

One word of caution, not all "current fed" converters are self protecting. For example, the circuit shown in Figure 21 will have very high switch current due to the photocurrent in the diode.

The total shift in V_{th} may be greatly reduced in long life, relatively low irradiation missions by operating the FETs at elevated temperatures (125 to 140°C) to accelerate annealing. Obviously higher operating temperatures compromise reliability therefore this technique must be applied judiciously.

The increase in $R_{DS(on)}$ due to neutron damage is essentially a thermal problem. The power dissipated (P_T) in the FET is proportional to:

$$P_T \alpha \ I^2 D(rms) \ R_{DS(on)} \tag{1}$$

The junction temperature (T_j) may be expressed as:

$$T_j = Ta + R_{\theta ja} P_T \tag{2}$$

Where:

T_a = ambient temperature
$R_{\theta ja}$ = thermal impedance from junction to ambient.

As $R_{DS(on)}$ increases, T_j, will increase. The effect is, however, not linear because $R_{DS(on)}$ is also a positive function of T_j so the increase in $R_{DS(on)}$ due to radiation is magnified. The thermal system has positive regeneration and can go into thermal runaway if $R_{DS(on)}$ becomes too high. This dependence and the determination of thermal stability are treated in detail in a Siliconix Application Note (4).

There are several things that can be done to minimize this problem:
1) Use the lowest voltage MOSFET consistent with the circuit requirements. In some cases, the use of series low voltage devices may be helpful. The direct series of several devices is not usually attractive from a circuit point of view, but there exist (5) power converter circuits which automatically divide the voltage across several switches. In these topologies, the use of lower voltage devices is quite practical.
2) Use a device with an initial $R_{DS(on)}$ much lower than the power level requires. If one device is not adequate, several can be paralleled.
3) Minimize the ambient temperature.
4) Minimize $I_{D(rms)}$. In power converter circuits, this can be accomplished by making the averaging inductor larger or by using alternate topologies (4). In particular, the family of converter circuits which includes that shown in Figure 19 are very useful (5).
5) Minimize $R_{\theta ja}$. This may be done by using a lower impedance heatsink, a non-isolated device mounting to the sink, parallel devices (lower effective $R_{\theta ja}$), and in general, good mounting practices.

In a practical design, one or more of these techniques may be used to arrive at a satisfactory solution.

Some Thoughts on Radiation Testing

The radiation test data shown in earlier sections is, for the most part, typical. However, the data from some devices, even of the same type, can be very different because the degree of parameter shift in a given device at a given irradiation level is strongly dependent on small variations in processing, the specific operating conditions during irradiation, and the character of the radiation source.

Wide variations in test results for the same type of device done at different times or in different facilities is unfortunately typical of radiation testing of all semiconductors.

If high levels of irradiation are anticipated and there is doubt concerning the adequacy of the design the best course of action is to install devices, from the same lot as those to be used in the flight equipment, in the actual circuit-then irradiate them while the circuit is operating.

Conclusion

The electrical characteristics of power MOSFETs are altered by radiation. The degree and type of characteristic change depends on the type, intensity, and duration of the radiation. In general, the effect of changing characteristics can be circumvented by proper circuit design. Successful power converter designs have been demonstrated for total doses greater than 200 Krad(Si). When radiation hardened MOSFETs become available, circuits capable of surviving total doses approaching ten Mrad(Si) should be practical.

Clearly the MOSFET is suitable for use in high radiation environments.

We recommend that the designer read the papers listed in the bibliography to obtain a more detailed understanding of radiation effects in power MOSFETs.

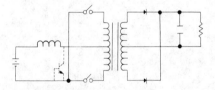

Radiation Activated Crowbar Examples
Figure 20

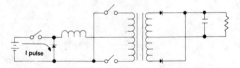

Unsuitable Current Fed Converter Example
Figure 21

Acknowledgements

Several of the figures used in the text were derived from the work of others:

Figures 2, 3, 4, 10 and 11	Reference 1
Figure 12	Reference 9
Figures 13, 14 and 15	Reference 10
Figure 16	Reference 2
Figure 18	Reference 6

5

References

1) H. Volmerange and A. Witteles, "Radiation Effects on MOS-POWER Transistors," IEEE Transactions on Nuclear Science, Volume NS-29, No. 6, December 1982.

2) D. Blackburn, et. al., "VDMOS POWER Transistor Drain-Source Resistance Radiation Dependence," IEEE Transactions on Nuclear Science, Volume No. 28, No. 6, December 1981.

3) G. Roper and R. Lowis, "Development of a Radiation Hard N-Channel Power MOSFET," IEEE 20th Annual Conference on Nuclear and Space Radiation Effects, July 18-20, 1983.

4) R. Severns, "SOA and Thermal Design for MOSPOWER Transistors," Siliconix Application Note AN83-10.

5) R. Severns, "Switchmode and Resonant Converter Circuits," International Rectifier Application Note, 1981.

6) D. Blackburn, et. al., "Ionizing Radiation Effects on Power MOSFETs During High Speed Switching," IEEE Trans. on Nuclear Science, Volume NS-29, No. 6, December 1982.

7) W. Baker, Jr., "The Effects of Radiation on the Characteristics of Power MOSFETs," Proceeding of Powercon 7, 1980, p. 3.

8) S. Rattner, "Additional Power VMOS Radiation Effects Studies," IEEE Trans. Nuclear Science, Volume NS-27, 1980.

9) S. Seehra and W. Slusark, "The Effect of Operating Conditions on the Radiation Resistance of VDMOS Power FETs," IEEE Trans. Nuclear Science, Volume NS-29, December 1982.

10) Blackburn, Benedetto and Galloway, "The Effect of Ionizing Radiation on the Breakdown Voltage of Power MOSFETs," IEEE Trans. Nuclear Science, Volume NS-70, No. 6, December 1983.

6.1 Power Supply Applications
6.1.1 High Frequency Power Conversion with FET-Controlled Resonant Charge Transfer

Abstract

The author discusses FET-controlled, high frequency, sinusoidal, power conversion. An off line multi-output power converter with output power exceeding 250 watts up to 500 kHz has been demonstrated. System operation is such that current flow occurs as successive packets of charge made from individual sinusoids of current. The resonant charge transfer process where switching loss is independent of frequency is described along with FET circuit implementation.

Introduction

Pressure to design cost effective reliable switching power supplies in smaller space points to high frequency operation. Further, the thermal impact on component reliability requires that supplies have higher efficiency which implies reduced switching loss.

Power conversion employing square waves has a fundamental disadvantage, vis-a-vis sinusoidal operation, because squarewave operation requires that the power switch (or its associated snubbing components) dissipate energy whenever (each time) voltage and current are handled (interrupted). Therefore, switching loss is directly related to the operating frequency; which is not the case using sine waves.

This paper presents a method that uses resonant charge transfer for DC to DC conversion. Resonant charge transfer gives low switching loss because each sinusoidal wave of current is allowed to terminate before the voltage forcing function is withdrawn. As a consequence, with resonant operation, the controlling switch turns both on and off at zero current to give

lower component loss and reduced component stress. Also there is no voltage or current overshoot.

Resonant Charge Transfer (RCT)

Figure 1 shows a diagram of the power-train for an RCT converter. Operation is as follows:

Filtered AC is converted to DC between $(160 \sim 250)$ V. FET Q_1 and Q_2 are alternately turned on for a fixed period of 6 μs but with a load-dependent variable PRF. The primary of the power transformer T is connected to the common point of Q_1 and Q_2 and in series with the resonant circuit consisting of L_1 and $C_3 + C_4$. Diodes D_3 and D_4 constrain the voltage excursion across C_3 and C_4 thereby stabilizing the resonant tank.

The secondary sections of T are auto-connected to give best cross-regulation and the main +5 V output is filtered and fed back (Vsense) to control the switching PRF of Q_1 and Q_2.

Low voltage differential series-pass FET regulators are used to regulate the auxiliary outputs.

A separate low power transformer is used to supply logic voltage but only during turn-on. After the system is up and running the rectified $+(13 \sim 15)$V secondary voltage is fed back to the input of a +5V regulator chip in the bias supply. This is done not only to improve the system efficiency but in addition to allow the logic system to be fed through the converter to utilize the energy stored on C_1 and C_2 to improve hold-up-time after loss of the AC input.

6

Reprinted by courtesy of PCI, April 1983 Proceedings.

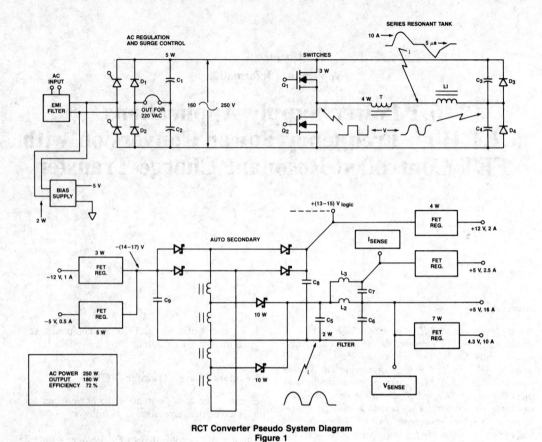

RCT Converter Pseudo System Diagram
Figure 1

	E	Ip	Ī	PIN	P_O (MAX)

AC POWER 250 W
OUTPUT 180 W
EFFICIENCY 72 %

FIXED
$E_H = 2E - N5 V$

$I_P = E_H \sqrt{\dfrac{C}{L}}$

$W = \Pi \sqrt{LC}$

$\bar{I}_{Pri} = \dfrac{2}{\pi} I_P \dfrac{W}{W + D}$

$POWER = E\bar{I}_{Pri}$

	E	Ip	Ī	PIN	P_O (MAX)
MAX →	125	12	6.0	750	420
NORMAL →	110	10.0	5.0	550	375
MIN →	100	8.0	3.6	360	260
HOLDUP →	80	6.0	2.6	200	150(80)
RUN →	110	10	2.3	250	180

W = 5   D = 1   n = .72

Resonant Waveforms
Figure 2

Figure 2 shows the current waveforms for the resonant tank. Relevant operating values are obtained as follows:

Assuming a minimum operating DC bus voltage of $E = \pm 80$ V and that the junction of the two 0.05 μf resonating capacitors is at -80 V, then when FET S^+ is turned on a sinusoidal pulse of current will flow. The peak of the sinusoid is equal to 6 A. This is because the reflected voltage across the primary winding of the 14 to 1 turns ratio transformer is about 80 V, and therefore, the voltage head E_H for the resonant loop is also equal to 80 V (the effective primary loop inductance including reflected secondary leakage is $\approx$ 18 μH). At full power output S^+ and S^- are alternately triggered at 6 μs (W+ D) intervals and the 4 μs sinusoids are separated by 2 μs intervals (D). Under this condition the average current is 2.5 amperes and the power drawn from the DC source is 200 watts.

At the maximum DC bus voltage of 250 V ($E = \pm 125$ V) the voltage head is 170 V which gives a 12 A peak pseudo-sinusoid pulse of current with a linear fall rate of about 4.5 A/μs (80 V $\div$ 18 μH). With 12 A pulses the maximum average current flowing in the resonant tank is now about 6 A and the power taken from the DC source is 750 watts.

RCT Control

Figure 3 shows how the RCT converter is controlled. The resonant primary sinusoids are transformed and rectified to give approximately 80 A 5 μs (nominal width) current pulses. Using the values shown for the capacitive input Pi-filter, V(A) and V(B) have the wave shape indicated. At 10 A output current the 1 μH choke has a current variation of ± 0.5 A.

In order to control the output voltage at $\textcircled{B}$, V(B) is compared to a stable reference voltage such that when V(B) goes below the reference the one-shot multivibrator (OS) is triggered. The OS produces a 6 μs (fixed width) pulse which retriggers the primary power switches. Accordingly, a new packet of charge is added to the 15 mf filter input capacitor whenever it is required (up to a maximum PRF equivalent to 6 μs). This control strategy is termed fixed-pulse-adaptive-variable-PRF because the PRF automatically adjusts to compensate for input DC bus variations, as well as load changes. This discontinuous format, where the control loop gain is zero between pulses, is quite stable because the 6 μs pulses that drive the primary switches are inverted and fed back to the reference side of the comparator. This assures that each pulse "uncrosses" the comparator input.

The measured output regulation is 1 mV/A ($R_0 = 1$ mΩ) from zero to 16 A. Also, it has been found that by adding a feedback path from V(A) to the reference side of the comparator the output impedance R_0 can actually be made negative. The reason is that with V(A) feedback, the current flowing through the input resistance in the filter choke causes the average value of V(A) to increase. Thus the output voltage is a stable positive function of load current. One last item, with 5 μs sinusoids (100 kHz) the 1 μs choke has an impedance of 500 mΩ. This is sufficient for effective voltage filtering yet allows an output current slew rate of 1.0 A per mS per mV (very good transient response).

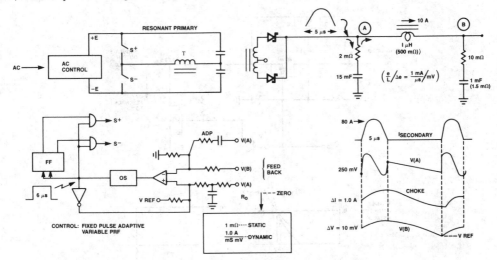

System Waveforms and Control
Figure 3

FET Regulator

In the future the FET will find wide application in the design of low voltage DC linear regulators. Figure 4 shows an example of a very low drop self-protected switch for an isolated 2 A output.

Two individual 50 mΩ n channel FETs are connected directly in parallel and controlled by shut down signal $\overline{SD}$ which is obtained from a logic chip with open collector output. When $\overline{SD}$ is high the two FETs conduct current with a $r_{DS(on)}$ of about 40 mΩ (at 100°C). The maximum gate to source voltage is set by the Zener diode to give current limiting. Note that the FET type and Zener voltage can be selected to give current limit (I_L) at the temperature stable point of the FET transconductance. When $\overline{SD}$ is low a third FET, used as an output shunt, clamps the output terminal to ground.

Figure 5 shows an example of a low drop, low output impedance, high current FET regulator.

The design provides for an isolated +4.3 V output with a current limit at ≈ 11 A. Two low voltage 50 mΩ FETs are connected in parallel and their source voltage is compared to REF$_V$. Feedback through amplifier A (type LM324) controls the FET gate voltage to give an output regulation of ≈ 1 mV/A. In order to obtain a current limit function the output of A is also fed back through B. Accordingly, when the FET gate voltage exceeds REF$_I$, the voltage level of the REF$_V$ input to A is reduced. The values for REF$_V$ and REF$_I$ are set to current limit at about 11.5 A. Also, REF$_C$ is set such that the negative going output of B "trips" C to crowbar the output at a level just above current limit.

Signal $\overline{SD}$, controlled from under/over voltage (uV/oV) circuits, which are not shown, serves to turn off the two series-pass FETs and crowbar the 4.3 V output. When low, $\overline{SD}$ turns off the regulator. When $\overline{SD}$ goes high the regulator will turn on, provided the output current is below I_L.

The nonlinear connections to capacitor C_2 and to the gate of the crowbar FET serve to ensure that when $\overline{SD}$ changes state the crowbar FET turns on after the series-pass FETs turn off and turns off before the series-pass FETs turn-on.

Capacitor C introduces a small delay which provides a transient over-current capability.

The regulator configuration of Figure 5 is simple (mainly three FETs and one LM324 chip), has low drop, and is flexible and functional in design (i.e., additional FET units can be paralleled to give higher current). Further, when using n-FETs for positive voltage regulation or p-FETs for regulating negative voltages, the basic regulator loop is unconditionally stable and gives good static regulation (≈ 0.25 A/μs−mV). The regulator loop is unconditionally stable because the series-pass FETs are used as a source follower (voltage gain < 1.0) yet they have high frequency response (even at low drawn-source voltage) and therefore the LM324 amplifier provides the low frequency dominant pole.

The reliability of FET low voltage regulators should prove quite high; the power components have low internal voltage stress, the FET has inherent ability to handle large short duration power peaks, and for DC regulator application large device leakage currents

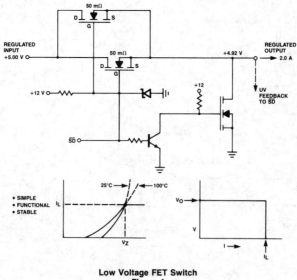

Low Voltage FET Switch
Figure 4

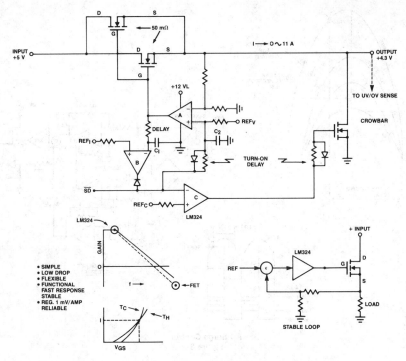

Low Voltage FET Regulator
Figure 5

are permissible. In fact in "test regulator circuits" the FETs have been run so hot "the leads turn blue" with no performance degradation.

Surge Protection; Feedforward Phase Control Rectifier Control

The goals for the power supply using RCT are shown in Table 1. From the outset it was appreciated that this is a challenging specification especially for volume production. The design requires low power component stress, and AC surge control is mandatory.

Table I
Resonant Regulator Specification

Vin $\begin{cases} (80 \sim 150) \text{ V}_{rms}, 200 \text{ V}_{rms} \text{ surge} \\ (160 \sim 300) \text{ V}_{rms}, 400 \text{ V}_{rms} \text{ surge} \end{cases}$

Po: 180 W; ± 12 V, ± 5 V, $+5$ V (16 A), $+4.3$ V (11 A)

Regulation: ± 50 mV (total static + dynamic)

Hold Up Time: 30 mS, +5 V (16 A)

B & W: All Standard, Remote Sense, Sequenced Turn On/Off, Self Protecting, uV/oV Clamp

Size: 254 in^3 (.147 ft^3)

Reliability: 5 yrs.

Cost: minimum

AC input-to-DC bus control is accomplished using a conventional SCR half-bridge with a perhaps unconventional but flexible SCR controller. The control method is shown in Figure 6.

For an input sinusoid of AC voltage a priori amplitude information is not available. Never use the conventional zero crossing time delay method for SCR trigger if it is necessary to isolate AC surges for the DC bus; instead, use a post-peak level-recross logic system where AC surges are automatically eliminated. The control system operates as follows:

Negative AC sinusoidal loops from the secondary of the bias transformer are used as input AC reference signals (the positive loops are distorted by the +5 V logic load) and inverted to give positive AC loops at Ⓐ. The AC waveform at Ⓐ is then compared to a DC reference level Ⓒ to give waveform Ⓓ. When the $\overline{\text{ISD}}$ signal is high waveform Ⓓ causes the SCR to be triggered at the trailing (rising) edge of Ⓓ. Noise protection is provided by setting the values of R and C so that insufficient trigger energy is available to the SCR below a minimum width of waveform Ⓓ.

The reference level at Ⓒ is adjusted by the peak synchronized squarewave Ⓑ where the resistor values are set to allow the SCR to fire just after the peak

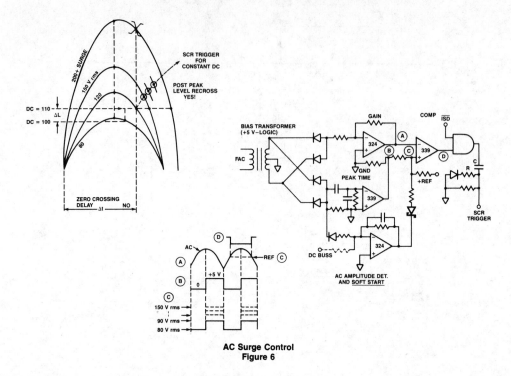

AC Surge Control
Figure 6

when the AC input is between 80 V_{rms} and 90 V_{rms}. This is done to give maximum DC bus voltage at low AC line.

Finally, because the output of a phase control rectifier is sensitive to amplitude dependent voltage slope (dV/dt); the trigger point is adjusted upward as the input AC voltage increases. This is accomplished by detecting the average AC peak amplitude and adjusting the reference level at Ⓒ. The 324 integrator also gives a soft start function at "Power-up".

Conclusions and Recommendations

Several authors have written excellent papers regarding the do(s) and don't(s) of high power, high frequency design so this subject is not discussed here. It suffices to say that the selection of magnetics and filter capacitors are important as is circuit layout and grounding. However, the following items should be emphasized.

1. Sinusoidal operation at high frequency is an important way to reduce the size of power supplies.

   Resonant operation, where the sinusoidal current pulses are allowed to naturally terminate, greatly reduces switching loss in the power handling components, especially in the power switches and power diode rectifiers. Accordingly, in the future

resonant conversion should allow much higher operating frequency than will square wave operation.

2. Reduced component stress.

   In low power systems (say less than 100 W output) component count may be a meaningful, perhaps dominant measure of reliability. However, in higher power ranges, especially when operating in harsh environment, component stress is dominant. The resonant mode, operating without current discontinuities, has little voltage overshoot, much reduced di/dt, and requires much less snubbing compared to square wave operation.

3. Use field effect transistors.

   In applications below 500 V and 50 A the FET – bipolar race is about over; the FET is fast, more rugged (where it counts, i.e., SOA), more efficient and cost effective. Indeed, for sinusoidal operation the simple low cost proportional drive long used with bipolar power switches is unavailable.

   The ease of design with FETs, their speed, voltage-activated low power drive requirements, uniformity and stability now allows them to be an item of commerce (almost anyone's equivalent type will work without circuit tweaking).

4. Use FETs for low voltage linear regulation.

   Low $r_{DS(on)}$ gives low voltage drop without saturation. The resulting high frequency response at low voltage coupled with the ease of drive, ability to parallel devices, the stability and uniformity of electrical parameters (especially transconductance), make them ideally suited for low voltage regulator applications.

5. Technology.

   The RCT system described has been operated with 1 μs sinusoids (it did not work as well as with 5 μs sinusoids—the magnetics require redesign).

However, the trend is clear; the FET switching speed is quite adequate at 1 MHz sinusoidal operation, as are film capacitors.

In the future the Gallium Arsenide FET technology offers potential for factors of 5 to 10 improvement in both speed and $r_{DS(on)}$. If this materializes, coupled with superior high frequency magnetic material such as Metglass, then sinusoidal operation in the $1.0 \sim 5.0$ megahertz range appears practical.

For higher power systems, resonant charge transfer (where current flow naturally terminates) is compatible with the SCR switch.

6

6.1.2 Practical Design Considerations For a Multi-Output ĆUK Converter

© 1979 IEEE. Reprinted, with permission, from PESC '79 IEEE POWER ELECTRONICS SPECIALISTS CONFERENCE, June 18–22, 1979, San Diego, CA, pp. 133–146.

Abstract

An experimental high frequency three output DC-DC converter design is presented in circuit detail, emphasizing the practical design needs of its contemporary optimum topology (ĆUK) output stages. Power component selection criteria are also given along with suitable load protection methods against output voltage reversals at turn-on. Inductor coupling criteria for ripple current reductions at input/output ports are developed including an alternate method for external tuning inductance insertion. Results of corresponding empirical evaluations of a 200 kHz, 55-W representative design are discussed.

1. Introduction

This decade has seen many significant advances in the art of power conditioning system design and contemporary electronic component technology. In the area of DC-to-DC conversion methods, the designer now has LSI components at his disposal for converter control, amplification, and supervisory functions. Yet, with all the current progress made in DC power system design, the main power translation stages have remained relatively untouched. This lag can be easily attributed to the need for this part of a DC-to-DC network to control relatively large amounts of power to perform its function. Consequently, the main power conversion stages must be formed from discrete active and passive circuit devices that contribute significantly to the total design's complexity, size, reliability, and power loss.

For these reasons, a great deal of effort has been expended on the search for a DC-to-DC power stage topology that has the *simplest possible circuit structure* with the *minimum* number of components while

yielding *maximum performance*. The current work to date has been spearheaded by a team of researchers at the California Institute of Technology who, in 1977, revealed their findings [1], to the design community. The topology was appropriately termed *optimum* as it achieved all the desired goals stated above.

Since 1977, further extensions [2, 3, 4], modeling methods [5, 6] and practical applications [7, 8, 9] have been suggested by the developers of the new topology which now make it an even more viable and attractive circuit structure for power conditioning designs. It is the intent of this paper to present a practical application of many of the features of the optimum topology (ĆUK) converter, including multiple outputs, isolation of input/output grounds and coupling of output filter inductances. Section 2 contains a short review of the basic topology operation and evolution of extensions for the reader's benefit. While total output power control achieved is less than 100 watts, the design presented does demonstrate the *compelling simplicity* of the topology and its many advantages.

Because of its promise of significant and controllable current ripple reduction at chosen circuit ports, a three-output ĆUK converter power stage was included in an experimental pulse-width modulation system undergoing evaluation for high-frequency operation under an in-house R&D study program. It must be emphasized that the resultant total system design to be described herein does not represent a truly optimum network from power loss or bandwidth standpoints, as the main goal of the study was design information and education for system practicality and producibility. Nevertheless, it was proven that it is *feasible* to utilize the ĆUK topology in converter applications requiring switching frequencies *approaching 200 kHz* with predictable circuit responses assuming proper layout and noise prevention practices [10] are followed.

Perhaps the most important achievement of this particular design is the *simultaneous reduction of input and one selected output port ripple currents to zero,* utilizing *tightly coupled* inductor windings and *small external inductances* for ripple current adjustments. From a system design standpoint, this implies *no need* for EMI filtering for switching current noise at the input to the converter. In addition, *no output filter capacitance* for ripple voltage reduction is necessary at the selected output port.

Section 3 presents the details of the total power conditioning system, including a somewhat *unusual ramp slope control* method of pulse-width modulation. Section 4 is devoted to power stage component selection for proper ratings and electrical values. In Section 5, the problem of output voltage reversal with power application is explored when input and output inductors are magnetically coupled, along with simple preventive circuit measures.

Ripple current projections in a four winding inductor assembly are considered in Section 6 with emphasis placed on achievement of predictable coupling coefficients to increase the effective inductances of selected windings for current ripple reduction. This subject is continued in Section 7 where it is demonstrated that the desired ripple current reduction can be accomplished by selection of small external inductances working in conjunction with *tightly-coupled* multi-inductor components.

Laboratory evaluations of a representative 200 kHz converter are presented in Section 8 as verifications of the analytical predictions and circuit responses given in earlier sections.

2. Background

The operation of the basic ĆUK converter stage has been covered in excellent detail in the references [1 through 4]; however, for those not familiar with the topology, it is worthwhile to briefly review the fundamental design features and extensions.

Figure 1 illustrates the evolution of the new converter topology from its basic form (1a) to coupling of input/output inductances (1b), introduction of transformer isolation (1c), and finally, recoupling of isolated input/output chokes (1d). The principal electrical features of all of these converters are nonpulsating input/output currents as shown graphically in Figure 1(a). It is easy to demonstrate that the voltage and

average current transfer ratio magnitudes for each of the circuits of Figure 1 are simply (continuous inductor currents assumed)

$$\frac{V}{V_\ell} = \frac{\eta D}{1-D} = \frac{\eta D}{D'} = \frac{I_1}{I_2} \quad , D' = 1 - D \quad (1)$$

where D is the duty cycle of conduction of transistor Q1 with respect to a constant timing period, T_S, and η is the converter circuit efficiency.

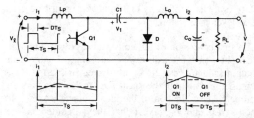

(a) Basic Inverting Topology with Nonpulsating Input/Output Current Waveforms

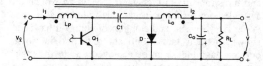

(b) Coupling of Input/Output Inductances (Single Core Construction)

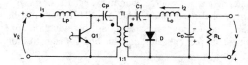

(c) DC-Isolated Extension with Separated Coupling Capacitance

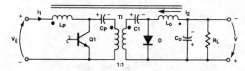

(d) Coupled-Inductor, DC Isolated Extension

Circuit Evolutions of the ĆUK Converter Figure 1

Capacitor C1 serves as the single energy transfer device between the input and output ports of the circuit in Figure 1(a). During the interval $D'T_S$ when Q1 is OFF, diode D is forward-biased and C1 is charging in the positive direction through inductance Lp. The voltage across the collector-emitter junction of the transistor is therefore positive and can be turned ON for the subsequent interval DT_S. As soon as Q1 turns ON, C1 becomes connected across the diode, thus reverse-biasing it. Thus, during the time DT_S,

C1 discharges through the load R_L and inductance L_O, charging the output capacitance C2 to a negative voltage. Circuit operation is repeated when interval $D'T_S$ is reached.

Because inductances are present on the input and output lines, the currents are nonpulsating with no sharp transitions that contribute to increased EMI as are present in conventional buck and boost designs. Output voltage magnitude can be less than, equal to, or greater than the input voltage value, depending on duty cycle as implied by (1), thus achieving true general level conversion.

The first extension of the converter topology to come to light was the feasibility of coupling of input and output inductances [2], as illustrated in Figure 1(b), without impact on the basic DC-to-DC conversion property. Examination of the ideal voltage waveforms across L_P and L_O reveal that they are identical in magnitude and timing, as diagrammed in Figure 2. Therefore, with proper choice of coupling coefficient and turns ratio, both inductances could be wound on a single core resulting in obvious reductions in converter size, weight, and component count. It has been shown, in fact, that significant reduction in ripple current magnitudes can be achieved due to the magnetic coupling between the chokes if the turns ratio and coupling coefficient maintain specific mathematical relationships [3]. DC bias effects on the core material are, of course, additive because of the required phasing between windings. This fact must be taken into account when selecting the induction core material to prevent saturation due to DC magnetization effects.

The problem of an inverted output not isolated from input ground potential was solved by a later extension of the topology [4]. As shown in Figure 1(c), the energy-transfer capacitor is simply broken into two parts, Cp and C1, with an isolation transformer separating the two halves of the circuit. Since no average or DC voltage can exist in the ideal sense across either transformer winding or the two inductances, the voltage on Cp is V_g and that on C1 is V. Note that the sum of these two capacitor voltages is $V_g + V = V_g/D' = V/D$, the same potential as that across the single coupling capacitance C1 of Figures 1(a) and (b).

Since both windings of the transformer in Figure 1(c) are isolated by the two energy-transfer capacitors, no DC transformer core magnetization can take place. Automatic volt-second balance is therefore achieved in the topology, requiring no special magnetic or circuit designs for balancing, as might be the case in a conventional "push-pull" converter. A smaller ungapped, square-loop core can be used for the transformer as a result of this circuit feature, with lower core and winding copper losses than would be possible with comparable converter topologies, as discussed in length in [4].

Because the basic converter operation of Figure 1(a) is unchanged from that of Figure 1(c), the desirable feature of inductance coupling is still possible, as shown in Figure 1(d), reducing the number of circuit magnetic assemblies to two.

The extension to multiple and isolated outputs [4] should now be obvious, as shown in Figure 3 for a three-output design. Again, by coupling of circuit inductances on one magnetic structure, only two magnetic assemblies are required for the total power stage. However, one must judiciously choose the coupling coefficients and turns ratio relationships be-

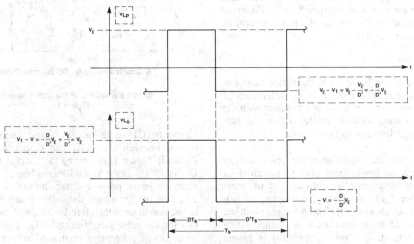

Ideal Lp, L$_O$ Voltage Waveforms for the Circuit of Figure 1a
Figure 2

tween windings in order to steer and reduce ripple currents in the directions chosen for design needs. Transformer turns ratios are not restricted to 1:1, but the ideal transfer voltage relationship specified by (1) must be modified to include the transformer turns ratio factor. For example, for the second output V_{o2} in Figure 3,

$$\frac{V_{o2}}{V_\ell} = n_2 \times \left(\frac{\eta D}{1-D}\right), \quad n_2 = N_{s2}/N_p \qquad (2)$$

One further observation about the topology of Figure 3 is needed. Note that the primary and secondary coupling capacitances have been relocated to the other side of the respective primary and secondary transformer windings in contrast to those shown in Figure 1(c) or 1(d). Because the capacitances are transformer coupled and isolated by T1, they can be placed on either side of their respective winding with no change in circuit operation. Since the cases of electrolytic capacitors are usually common to their negative terminals this placement effectively grounds the cases of these capacitors, thus reducing the high frequency radiated noise due to the pulsating switching currents in these capacitors.

3. Experimental Circuit Designs

In order that the results of the development study of the ĆUK converter operation and its control electronics could be applied to representative production program needs, a three-output configuration was selected, each isolated from one another and from the input DC power ground. Voltage levels were set at +5 V, +15 V, and −15 V DC, with corresponding maximum current capabilities of +5 A, +1 A, and −1 A DC, respectively. Since the +5 V load was the highest and desired tolerance, the lowest of the three outputs, the +5 V output was chosen for converter pulse-width modulation (PWM) control with the ±15 V lines controlled by cross-regulation through the converter transformer secondary turns ratio relationships.

Figure 4 is a schematic of the overall converter system, with each functional area identified. Like its ĆUK output power stage, the PWM control system is somewhat *unique* with respect to its control method. Because currently available PWM LSI integrated circuits do not use this control and isolation approach, SSI logic and discrete parts were employed.

The system as shown consists of an error amplifier, individual clock and ramp generators, voltage comparator, amplification and buffer stages for correct gate drive levels for the power VMOS converter switch. Isolation between input/output grounds is achieved by allowing the error amplifier, which is powered by the +5 V output, to interface with the ramp generator control through a *high speed* optical coupler. A simple shunt regulator, consisting of a temperature-compensated reference diode (VR3) and bias resistor (R4) provides the single regulated voltage needed for all remaining PWM Controls. The input +28 V is used as the unregulated voltage source for this low power regulator.

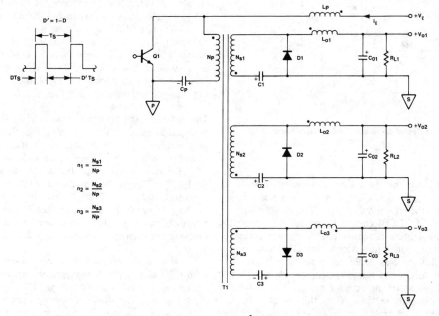

Three-Output, DC Isolated, Scaled Version of the ĆUK Converter (Coupled-Inductors)
Figure 3

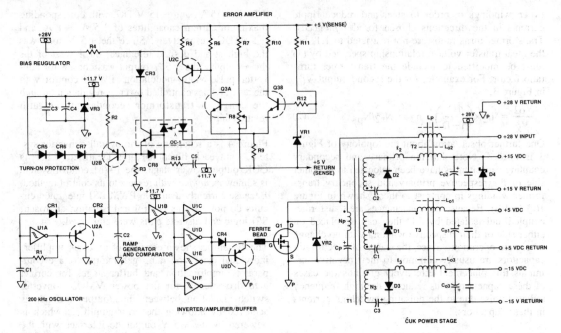

Experimental Converter System
Figure 4

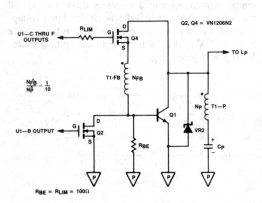

Alternate Design Approach for Converter Power Drive
Figure 4A

The PWM operates as a fixed frequency, *variable OFF* time system. A change in +5 V potential is detected by the error amplifier which adjusts the light-emitting diode (LED) current in the optical coupler, OC-1. This current change alters the emitter current of the light sensitive transistor within OC-1 changing the base voltage bias of the ramp control transistor, U2B. This action, in turn, *changes the slope* of the ramp voltage appearing across capacitor C2, causing a corrective shift in time when this ramp voltage reaches the fixed high level input voltage threshold of a CMOS Schmidt trigger inverter gate (U1B), designated as the comparator for the PWM system. This

ramp slope control method is diagrammed in Figure 5 for duty cycle adjustment.

U1A, another Schmidt trigger gate, in conjunction with R1, C1, U2A, and diode CR1, form a very simple 200 kHz astable oscillator for fixed-frequency PWM synchronization. When the output of U1A is in a low state, U2A discharges both the oscillator capacitor C1 and the ramp capacitor C2 through diodes CR1 and CR2 respectively, until the lower input voltage threshold point of U1A is reached. U1A's output then goes to the logic one state, allowing these capacitors to recharge.

The output of the system comparator, U1B, is buffered and inverted by the remaining four Schmidt triggers within U1, all connected in parallel, for lower output impedance drive through CR4 for rapid charge of the gate capacitance of the power stage VMOS device, Q1. This ensures fast turn-on of Q1 to minimize switching losses. U2D discharges the stored charge in the gate capacitance of Q1 when the driver gates return to the logic zero state concurrent with the discharge of the ramp capacitance, C2.

The error amplifier itself is a conventional design with U2C as a buffer for current control of the LED within OC-1. VR1 is a low voltage (1.23 V) reference diode selected for stable temperature characteristics with low bias current. Variable resistor R8 is used to remove output voltage variations due to initial component tolerance within the control loop.

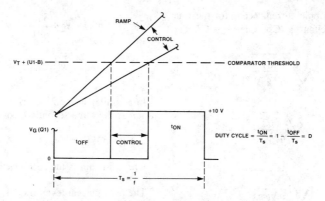

Fixed-Frequency, PWM Converter Control Concept
Figure 5

Three diodes in series (CR5, CR6, CR7) clamp the ramp control transistor current level to a fixed maximum value at converter input power application, thus limiting the duty cycle of drive to Q1 to prevent excessive power dissipation from high inrush currents. These diodes, along with CR8, also provide a path for discharge of capacitance C5 in the lead-lag frequency compensation network for the converter on input power removal. Resistor R13 is the companion component in this compensation.

Capacitors C3 and C4 are small noise filter capacitors for the bias regulator. A small ferrite bead on the gate lead of Q1 reduces high frequency noise problems due to gate drive. A suppression zener diode clamp (VR2) limits the turn-off transient voltage magnitudes across the drain-to-source terminals of Q1 to safe levels.

The configuration of the ĆUK power stage shown in Figure 4, including component rating considerations, are the subjects of Sections 4, 5, and 6 to follow.

3.1 Alternate Converter Power Switch

The limited number of vendor suppliers of high current VMOS power FETs with low drain-to-source resistance plus their current high costs may prove to be uneconomical in some high-volume or militarized power system applications. Lower power VMOS is currently more available at a proportionally lower cost.

In our early experiments with the breadboard of the multi-output converter, we used an alternate design in the area of buffer stage and power control. This alternate scheme is shown in Figure 4A. A high-frequency power transistor (2N6277) is used in place of the high current VMOS (Q1 in Figure 4) and lower current VMOS devices used as base drive and

removal controls. Note that an additional winding (T1-FB) was added to the power transformer to aid in rapid turn-on and saturation of Q1. The position of the winding in the source of Q4 also increases the effective V_{GS} (ON) of Q4, thus increasing the available base current for Q1. The lower VMOS FET (Q2) turns on when Q4 is commanded off and removes the stored charge in the B-E junction rapidly through its low ON resistance. Resistance R_{LIM} and R_{BE} provide for current limiting in the drive circuitry when the voltage across T1-FB reverses as a result of Q1 being turned OFF to prevent possible upset and/or damage in the controlling CMOS logic.

Switching losses in this alternate buffer and power control method are limited primarily to those due to Q1. If a square loop core material is used for T1, one must ensure that the DC imbalance produced by this non-capacitively coupled winding does not produce saturation. This scheme for the power switch of the converter system is very attractive for switching frequencies in the 40 kHz range should one choose to reduce the converter frequency of operation.

4. Power Stage Component Selection

Because the ĆUK power stage and its components were the key to an optimal converter design, much emphasis was placed on a proper understanding of the component values and ratings needed to achieve reliable and predictable operation.

In the matter of the energy transfer capacitances, because they are asked to transfer the entire power in AC form from input to outputs, the rms ripple current capability of these parts is a prime consideration. This subject was covered in [4] and [8] for single output ĆUK converters; however, the results can be used to predict the needs for a multi-output situation. Referring to the scaled and triple output design of Figure 3, ripple current needs for the four energy transfer capacitances (Cp, C1 ⟶ C3) can be estimated to be

$$i_{C1} \text{ (rms)} = I_{o1} \sqrt{\frac{V_{o1}}{\eta V_\ell n_1}} \qquad (3)$$

$$i_{C2} \text{ (rms)} = I_{o2} \sqrt{\frac{V_{o2}}{\eta V_\ell n_2}} \qquad (4)$$

$$i_{C3} \text{ (rms)} = I_{o3} \sqrt{\frac{V_{o3}}{\eta V_\ell n_3}} \qquad (5)$$

$$i_{Cp} \text{ (rms)} = \sum_{s=1}^{3} n_s i_{Cs}, \text{ or}$$

$$i_{Cp} \text{ (rms)} = \frac{1}{\sqrt{\eta V_\ell}} \sum_{s=1}^{3} \sqrt{n_s V_{os}} \times I_{os}, \qquad (6)$$

$$V_{Cp} = V_\ell, \quad V_{C1} = V_{o1},$$
$$V_{C2} = V_{o2}, \quad V_{C3} = V_{o3} \qquad (7)$$

Capacitance values of these parts are chosen to keep their voltage ripple magnitudes to reasonable values to minimize OFF voltage stresses on Q1 and the secondary commutating diodes. A good first approximation is 1% of the corresponding average voltage stress for ripple voltage peak-to-peak magnitude. For example, for capacitor Cp in Figure 3, using the basic expression $i_\ell = C_p dV/dt$,

$$C_p > \frac{I_\ell DT_S}{0.01 V_\ell}, \text{ or}$$

$$C_p > 100 \frac{P_\ell DT_S}{V_\ell^2} \qquad (8)$$

where P_ℓ is the total converter input power, and I_ℓ is the average input current at that power level. Similarly, for the secondary transfer capacitances,

$$C_s > 100 \frac{P_{os}}{V_{os}^2} DT_S \qquad (9)$$

where s is the secondary number in question (1, 2, or 3).

Peak OFF voltage Q1 of Figure 3 is simply the line voltage magnitude plus the reflected secondary capacitance voltage of T1.

$$V_{Q1} \text{ (OFF)} = V_\ell + \frac{V_{o1}}{n_1} = V_\ell + \frac{V_{o2}}{n_2}$$

$$= V_\ell + \frac{V_{o3}}{n_3} \qquad (10)$$

In a similar fashion, reverse voltage levels appearing across the three secondary commutating diodes are

$$V_{Ds} \text{ (OFF)} = V_{os} + n_s V_\ell \qquad (11)$$

where s again is the secondary number (1, 2, or 3).

The peak current level in Q1 is the sum of the average line current, I_ℓ, and the reflected secondary load currents.

$$I_{Q1} \text{ (ON)} = I_\ell + \sum_{s=1}^{3} n_s I_{os} \qquad (12)$$

In terms of output voltages, (12) can be restated as

$$I_{Q1} \text{ (ON)} = \sum_{s=1}^{3} \left[\frac{V_{os}}{\eta V_\ell} + n_s \right] I_{os} \qquad (13)$$

Forward current stress levels in the secondary commutating diodes is related to the corresponding secondary load current by the following relationship.

$$I_{FDs} \text{ (ON)} = \left[\frac{V_{os}}{\eta V_\ell n_s} + 1 \right] I_{os} \qquad (14)$$

s = 1, 2, or 3

Since the output filter portions of the converter of Figure 3 are essentially that of a *buck* regulator, inductance and capacitance values for $L_{o1} \rightarrow L_{o3}$ and $C_{o1} \rightarrow C_{o3}$ can be formulated from standard formulae [11] based on ripple current and voltage needs for each output. Ripple current reduction by inductive coupling between the input and output chokes of Figure 3 will be considered in Sections 6 and 7. Like the output filter networks, the design of the isolation and scaling transformer (T1 in Figure 3) can follow conventional techniques. For operation at high frequencies, *ferrite* toroids are recommended for the small core areas and low loss. Winding techniques for leakage inductance reduction are mandatory to reduce transient voltage spikes across semiconductor junctions during circuit transitory periods. Choice of turns ratios are based on desired limits of duty cycle variations with line and load changes, stress limitations on the surrounding semiconductors and energy transfer capacitances, and, of course, *cross-regulation needs* for the unregulated outputs.

Even though gapping of the core material is not required in the ideal sense, as mentioned in Section 1, *rapid* load demand changes or *sharp* excursions in line voltage may produce transient conditions such that core movement into *saturation* is feasible. One practical application [8] did experience such situations; therefore, precaution must be exercised in evaluating the need for a gapped magnetic core for T1 in light of the intended converter application and *dynamic* input/output electrical specifications.

5. Transient Output Reversal

In ĈUK converter designs where magnetic *coupling* of *input* and *output* inductances is desirable for ripple reduction, one undesirable feature of operation is introduced; namely, that of *transient voltage polarity reversal* on power application. For systems where short term reversals of outputs could produce damage to output capacitances or loads (e.g., monolithic ICs), this phenomenon must be circumvented or reduced to acceptable magnitudes.

Figure 6 is a diagram of this temporary reversal for a single output ĈUK converter with transformer isolation and coupling between input and output chokes. Inrush current through Lp due to first turn-on of Q1 or by charging of the primary energy transfer capacitance Cp when V_ℓ is applied produces a *secondary transient current* in the direction shown in Figure 6. This direction is *opposite* to that of normal current flow in converter steady-state operation, and results in a *reversal* of output voltage polarity for the transient time period.

This condition will persist until the potential levels of Cp and C1 reach values to counteract the voltage across L01 produced by the coupled primary inductance voltage. The reversal period will be determined by the time constants and magnitudes of the power stage impedances and output load values.

Figure 7 suggests some solutions to reversal protection for loads and output filter capacitors. In Figure 7(a), a blocking diode is placed in line with the secondary inductance to ensure a proper undirectional current path for steady-state output current, but prevents reverse current flow at turn-on. However, its presence does add to the power waste in the converter during normal operation. Figure 7(b) adds an output clamp diode to limit reverse transient potential to less than a volt, but does not conduct during normal operation. Use of a Schottky diode here is recommended due to the gray areas presently in prediction of damage thresholds for ICs or polarized filter capacitors.

The solution of Figure 7(a), while lowering converter efficiency, may be more attractive when the problem of transient currents in T1 producing core saturation is considered as mentioned in Section 4. There may be more sophisticated methods for reversal prevention; however, the two noted in Figure 7 seem to remain the simplest found to date.

6. Ripple Reduction by Inductor Coupling

The proportionality of the two inductor voltage waveforms of the circuits of Figure 1(a) led in [2] to the idea of *inductive coupling* and to the subsequent coupled-inductor extension of the ĈUK converter shown in Figure 1(b). At first thought, the equal amplitudes of the inductor voltages implies a need for 1:1 turns ratio only (or Lp = L_O for perfect coupling). However, in [3], it was shown that there is really no need for such restriction and the coupling of inductors may be achieved at *any inductor values*. In fact, proper choice of the relationship between turns ratio of self inductance and the coupling coefficient can result in

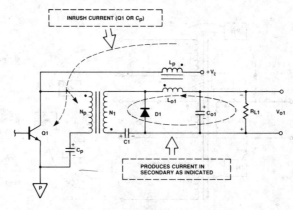

Illustration of Output Reversal on Turn-On
Figure 6

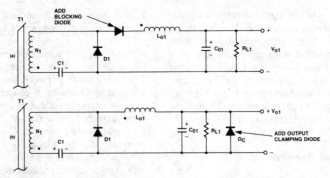

Suggested Methods of Protection for Output Polarity Reversal on Turn-On
Figure 7

zero current ripple at *either input or output* ports of the circuits of Figure 1(b) or 1(d), as dramatically demonstrated in [3] for a single output design.

In our case, however, the problem of ripple current steering and reduction is magnified due to the presence of more than one secondary inductance and the corresponding complex interrelationships between turns ratios and coupling coefficients. However, this problem is simplified if a first-order model, such as that shown in Figure 8, is utilized for analysis. As shown, all windings are mutually coupled (M's), and the assumption is made that the primary and secondary voltages are related to the primary potential by proportionality factors a, b, and c. In the design of Figure 4, these factors would be the secondary turn ratios of T1, or $a = n_1$, $b = n_2$, and $c = n_3$, assuming an ideal transformer model for T1. Proceeding with the voltage matrix for the model of Figure 8, we find

$$\begin{bmatrix} v_p \\ av_p \\ bv_p \\ cv_p \end{bmatrix} = \begin{bmatrix} L_P & M_{12} & M_{13} & M_{14} \\ M_{12} & L_{o1} & M_{23} & M_{24} \\ M_{13} & M_{23} & L_{o2} & M_{34} \\ M_{14} & M_{24} & M_{34} & L_{o3} \end{bmatrix} \begin{bmatrix} di_p/dt \\ di_1/dt \\ di_2/dt \\ di_3/dt \end{bmatrix} \quad (15)$$

where L_P, L_{o1}, L_{o2}, and L_{o3} are the self or open-circuit inductances of the windings, and the various M's represent mutual inductance values between the four windings. We now define effective turns ratios ($n_{o1} \longrightarrow n_{o3}$) and coupling coefficients ($k_1 \longrightarrow k_6$) as

$$n_{o1} \triangleq \sqrt{\frac{L_P}{L_{o1}}}, \; n_{o2} \triangleq \sqrt{\frac{L_P}{L_{o2}}}, \; n_{o3} \triangleq \sqrt{\frac{L_P}{L_{o3}}},$$

$$k_1 \triangleq \frac{M_{12}}{\sqrt{L_P L_{o1}}}, \quad k_2 \triangleq \frac{M_{13}}{\sqrt{L_P L_{o2}}},$$

$$\hspace{8cm} (16)$$

$$k_3 \triangleq \frac{M_{14}}{\sqrt{L_P L_{o3}}}, \quad k_4 \triangleq \frac{M_{23}}{\sqrt{L_{o1} L_{o2}}},$$

$$k_5 \triangleq \frac{M_{24}}{\sqrt{L_{o1} L_{o3}}}, \quad k_6 \triangleq \frac{M_{34}}{\sqrt{L_{o2} L_{o3}}}$$

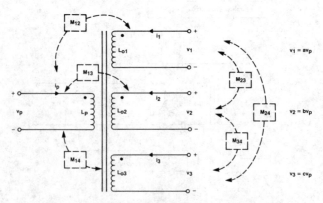

First-Order Inductor Model for Ripple Current Predictions
Figure 8

Substitution of these definitions into the matrix of (15) and solving for the various current derivatives will then allow us to solve for each winding's *effective inductance* value in relation to its self-inductance. After a few lines of algebra and the aid of Cramer's rule,

$$\frac{di_p}{dt} = \frac{v_p}{L_{efp}} , \quad \frac{di_1}{dt} = \frac{av_p}{L_{efs1}} ,$$

$$\frac{di_2}{dt} = \frac{bv_p}{L_{efs2}} , \quad \frac{di_3}{dt} = \frac{cv_p}{L_{efs3}} \tag{17}$$

where the four effective inductances can be shown to be

$$L_{efp} = \frac{N(k) \times L_p}{(1 - k_5{}^2)(1 - bn_{o2}\, k_2) - (cn_{o3}\, k_4 - an_{o1}\, k_6) \times D(k)} \tag{18}$$

$$L_{efs1} = \frac{N(k) \times L_{o1}}{(1 - k_2{}^2)\left(1 - \dfrac{cn_{o3}\, k_5}{an_{o1}}\right) - \left(\dfrac{bn_{o2}}{an_{o1}} - \dfrac{k_6}{an_{o1}}\right) \times D(k)} \tag{19}$$

$$L_{efs2} = \frac{N(k) \times L_{o2}}{(1 - k_5{}^2)\left(1 - \dfrac{k_2}{bn_{o2}}\right) - \left(\dfrac{an_{o1}}{bn_{o2}} - \dfrac{cn_{o3}\, k_1}{bn_{o2}}\right) \times D(k)} \tag{20}$$

$$L_{efs3} = \frac{N(k) \times L_{o3}}{(1 - k_2{}^2)\left(1 - \dfrac{an_{o1}\, k_5}{cn_{o3}}\right) - \left(\dfrac{k_4}{cn_{o3}} - \dfrac{bn_{o2}\, k_1}{cn_{o3}}\right) \times D(k)} \tag{21}$$

$$N(k) = (1 - k_5{}^2)(1 - k_2{}^2) - (k_3\, k_4 - k_1\, k_6)^2, \text{ and}$$

$$D(k) = k_3\, k_4 - k_1\, k_6$$

It now remains to investigate some of the interesting values for the effective inductances above such that one will approach *infinity,* thus producing *zero* ripple at that port.

Table 1 is a listing of four cases where the proportionality factors a, b, and c, are set equal to combinations of coupling factors and turns ratios of self inductances as defined by (16). Note that it is possible for one winding to approach *infinite* effective inductance while another will equal its *open-circuit* inductance value at that point.

The four cases illustrated do point out *very important* restrictions on the coupling coefficients. For a tightly coupled multi-inductor design (all k's $\approx$ 1), both the denominator and the numerator of the expressions in

(18) through (21) vanish, resulting in undeterminate effective inductances. The coupled inductors are thus reduced to a multiple winding ideal transformer, for which predictable current ripple reduction performance becomes virtually impossible.

Another problem that comes immediately to mind is the difficulty of design of the actual magnetic structures needed to attain predictable coupling coefficient relationships between windings. Such structures imply complex flux paths, each of which would have to be individually tuned for desired interactions. Such assemblies would be expensive, difficult to manufacture for producibility from unit-to-unit, and would need rigid controls to prevent coupling coefficient changes with physical environments, such as temperature, vibration, or shock.

6

Table 1
Matrix of a, b, c versus Effective Inductances for Multiple-Inductor Coupling Analysis
Note: $N = (1 - k_5^2)(1 - k_2^2) - (k_3 k_4 - k_1 k_6)^2$

a	b	c	L_{efp}	L_{efs1}	L_{efs2}	L_{efs3}
(A) $\dfrac{k_1}{n_{o1}}$	$\dfrac{k_2}{n_{o2}}$	$\dfrac{k_3}{n_{o3}}$	L_p	$\dfrac{N \cdot L_{o1}}{\left(1 - k_2^2\right)\left(1 - \dfrac{k_3 k_5}{k_1}\right) - \left(\dfrac{k_2 k_3}{k_1} - \dfrac{k_6}{k_1}\right)(k_3 k_4 - k_1 k_6)}$	$\longrightarrow \infty$	$\dfrac{N \cdot L_{o3}}{\left(1 - k_2^2\right)\left(1 - \dfrac{k_1 k_5}{k_3}\right) - \left(\dfrac{k_4}{k_3} - \dfrac{k_1 k_2}{k_3}\right)(k_3 k_4 - k_1 k_6)}$
(B) $\dfrac{1}{k_1 n_{o1}}$	$\dfrac{k_4}{k_1 n_{o2}}$	$\dfrac{k_5}{k_1 n_{o3}}$	$\dfrac{N \cdot L_p}{\left(1 - k_5^2\right)\left(1 - \dfrac{k_4 k_2}{k_1}\right) - \left(\dfrac{k_4 k_5}{k_1} - \dfrac{k_6}{k_1}\right)(k_3 k_4 - k_1 k_6)}$	L_{o1}	$\dfrac{N \cdot L_{o2}}{\left(1 - k_5^2\right)\left(1 - \dfrac{k_2 k_1}{k_4}\right) - \left(\dfrac{k_3}{k_4} - \dfrac{k_5 k_1}{k_4}\right)(k_3 k_4 - k_1 k_6)}$	$\longrightarrow \infty$
(C) $\dfrac{k_4}{n_{o1} k_2}$	$\dfrac{1}{k_2 n_{o2}}$	$\dfrac{k_6}{n_{o3} k_3}$	$\longrightarrow \infty$	$\dfrac{N \cdot L_{o1}}{\left(1 - k_2^2\right)\left(1 - \dfrac{k_5 k_6}{k_4}\right) - \left(\dfrac{k_3}{k_4} - \dfrac{k_2 k_6}{k_4}\right)(k_3 k_4 - k_1 k_6)}$	L_{o2}	$\dfrac{N \cdot L_{o3}}{\left(1 - k_2^2\right)\left(1 - \dfrac{k_4 k_5}{k_6}\right) - \left(\dfrac{k_4 k_2}{k_6} - \dfrac{k_1}{k_6}\right)(k_3 k_4 - k_1 k_6)}$
(D) $\dfrac{k_5}{n_{o1} k_3}$	$\dfrac{k_6}{n_{o2} k_3}$	$\dfrac{1}{n_{o3} k_3}$	$\dfrac{N \cdot L_p}{\left(1 - k_5^2\right)\left(1 - \dfrac{k_2 k_6}{k_3}\right) - \left(\dfrac{k_4}{k_3} - \dfrac{k_5 k_6}{k_3}\right)(k_3 k_4 - k_1 k_6)}$	$\longrightarrow \infty$	$\dfrac{N \cdot L_{o2}}{\left(1 - k_5^2\right)\left(1 - \dfrac{k_2 k_3}{k_6}\right) - \left(\dfrac{k_3 k_5}{k_6} - \dfrac{k_1}{k_6}\right)(k_3 k_4 - k_1 k_6)}$	L_{o3}

The question may then be asked in light of the above discussion. *Is it feasible to simulate predictable coupling coefficients by external means using a tightly-coupled inductor assembly?* The answer is *yes* and can be accomplished with *small external inductances* working in conjunction with the tightly-coupled inductor as shown in the next section of this paper.

Before turning to this ripple current reduction by external impedance, a few comments are needed about special cases of effective inductances that may not be completely obvious from those given in Table 1. If it is possible to maintain

$$k_3 k_4 = k_1 k_6; \quad k_2, k_5 < 1 \qquad (22)$$

then a simplified table of effective inductance values can be constructed, as is done in Table 2, for specified equalities between remaining coupling coefficients and effective turns ratios of windings.

Cases A and B of Table 2 are very significant, for they imply that it is possible to *simultaneously* increase the effective inductances of two windings such that zero current ripple can exist in *both*. For example,

in Case A, *zero* current ripple on *input* and one *output* is achievable, while the remaining windings approach their self inductance values.

Cases D1 and D2 are included as demonstrations of the *"negative"* ripple situations which implies a change in current slope from that which would normally appear with a known voltage polarity across an inductance. This phenomenon was explained in [2] and [3] and will not be considered further here.

7. Ripple Reduction with External Chokes and Inductor Coupling

Faced with the serious problem of internal coupling coefficient control of the inductor assembly to properly and reliably reduce ripple currents at selected ports of the ĆUK converter, attention is turned to an analytical investigation of the feasibility of external simulation of leakage inductances of windings for control. To begin, the first-order inductor model of Figure 8 is modified to include such inductances, as shown in Figure 9, and a redefinition of inductor effective turns ratios and coupling coefficients is made as shown in (23).

CASE	CONDITIONS	L_{efp}	L_{efs1}	L_{efs2}	L_{efs3}
A	ZERO CURRENT RIPPLE (INPUT, ONE OUTPUT) $bn_{o2} k_2 = 1$, $an_{o1} = cn_{o3} k_5$	$\longrightarrow \infty$	$\longrightarrow \infty$	L_{o2}	L_{o3}
B	ZERO CURRENT RIPPLE (TWO OUTPUTS) $bn_{o2} = k_2$ $an_{o1} k_5 = cn_{o3}$	L_p	L_{o1}	$\longrightarrow \infty$	$\longrightarrow \infty$
C	BALANCED RIPPLE CURRENT REDUCTIONS (INPUT, ALL OUTPUTS) $bn_{o2} = 1$ $an_{o1} = cn_{o3}$	$(1 + k_2) L_p$	$(1 + k_5) L_{o1}$	$(1 + k_2) L_{o2}$	$(1 + k_5) L_{o3}$
D1	"NEGATIVE" CURRENT RIPPLE CONDITIONS $bn_{o2} k_2 > 1$ $cn_{o3} k_5 > an_{o1}$	NEGATIVE	NEGATIVE	POSITIVE	POSITIVE
D2	$bn_{o2} < k_2$ $an_{o1} k_5 > cn_{o3}$	POSITIVE	POSITIVE	NEGATIVE	NEGATIVE

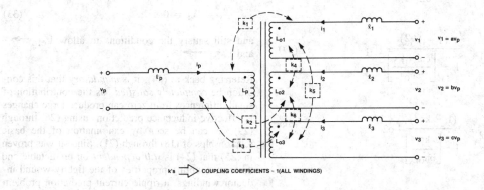

$v_1 = av_p$

$v_2 = bv_p$

$v_3 = cv_p$

$k's \Longrightarrow$ COUPLING COEFFICIENTS ~ 1(ALL WINDINGS)

First-Order Inductor Model with External Inductances
Figure 9

6

$$n'_{o1} = \sqrt{\frac{L'_P}{L'_{o1}}} \quad , \quad n'_{o2} = \sqrt{\frac{L'_P}{L'_{o2}}} \quad , \quad n'_{o3} = \sqrt{\frac{L'_P}{L'_{o3}}} \quad ,$$

$$k'_1 = k_1 \sqrt{\frac{L_P L_{o1}}{L'_P L'_{o1}}} \quad , \quad k'_2 = k_2 \sqrt{\frac{L_P L_{o2}}{L'_P L'_{o2}}} \quad , \quad k'_3 = k_3 \sqrt{\frac{L_P L_{o3}}{L'_P L'_{o3}}} \quad , \tag{23}$$

$$k'_4 = k_4 \sqrt{\frac{L_{o1} L_{o2}}{L'_{o1} L'_{o2}}} \quad , \quad k'_5 = k_5 \sqrt{\frac{L_{o1} L_{o3}}{L'_{o1} L'_{o3}}} \quad , \quad k'_6 = k_6 \sqrt{\frac{L_{o2} L_{o3}}{L'_{o2} L'_{o3}}}$$

where $L'_P = L_P + \ell_P$, $L'_{o1} = L_{o1} + \ell_1$, $L'_{o2} = L_{o2} + \ell_2$, and $L'_{o3} = L_{o3} + \ell_3$. Repeating the voltage matrix and solution for current rates, the effective inductance expressions for the four windings become the *same* equations as (18 $\longrightarrow$ 21), except of course, for replacement of all k, n, and L symbols with primes as indicated in (23) above. For the condition

$$k'_3 k'_4 = k'_1 k'_6 \tag{24}$$

substitution of the values from (23) into (24) shows that

$$k_3 k_4 = k_1 k_6 \tag{25}$$

still must be satisfied within the multi-winding inductor itself to reduce the effective inductances to the following simplified equations.

$$L'_{efp} = \frac{(1 - k_2'^2)}{1 - bn'_{o2} k'_2} L'_P \tag{26}$$

$$L'_{efs1} = \frac{(1 - k_5'^2)}{1 - \left[\dfrac{cn'_{o3} k'_5}{an'_{o1}} \right]} L'_{o1} \tag{27}$$

$$L'_{efs2} = \frac{(1 - k_2'^2)}{1 - \left[\dfrac{k'_2}{bn'_{o2}} \right]} L'_{o2} \tag{28}$$

$$L'_{efs3} = \frac{(1 - k_5'^2)}{1 - \left[\dfrac{an'_{o1} k'_5}{cn'_{o3}} \right]} L'_{o3} \tag{29}$$

A case matrix similar to that of Table 2 can be prepared using (26) through (29); however, that case concerning itself with *zero* input and *zero* output ripple on the first secondary is of direct interest to this work. This situation according to (26) and (27) above should occur for

$$bn'_{o2} k'_2 = 1 \quad \text{and} \quad an'_{o1} = cn'_{o3} k'_5 \tag{30}$$

when $L'_{efp} \longrightarrow \infty$ and $L'_{efsl} \longrightarrow \infty$. Substitution for prime values as defined by (23) into (30) and solving for the external inductances yields

$$\ell_2 = b \sqrt{L_P L_{o2}} - L_{o2} \tag{31}$$

$$\ell_3 = \left[\frac{c}{a} \right] \sqrt{L_{o1} L_{o3}} - L_{o3} \tag{32}$$

for k_2, $k_5 \approx 1$ (tightly coupled windings).

Since the relationships of (26) and (27) do not include L'_{o1} or L'_P in their denominators, then we can set

$$\ell_P = \ell_1 = 0 \tag{33}$$

and still satisfy the conditions to allow $L'_{efp} \longrightarrow \infty$ and $L'_{efsl} \longrightarrow \infty$.

Referring back to (24), it is *mandatory* that this condition be *completely satisfied,* as the contribution of small differences from zero can produce large changes in effective inductance predictions using (26) through (29), as can be seen by examination of the basic relationships of (18) through (21). Since it was proven in (25) that (24) is *still dependent* on predictable and uniform coupling properties of the tightly-wound inductor windings, a ripple current prediction problem can still exist.

What is interesting, however, about condition (25) is that it can be *absolutely* satisfied if

$$k_1, k_3, k_4, k_6 = 0 \tag{34}$$

which implies *two non-coupled* inductor assemblies, each with two windings, such that L_p and L_{o2} are wound on the same core and L_{o1} and L_{o3} on the other. External inductances ℓ_2 and ℓ_3 are then added in series with L_{o2} and L_{o3}, respectively, to reduce both *input* and the *first secondary* (in our case, the high current $+5$ V section) ripple currents to *zero*. In other words, the coupled-inductor analysis prediction problem reduces to that of two ĈUK single output converters operating *independently* of each other.

For the above reasons, two inductor assemblies were used in the experimental design as shown in the schematic of Figure 4, and the small inductors in the $+$ and -15 V lines were selected for *zero* $+28$ V ripple current and *zero* $+5$ V ripple current conditions. The effective inductances in the $+$ and -15 V output filters, including the external inductors, are easily derived from (31) and (32) as

$$L'_{efs2} = b \sqrt{L_p L_{o2}}$$

$$L'_{efs3} = \left[\frac{c}{a}\right] \sqrt{L_{o1} L_{o3}} \qquad (35)$$

Note that, in Figure 4, only *one* transient output polarity reversal diode is required (D4). Since the $+5$ V and -15 V inductor windings are coupled together but *not* coupled to the input, *no* protection is needed on these outputs for this phenomenon as described earlier in Section 5.

8. Experimental Results

To correlate the analytical response predictions of ripple current reduction and expected system performance, the circuitry shown in Figure 4 was breadboarded using the components types and values given in Table 3.

The core material for the power transformer (T1) was of the ferrite variety (F-42207-TC), with good high frequency and tight coupling practices followed in its construction. Primary-to-secondary turns ratio was set at 15:7 for the $+5$ V winding, while the ±15 V windings were identical and turns ratios selected to be 15:17 with respect to the primary.

The two output inductor assemblies (T2, T3) were built using standard Magnetics Inc. 54167-0.5R cores, with 1:1 turns ratios between their respective windings for evaluation simplicity. Each winding on T2 had 33T, with 23T on those of T3. Measured open-circuit inductances of T2 were 107 μH with no DC bias, while those of T3 were very close to 54.6 μH. Under representative DC bias currents, the open-circuit inductances of T2 dropped to 89.5 μH, at 200 kHz. Under bias, those of T3 reduced to 46 μH at the same frequency. Both magnetics were constructed with tightly coupled windings.

Table 3
Component Values/Types
(Reference — Figure 4)

CONTROL ELECTRONICS

R1	6.56 K	C1	1000 pF
R2	1.4 K	C2	200 pF
R3	11 K	C3	0.1 μF
R4	1K, 1 W	C4	1.0 μF
R5	100 ohm	C5	0.1 μF
R6	51 K		
R7, R11	7.5 K	CR1–CR8	IN4150
R8	5 K	VR1	ICL8069
R9	6.19 K	VR2	IN4978
R10	10 K	VR3	IN941
R12	1.8 K	OC-1	HP-5082-4350
R13	2.4 K	U1	CD-40106BE
		U2	MPQ2907
		Q3	2N2920A

ĈUK POWER STAGE

C1	360 μF, 10 V	D1	SD-41
C2, C3	47 μF, 35 V	D2, D3	IN6077
C01	15 μF, 10 V	D4	MBR040
C02, C03	12 μF, 35 V	Q1	IRF-100
Cρ	160 μF, 50 V	VR2	IN6163

The two external chokes (ℓ_2, ℓ_3) were of the RF variety from the Dale IH-3 series. Since the open circuit inductances L_p and L_{o2} are known and a $= 7/15$, b $=$ c $= 17/15$ (31), (32) and (35) can be used to calculate the values of ℓ_2, ℓ_3, L'_{efs2} and L'_{efs3}. Therefore, for both zero ripple current on the $+28$ V line and the $+5$ V output,

$$\ell_2 \cong 12 \,\mu H, \; \ell_3 \cong 65 \,\mu H,$$

$$L'_{efs2} = 102 \,\mu H, \; L'_{efs3} \cong 111 \,\mu H \qquad (36)$$

The testing performed on the experimental breadboard of the system shown in Figure 4 can be categorized into the following three groups.

8.1 Group One: Ripple Current Adjustments

In these tests, first the values of ℓ_2 and ℓ_3 were adjusted respectively to ℓ_{20} and ℓ_{30}, for which the current ripples in the input port and in the $+5$ V output port were reduced to approximately *zero*, as shown in the waveform photograph of Figure 10. When the zero ripple current condition was obtained, the effective inductance values (L'_{efs} and L'_{efs3}) of the $+$ and -15 V outputs were measured from the relevant voltage and current wave shapes. Then the values of ℓ_2 and ℓ_3 were *increased* from their zero ripple values of ℓ_{20} and ℓ_{30} and *positive* slope ripple currents in the input and in the $+5$ volt output were obtained as shown in Figure 11. When the values of ℓ_2 and ℓ_3 were *decreased* to levels below ℓ_{20} and ℓ_{30}, *negative* slope ripple currents of Figure 12 were observed.

6

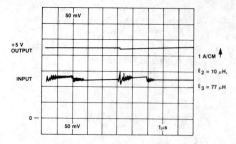

"Zero" Ripple Case
Figure 10

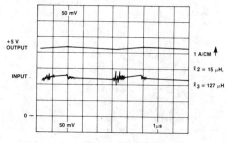

"Positive" Ripple Case
Figure 11

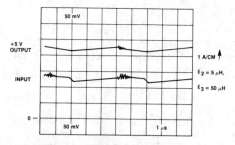

"Negative" Ripple Case
Figure 12

The values of ℓ_2 and ℓ_3 for different current ripple conditions are summarized in Table 4 along with the theoretical values from (36). Also shown in Table 4 are experimental and theoretical values for L'_{efs2} and L'_{efs3} under the zero ripple current condition.

Examination of Table 4 shows excellent correlation between experimental and calculated values of circuit inductances. Variations noted are in part due to the differences between assumed and actual inductance values of L_p, L_{o1}, L_{o2}, and L_{o3}, and to the measurement accuracy in determining ℓ_{20} and ℓ_{30} of the fixed value, discrete chokes.

The slight change in the magnetization current of T1 along with the input current level of Figures 10 through 12 is primarily caused by differences in switching times of Q1 and the three secondary commutation diodes. During these switching periods, the

Table 4
Experimental Results versus Calculated Values of External and Effective Secondary Inductances

	EXPERIMENTAL VALUE (μH)	CALCULATED VALUE (μH)
ℓ_{20}	10	12
ℓ_{30}	77	65
ℓ_2 ("+" slope)	15.0	>10.0
ℓ_3 ("+" slope)	127.0	>77.0
ℓ_2 ("−" slope)	5.0	<10.0
ℓ_3 ("−" slope)	50.0	<77.0
L'_{efs2}	87.4	102.0
L'_{efs3}	108.0	111.0

secondary voltages of T1 and T2 become unbalanced (v_p, v_1, v_2, and v_3 of Figure 9), and the equations (26 $\longrightarrow$ 29) become invalid since they assume ideal time phasing between all potentials.

The noise and ringing observed in the current waveforms are caused in part by the presence of inductive and capacitive parasitics in the breadboard layout that become significant at high switching frequencies which modulate current levels, and by the non-ideal nature of instrumentation with regard to grounding and noise.

8.2 Group Two: Line/Load Variations

These tests were mainly performed to evaluate the steady state operating properties of the converter. The tests were conducted while maintaining zero current ripple in the input and the +5 V output. These tests included line and load regulation and system efficiency measurements.

Table 5
Line Regulation with R_{L1} (5 V Output Load) = 1 Ω, R_{L2}, R_{L3} ($\pm$15 V Output Loads) = 15 Ω

VIN VOLTS DC	+5 V VOLTS DC	+15 V VOLTS DC	−15 V VOLTS DC
22	4.98	15.02	−14.92
28	5.00	14.58	−14.50
32	5.00	14.37	−14.29

Table 6
+5 V Output Load Regulation with V_{in} = 28 V DC, $R_{L2} = R_{L3} = 15 \Omega$

R_{L1} (Ω)	+5 V (VOLTS DC)	+15 V (VOLTS DC)	−18 V (VOLTS DC)
10	5.00	12.05	−12.06
2	4.99	13.15	−13.14
1	5.00	14.54	−14.46

In Tables 5 and 6 we observe that the load and line regulation of the +5 V output are very satisfactory and do not need any comments. It is interesting to note that, although the line regulation of the + and

−15 volt outputs are well within required ranges, their absolute value *decreases* with *increasing* line voltage. This change is enhanced with respect to +5 V load variations as noted in Table 6. This phenomenon can be explained when it is observed that + and −15 volt output voltages are determined mainly by the voltage impressed across the N_1 winding of T1 during the off time of Q1. This voltage is made up of the voltage on C1 and the forward drop of diode D1. The voltage on C2 does not significantly change with line variation; however, the forward voltage drop of D1 increases with decreasing line voltage due to increased value of the reflected current from the primary, thus increasing the total impressed voltage across N_1. With +5 V load variation, this effect becomes more pronounced. Since now, in addition to the forward voltage drop of D1 which changes with load, the steady-state value of the voltage on C1 also changes to make up for the changing drops in the winding resistance of T1, L_{o1} and in its own ESR. As a net result, the voltage impressed across N_1 during the off time of Q1 *increases* with increasing +5 V load and *decreases* with decreasing +5 V load.

In Table 7, efficiency figures are calculated for minimum, nominal and maximum line voltages while the converter is delivering full output power. These figures compare well with practical efficiencies that are presently being obtained from conventional switching regulators operating at much lower switching frequencies.

Table 7
Efficiency Measurements at Full Output Loads

V_Q (VOLTS DC)	I_Q (AMP DC)	P_Q (WATTS)	P_o (TOTAL) (WATTS)	% EFFICIENCY
22	3.5	77.0	55.0	71.4
28	2.6	72.8	53.2	73.0
32	2.3	73.6	52.4	71.2

Power losses are attributed to those produced by semiconductor switching actions, dissipation in the energy transfer capacitances, IR drops in Q1, and resistances of magnetic windings. As mentioned earlier, the experimental design did not include optimization measures to minimize these losses.

8.3 Group Three: Reversal Protection

In this test, output voltage polarity reversal of the +15 V output during turn-on was investigated. This voltage polarity reversal occurs only at the +15 V output during turn-on due to magnetic coupling between Lp and L_{o2}. The photograph of Figure 13(a) shows that negative going spike went as high as −20 volts and recovered in about 60 μsec after turn-on. Figure 13(b) demonstrates that a diode at the +15 V

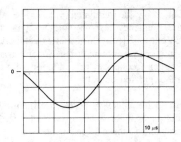

(a) Without Diode Clamp
+15 V Output

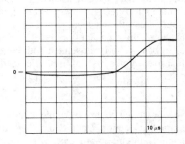

(b) With Diode Clamp
+15 V Output

Output Waveforms at Turn-On
Figure 13

output effectively clamps this negative spike to a diode drop below the +15 V return line.

9. Conclusions and Remarks

A high frequency triple-output DC-to-DC converter design utilizing a ĆUK output power stage is proposed herein, with magnetic coupling between selected output filter inductances for attaining *zero* ripple currents at *both* the *input* power port and *one output*. The design also includes *transient polarity reversal* protection at the output port magnetically coupled to the input line.

Although the control electronics for converter pulse-width-modulation were somewhat unconventional due to the use of *ramp slope* control, standard PWM LSI integrated control circuitry could be employed if desired. By the use of a VMOS power device instead of a bipolar transistor for the ĆUK converter primary switch, the interface electronics are greatly simplified and overall converter efficiency improved due to the *micropower* gate drive required for the VMOS field effect transistor.

Obviously, the converter could be designed to operate at lower switching frequencies with corresponding increases in energy transfer capacitor values and larger magnetic cores for the isolation transformer and inductor assemblies. In fact, the need for *no* output

6

filter capacitor on the regulated +5 V line should simplify the frequency response of the converter as was pointed out in [3]; therefore, an extensively high switching frequency is *not* required in comparison to the filter corner frequency for reduction of voltage ripple on the regulated output.

With regard to frequency response, no mention was made in this paper of breadboard converter performance in this area as these experiments are currently in process. Modeling of the converter for response predictions can follow published techniques [1, 2, 5] with modifications made to include nonregulated output loads and their filter networks reflected to the controlled output and the magnetic coupling of input/output inductances. Investigations are being performed in the areas of bandwidth improvement with increased stability margins by the use of multiple feedback control techniques such as those described in [6] and [8].

The success of external control of ripple current magnitudes reported in Section 7 and 8 has *implications* of corresponding circuit improvements to the basic single output ĆUK converters of Figure 1 that could *simultaneously* eliminate both *input and output* ripple currents. In addition, it should be possible for converter design with more than three outputs to utilize the techniques described in Sections 6 and 7 to attain *zero* ripple currents on the *input* and *more than one output* by proper choice of output inductances to be magnetically coupled.

Acknowledgements

The authors would like to thank Morris Rauch for his diligent work on the evaluation of the experimental breadboard and Karl Dakteris for the many hours spent in prototype magnetic designs. Thanks are due also to Dusan Jocic for excellent magnetic prototype construction work.

Most of all, special personal gratitude is sent to Drs. Slobodan Ćuk and R.D. Middlebrook for their helpful advice and inspiration in putting this work together.

References

[1] Slobodan Ćuk and R.D. Middlebrook, *"A New Optimum Topology Switching DC-to-DC Converter,"* IEEE Power Electronics Specialists Conference, 1977 Record, pp. 160–179 (IEEE Publication 77CH1213-8 AES).

[2] Slobodan Ćuk and R.D. Middlebrook, *"Coupled-Inductor and Other Extensions of a New Optimum Topology Switching DC-to-DC Converter,"* IEEE Industry Applications Society Annual Meeting, 1977 Record, pp. 1110–1126 (IEEE Publication 77CH1246-8-IA).

[3] Slobodan Ćuk, *"Switching DC-to-DC Converter with Zero Input or Output Current Ripple,"* IEEE Industry Applications Society Annual Meeting, 1978 Record, pp. 1131–1146, (IEEE Publication 73CH1346-61A).

[4] R.D. Middlebrook and Slobodan Ćuk, *"Isolation and Multiple Output Extensions of a New Optimum Topology Switching DC-to-DC Converter,"* IEEE Power Electronics Specialists Conference, 1978 Record, pp. 256–264 (IEEE Publication 78CH1337-5 AES).

[5] R.D. Middlebrook, *"Modelling and Design of the ĆUK Converter,"* Proc. Sixth National Solid-State Power Conversion Conference (Powercon 6), pp. G3.1–G3.13, May 1979.

[6] Shi-Ping Hsu, Art Brown, Loman Rensink, and R.D. Middlebrook, *"Modeling and Analysis of Switching DC-to-DC Converters in Constant-Frequency Current-Programmed Mode,"* IEEE Power Electronics Specialists Conference, 1979 Record, paper 4A.6.

[7] Slobodan Ćuk and Robert W. Erickson, *"A Conceptually New High-Frequency Switched-Mode Amplifier Technique Eliminates Current Ripple,"* Proc. Fifth National Solid-State Power Conversion Conference (Powercon 5), pp. G3.1–G3.22, May 1978.

[8] Loman Rensink, Art Brown, Shi-Ping Hsu, and Slobodan Ćuk, *"Design of a Kilowatt Off-Line Switcher Using a ĆUK Converter,"* Proc. Sixth National Solid-State Power Conversion Conference (Powercon 6), pp. H3.1–H3.26, May 1979.

[9] R.D. Middlebrook, Slobodan Ćuk, and W. Behen, *"A New Battery Charger/Discharger Converter,"* IEEE Power Electronics Specialists Conference, 1978 Record, pp. 251–255 (IEEE Publication 78CH1337-5 AES).

[10] R.D. Severns, *"High-Frequency Switching Regulator Techniques,"* IEEE Power Electronics Specialists Conference, 1978 Record, pp. 290–298, (IEEE Publication 78CH1346-61A).

[11] A.I. Pressman, *"Switching and Linear Power Supply, Power Converter Design,"* Hayden Book Co., Inc., 1977, pp. 289–315.

6.1.3 The Generalized Use of Integrated Magnetics and Zero-Ripple Techniques in Switch Mode Power Converters

Abstract

Methods are presented that allow the discrete transformers and inductors of switch mode power converters to be unified in single magnetic structures. It is demonstrated that unified magnetics and zero ripple operation are general phenomena applicable to all types of switch mode power converters.

Introduction

Seemingly, a day does not pass now without an announcement of a new advance in microcircuit technology, bringing us closer to the age of ultra miniaturization of electronic products. For those of us who must design and develop power processing systems it is a time of frustration and reflection, for as product sizes shrink, so must their power conditioners.

Twenty-five years ago, when the sizes of electronic systems were measured in terms of room dimensions, their power supplies could afford to be large. Today, with electronic products, such as calculators smaller than the dial of a watch, the power supply subsystems are often as large or larger than the electronics. Another example is the personal computer, where power supply areas often occupy up to 50% of enclosure volume, with a cost approaching 45% of product price.

In a concerted effort to reduce power supply sizes, recent years have seen switch mode conversion and processing designs pushed higher and higher in operating frequency. Theoretically, at least, converters with high switching rates imply that their circuits will have smaller magnetic components. There are however, very practical limits to the size reduction obtainable from high frequency operation [1].

There is little disagreement among power converter designers that the magnetic parts of any switch mode design are the major contributors to supply cost, weight, and size. For these reasons, it is not uncommon for designers to select converter approaches that have a minimum magnetic content, even though, in many cases, one with more magnetic components would be more suitable from an overall circuit performance standpoint.

Realizing that moving to higher switching frequencies to reduce magnetic size has practical boundaries, designers are now turning to another avenue of investigation — *magnetic integration*. This rather innovative-sounding, but accurate, term is used to describe magnetic design techniques whereby various inductive and transformer elements of a power converter can be combined on a single core structure.

If one accepts the above definition of magnetic integration as applied to switch mode DC-to-DC power converters, then two converter topologies can be readily identified as integrated magnetic circuits, namely, the buck-boost-derived *flyback converter* and a special variation of the "boost-buck-derived" ĆUK converter[2].

Both of these circuits, together with their discrete magnetic counterparts, are shown in the lower portions of Figure 1. In the case of the flyback converter, the required inductance for energy storage is simply "built-into" the isolation transformer by proper choice magnetization inductance. Thus, the transformer of the flyback converter serves two important purposes — as an isolation element for input and output grounds and as a means of primary energy storage for supplying load power needs. In the case of the integrated-magnetic version of the transformer-isolated ĆUK converter, its single magnetic houses all inductive functions of the converter and, by selecting proper amounts of mutual inductance that exist between windings, input and output ripple current magnitudes can be controlled, and even be reduced to *zero* in special instances [3].

Both the buck-boost converter and its dual circuit, the ĆUK converter, can be evolved from the appropriate cascade arrangements of basic buck and boost circuits, as demonstrated in [2]. It is interesting to note that integrated-magnetic versions of isolated buck-boost and ĆUK converters are perfectly feasible and well-documented circuit possibilities, but little is publicly known about similar integration methods for transformer-isolated versions of their parent circuits, as visually emphasized by the vacant sections of Figure 1.

In this paper, we will explore design techniques for the integration of the transformer and inductive components of buck and boost-derived converter topologies. To set the stage for this examination, a brief introductory section on basic electromagnetic modeling and analysis is included. Discussions of magnetic integration of the inductive components of a forward converter are then presented, with extensions for reducing the output ripple current significantly by external means. By duality integrated-magnetic boost converter circuits are then evolved and extensions illustrated for input ripple current reductions.

Integrated-magnetic "push-pull" DC-to-DC converter structures are then examined, including one based on a special variation of the Weinberg circuit. Results of laboratory tests of a representative integrated magnetic converter are then shown, followed by discussions of small-signal averaged modeling for stability and control analysis.

Tools for Magnetic Circuit Modeling

Integrated magnetics for converters brings to mind a picture of magnetic structures that are highly complex and unwieldy from design, analysis, and construction viewpoints. To some degree, this concern is understandable, for most engineers today are only familiar with magnetic design methods that address inductors and transformers as individual converter components. Just the thought of having to deal with a magnetic assembly with more than one major flux path is often a deterrent to an engineer to attempt such a design.

The attitude of designers in this regard is now changing. However, for many converter designers, the tools to properly design and model integrated-magnetic components have been unused over the years. Thus, before beginning the exploration of integrated-

magnetic buck and boost converters, it is worthwhile to digress briefly and review some not-so-familiar electromagnetic fundamentals and magnetic circuit modeling methods using electric circuit analogs. For those readers interested in a more comprehensive review than the one to follow, reference [4] is highly recommended.

Recall that the similarity between Kirchoff's voltage and current laws for *linear electric circuits* and Ampere's circuital laws related to magnetomotive force and flux continuity in *linear magnetic circuits* permit the use of electric circuit analogs for analysis purposes. Such analysis makes voltage, V, analogous to magnetic potential, f; current, i, analogous to magnetic flux, ϕ; electrical resistance, R, analogous to magnetic reluctance, R. Furthermore, because the electric circuits derived by the use of these analogs are *linear,* they can be manipulated into even more useful forms by established duality relationships [5]. Transformation by duality then produces electric circuit models that relate magnetic reluctances, R's, to inductances, L's; flux linkages in windings, λ's, to voltages, v's; and flux levels in magnetic paths, ϕ's, to currents, i's.

The rate of change of flux with time within a coil of wire (with or without a ferromagnetic core) of N turns can be related to λ, as

$$V = \frac{d\lambda}{dt} = N \frac{d\phi}{dt} = L \frac{di}{dt} \qquad (1)$$

by Lenz's and Faraday's Law. As shown in (1), a similar relationship can be stated in terms of the inductance (L) of the coil and the instantaneous current (i) thru it. In the case of a single coil of wire, ϕ is the measure of flux linkage within the core produced by *self-induction*. In an instance where there are multiple coils with common magnetic paths, the total flux linkage of one coil would be the *sum* of that produced by self-induction plus those produced by mutual interaction, or *mutual induction*, with the others.

The *magnetomotive force*, F, of an excited coil of wire (with or without a ferromagnetic core) is defined as the product of the instantaneous current through it, i, and the number of turns, N, of the coil. In equation form, F can be directly related to *magnetizing force*, H, and its magnetic path length, 1, as

$$F = Ni = Hl \qquad (2)$$

The *self-inductance*, L, of a single coil of wire of N turns, can be equated to the rate of change of flux with current from (1) as

$$L = N \frac{d\phi}{di} \qquad (3)$$

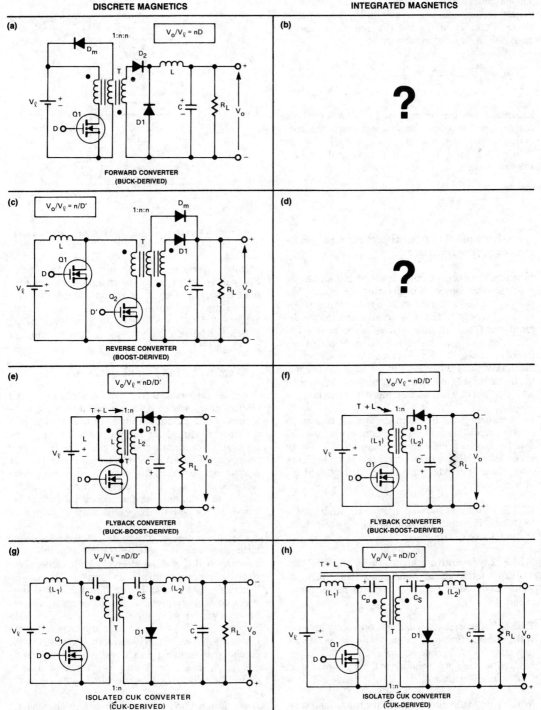

Discrete and Integrated-Magnetic Converter Possibilities—What About the Forward and Reverse Designs?
Figure 1

Assuming that a *linear* relationship between flux level and magnetizing force is always maintained, R can be defined as the ratio of a change in F produced by a change in ϕ. From (2) and (3), reluctance can be expressed in terms of inductance as

$$R = \frac{dF}{d\phi} = N\frac{di}{d\phi} = \frac{N^2}{L} \qquad (4)$$

Reluctance can also be expressed in terms of related magnetic path length, cross-sectional area of the magnetic material, A_C, and the *permeability*, μ, of the path in question. If A_C is uniform in value throughout the path, then

$$R = \frac{1}{\mu A_C} = \frac{1}{P} \qquad (5)$$

where P is defined as material permeance, the reciprocal element of reluctance.

In magnetic circuits, arrows are used to indicate the assumed directions of winding currents, rather than polarity marks for magnetomotive forces (often shortened to "mmfs"). The most popular method for determining flux direction is the "right-hand rule" illustrated in Figure 2(a). With the right hand positioned as shown, the flux direction will be indicated by the direction of the curvature of the fingers. Note that the thumb must be pointed in the assumed direction of winding current when making this determination.

In representing multi-winding magnetic circuits in electric circuit diagrams, it is customary to use *dot notation* to convey the voltage polarity relationships of each winding relative to the others, as shown in Figure 2(b). Three basic rules are followed in the "dotting" of windings:

RULE 1. Voltages induced in any two windings due to changes in mutual flux will have the *same* polarity at "dot-marked" terminals.

RULE 2. If positive currents flow into the "dot-marked" terminals of related windings, then the mmfs produced in each winding will have *additive* polarity.

RULE 3. If any related winding is open-circuited, and if the currents flowing into the "dot-marked" terminals have a *positive* rate of change, then the voltage induced in the open winding will be *positive* at its "dot-marked" terminal.

When a magnetic circuit arrangement has more than two windings and contains *more* than one major flux path, then multiple "dotting" windings is necessary to

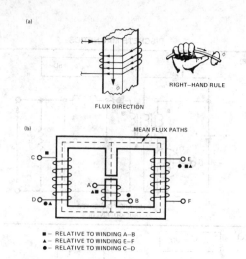

RIGHT–HAND RULE

FLUX DIRECTION

MEAN FLUX PATHS

■ – RELATIVE TO WINDING A–B
▲ – RELATIVE TO WINDING E–F
● – RELATIVE TO WINDING C–D

Determining Flux Directions (a), and Dotting of Windings (b)
Figure 2

visually express voltage polarities. However, this is easily done by the use of the right-hand rule and the three dotting rules given above. An example of multiple "dotting" is shown in Figure 2(b), where three sets of "dots" are needed to express winding polarities relative to each one of the three windings of the magnetic.

A Magnetic Circuit Modeling Example

All of the fundamental definitions related to magnetic circuits given above, when combined with the techniques of electric circuit modeling and duality, produce a powerful set of analysis tools for quick and accurate examination of any magnetic circuit arrangement, no matter how complex it may be. The electric circuit models that result are just that — *circuit models* — and do not depend on abstract mathematical relationships (such as mutual inductance expressions) for performance evaluations.

As an illustration of the power of these modeling techniques, consider now an example. For this exercise, we will derive an electric circuit model for a two-winding transformer with parasitic leakage inductances, as shown in Figure 3(a). Here, an ungapped toroidal core provides the major magnetic path between windings (ϕ_m), and each winding has a leakage path (ϕ_1, ϕ_2) for flux that is not contained in the material path. We will assume for this exercise that the core's cross-sectional area is uniform throughout its body, and that a mean path length can be used to define the reluctance of the core.

If we designate N_p as the primary exciting winding, we can first determine the voltage polarity of the secondary winding using the right-hand rule. With a primary current direction and the winding dotted as shown in Figure 3(a), fluxes ϕ_1, and ϕ_2 must have positive directions as indicated in this figure. The dot for the secondary therefore must be at the top of this winding to produce a flux in the same direction as ϕ_m.

With winding polarities established, we note that the secondary voltage, V_0, must have a polarity as shown, for the given polarity of exciting voltage, V_s. Because the resulting secondary current is "out of the dot", we also note from our earlier transformer rules that this current, i_0, will produce an mmf that opposes that of the primary.

Next, we draw the equivalent magnetic circuit diagram for the arrangement of Figure 3(a). This is done in Figure 3(b), using the analogous relationships defined earlier. Note that ϕ_m and ϕ_1 have the same polarity in the primary winding, since they are caused by i_s. However, in the secondary winding, ϕ_m and ϕ_2 have opposite polarity, since one flux (ϕ_m) produces i_0 while the other (ϕ_2) is a consequence of i_0.

From the magnetic circuit of Figure 3(b), we can now develop a dual permeance network for it, resulting in the equivalent circuit of Figure 3(c). Our next step is to scale the network of Figure 3(c) by the number of turns of the winding we select as reference for the final electric circuit representation. This is accomplished in Figure 3(d). The major purpose of this scaling step is to place all circuit permeances in a form that can be directly related to inductance. Note that the scaled model of Figure 3(d) also permits easy conversion of flux linkages to primary and secondary voltage values.

Finally, we convert the scaled permeance network to one involving voltages and inductances. This conversion is shown in Figure 4(a). Note that we have added an ideal transformer to this network to properly scale terminal secondary voltage and current values. Using impedance translation methods, we can also "move" all or some primary inductances to the secondary side of the final electric circuit model if desired. Two versions of such impedance movements are shown in Figures 4(b) and 4(c).

In just four easy steps, we have been able to develop a realistic electric circuit model for a somewhat complex magnetic circuit arrangement. Note that the final models of Figure 4 do not involve mutual inductances and are in a form that a designer can easily relate physical properties of the corresponding magnetic system of Figure 3(a) to familiar electric circuit quantities.

Looking at Figure 4(c), we see that inductances L_1 and L_2 represent those produced by flux leakages, and are often called *leakage inductances*. Inductance L_c in the networks of Figure 4 represents that produced by the primary turns (N_p) wound on the ferromagnetic core of the transformer, and is not the mutual inductance (M) shared by primary and secondary windings. Recall that mutual inductance is a mathematical measure of the degree of coupling between two windings of a magnetic. In this case, we can easily find the value

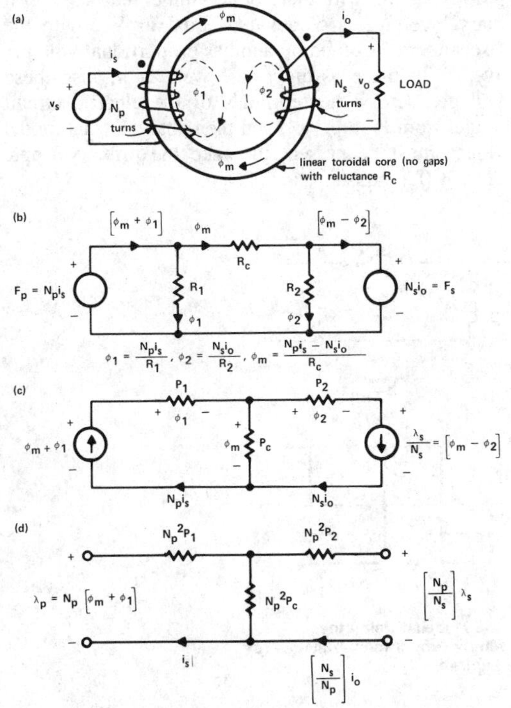

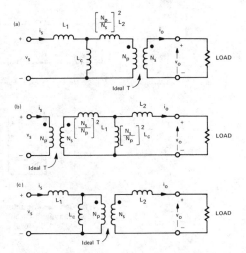

Developing Reluctance and Permeance Circuit Models for a Two-Winding Transformer with Leakage Fluxes
Figure 3

Electric-Circuit Equivalent Networks for the Two-Winding Transformer of Figure 3(a)
Figure 4

of M by writing the two nodal equations that relate input and output currents of the circuit models, and then isolate the common inductance terms within them. If this is done, then M is found to be

$$M = \frac{N_S}{N_P} \times L_C \qquad (6)$$

Achieving Zero-Ripple Currents

Although the techniques of achieving zero-ripple currents in the windings of selected magnetic designs have recently been thoroughly explored [3], there is much historical evidence that the use of magnetic means to lower ripple currents in converter outputs is not a new discovery. For example, G.B. Crouse disclosed in his 1933 United States patent [6] a magnetic method for lowering the ripple voltage of an L-C filter for use in radio receiver power supplies. This method, upon close scrutiny, appears to be identical to one reported some years later by S. Feng [7] for reducing the size of filter capacitors used across the outputs of power converters. Several other examples [8,9,10] also exist.

In both cases, the method entailed the addition of another winding to the filter inductor, whose mutual inductance relationship to the original inductor winding was selected by design to "steer" the ripple current in the main winding to the added "secondary" winding and, therefore, away from the output of the filter. The resulting circuit arrangement is shown schematically in Figure 5, for application as the secondary filter of a forward converter. Note that a filter capacitor (C) is necessary for DC isolation of the "inter" winding and is selected to maintain an aver-

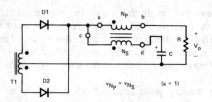

"Coupled-Inductor" Output Filter Network of a Quasi-SquareWave Converter
Figure 5

age voltage across it equal to V_O in the presence of the ripple current. Ideally, the mutual inductance shared by the inductor windings is then chosen to *completely* remove the ripple current from the load (R) and to steer it to the inner winding (N_3). For this reason, no output filter capacitor is shown across R in Figure 5. However, in practical designs, some capacitance is usually added across R for decoupling of noise and for additional energy storage for instantaneous load demands.

To understand how the inductor arrangement of Figure 5 can magnetically reduce ripple currents, we can use the electric circuit model developed in Section 2.1. Looking at Figure 5, we can assume that the voltages impressed across the two windings of the inductor are proportional in amplitude and equal in dynamic periods, such as those shown in Figure 6(a). For our purposes, we will therefore assume that the "primary" voltage is of a value equal to V_S, with the "secondary" excited by another proportional voltage, av_S. With these assumptions, we can impose these voltages across the terminals of the electric circuit model from Figure 4(c), and then analyze what model values must be present to make the primary ripple current (i_S) vanish.

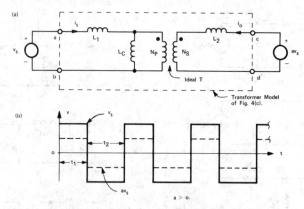

The Coupled-Inductor of Figure 5 Modeled Using the Transformer Equivalent Circuit (a), Driven by Proportional Voltages (b)
(Note: a = 1 for Figure 5)
Figure 6

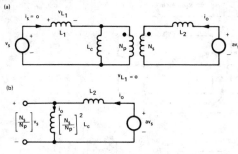

Circuit Conditions in the Model of Figure 6 for $i_s + 0$
Figure 7

This analysis task is easily performed, as illustrated in Figure 7. Writing the circuit equations for Figure 7(b) gives

$$av_S = \left[L_2 + \left(\frac{N_S}{N_P}\right)^2 L_C\right] \frac{di_O}{dt} \qquad (7)$$

$$\left(\frac{N_S}{N_P}\right) v_S = \left(\frac{N_S}{N_P}\right)^2 L_C \left(\frac{di_O}{dt}\right) \qquad (8)$$

$$L_2 = \left(\frac{N_S}{N_P}\right)^2 L_C \left[\frac{aN_P}{N_S} - 1\right] \qquad (9)$$

Therefore, if we select the leakage inductance of the secondary winding to meet the needs of (9), then the primary ripple current will be reduced to zero! Although not shown here, a similar analysis of the circuit model of Figure 6(b) for zero-ripple secondary current can be made. The value of primary leakage inductance for $i_O = 0$ is then found to be

$$L_1 = L_C \left(\frac{N_S}{aN_P} - 1\right) \qquad (10)$$

Comparing the constraints imposed on the values of L_2 and L_1 by (9) and (10), respectively, one finds that it is not possible to achieve zero values of i_S and i_O simultaneously. This point is also made by the current waveforms for i_S and i_O illustrated in Figure 8. Note also from the equations accompanying Figure 8 that the *effective* inductance seen at the "primary" terminals of the two-winding magnetic will be equal to its open-circuit value (i.e. "secondary" side open) when L_1 is selected to reduce i_O to zero. Conversely, the effective inductance seen at the "secondary" terminals will be equal to its open-circuit magnitude when L_2 is set to reduce i_S to zero.

In actual practice, it is difficult to design and manufacture magnetic assemblies with consistent and specific values of parasitic leakage inductances to achieve the zero-ripple current conditions as defined by (9) or (10). One viable solution to this producibility problem is to *tightly* wind primary and secondary turns to reduce L_1 and L_2 to essentially zero, and then to insert a small *external* "trimming" inductor [8] in series with the input or output to emulate the required inductance needed for L_1 or L_2. This trimming method is shown in Figure 9, along with corresponding values of L_{ext} necessary for zero-ripple current conditions.

Because of recent emphasis on application of zero-ripple-current filters in ĆUK converter variations [9], [10], it is now generally believed that this particular family of converter topologies is the only one that can benefit by their use. However, as we have just seen, this is not the case. In boost converters, these same principles of ripple current reduction can be applied to their input inductor arrangements, one example of which is shown in Figure 10. And, as we will soon see, this method of ripple current reduction can also be advantageously used in both buck and boost derived *integrated-magnetic* converter systems.

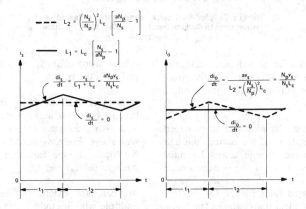

Input and Output Current Waveforms of Figure 6(a)
When Selected Values of L_1 or L_2 are Present (Note: a = 1 for Figure 6(a))

Figure 8

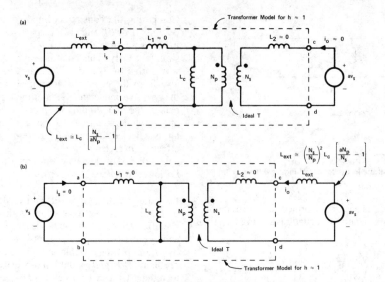

**Using Small External Inductances to Trim Ripple Current
Magnitudes in the Circuit of Figure 6(a)
Figure 9**

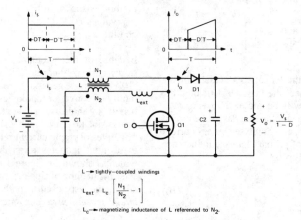

**Using a Coupled-Inductor Arrangement in a Basic Boost
Converter to Reduce Input Ripple Current Magnitude
Figure 10**

The preceding explanation for zero ripple, while correct, may leave one wondering what is really going on and if the results are truly general or just a bit of mathematical serendipity.

A very simple intuitive explanation can be provided by looking again at Figure 6. If the current through L_1 is to be zero, then the voltage across L_1 must also be zero. To the left of L_1 is the voltage source V_S. To the right of L_1 is a voltage divider formed by L_C and the transformed value for L_2: $L_2' = (N_S/N_P)^2 \cdot L_2'$ as well as the ideal transformer. The game being played is to adjust the values for L_C, L_2', and N_P/N_S so that the voltage source aV_S is transformed to a value equal to V_S at the right side of L_1. When

this is accomplished, the voltage across L_1 is zero and so is the ripple!

Remember that, in this particular example, a = 1 because both windings are excited by the same source. This is not the case in general as the next example will show. For the special case where a = 1, then $N_P/N_S > 1$ (see Equation 9) if L_2 is to have a positive value.

We can now extend our zero-ripple example to a multiple winding inductor in a multiple output quasi-square wave converter in which more than one output has zero ripple. The use of a single multiple winding inductor reduces the total weight and volume of the

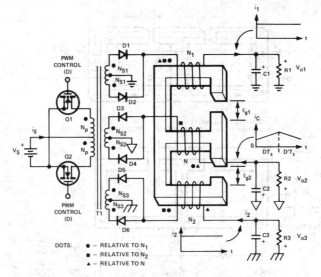

DOTS: ● − RELATIVE TO N₁
       ■ − RELATIVE TO N₂
       ▲ − RELATIVE TO N

**Coupled-Inductor Arrangement for a Quasi-Square Wave
DC-DC Power Converter
Figure 11**

inductive portion of the system, and also lowers cost and enhances the dynamic cross-regulation properties between the outputs of the converter. One of several possible versions of a converter with "coupled" output inductances is given in Figure 11.

Those familiar with the design aspects of coupled-inductors for converters such as this one will note that the winding arrangements, as well as the core structure itself, are somewhat unconventional. Here, a three-legged core is employed, with two inductor windings placed on the outer portion of the structure, and the third winding contained on the "center" leg. Air gaps are placed in each of the two outer legs. As illustrated by the inductor current waveforms of Figure 11, it is postulated that the ripple currents in two of the three inductor windings can be made to disappear!

At first glance, it may seem that the analysis effort necessary to understand how the magnetic arrangement can achieve significant current ripple reductions in selected windings is overwhelming. However, if we use the modeling techniques just demonstrated, the analysis becomes straightforward and its results illuminating.

Note that the dynamic voltages impressed across each of the three windings of the magnetic are proportional, always in phase with one another, and have the same frequency. Voltage proportions are ideally set by the ratios of the transformer secondary turns relative to primary turns, and for this analysis, we will assume

ideal conditions exist (i.e., diode drops may be neglected and perfect coupling of transformer windings has been achieved). With this assumption, we define

$$V_{N1} = av_S, \ a = N_{S1}/N_P$$

$$V_{N2} = bv_S, \ b = N_{S3}/N_P \tag{10A}$$

$$V_N = cv_S, \ c = N_{S2}/N_P$$

where V_S is the dynamic waveform produced across either of the primary windings of T1 by the switching actions of Q1 and Q2. By our earlier assumption, we know that this converter is being operated at a constant switching frequency, and that Q1 and Q2 are being alternately turned on for a portion, D, of a switching cycle.

The first three steps in evolving the model are shown in Figure 12. In part (a) flux directions are found in each of the three magnetic paths, using the right-hand rule method and assuming that the center leg is the controlling and dominant source of mmf.

Three other minor flux paths are shown in Figure 12(a). These represent "leakage" fluxes, i.e., not contained within the core material and are to be expected in any practical magnetic winding arrangement. Therefore, we have added them to our magnetic system to ensure that our final electrical model will include leakage inductance effects associated with each of the three windings.

6

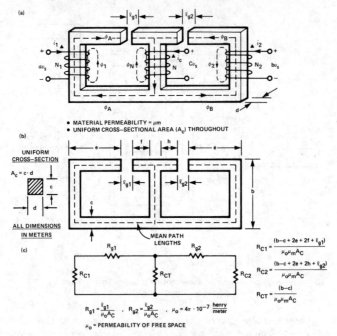

(a)

• MATERIAL PERMEABILITY = μ_m
• UNIFORM CROSS–SECTIONAL AREA (A_c) THROUGHOUT

(b)

UNIFORM
CROSS–SECTION

$A_c = c \cdot d$

ALL DIMENSIONS
IN METERS

(c)

MEAN PATH
LENGTHS

$$R_{C1} = \frac{(b-c + 2e + 2f + \ell_{g1})}{\mu_o \mu_m A_C}$$

$$R_{C2} = \frac{(b-c + 2e + 2h + \ell_{g2})}{\mu_o \mu_m A_C}$$

$$R_{CT} = \frac{(b-c)}{\mu_o \mu_m A_C}$$

$$R_{g1} = \frac{\ell_{g1}}{\mu_o A_C} \quad , \quad R_{g2} = \frac{\ell_{g2}}{\mu_o A_C} \quad , \quad \mu_o = 4\pi \cdot 10^{-7} \frac{henry}{meter}$$

μ_o = PERMEABILITY OF FREE SPACE

**Reducing the Inductor Assembly of Figure 11 to a
Reluctance Network
Figure 12**

Next, in part (b) a two-dimensional view of the structure of (a) is given in order to be able to relate core dimensions and air gap lengths to values for reluctance. Because winding details are not important for these determinations, they are omitted for clarity.

In part (c) an "electrical" network of reluctances is constructed from the magnetic path details of part (b). Also shown in part (c) are the equations relating reluctance values to core dimensions, permeabilities, and cross-sectional area.

We now add to our reluctance network of Figure 12(c) the mmf's produced by the three winding currents as well as surrounding leakage reluctances. This is done in Figure 13(a). The "polarity" of each of the winding mmf's in our model is established using **Rule 2** in conjunction with the assumed flux directions in each leg of the core in Figure 12(a). Note that, because each winding current is directed into its respective "dot," the mmf polarities are such that they will produce fluxes that are in the *same* direction as those assumed in Figure 12(a) in their respective legs.

Although the magnetic circuit model of Figure 13(a) is perfectly suitable for translation into an electrical equivalent network, some simplification may be made.

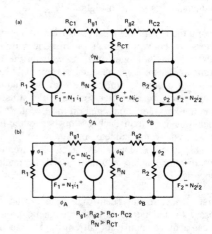

$$R_{g1}, R_{g2} \gg R_{C1}, R_{C2}$$
$$R_N \gg R_{CT}$$

**(a) Adding mmf's to the Network of Figure 12(c),
(b) Simplification of (a)
Figure 13**

For the most part, air gap and leakage reluctances will always be much greater than those presented by magnetic material paths, since material permeabilities are much greater than that of air. Therefore, the magnetic model of Figure 13(a) may be reduced in complexity to that shown in Figure 13(b) with little loss in accuracy.

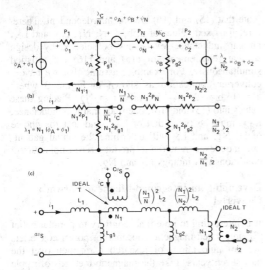

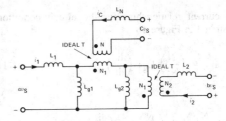

Final Electric Circuit Model for the Magnetic Arrangement of Figure 12(a)
Figure 15

Developing an Equivalent Electrical Circuit Model for the Reluctance Network of Figure 13(b)
Figure 14

The next step in our model development is establishing a dual network for the reluctance and mmf system of Figure 13(b). Using the duality procedures outlined in [5], the permeance circuit of Figure 14(a) is easily obtained. A choice must now be made as to which of the three windings we want the final elements of the electrical model to be "referenced." We arbitrarily chose N_1, keeping in mind that we can always alter our final electrical model later to refer to N_2 or to N if we so desire.

We now scale all of the elements of the model of Figure 14(a) by N_1. This process gives us a model relating flux linkages to winding current values, as illustrated in Figure 14(b). We then convert the permeance model of Figure 14(b) to one involving inductances and winding voltages. This is done in Figure 14(c), with ideal transformers added to account for the model reference to N_1.

Finally, using conventional impedance transformations, we move the leakage inductances associated with N and N_2 to the "primary" sides of their respective ideal transformers. The result of this last model manipulation is shown in Figure 15.

In a few basic steps, we have been able to reduce a rather complex magnetic system to an equivalent electrical circuit. In reviewing our steps, we find that we can easily relate all of the elements of the final electrical model to the parameters of the magnetic assembly, including physical dimensions and perme-

abilities of various parts of magnetic paths. The final model gives excellent visibility into the effects of winding voltages and currents; also, how we can reduce ripple currents in selected windings to zero!

Referring to the converter circuit of Figure 11, we remind ourselves that we are interested in the model conditions that must exist if the dynamic currents, i_1 and i_2, are zero. Looking now at our final model in Figure 15, we simply set i_1 and i_2 to zero, and then establish what circuit conditions prevail as a result of their absence. These conditions are shown in Figure 16.

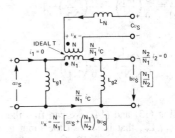

Circuit Conditions in the Model of Figure 15 for Ripple Currents in N_1, N_2 Equal to Zero
Figure 16

With i_1 and i_2 at zero value, the voltages impressed across the gap inductances, L_{g1}, and L_{g2} will be simple reflections of associated winding voltages, av_s and bv_s, respectively. Also, the dynamic voltage drops across the leakage inductances, L_1 and L_2, must be zero, because their respective branch currents are zero. At this point, we can make an important observation — the values of the leakage inductances, L_1 and L_2, are *not* important in establishing a zero-ripple current condition in either winding N_1 or in winding N_2!

In order to discover what remaining parameters of the magnetic arrangement are important in making i_1 and i_2 zero, we now determine the describing voltage

6

and current relationships for the circuit conditions illustrated in Figure 16

$$av_S = \left(\frac{N}{N_1}\right) L_{g1} \left(\frac{di_c}{dt}\right) \tag{11}$$

$$b \frac{N_1}{N_2} v_S = \left(\frac{N}{N_1}\right) L_{g2} \left(\frac{di_c}{dt}\right) \tag{12}$$

$$cv_S - v_x = L_N\left(\frac{di_c}{dt}\right) \tag{13}$$

$$v_x \quad \frac{N}{N_1} \left[av_S + \left(\frac{N_1}{N_2}\right) bv_S\right] \tag{14}$$

Combining (11) and (12) with (13) and (14) to eliminate the common di_c/dt term gives

$$\frac{L_N}{L_{g1}} = \frac{c}{a}\left(\frac{N_1}{N}\right) - \frac{b}{a}\left(\frac{N_1}{N_2}\right) - 1 \tag{15}$$

$$\frac{L_N}{L_{g2}} = \frac{c}{b}\left(\frac{N_2}{N}\right) - \frac{a}{b}\left(\frac{N_2}{N_1}\right) - 1 \tag{16}$$

Also, using (13) and (14), we can find the effective inductance, L_e, that is seen at the input to winding N when i_1 and i_2 are zero. Performing this operation gives

$$L_e = L_N + \left(\frac{N}{N_1}\right)^2 \left(L_{g1} + L_{g2}\right) \tag{17}$$

Looking at (15) and (16), we now realize that if their relationships are satisfied, the ripple currents in both N_1 *and* N_2 will be entirely eliminated! These equations also predict that the zero-ripple current conditions are *not* dependent on external converter operational conditions, such as line, load, or switch duty cycle magnitudes! The phenomenon will be dependent, however, on winding turns, proportional constants set by the transformer secondary-to-primary turns ratios, leakage inductance of the center leg of the magnetic assembly and the values of the two air gap inductances.

Note that (15) and (16) are *not* interdependent. Therefore, for fixed values of a, b, c, N, N_1, N_2, and L_N, air gap inductance levels may be selected by design so that requirements of (15) and (16) are satisfied simultaneously. For example, assume that we have a converter application as shown in Figure 3 where a = 1, b = 2, c = 3, and $N_1 = N_2$. Placing these values into (15) and (16), one finds that gap inductance, L_{g2}, must be twice that of L_{g1}, and that the choice of N, N_1, and L_N will determine the actual gap inductance levels required to produce zero-ripple current conditions in windings N_1 and N_2.

Developing an Integrated-Magnetic Forward Converter

It is interesting to note that, contrary to popular belief [3], integrated-magnetic concepts for power converters are not innovations of research performed over the past seven years. Like the magnetic methods of ripple current reduction described in the last section of this paper, the concept of integration of the magnetic functions of power processing circuits has historical roots in much earlier periods. Perhaps the best documented evidence of this fact is an obscure (and mistitled) United States patent by Cielo and Hoffman of the IBM Corporation [14] in 1971, which discloses possible circuit methods for integration of the transformer and inductor of so-called "push-pull" DC-to-DC converters. Later in this paper, we will examine the methods of this interesting patent by extending the integrated-magnetic forward converter concepts to include "push-pull" switch arrangements.

With a brief exposure to magnetic circuit modeling methods and a review of inductive ripple current reduction techniques behind us, let us proceed to systematically construct a *single* magnetic assembly wherein the transformer and the inductor of a forward converter can be contained. No rigorous synthesis procedure will be followed here, but rather a path of *deduction and intuition* based on the general operation of this converter and knowledge of the flux change relationships that must exist in its inductive components.

The first step in our integration process will be to redraw the forward converter circuit of Figure 1 in a manner so as to emphasize both the electrical and magnetic aspects of the converter. Then, assuming that the converter is operating in the continuous mode of inductor energy storage, the equations describing circuit conditions for each of the two switching states are found, and then placed in a format relating corresponding flux changes in the transformer and inductor. Because we are interested at this time in only major flux relationships, the describing circuit equations can be ideal, thus ignoring all parasitics including switch and diode drops, switching losses, etc. The elimination of parasitics is not a requirement, but it is a convenience for this discussion.

The redrawn forward converter is illustrated in Figure 17. The core of the transformer is shown ungapped, while the inductor core is shown with an air gap, as is the usual case for a magnetic that must withstand DC bias. Transformer windings are drawn and dotted to emphasize an assumed counter-clockwise direction of flux in its core and to produce a positive voltage (V_1) across the secondary winding (N_S) when the primary switch (Q1) is ON. Note also that we have included a "core-reset" winding (N_{p2}) on the transformer for energy removal during intervals when Q1 is OFF. A core material with low residual flux is presumed. For the inductor with its winding position as shown, flux direction will be clockwise.

Next, the ideal voltages that appear across the transformer and inductor windings are found by simple circuit analysis for each of the two switching states of the converter, DT_S and $(1-D)T_S = D'T_S$. Using (1), we then relate the rate of flux change in the transformer core (ϕ_T) and in the inductor (ϕ_L) to these voltage magnitudes.

During DT,

$$\dot{\phi}_T = \frac{d\phi_T}{dt} = \frac{V_s}{N_{p1}} = \frac{V_1}{N_S} \qquad (18)$$

$$\dot{\phi}_L = \frac{d\phi_L}{dt} = \frac{V_1}{N_L} - \frac{V_o}{N_L} \qquad (19)$$

During D'T,

$$\dot{\phi}_T = \frac{V_S}{N_{p2}} \qquad (20)$$

$$\dot{\phi}_L = \frac{V_o}{N_L} \qquad (21)$$

Looking at the equations for interval DT_S, we see that we can combine them to remove the intermediate secondary voltage value, V_1:

$$\dot{\phi}_T = \left(\frac{N_L}{N_S}\right)\dot{\phi}_L + \frac{V_o}{N_S} \qquad (22)$$

Note that the last term of (22) is of a form that could be considered as defining a flux change in a magnetic medium that is *dependent* on the value of the output voltage, V_o, of the converter and the number of secondary turns on the transformer, N_S. It follows then, that if we make this consideration, that such a magnetic medium must be a part of the transformer assembly to satisfy the conditions of (22). Therefore, we can rewrite (22) as

$$\dot{\phi}_T = \left(\frac{N_L}{N_S}\right)\dot{\phi}_L + \dot{\phi}_o \qquad (23)$$

where $\dot{\phi}_o = V_o/N_S$. Turning now to the first right-hand term of (22) and (23), we note that its contribution to ϕ_T is dependent on a fraction, N_L/N_S, of the flux change rate in the inductor of the converter. Since our ultimate goal is to make the inductor an integral part of the same magnetic assembly that also contains the transformer windings, it is logical to assume that N_L should be made equal to N_S, so as to contain all inductor flux in a single magnetic path.

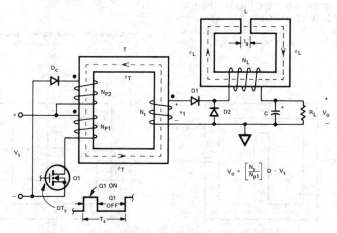

Conventional Forward DC-DC Power Converter (Idealized)
Drawn to Emphasize Magnetic Operations
Figure 17

As we will see later when we analyze the integrated magnetic that results from this "synthesis" exercise, N_S must be equal to N_L to realize a *non-pulsating* output current waveform and to achieve an input-to-output voltage transfer function that *matches* that of a forward converter.

Setting $N_L = N_S$ in (23), we arrive at a final expression for $\dot\phi_T$ during interval DT_S:

$$\dot\phi_T = \dot\phi_L + \dot\phi_0 \qquad (24)$$

Remembering our previous magnetic modeling, we can interpret (24) as defining a magnetic assembly in which there are three major flux paths. This equation also tells us that the flux change in an input source related path ($\dot\phi_T$) contributes to the change in another path ($\dot\phi_L$) associated with the "inductor" portion of the assembly, as well as to flux change ($\dot\phi_0$) in a third path. From the assumption made in (23), this third path is related to the output voltage value of the converter.

These general observations permit us now to sketch a possible magnetic path arrangement that satisfies the conditions of (24) for the switching interval DT_S. This sketch is shown in Figure 18(a). Note that the "inductor" path includes an air gap, as we expect that this leg of the magnetic core will need one to establish the required amount of storage inductance and to sustain DC bias without material saturation. To the outer legs of the core arrangement, we have added windings for primary and secondary in accord with (18) and (19), dotted properly to produce the required polarity of V_0 for the indicated directions of i_S and i_0.

Using the magnetic arrangement of Figure 18(a) as a baseline, we now look at what must be added to permit the conditions of (20) and (21) to be satisfied for the other switching interval $D'T_S$. First, a winding

to release the energy stored in the center leg is needed, and its turns (N_L) must equal those of N_S from our earlier discussions. Second, a winding (N_{P2}) is needed in the same leg of the core as that of the primary for "reset" purposes. The magnetic arrangement that results is shown in Figure 18(b). Note that these two windings are dotted so as to produce the *same* flux directions in their respective core legs as those in Figure 18(a).

Our final task in this deductive "synthesis" process is to combine the two magnetic arrangements of Figure 18 by the addition of switches and diodes to establish the required winding voltage values for each switching interval of the converter. From our knowledge of the forward converter, we would expect that no more than one switch and three diodes would be needed, and this is indeed the case.

Figure 19 is the schematic of the integrated-magnetic version of the forward converter that results from our efforts. Reflecting now on the steps that were necessary to place the transformer and inductor on a single-core structure, we find that no exotic efforts were really required and that, in reality, the deductive procedure followed was rather straightforward and ele-

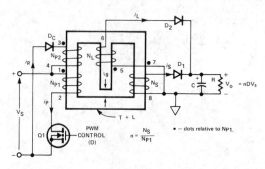

A Forward Converter with Integrated Magnetics
Figure 19

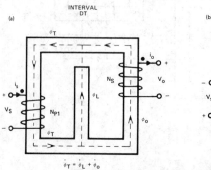
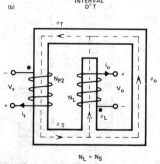

Developing Magnetic Core Arrangements for Each State of the Converter in Figure 11
Figure 18

gantly simplistic. It now becomes clear that we can use similar deductive methods to evolve integrated-magnetic versions of any transformer-isolated buck-derived converter. Once this is done, we can then use the principles of duality between converter circuits [15,16] to formulate complementary boost topologies.

Verification by Analysis

Following the evolution of an integrated-magnetic arrangement for a converter, it is worthwhile, if not mandatory, that its structure be verified by analysis. This is necessary in order that no major design aspect has been overlooked, as well as to examine the effects of any circuit parasitics (such as leakage inductances, etc.) that may have been introduced by the integration process.

For the integrated-magnetic forward design of Figure 19, the analysis procedure begins with the extraction of the magnetic from the overall circuit topology, labeling each winding as to voltage polarities as well as current directions that are imposed by the converter. It is also advantageous to mark each winding terminal and carry the identifying marks through the modeling procedure, so that the final electric circuit model found by the analysis can be directly substituted for the magnetic assembly in the converter topology.

An isometric sketch of the extracted magnetic from Figure 19 is shown in Figure 20(a). To this sketch, we add flux directions in each leg, assuming that the primary winding, N_{P1}, is the exciting and dominant winding of the magnetic. From this point on, the analysis procedure follows the same steps of the two-winding transformer example of Figure 3(a), keeping in mind that there are three magnetic paths and four sources of mmf in this case. Also, since we are only interested in major flux paths, we have omitted all leakages for this examination. However, they can be easily added later and the corresponding electric circuit model changed to reflect their presence.

The electric circuit model of the magnetic of Figure 20(a) is shown in Figure 20(b). For the sake of brevity, we have not included the intermediate steps of the modeling effort here, but they can be found in [17]. This particular model is referenced to the main primary of the magnetic, N_{P1}, as evidenced by the turns ratios of the three ideal transformers within it. Inductance L_g represents that presented by the center-leg-winding, while the two inductances, L_C, represent those presented by the outer-leg windings of the core (referenced to N_{P1}).

The next step is to place the electric circuit model into the converter circuit, as shown in Figure 21. In making this placement, we have also "moved" L_g and one of the two L_C inductances "through" the

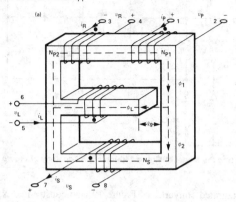

- UNIFORM CROSS–SECTIONAL AREA (A_C) THROUGHOUT
- MATERIAL PERMEABILITY OF ALL SECTIONS IS THE GAME
- DOTS RELATIVE TO N_{P1}

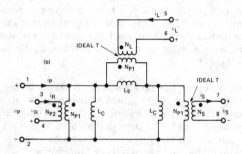

An Electric Model (b) for the Magnetic System of (a)
Taken from Figure 13
Figure 20

ideal transformer of turns ratio $N_{P1}:N_S$, using standard impedance translation methods. This is done to position most of the inductances to the "right" of this transformer in order that the final equivalent converter circuit can be compared to that of a forward converter with discrete magnetics. Note that this impedance movement also required a change in the turns ratio of the ideal transformer across L_g.

In practical designs, we would expect that L_C would be much larger in inductance value than L_g, since the permeability of free air is much less than that of a ferromagnetic material. Therefore, we can simplify our analysis at this point by eliminating the reflected L_C across terminals 7 and 8 of the circuit model in Figure 21. Also, we can assume that the L_C across terminals 1 and 2 of the model can be viewed as the "magnetizing inductance" of a real transformer of ratio $N_{P1}:N_S$. It is also apparent that winding N_{P2} can be a part of this same transformer, since it parallels L_C.

With these assumptions in mind, we can now redraw the circuit of Figure 21 in a slightly simpler form shown in Figure 22(a). We also remember that the

6

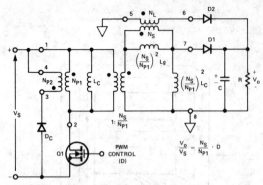

Placing the Model of Figure 14(b) Into the Topology of Figure 13
Figure 21

integrated converter of Figure 19 also constrained N_L to be equal to N_S and, therefore, the turns ratio of the ideal transformer across L has been changed accordingly in Figure 22(a). With the turns ratio of this transformer now 1:1, it can be completely removed by simply reorienting the circuit positions of diodes D1 and D2.

This final analysis step leads to the converter configuration of Figure 22(b), which matches that of a conventional forward converter! We see also that the equivalent output inductance of this converter is equal to that set by the gap of the integrated magnetic multiplied by the squared ratio of "transformer" secondary-to-primary turns. If we had not chosen N_L to be equal to N_S, this equivalent circuit and its predecessor in Figure 22(a) tells us that the equivalent inductance would not be of the same value for each switching state of the converter and, therefore, we could expect the output current to be somewhat pulsating, just as would be experienced in a forward converter with a "tapped" output inductor [17]!

It is evident from the results obtained above that the analysis of an integrated magnetic by the use of electric model equivalents has great value and provides much valuable design information. It is particularly valuable in designing integrated-magnetic versions of existing converter designs, where filter inductances and transformer characteristics (turns ratios, etc.) are known quantities. Using the equivalent circuit, such as the one in Figure 22(b), these quantities can then be directly related to the parameters of the integrated magnetic, such as core dimensions and permeabilities, winding turns, etc.

Voltage and Current Waveforms

From a "black box" standpoint, we would expect an integrated-magnetic version of a converter to have the same voltage and current characteristics as its discrete-magnetic counterpart. For the forward converter design of Figure 19, this is indeed the case,

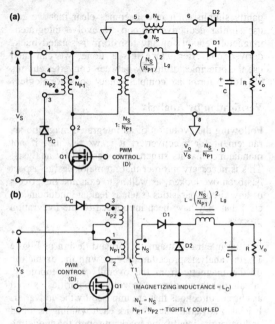

Further Circuit Manipulation (a) to Derive the Discrete Magnetic Version of Figure 13 Shown in (b)
Figure 22

as is evidenced by the dynamic current and voltage waveforms of Figures 23 and 24, respectively. These waveforms and their magnitudes were derived by inspection of the current and voltage conditions that must exist in the electrical circuit models of Figures 21 and 22 for each converter state (continuous mode of energy storage assumed). As we see from both sets of waveforms, current and voltage stresses on the switches and diodes remain the same as those that would be experienced in a conventional forward design.

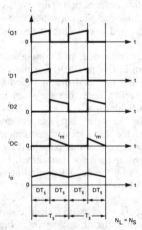

Idealized Currents in the SPC Topology of Figure 13
(Continuous mode)
Figure 23

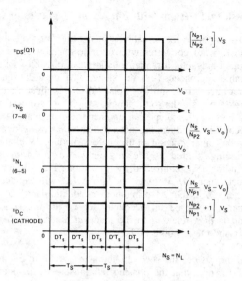

Voltage Waveforms in the SPC of Figure 19 (Continuous Mode) Figure 24

Developing Integrated-Magnetic Boost Converters

It is possible to follow similar procedures of deductive synthesis and analysis from the last section of this paper to evolve integrated-magnetic versions of various boost-derived converters, such as the *reverse converter* [16] in Figure 1. However, given a dual buck-derived converter approach, it is much easier to use duality methods [15,16] to derive these complementary circuits.

For example, given the integrated-magnetic forward converter, duality manipulations then produce the integrated-magnetic boost equivalent of the reverse converter of Figure 1. The unusual converter that results is shown in Figure 25. In this case, when Q2

is ON, energy is stored in the center-leg of the magnetic. During this same time period (DT$_S$), load needs are supported by the output capacitor, C, and by magnetizing energy stored in the outer leg of the magnetic from the previous switching cycle (via D3). When Q2 turns OFF, Q1 is turned ON and the energy stored in the center leg is magnetically routed to the output load, R, via winding N$_S$, with diode D1 now forward-biased and the other diodes non-conducting. Like its discrete-magnetic contemporary in Figure 1, the ideal input-to-output voltage transfer function is simply

$$\frac{V_O}{V_S} = \frac{N_S}{N_P} \times \frac{1}{1 - D} \tag{25}$$

with the "inductor" turns equal to that of the "primary" winding of the magnetic assembly.

Adding Zero-Ripple Current Features

Magnetic integration of the inductors and the transformers of buck and boost-derived converters does *not* lessen the possibility of adding additional windings to control ripple current magnitudes on output or input lines.

As we have seen in prior examples of integrated-magnetic forward and reverse converters, the inductive part of their magnetic assemblies is isolated to one path of the core arrangement. Therefore, by adding another winding in these same paths and impressing a voltage across it that is proportional in amplitude and of the same frequency as that appearing across the original inductor winding, we can steer the ripple current from the inductor winding to the other winding and, therefore, away from the output. We also have the option of "trimming" the ripple current magnitude by external means [11] if we so desire.

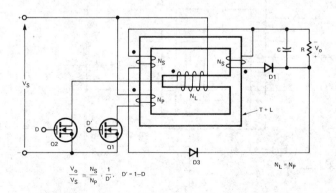

$$\frac{V_O}{V_S} \cong \frac{N_S}{N_P} \cdot \frac{1}{D'}, \quad D' = 1-D$$

An Integrated-Magnetic Version of the Reverse Converter Shown in Figure 1 Figure 25

Figure 26 shows an integrated-magnetic forward converter with output ripple current control capability. Note that another winding has been added to the center-leg section (N_{LR}), along with a small series inductor (L_X) and DC-isolation capacitor (C_R). Winding N_L is the original inductor winding and the new winding, N_{LR}, is wound in close proximity with it. This is done to minimize leakage inductance effects and to maximize the mutual inductance that will exist between the two windings.

With the new winding positioned in the converter circuit as shown in Figure 26, the value of inductance L_X needed to cause the output ripple current to vanish is found easily using (9) with a = 1:

$$L_X = \left(\frac{N_{LR}}{N_L}\right)^2 Lg \left(\frac{N_L}{N_{LR}} - 1\right) \qquad (26)$$

where Lg is the inductance presented by the air gap of the center leg with N_L turns.

In a similar manner, a ripple-control winding can be added to the integrated magnetic of a reverse converter, as is done in Figure 27. In this case, the value of L_X to reduce the input ripple current to zero is again the *same* as given by the solution of (26), assuming that windings N_L and N_{LR} are tightly coupled together.

In actual designs, there will be some slight slope in the "zero" ripple currents of these two converters due to the presence of the inductances (L_C) of the outer-leg primary and secondary windings. However, since L_C is usually much greater in value than that posed by the "inductor" winding, Lg, this deviation from an ideally flat current waveform is normally so small that it can be neglected for all practical purposes.

Push-Pull Designs with Integrated Magnetics

Just as there are so-called "push-pull" versions of the forward converter with discrete magnetics, integrated-magnetic arrangements are also possible. Two viable approaches [14] are shown in Figure 28. The first alternative, shown in part (a) of this illustration, looks very similar to the single-switch forward design of Figure 19, except that another set of primary and secondary windings has been added on opposing sides of the outer portion of the core structure. With the windings dotted as shown (relative to the primary winding, N_p, controlled by Q1), diode D1 will conduct when Q1 is ON, and diode D2 will conduct when Q2 is ON. During both conduction intervals, energy is stored in the center leg of the magnetic, and is then released to the load via diode D3 when Q1 and Q2 are both OFF.

In the case of the second alternative shown in Figure 28(b), note that the phasing relationships (i.e., dots) of the two secondary windings (N_S) have been changed from those of Figure 28(a). This change now prevents significant energy from being stored in the center leg of the magnetic, since conducting secondaries lie on the same magnetic path as their corresponding primary windings. For example, in Figure 28(b), when Q1 is ON, D1 will conduct and when Q2 is ON, D2 conducts. Therefore, another winding (N_{L1}) has been added to the center leg to provide a means of energy storage in the center leg when either Q1 and Q2 is conducting.

Ideal Voltage Transfer Functions

The input-to-output voltage relationships for the continuous mode of inductive energy storage for the converters of Figure 28 can be easily found by equating the volt-seconds appearing in the center leg of their

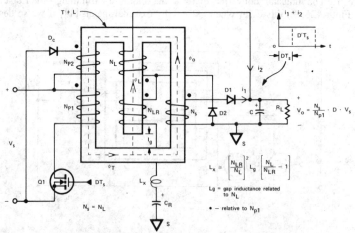

Adding Ripple Control Features to the Integrated-Magnetic Forward Converter of Figure 19 (N_L, N_{LR} tightly coupled)
Figure 26

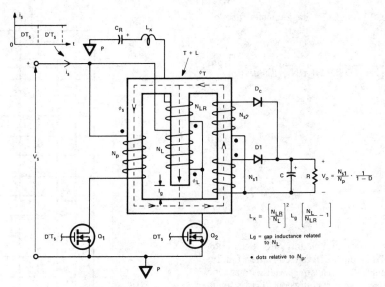

**Adding Ripple Control Features to the Integrated-Magnetic
Reverse Converter of Figure 25 (N_L, N_{LR} tightly coupled)
Figure 27**

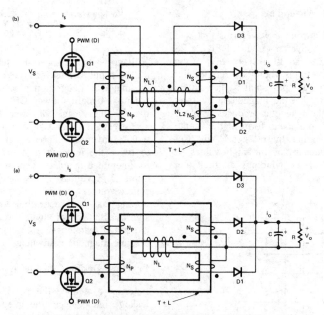

**Two "Push-Pull" Versions of the Forward Converter
of Figure 19
Figure 28**

integrated-magnetic assemblies during each of the two switching intervals, DT_S and $D'T_S$. In the case of Figure 28(a), the DC voltage transfer is ideally found to be

$$\frac{V_O}{V_S} = \frac{N_L}{N_P} \left\{ \frac{D}{1 - D\left[1 - (N_L/N_S)\right]} \right\} \qquad (27)$$

For N_S equal to N_L, (20) reduces to

$$\frac{V_O}{V_S} = \frac{N_S}{N_P} \times D \qquad (28)$$

which confirms that this integrated-magnetic converter design is buck-derived. For the other circuit arrangement in Figure 28(b), the DC voltage transfer function is ideally

$$\frac{V_O}{V_S} = \frac{N_S}{N_P}\left[\frac{1}{1 + (D'/D)(N_{L1}/N_{L2})(N_S/N_P)}\right] \quad (29)$$

In this instance, if

$$\frac{N_P}{N_S} = \frac{N_{L1}}{N_{L2}} \quad (30)$$

then (22) becomes

$$\frac{V_O}{V_S} = \frac{N_S}{N_P} \times D \quad (31)$$

which, like (28), confirms that this circuit variation is a member of the buck-derived converter family. Interestingly enough, this converter, with its magnetic replaced by its electric-circuit equivalent [17], is found to be an integrated-magnetic version of the Weinberg circuit [18]!

Voltage and Current Waveforms

The typical current and voltage waveforms that will be observed in either converter circuit of Figure 28 are shown in Figures 29 and 30, respectively. In the case of the current waveforms, they look very similar to those of a conventional "push-pull" quasi-square wave converter, except that the magnetizing current of the outer legs of the magnetic produces a minor step in the otherwise continuous form of the output current. As discussed earlier, these "steps" will usually be very small because of the high values of effective inductances of the windings on the outer legs.

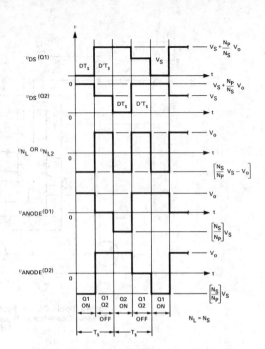

Voltage Waveforms in the Converters of Figure 28
(Continuous Mode)
Figure 30

Voltage waveforms also are quite similar to those of a "push-pull" buck-derived converter, except that the magnetic integration process has yielded another potential benefit—lower OFF voltage stress on the two switches of the converter. Ideally, the OFF voltage stress on either switch in a conventional "push-pull" converter will be equal to a maximum value of twice the input voltage magnitude. However, in its integrated-magnetic version, Figure 28(a), maximum OFF voltage stress is equal to the value of source voltage plus the reflected output voltage. This implies that, by proper choice of turns ratio, we can significantly reduce switch voltage stress by using integrated magnetics as compared to a conventional discrete-magnetic equivalent!

Practical Design Considerations

In order to reduce the possible harmful effects of leakage inductances, the integrated-magnetic circuit variation of Figure 28(b) is often preferred over that of its contemporary approach in Figure 28(a). Note that, in Figure 28(a), corresponding primaries and secondaries are positioned on the core legs so that they cannot be tightly coupled together (i.e., wound tightly together). However, in the converter of Figure 28(b), because corresponding primaries and secondaries lie on the same legs of the core structure, they can be wound tightly together to maximize their magnetic coupling and to minimize parasitic leakage inductances. Therefore, for this reason, the design of

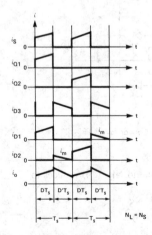

Ideal Current Waveforms in the Converters of Figure 28
(Continuous Mode)
Figure 29

Figure 28(b) is usually chosen rather than the circuit of Figure 28(a), even though the latter approach requires an additional winding to be added to the center leg for energy storage.

Another advantage of the converter in Figure 28(b) is the presence of an inductor in series with both primary switches. Thus, high instantaneous current due to conduction time overlap is automatically eliminated. Also, this converter, like its discrete-magnetic counterpart, can be made to operate as a boost converter, if the primary switches are purposely forced to have overlapping conduction intervals [16].

Small-Signal Modeling Considerations

Even though the magnetics of a buck or boost-derived converter have been magnetically integrated, modeling and analysis for stability and other dynamic studies is straightforward, using proven techniques such as the state-space averaging method [2]. All that is necessary is to derive an electric circuit model for the integrated-magnetic system, and then to add it in place of the magnetic in the converter topology. Following this, the normal procedures for deriving small-signal averaged models are followed, accounting for additional inductive elements [19] as presented by the integrated-magnetic arrangement.

For example, the averaged small-signal model for the integrated-magnetic version of the forward converter of Figure 16, using its equivalent electric-circuit topology from Figure 22, is illustrated in Figure 31. Note that the averaged model accounts for all of the inductive elements of the electric-circuit model, as well as its isolation transformer. If we had chosen to add leakage inductances to the electric-circuit representations in Figure 22, they could also have been easily accounted for in the equivalent small-signal model.

Laboratory Investigations

All of the integrated-magnetic versions of the various converters shown in this paper have been built and tested by the authors over the past two years. Major areas of interest for these empirical examinations were verification of expected DC voltage transfer functions and anticipated dynamic waveforms within the converters. In all cases, the results obtained were very close to those predicted.

Typical of the test results are the voltage waveforms taken for a low-power version of the integrated-magnetic Weinberg converter of Figure 28(b), shown in Figures 32(a) through (c). In this instance, the supply voltage was set at 40 VDC, and PWM duty cycles of Q1 and Q2 manually adjusted to obtain an output voltage of 5 VDC, across a load of 5 ohms and an output capacitor, C, of 100 microfarads. PWM clock frequency was set very close to 100 kHz, resulting in a 50 kHz switching frequency of each primary power switch. Also, to minimize voltage "spikes" produced by leakage inductances, small R-C snubbers (1.2 K ohm $\times$ 0.01 μF) were added across each primary winding and across N_{L2}.

The integrated magnetic used for this particular experiment was composed of two joined Ferroxcube E625 E-cores, with each of their center legs ground to produce a total air gap length of 0.01''. Three bobbins were then placed on each leg of the core structure, each with two sets of windings. Each winding set was composed of 30 turns of #20 AWG wire and 15 turns of #20 AWG wire. The completed magnetic was then connected into the converter arrangement of Figure 28(b), with windings phased as shown. Windings with 30 turns were used as primaries and for N_{L1}, while the 15-turn windings were used for the secondaries and for N_{L2}.

Using the relationships developed in Section 2, the inductance of the 15T center-leg winding was calculated to be about 1 mh for an air gap length of 0.01''. Inductances of the primary windings on the outer legs were also calculated to be about 10 times that of this center-leg winding, confirming the earlier assumption that these latter inductances are indeed much larger in value.

6

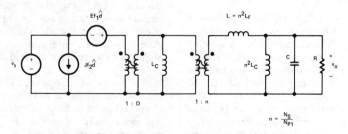

Averaged Small-Signal Model of the Integrated-Magnetic
Forward Converter of Figure 19
Figure 31

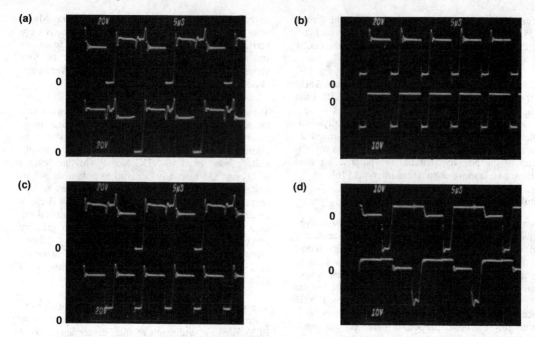

Breadboard Test Photos of Voltages Within the Integrated-Magnetic Weinberg Converter of Figure 28(b)
Figure 32

Comparing the lab photos of Figure 32 to the ideal waveforms of Figure 30 expected in this design, we find that they closely match in all instances. Also, the observed duty cycle of each primary switch is in close agreement with that predicted earlier by (31) when correction is made for converter efficiency.

Concluding Remarks

In this paper, we have presented a broader and more general insight into integrated-magnetic concepts for switch mode DC-to-DC power converter circuits and, hopefully in doing so, have dispelled a common belief among designers that the transformers and inductors of buck and boost-derived converters cannot be magnetically blended on a single core arrangement.

Also contrary to popular thought, the demonstrated ability to integrate the magnetics of these converters also removes the restriction that impressed winding voltages be completely proportional in all respects. In general, this particular restriction is correct, if one thinks only in terms of winding arrangements placed on a single magnetic path. However, as we have seen, multiple-path arrangements are perfectly viable design approaches. It must be remembered that, even though the winding voltages of the transformers and inductors in converters may not be completely proportional in frequency and amplitude, they share a common operational property—flux change. Therefore, if relationships can be established between their flux changes from a knowledge of their position in a converter circuit, then it is possible to use them to deductively synthesize a single magnetic system to house all transformer and inductive functions.

With integrated-magnetic versions of first-order buck and boost converters established, the possibility of magnetic integration of all of their family member topologies becomes immediately evident.

Many of these possibilities are shown in [17], while others are still to be disclosed in the future. Although we have presented integrated-magnetic versions of single-output buck and boost converter arrangements, the idea is easily extended to accommodate multiple outputs, or multiple inputs as the case may be. This is easily done by the addition of more secondary or primary windings on appropriate core legs, together with contemporary switches and diodes.

In this paper, we have also shown methods whereby more windings can be added to inductive core paths to control input or output ripple current magnitudes, and significantly reduce them by proper internal and/or external magnetic means. Even though we have used single inductive paths in making this demonstration, it is also feasible to add more core legs with appropriate air gaps to the integrated-magnetic systems to control ripple current characteristics of contemporary windings, much in the same manner as was shown in [13] for the two inductors of a ĆUK converter.

The discussion here has been restricted to switch mode power converters. Work presently in progress indicates that these techniques can also be applied to resonant converter circuits. Take for example the series resonant converter, with an inductor in series with the transformer primary. This inductor can be absorbed into the transformer simply by increasing the primary-secondary leakage inductance. This produces an integrated magnetic.

Finally, we conclude our paper with the following reflective thoughts. Although recent years have seen an increased interest in integration concepts for minimizing the magnetic content of converters or for enhancing their conversion properties, such techniques have historical roots, reaching back into the early years of this century. It is interesting to note that many advances in magnetic concepts for electronics were made in these earlier periods, with years following the introduction of the first electronic digital computer seeing little or no new work being continued in this important aspect of power electronics design. Shortly thereafter, a sad decline in educational opportunities for engineers in the field of practical magnetics design began, and continues today. Ironically, it appears now that magnetics could hold the key to achieving smaller and more efficient power conversion systems. As was noted in the introduction of this paper, it is truly a time of frustration and reflection for the power electronics engineer.

Acknowledgements

The authors would once again like to thank Morris Rauch for his help in constructing the breadboards of the various integrated-magnetic converters shown in this paper. Also, a special thank-you is sent to Colonel Wm. T. McLyman of CALTECH's Jet Propulsion Laboratory for constructing one of the integrated magnetics for the breadboards, and to the engineering directorate of Christie Electric Corporation for the use of their laboratory facilities during the breadboard testing efforts.

To J. Cielo and H. Hoffman of IBM, we send a very special inspiration thanks for their exceptional 1971 patent concerning integrated magnetics for "push-pull" DC-to-DC power converter circuits.

References

[1] R. D. Middlebrook, *"Reduction of Switching Ripple in Power Converters,"* Proceedings of the Sixth International PCI Conference, April 19–21, 1983, p. 6–14.

[2] Slobodan Ćuk, *Modelling, Analysis, and Design of Switching Converters*, Ph.D. Thesis, California Institute of Technology, Pasadena, CA, 1977.

[3] Slobodan Ćuk, *"A New Zero-Ripple Switching DC-to-DC Converter and Integrated Magnetics,"* 1980 IEEE Power Electronics Specialists Conference Record, p. 12–32 (IEEE Publication 80CH1529-7 AES).

[4] J. K. Watson, *Applications of Magnetism*, John Wiley and Sons, 1980.

[5] C. Desoer and E. Kue, *Basic Circuit Theory*, McGraw Hill, 1969, p. 453–457.

[6] G. B. Crouse, *"Electrical Filter,"* U.S. Patent 1,920,948 (August 1, 1933).

[7] S. Feng et.al., *"Small-Capacitance Nondissipative Ripple Filters for DC Supplies,"* IEEE Transactions on Magnetics, Vol. MAG-6, No. 1, March 1970.

[8] A. Llodye, *"Choking Up on L-C Filters,"* Electronics Magazine, August 21, 1967 issue.

[9] G. C. Waehner, *"Switching Power Supply Common Output Filter,"* U.S. Patent 3,916,286 (October 28, 1975).

[10] W. J. Hirschberg, *"Improving Multiple Output Converter Performance with Coupled Output Inductors,"* Proceedings of POWERCON 9, July 1982.

[11] G. E. Bloom and A. Eris, *"Practical Design Considerations of a Multi-Output ĆUK Converter,"* 1979 IEEE Power Electronics Specialists Conference Record, p. 133–146 (IEEE Publication 79CH1461-3 AES).

[12] L. Rensink, *Switching Regulator Configurations and Circuit Realizations*, Ph.D. Thesis, California Institute of Technology, Pasadena, CA, 1979.

6

[13] Slobodan Ćuk, *"Analysis of Integrated Magnetics to Eliminate Current Ripple in Switching Converters,"* Proceedings of the Sixth International PCI Conference, April 19–21, 1983, p. 361–386.

[14] J. Cielo and H. Hoffman, *"Combined Transformer and Indicator Device,"* U.S. Patent 3,553,620 (January 5, 1971).

[15] Slobodan Ćuk, *"General Topological Properties of Switching Structures,"* 1979 Power Electronics Specialists Conference Record, p. 109–130 (IEEE Publication 79CH1461-3 AES).

[16] Rudy Severns, *"Techniques for Designing New Types of Switching Regulators,"* Proceedings of the International PCI Conference, September 1979, paper 5.5.

[17] G. E. Bloom and Rudy Severns, *Modern DC-to-DC Switchmode Power Converter Circuits,* to be published in 1984 by Van Nostrand-Reinhold, Inc. (Catalog #0-442-21396-4), also available from e/j BLOOM Associates, 1312 Lovell Avenue, Arcadia, CA 91006.

[18] A. H. Weinberg, *"A Boost Regulator with a New Energy Transfer Principle,"* Proceedings of the 1974 Spacecraft Power Conversion Electronics Seminar (ESRO Publication SP-103).

[19] R. D. Middlebrook, *"Design Techniques for Preventing Input Filter Oscillations in Switched-Mode Regulators,"* Proceedings of POWER-CON 5, May 1978, p. A3-1 to A3-16.

6.1.4 Siliconix Switch Mode Power Supply Kit

Introduction

The Siliconix power supply kit is an educational tool which will give you a hands-on introduction to both switch mode power converters (SPC) and devices from Siliconix's expanding line of power converter components. The circuit operates at a switching frequency of 150 kHz and a ripple frequency of 300 kHz. The use of switching frequencies greater than 100 kHz represents the present trend in SPC design. The benefits of high frequency operation are reduced size, weight, and cost along with improvements in transient response and input noise rejection.

Circuit Description

There are literally hundreds of different SPC circuits from which to choose. For the kit we have chosen a buck-derived circuit which is commonly used in off-line applications. Normally, this type of circuit would be operated from a 100 to 350 volt source, but for reasons of safety and convenience, we have scaled the source voltage down to 24 to 32 VDC. The waveforms and circuit operation remain the same. It is important to emphasize that the kit is designed as a *learning tool* and is *not* intended to be used as a production power supply.

A block diagram of the power supply is shown in Figure 1. The power converter uses a control loop which compares the SPC output voltage (V_O) to a regulated reference voltage (V_{ref}) at the error amplifier. The error amplifier generates an error voltage proportional to the difference between V_O and V_{ref}. The error voltage is then compared to a ramp signal and converted to a pulse-width-modulated (PWM) control signal for the power converter (SPC).

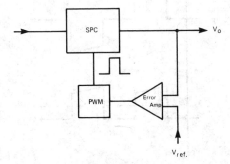

Power Supply Block Diagram
Figure 1

A simplified schematic for the power converter is S2) are connected in a half-bridge configuration. In this configuration the switches see a maximum voltage of V_S and a peak current of twice the input current (I_i). The relatively low voltage stresses on the power switches make this circuit suitable for high input voltage (100 to 400 volt) operation; however, for a given type of switch, the high switching currents become a limiting factor, so this design is rarely used at power levels above 500 watts.

As stated earlier the kit SPC operates from 24 to 32 volts, a much lower than normal voltage for this configuration for reasons of safety. The effect of low input voltage and high switching currents is a reduction in the circuit efficiency of 5 to 10%. This should be kept in mind when testing the circuit.

6

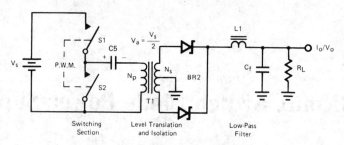

Simplified Power Supply Schematic
Figure 2

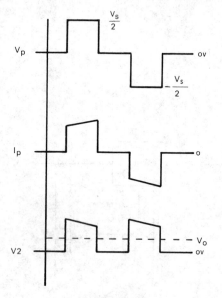

Idealized Waveforms
Figure 3

Switches S1 and S2 conduct alternately for equal periods of time. The actual "on" time (t1) of the switches is controlled by the PWM drive to each switch. The waveforms shown in Figure 3 illustrate how the circuit (in Figure 2) works. The action of the switches causes a pulsating DC voltage (V1), the AC component of which is applied across the primary of T1 (C5 blocks the DC component). The primary voltage (V_p) is stepped down to approximately 7.5 volts by T1 and rectified by BR2 producing a pulsating DC voltage (V2).

BR2 is a dual Schottky diode assembly. Schottky diodes are used in low voltage, high current output applications because of their low forward voltage drop (V_F). BR2 has a rated V_F of 0.5 volts and accounts for about 30% of the circuit losses. A normal PN junction diode has a V_F of 0.8 to 1.2 volts with

correspondingly higher losses. Schottky diodes also have a fast reverse-recovery-time (t_{rr}) which reduces the rectifier losses at very high switching frequencies.

Referring again to Figure 2, the average value of V2 is the output voltage (V_o), and the relationship to the PWM is:

$$V_O = \frac{V_s}{2} \left(\frac{N_s}{N_p} \right) \qquad (1)$$

Where D is the portion of the switching cycle during which the switches are conducting

$$D = \frac{t1}{T_s} \qquad (2)$$

where: t1 is the conduction time of the switches.
T_s is the switching period.

We can see from equation (1) that if V_O is to remain constant, D must become smaller as V_s increases and vice versa. This is the basis for the control action.

For most applications the switching noise and ripple must be very low, so in addition to averaging, the output filter must reject most of the AC component of V2. This is accomplished by making the cutoff frequency (f_c) of the filter lower than the ripple frequency (f_r). In the kit converter, f_c is so much less than f_r that the equivalent series resistance (ESR) of the output capacitors greatly reduces the available attenuation. Figure 4 shows an equivalent circuit for the filter and attenuation curves with and without ESR.

The filter inductor current waveform can have two possible shapes: either the current goes to zero during the switching interval, or it doesn't. Another way of stating this is either all of the energy stored in L1 while FET 1 or FET 2 are conducting is discharged into the load when FET 1 or FET 2 are off, or it is not. At first

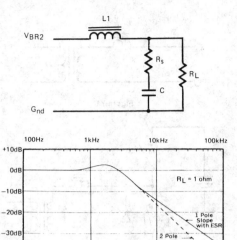

Figure 4

glance this seems like a trivial distinction, but it is not. The conduction mode of the inductor current defines the operating mode (continuous or discontinuous) of the converter. The open loop static and dynamic properties of the converter will differ dramatically from one mode to the other. This is characteristic of all SPC circuits. The conduction mode must be known to properly stabilize the loop and to predict the circuit

performance. This will be discussed again in the control loop stabilization section.

The schematic for the complete power supply is given in Figure 5. All control loop and drive functions are performed by the Siliconix PWM125 (IC1). The PWM125 is a high performance pulse-width-modulator IC designed specifically for high frequency power converter applications. The PWM125 has the same pinout and is functionally similar to its predecessor, the SG1525A. The PWM125 features separate analog and digital grounds, true TTL gate structures for precise metering of pulse modulated signals, and completely redesigned totem-pole output sections with very low crossover current (120 mA). In earlier totem-pole designs (i.e. 1525A), crossover currents of 950 mA caused catastrophic failure at high speeds. Other features unique to the Siliconix PWM series (PWM 25/27, 125) are high operating temperature and high voltage capability. The PWM125 will be discussed in greater detail later in this text.

Actual Circuit Waveforms

The waveforms shown in Figure 3 are idealized. In the actual circuit, the waveforms are somewhat different, and it is worthwhile knowing why. Figure 6 shows the actual waveform (V_p) across the primary of T1. What interests us are the spikes and ringing when one of the FETs turns off.

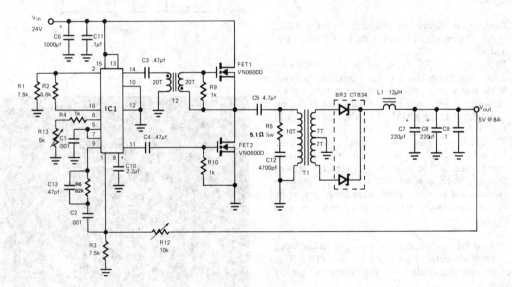

Siliconix SMPS Kit Schematic
Figure 5

6

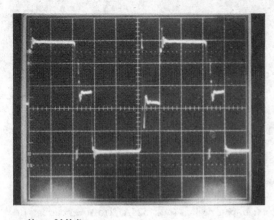

V$_{in}$ = 24 Volts
I$_{out}$ = 8 A
Vertical Scale — 5 V/Div.
Horizontal Scale — 1 μS/Div.

**Primary Waveform
Figure 6**

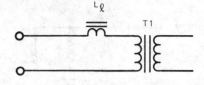

Figure 7a

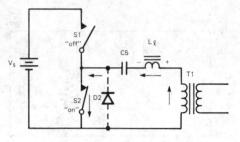

Figure 7b

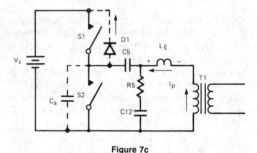

Figure 7c

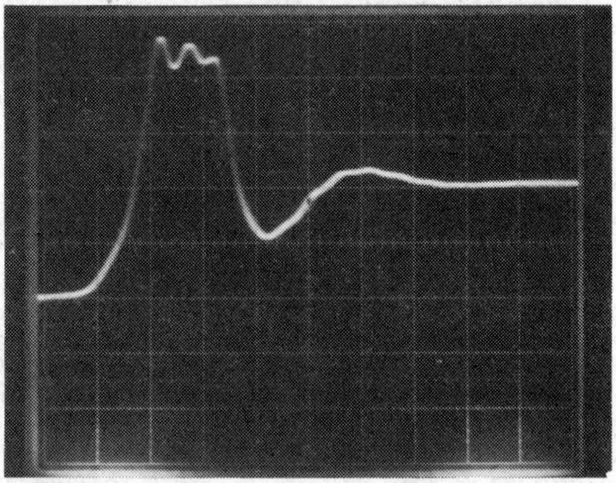

In a real transformer, the primary and secondary windings are not perfectly coupled, and some of the primary flux linkages will not be shared by the secondary. This creates an equivalent series inductance (referred to as leakage inductance) Lϱ shown in Figure 7a.

The mechanism for generating the voltage disturbance is as follows:

1. Referring to Figure 7b, assume S1 is off and S2 is on. The primary current is flowing through Lϱ and S2.
2. Referring to Figure 7c, if we turn S2 "off" suddenly, the current in Lϱ must continue to flow. What happens is that the voltage across Lϱ reverses and rises until D1 conducts (D1 is the integral diode in the MOSFET). When the energy in Lϱ has been discharged, D1 will cease to conduct. The circuit now rings at a frequency determined by Lϱ and C$_a$, where C$_a$ is the total device and stray capacitance at that node. If D1 were not present then the voltage across S2 would rise until the switch went into avalanche breakdown. D1 limits the voltage rise to just above V$_S$ and acts as a clamp.

To reduce the ringing, there is a simple RC network made up of R5 and C12 across the primary of T1. This does not completely eliminate the ringing but greatly reduces it.

Another feature of this waveform which may appear puzzling is the apparent change in turn-off switching between heavy and light loads. This effect is shown in Figures 8 and 9. Figure 8 is the turn-off waveform for FET 2 with an 8A output load. The turn-off transition

V$_{in}$ = 24 Volts
I$_{out}$ = 8 A
Vertical Scale — 5 V/Div.
Horizontal Scale — 50 nS/Div.

**S2 Turnoff Voltage Waveform
Figure 8**

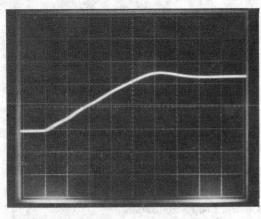

V_{in} = 24 V
I_{out} = 0.5 A
Vertical Scale — 5 V/Div.
Horizontal Scale — 50 nS/Div.

S2 Turnoff Voltage Waveform
Figure 9

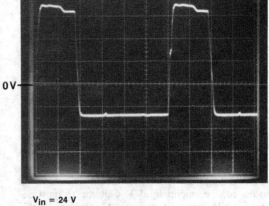

V_{in} = 24 V
I_{out} = 0.5 A
Vertical Scale — 5 V/Div.
Horizontal — 1 μS/Div.

FET 2 Gate
Figure 11

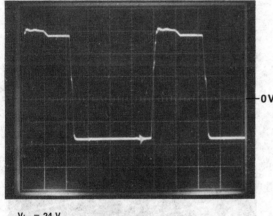

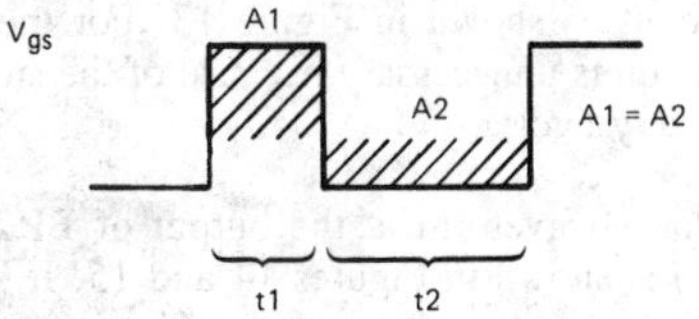

V_{in} = 24 V
I_{out} = 8 A
Vertical Scale — 5 V/Div.
Horizontal — 1 μS/Div.

FET 2 Gate Voltage Waveform
Figure 10

Figure 12a

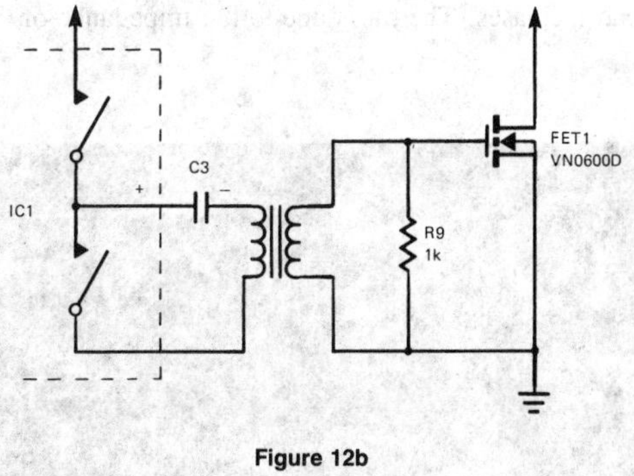

Figure 12b

6

is 50 nsec. Figure 9 is the same waveform with the output load reduced to 0.5 A, and the *apparent* switching time is 200 nsec. In fact, the switching time of the FET has not changed at all; it is 50 nsec in both cases. The reason for the slower transition time at light loads can be explained by referring back to Figure 7c. At turn-off the current flowing in S2 will be available to charge C_a, and the rate of voltage rise across S2 will be dependent on the current ($I_{L\varrho}$) available to charge C_a. When the output current is 8 A, ample current is available to charge C_a, and the rate of voltage rise is determined by the FET switching characteristics. However, when the output load is reduced to 0.5 A,

the current available to charge C_a is reduced by a factor of 16. In this case, the charging of C_a completely dominates the voltage rise time.

Another waveform of interest is the gate drive waveform (V_{gs}) shown in Figures 10 and 11. Ideally we should see symmetrical waveforms independent of D. A cursory examination of the oscillographs shows that this is not so; the positive and negative amplitudes vary with D. Since this is an AC coupled waveform (through capacitor C3 in Figure 12a), the enclosed areas are equal as is shown in Figure 12b. From a

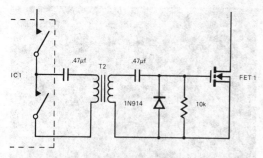

Transformer Drive Cir. for Wide Variation of Duty Cycle
Figure 13

practical point of view, this means that as D is increased, the amplitude of V_{gs} during t1 must decrease while the amplitude during t2 increases. In this particular application, the variation in V_{gs} is not large and is not a problem. In some circuits the variation in D may be much larger causing V_{gs} to be too large at small duty cycles and too small at large duty cycles. This problem can be eliminated by modifying the drive circuit as shown in Figure 13. For the kit this modification is unnecessary because of the small variation in duty cycle.

The voltage waveform at the output of BR2 (V2 of Figure 2) is shown in Figures 14 and 15. It is clearly not ideal. In addition to the ringing introduced on the primary leakage inductance, there is a distinct tilt to the top of the waveform which increases as the output load increases. The tilt is due to the impedance of C5

in series with the primary of T1. The AC voltage across C5, due to the transformer primary current, subtracts from the applied voltage and causes the tilt in the output waveform. As the power supply output load increases, the voltage drop across C5 increases which increases the tilt in V2. When this type of circuit is used directly from the AC mains at the same output power rating, the waveform tilt is insignificant. The higher primary voltage results in a lower primary current which, in turn, results in a smaller voltage drop across C5. By scaling down the input voltage for the kit, the effect of C5 is magnified. The waveform tilt could be reduced by increasing the value of C5, but this is not necessary for the purposes of this kit.

Control Loop Stabilization

For the power converter to function as a regulated power supply, some means is needed to maintain a constant output voltage, as the output load and input voltage vary. This function is provided by the feedback control loop. As was pointed out earlier, most of the control loop functions are provided by the PWM125 (IC1). The issue to be discussed now is that of loop stability and transient response performance.

Figure 16 shows an equivalent block diagram of the power supply control loop. To analyze the stability of the loop, we will calculate the gain and phase characteristics of the loop and then graph them using a Bode plot.

The total loop gain (A_T) is: $A_T = G1\ G2\ G3$ (3)

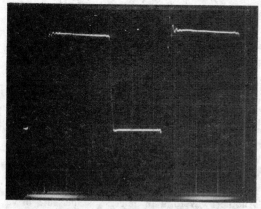

V_{in} = 24 V
I_{out} = 1 A
Vertical Scale — 2 V/Div.
Horizontal Scale — 500 nS/Div.

Output of BR2 @ 1 A Load
Figure 14

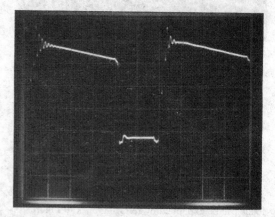

V_{in} = 24 V
I_{out} = 8 A
Vertical Scale — 2 V/Div.
Horizontal Scale — 500 nS/Div.

Output of BR2 @ 8 A Load
Figure 15

NOTE: *Current Goes Slightly Negative During Transition Period.*

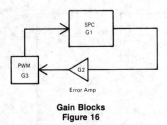

Gain Blocks
Figure 16

G1 represents the gain of the power converter. The AC part of G1 is determined by the output filter which, as we saw in Figure 4, is a simple 2 pole LC filter with an ESR zero. The result is a single-pole characteristic with the cutoff frequency (f_c) equal to 2.4 kHz. The DC portion of G1 is determined by the duty cycle and the transformer (T1) turns ratio for D = 0.5. The DC value for G1 is:

$$G1 = 0.7 \tag{4}$$

This analysis presumes that the converter is operating in the continuous inductor current mode. If the load is reduced to the point where the discontinuous mode exists then the converter will have a single, rather than a two pole characteristic. Simultaneously, the loop gain will drop. Fortunately, these two changes work together in this application, and the loop remains stable.

G2 is the gain of the error amplifier, and we can calculate it from Figure 17. We will assume the amplifier gain is large. The DC value of G2 is very large (60 dB). There will be a pole at the origin due to C2, and a zero will appear in the transfer function at a frequency determined by R6 and C2:

$$f_0 = \frac{1}{2\pi RC} = 1.9 \text{ kHz} \tag{5}$$

The asymptotic gain at 1.9 kHz is:

$$G2 = \frac{R6}{R12} = 14.6 \tag{6}$$

The combination of C13 and R6 will produce a pole at 41 kHz. An asymptotic approximation for G2 is shown in Figure 17b.

G3 is the gain characteristic of the pulse-width-modulator. Given the particular modulator used in the PWM125 (IC1), the gain expression is very simple:

$$G3 = \frac{1}{V_{ramp}} \tag{7}$$

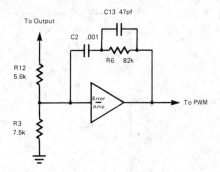

Figure 17a

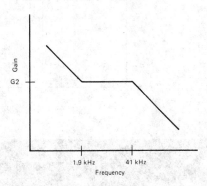

Figure 17b

Where V_{ramp} is the peak-to-peak amplitude of the ramp voltage which appears on pin 5 of IC1:

$$V_{ramp} = 3.2V \text{ p-p} \tag{8}$$

So that:

$$G3 = .31 \tag{9}$$

We can now plot A_T, as shown in Figure 18, in asymptotic form. We can see that gain crossover is about 9 kHz with a single pole roll-off which should give us a phase margin close to 90°. The ultimate test for stability is the transient response to a step load. The transient response of the kit is shown in Figure 19. The response is well damped with a very small step change (about 40 mV). Given that the load step is 1 A superimposed on a 5 A DC load, the transient response is excellent. The small high speed oscillations on the leading edge of the step response are inherent in the dynamic load and are not due to the control loop.

It is quite possible to have a much higher gain crossover frequency and an even better transient response,

6

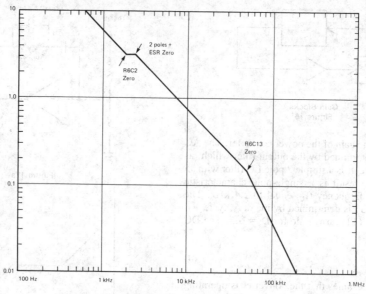

2 poles +
ESR Zero

R6C2
Zero

R6C13
Zero

Figure 18

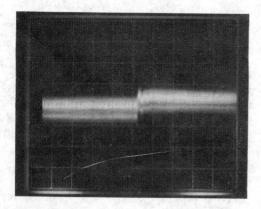

V_{in} = 24 V
I_{out} = 5 A
Vertical Scale — 50 mV/Div.
Horizontal Scale — 100 μS/Div.

Response Time to A 1 A Step Change with A 5 A Load
Figure 19

but for the purposes of this kit, the reduced bandwidth presents fewer problems for the user and still provides excellent transient response.

The PWM125 — How It Works

For ease of understanding the Siliconix PWM series (PWM25, 27, 125), the circuitry has been divided into 3 sections:

1. Clock and Control
2. Metering and Output
3. Regulation and Shutdown

Clocking and Control Circuit (Figure 20)

The ramp and timing signals are generated by charging a capacitor (Ct at pin 5) from an internal current source which is programmed by an external resistor (Rt at pin 6) using a current mirror technique. A comparator in the oscillator section is set to trip at 0.8 V (low) and 3.6 V (high). When capacitor Ct (pin 5) is charged to the high trip point, the discharge transistor (Qd) turns "on" and discharges Ct to ground through pin 7 (Qd collector). Pin 7 is normally connected to pin 5 on the PWM125. When the capacitor (Ct) is discharged to approximately 0.8 V, Qd is turned "off," and the cycle starts over again. The slow charge and fast discharge cycle generates the ramp signal for the pulse-width-modulator circuit. Other circuits in the oscillator generate the clock pulse for the metering circuits.

The Error Amplifier is a transconductance amplifier with a common-mode input range of 1.5 to 5.2 volts and a gain of 0.5 to 5.5. For positive output regulation the non-inverting (+) input can be referenced at any point within the common mode range. For regulated output voltages above 5.1 volts, no resistive divider is required, and a single resistor to the V_{ref} output (pin 16) can be used as the error amplifier reference. A voltage divider is then designed to reference the inverting input (− pin 1) to the same voltage as pin 2. In cases where the regulated output is to be below 5.1 volts, a voltage divider should be used to set the desired reference level. For negative output regulation, this process is reversed, and the inverting (−) input of the error amplifier can be monitored at pin 9 (Comp.). Open loop gain for the PWM series

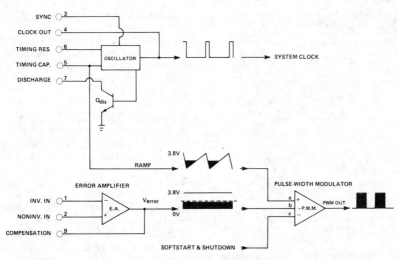

PWM125 Clock & Control Circuit
Figure 20

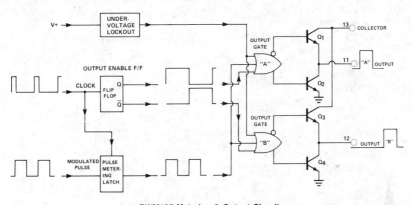

PWM125 Metering & Output Circuit
Figure 21

amplifiers, with a load resistance of $\geqslant 10$ megohms is 60 to 80 dB. Because the error amplifier is a transconductance design ($V_O = g_m R_L V_i$), as the output is loaded the gain is reduced. The details on how to use this amplifier were discussed in the section on control loop stabilization.

The pulse-width-modulator compares the incoming ramp and error voltages to form the pulse width modulated signal shown in Figure 20 (the shaded areas). As the error voltage changes, the active time on is varied according to the reference crossing points on the rising and falling edges of the ramp, and the pulse width is varied or "modulated" accordingly. As demand for output current increases, or as the power converter input voltage changes, the change is sensed by the error amplifier, and the error voltage changes. These changes cause the ramp crossing points to shift and the pulse width to vary proportionately. Changes

in input voltage or output current are met with little or no change in output voltage. Figure 20 shows the error voltage vs. ramp at about 50% duty cycle.

Metering and Output Circuits

The clock pulse generated by the oscillator circuit is used to trigger the metering flipflop which controls output gate selection. This ensures that the totem-pole outputs at pins 11 and 14 are never on simultaneously. The metering latch ensures that only one modulated pulse per clock period is allowed and prevents double-pulsing of the outputs.

The output control gates are 3 input NOR gates (4 input in the PWM25, 27) controlled by the two previously described signals and the undervoltage lockout/ shutdown circuits. When bipolar transistors are used as the power switches, a period of dead time is re-

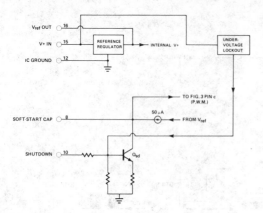

PWM125 Regulator & Shutdown Circuit
Figure 22

quired to get rid of the stored charge inherent to bipolar transistors. The maximum dead time allowed in the PWM125 is 120 nsec. When MOSPOWER FETs, which have no storage time problems, are used as the power switches, the dead time is set to minimum by connecting pin 7 (Qd collector) directly back to pin 5 (Ct). Do not use the PWM125 with bipolar transistors as there is no dead time control. Use the Siliconix PWM25, 27 which have the adjustable dead time feature.

Regulation and Shutdown Circuits

The regulation circuit provides a 5.1 volt $\pm 1\%$ source for the internal circuitry and a 20 mA output at pin 16 (V_{ref}). Also included is the undervoltage lockout which ensures a stable off condition when IC V+ (pin 15) is below 8 V.

The soft-start circuit provides protection by bringing the pulse width up slowly. This feature prevents cold start stresses on the SPC components when the power supply is turned on under heavy load conditions.

The shutdown circuit provides an external control for turning the outputs "on" or "off." When the shutdown transistor (Q_{sd}) is "off," a current source from the regulator circuit charges the soft-start capacitor at pin 8. The charge time determines the speed at which the modulated pulse comes up to the required width. When Q1 is turned "on" by the undervoltage lockout circuit or by an active high (2.4 V) on the shutdown pin, the soft-start capacitor is discharged to ground which disables the PWM circuit. When Q1 is turned "off," the soft-start capacitor charges again, and the soft-start cycle is repeated.

6.1.5 A 500 kHz Switching Inverter for 12 V Systems (AN79-1)

This note describes the design of a 12 V to ± 20 V inverter. The design is of the flyback type, made much more practical by the use of DMOS devices running at a high switching rate. It is an energy transfer circuit, not a voltage or current transfer circuit; the output power is maintained at a constant level for a given pulse width-modulator (PWM) operating point. If the load requires less current, the output voltage will soar. Conversely, if the output current demand increases, the output voltage will sag to maintain the constant output power ($V \times I$). This contrasts sharply with conventional circuits that deliver a constant output voltage per PWM operating point. Maintaining constant power permits simplified circuitry with reduced magnetic and filter requirements.

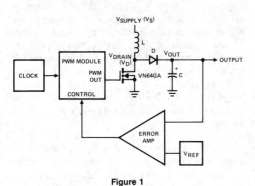

Figure 1

The principles of operation may be described by the basic circuit shown in Figure 1, which includes the pulse width modulator (PWM) control circuit; the high frequency MOSPOWER FET Switch; the flyback circuit inductor (L), diode (D) and capacitor (C); and an error amplifier.

At the heart of this design is the PWM control circuit which provides the control pulse to the DMOS Power Switch in the flyback circuit. The output of the PWM is a pulse whose width is proportional to the input control voltage and whose repetition rate is determined by an external clock signal. To provide the control input to the PWM and to prevent the output voltage from soaring or sagging as the load changes the error amplifier and reference voltage complete the design. They act as the feedback loop in this control circuit much like that of a servo control system. Pertinent waveforms are given in Figure 2; 2a describes the conditions that exist at 50% (maximum) duty cycle and 2b describes those at a low duty cycle.

Operation is as follows:

1. Between t_0 and t_1 the DMOS is turned on and applies the supply voltage across L. The drain current is closely approximated as $I_L = t \times V_S/L$ and the final current I_L (peak) $= (t_1 - t_0) V_S/L$. The energy stored in L is $E_L = (t_1^2 - t_0^2) V_S^2/2L$.

6

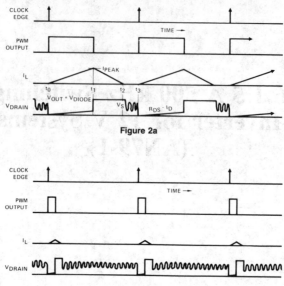

Figure 2a

Figure 2b

2. At the instant of t_1 the VN64GA is turned off and its drain voltage increases to the voltage on C plus the forward voltage drop of D because of the inductor L. This works well when V_{OUT} is much larger than V_S and permits nearly all of the energy stored in L to be transferred to C with no large current or voltage spikes.

3. Between t_2 and t_3, no net DC current flows through L, and V_{DRAIN} tends toward V_{SUPPLY}, although a ringing will occur between L and the capacitance of the FET and D. The cycle is repeated at a frequency f_{CLOCK}, such that the power drain from V_S is $(1/2(t_1 - t_0)^2\ V_S^2)/f_{CLOCK}$.

If we let the duty cycle

$$\delta = \frac{t_1 - t_0}{t_3 - t_0} = (t_1 - t_0) \times f_{CLOCK},$$

where $f_{CLOCK} \equiv \dfrac{1}{t_3 - t_0}$,

then $P_{IN} = \dfrac{\delta\ V_S^2 (t_1 - t_0)}{2}$

If $\delta = 1/2$ max, then P_{IN} max $= \dfrac{V_S^2}{8L\ f_{CLOCK}}$

The peak current through the DMOS will

closely be $\dfrac{4\ P_{OUT}}{\eta\ V_S}$; and the value of inductor

L will be $\eta\ V_S^2 / 8\ f_{CLOCK}\ P_{OUT}$, where η is the overall power efficiency of the circuit.

The power losses in the circuit are, in order of importance:

A. $(I_D)^2 R_{DS}$ loss in the MOSFET
B. $I_{LOAD}\ V_F$ loss in the catch diode D
C. Transient charge losses in D due to a slow turn-on characteristic. This actually allows many volts of forward bias on D just at time t_1 while the diode is attempting to turn on. Turn-off losses are not important in this circuit.
D. Loss in L due to hysteresis and saturation effects. The inductor is required to pass all of $I = 4 P_{OUT}/\eta\ V_{SUPPLY}$, which usually results in magnetic saturation and losses. Fortunately, the higher frequencies allow fewer turns on L, reducing the number of turns and I_{SUPPLY} product, and the core material of L can be high-frequency ferrite, which is less prone to saturation (for a given inductance).
E. Simple $C_{STRAY}\ V^2_{OUT}\ f_{CLOCK}$ losses, where C_{STRAY} includes C_{DS} of the FET and C_D of the rectifier.

Figure 3 is the schematic of a simple 35 watt inverter designed to produce ±20 V regulated outputs from 12–16 volt inputs. U_1 is a simple Schmitt-trigger oscillator with a nominal 50% duty cycle. This duty cycle waveform buffered into Q_5 by Q_4 and U_2 through U_6 runs the system at full output power.

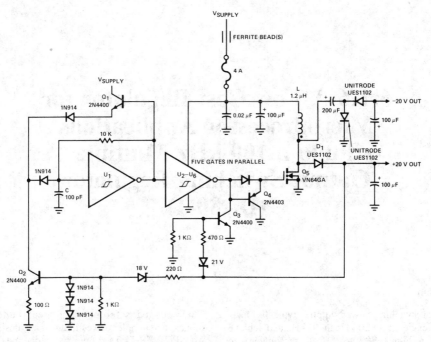

Figure 3

The duty cycle is reduced to stabilize the output voltage by the feedback amplifier Q_2 and its 18 V zener. Q_1 and its associated diodes switch Q_2's discharge current away from C during the discharge period of the transformer's field, and allow it to shorten the period in which primary current is drawn through the transformer. Q_3 and its 21 V zener prevent the output from soaring (if a fault condition occurs in the feedback regulator) by simply clamping Q_5's gate drive.

L is wound on a ferrite core; I used a wideband toroid 1 1/2" in diameter with a cross-sectional area of about $1/10$ in^2. The resistance of the winding should be held to less than $1/20$ Ω; #22 wire is adequately large.

6

6.1.6 A Low Cost Regulator for Microprocessor Applications Build a 100 kHz Multiple Output Switching Regulator (DA80-1)

Introduction

Commercial switching power supplies typically operate at frequencies from 20 kHz to 40 kHz and achieve efficiencies as high as 70% to 75% at a reasonable size and weight. These same efficiencies or better can be realized by increasing the operating frequency to 100 kHz and above when using MOSPOWER FETs as the power switching transistors. At these higher frequencies, much smaller reactive components are necsary, thus decreasing the cost, size, and weight of the power supply while maintaining the same output power. The main factor limiting the operating frequency of conventional switching supplies is the inherently slow switching times of the power bipolar transistors due mainly to minority carrier storage time. MOSPOWER FETs are majority carrier devices and therefore do not have storage time. The VN4000A series of 400 volt MOSPOWER FETs have maximum switching times of 100 ns, thus enabling efficient switching rates up to 500 kHz and above.

This higher operating frequency results in a reduction of the size and cost of ferrites and capacitors needed for the same power transfer and filtering capability. Since DMOS is a voltage controlled device, drive circuits are much simpler and consume less power than high current bipolar drives. DMOS' rectangular safe-operating area means that maximum rated voltage and current can be controlled simultaneously with no fear of second breakdown. Snubbers add extra cost and dissipate excess power in bipolar designs — none are needed to protect MOSPOWER FETs. Catch diodes are required for totem-pole switch configurations such as full-bridge and half-bridge power supplies to catch high voltage inductive spikes. These diodes must be added externally to bipolar designs at extra cost, but they are already built into MOSPOWER FETs.

Their rugged safe-operating area, built-in catch diodes, and simpler drive circuits make designing with MOSPOWER FETs simple and economical.

Power Supply Overview

The power supply presented here uses two VN4000A 400 volt MOSPOWER FETs in a half-bridge power switch configuration (Figure 1). Outputs available are +5 volts at 20 amperes and ±15 volts (or ±12 volts) at 1 ampere. Since linear three terminal regulators are used for the low current outputs, either ±12 or ±15 volts can be made available with a simple change in the transformer secondary windings (see Construction Details). A TL494 switching regulator IC provides pulse width modulation control and drive signals for the power supply. The upper MOSPOWER FET (Q_7) in the power switch stage is driven by a simple transformer drive circuit. The lower MOS (Q_6), since it is ground referenced, is directly driven from the control IC.

For initial start-up, a linear regulator (Q_3, R_4, and D_2 in Figure 1) supplies about 14 volts from the full-wave rectified line voltage for all the drive and control circuitry. Once the power supply starts up, the voltage from a separate secondary winding (#2) is rectified and filtered and used to supply all power to the control IC and drive circuitry. When this supply reaches full voltage (about 18 volts), diode D_1 is reverse biased, thus automatically turning off the less efficient linear regulator used for start-up. A minimum current of one ampere must be drawn from the +5 V output to assure turning off the linear start-up regulator. If less current is drawn from this output, the control circuitry will be powered by the linear regulator, and excess power will be dissipated in Q_3.

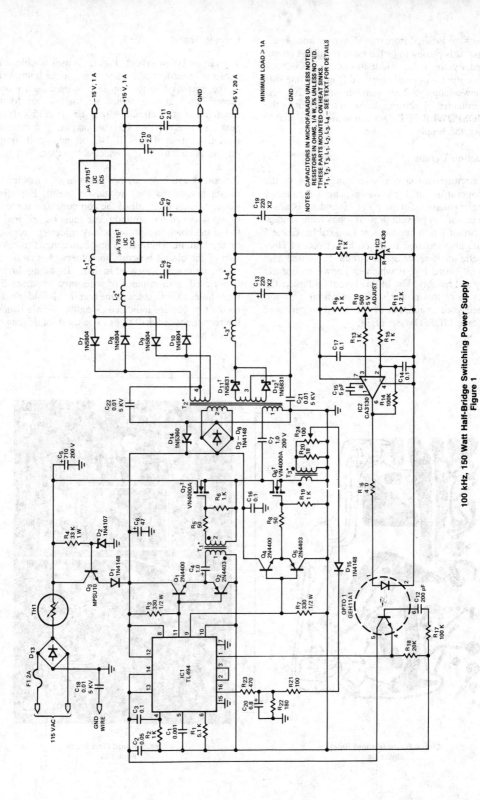

NOTES: CAPACITORS IN MICROFARADS UNLESS NOTED.
RESISTORS IN OHMS, 1/4 W, 5% UNLESS NOTED.
† THESE PARTS MOUNTED ON HEAT SINKS.
*T_1, T_2, T_3, L_1, L_2, L_3, L_4 — SEE TEXT FOR DETAILS

100 kHz, 150 Watt Half-Bridge Switching Power Supply
Figure 1

All outputs are isolated from the AC power line. The 5 volt output was chosen to be the main regulated output controlled by the pulse width modulator. Feedback from this output is optically isolated from the line side of the power supply. The complete supply is over-current protected by sensing the source current in the lower MOSPOWER FET (Q_6) and using this signal to shut down the supply.

Construction Details

Careful circuit board layout is very important for the proper operation of high power, high-frequency switching regulators. Single point grounding is absolutely necessary to prevent ground loops from rendering the circuit totally unstable or inoperable. Ground planes are also required to lessen the effects of electromagnetic interference on the circuit. Presented here is a circuit board layout which is known to operate correctly and reliably. Use of this layout will make the construction of this power supply much simpler and will speed your evaluation of the VN4000 series of high-voltage MOSPOWER FETs.

Circuit Board

The circuit board layout (Figure 2) uses double sided construction. Most of the traces are on the bottom side of the board while the top side is used as a ground plane. Three ground planes are used — one for the input and control circuitry, one for the ±15 volt outputs and one for the +5 volt output. If a common ground is desired for all DC outputs, both output ground planes may be connected together.

Plated-through holes are not necessary for making the circuit board, but they would be useful. If plating-through is not used, all of the components connecting to top traces or the ground plane must be soldered on top of the board in addition to any soldering necessary on the bottom. There are a few connections from one side of the circuit board to the other side that do not have components mounted in them. If plating-through is not used, a short piece of wire must be soldered in these holes to both sides of the board. All solder points on the top side are indicated by an 'X'. Table I shows the recommended drill sizes to be used for drilling the circuit board.

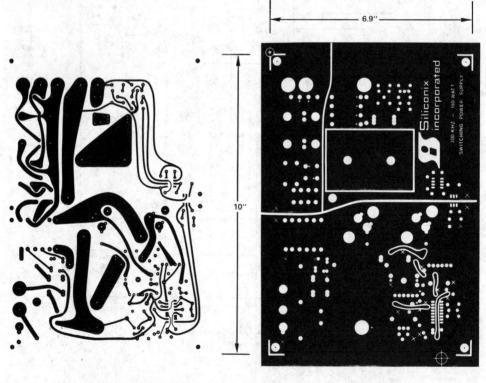

Circuit Board (Bottom Side)
Figure 2a

Circuit Board (Top Side)
Figure 2b

Table 1
Recommended Drill Sizes

The following drill sizes should be used on the circuit board:

#66   IC1-IC3, 1/4 W resistors, disc capacitors, opto-isolator, Q_1-Q_5, D_1-D_6, C_6, C_{13}

#60 T_1 bobbin leads, 1/2 W resistors, C_8, C_9, T_3 secondary, TH_1, F_1, C_{18}, C_{21}, C_{22}, D_2, D_{14}

#57   1 W resistors, T_2 bobbin leads, D_7-D_{10}, C_7, D_{13}, C_{13}, C_{19}

#54 R_{10}, IC4, IC5, L_1 – L_4, R_{24}, T_3 primary

#44   TO-3 lead sockets, line cord

#23 F_1, IC4, IC5 and TO-3 screw-mount holes, 5 V CT

3/16″ D_{11}, D_{12}

1/4″ C_5

5/16″   Banana jacks

Transformers

Three transformers are used in this power supply: 1) DMOS drive transformer, 2) power transformer and 3) current sense transformer. The winding details explained here should be closely followed, especially for the power transformer T_2.

T_1 — DMOS Drive Transformer

Using the correct bobbin and pot core (see parts list), wind the following:

Primary (1) — 20 turns of #24 enamel wire

Secondary (2) — 30 turns of #24 enamel wire

Make all connections to the bobbin according to the parts placement diagram (Figure 3).

T_2 — Power Transformer

Using the appropriate bobbin, wind the following (pin 1 of the bobbin has a notch for identification):

Primary (1) — The primary is made up of a type of litz wire using several strands of regular enamel wire (refer to Figure 4). Cut 8 strands of #28 enamel wire (about 6 feet long) and place them together in parallel. Twist the ends only together (not the whole length), but do not solder. Fold this twisted bundle of wires in half and wind 8 turns of this doubled over bundle onto the bobbin. Cut the folded over end of the bundle so that there are now 4 ends coming out of the bobbin. Twist the ends of each newly cut bundle. Next, connect one of the beginning bundles (D) to the end of the other (B). This effectively connects the bundles in series, wound in the same direction, to form a single 16 turn primary. The purpose of winding in this manner is to equalize the flux across the transformer core. There should now be two ends of the wire free and two ends connected to each other. Connect these free ends to the bobbin as shown in Figure 3. Make sure all windings are wound tight and neat — do not waste any space. Now put a layer of transformer tape to cover the primary.

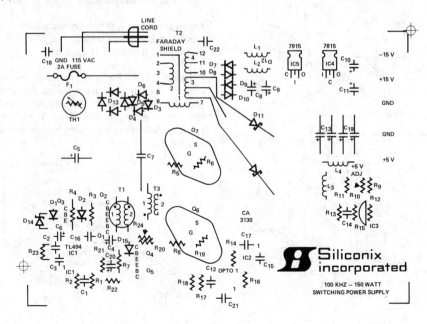

Parts Location Diagram
Figure 3

Secondary (2) — Start-up Winding: Wind 4 1/2 turns of #24 enamel wire (about 20 inches long) evenly on top of the primary winding. Connect the ends as shown in Figure 3. Put a single layer of transformer tape over the start-up winding.

Faraday Shield — This is a shield used to minimize radiated electromagnetic interference (EMI). Cut a piece of 5/8 inch copper tape about 3 inches long and wrap this around the existing windings (refer to Figure 5). Do not make a complete loop — leave about 1/4 inch between the ends so that they can't touch. Solder a small stranded wire (#20) onto the shield and connect it to the bobbin as shown in Figure 3. Put a layer of transformer tape over the Faraday shield.

Step 1 — Parallel 8 strands #28 wire

TWIST ENDS

Step 2 — Fold in half

Step 3 — Wind on bobbin

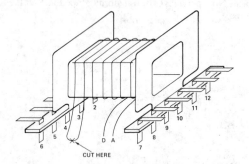

CUT HERE

Step 4 — Connect C and A to pins 6 and 7

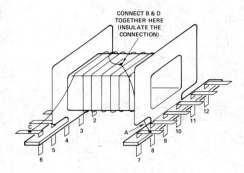

CONNECT B & D TOGETHER HERE (INSULATE THE CONNECTION)

Primary Power Transformer
Figure 4

Secondary (4) — ±15 volt secondary: Make another litz wire similar to the primary but this time parallel 6 strands of #28 enamel wire about 40 inches long (refer to Figure 6). Twist the ends together and double the bundle over itself. Wind 5 turns of this doubled over bundle neatly on the bobbin (4 1/2 turns for ±12 V outputs). Cut the double end and connect B and D together and connect to bobbin pin #11. Connect the two free ends of the bundle to the other bobbin pins. Put a layer of transformer tape over these windings.

Secondary (3) — 5 volt secondary: Make up some insulated copper tape by placing transformer tape on one side of a 10 inch long piece of 5/8 inch wide, 2 mil copper tape (refer to Figure 7). Make two of these insulated tapes. Make sure the transformer tape is slightly wider than the copper tape so that the windings don't short to each other. Wind both of these insulated tapes at once (like bifilar tape) for two turns. Connect the beginning of one tape (A) to the end of the other tape (D) — this is the 5-volt center tap. Connect three #18 stranded wires (or #18 ribbon cable) in parallel to each of the free copper tape ends and to the center tap. Spread out the stranded wires flat when soldering to the copper tape — this makes a much neater and less bulky connection. Connect the ends of these paralleled wires to the output rectifiers and P.C. boards as shown in Figure 3. Wrap a final layer of transformer tape to hold everything together.

T3 — Current Sense Transformer

Place the windings directly on the toroid:

Secondary (2) — Wind the secondary first. Wind 7 turns of #20 enamel wire (about 10 inches long) onto one side of the toroid (Figure 8).

Primary (1) — Form the primary by soldering 2 strands of #16 enamel wire to the circuit board connections (Figure 3). Run these strands through the center of the toroid. This forms the one turn primary. Solder the secondary into the board as shown (Figure 8).

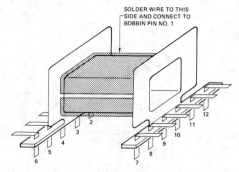

SOLDER WIRE TO THIS SIDE AND CONNECT TO BOBBIN PIN NO. 1

Faraday Shield
Figure 5

Step 1 — Parallel 6 strands #28 wire

TWIST ENDS

Step 2 — Fold in half

Step 3 — Wind on bobbin

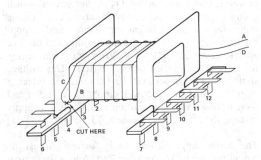

CUT HERE

Step 4 — Connect the free ends to bobbin pins #10, 12. Connect doubled over end to pin 11.

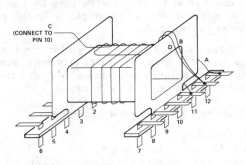

±15 V Secondary
Figure 6

Output Inductors

L_1 and L_2 are identical. Wind one turn of #18 enamel wire through each core and solder to the circuit board.

L_3 is also one turn, but use three strands of #18 in parallel.

L_4 is an air core inductor. Close wind 10 turns of #16 wire on a 5/16 inch diameter form.

Heat Sinks

Mount the TO-3 heat sinks off the board with 1/4 inch spacers (to make space for R_5, R_6, R_8, and R_{19}). No insulating washers are needed, but heat sink compound should be used.

Step 1 — Make two insulated copper straps

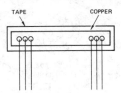

TAPE   COPPER

Step 2 — Place one strap on top of the other with leads on opposite sides.

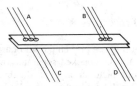

Step 3 — Wrap two turns of this double copper tape onto the bobbin. Connect A to D and solder this center tap into the circuit board. Connect the other free ends, B and C, to diodes D_{11} and D_{12}.

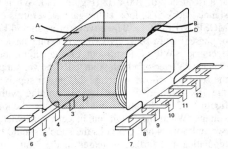

5 V Secondary
Figure 7

Mount IC4 and IC5 onto their heat sinks with thermal compound. IC5 should be insulated from the heat sink. Use metal screws for mounting IC4 and IC5 and use plastic screws for the other mounting screws for the TO-66 heat sinks. Cut off the center lead of each regulator and insert the other two pins into the circuit board.

Schottky rectifiers D_{11} and D_{12} mount directly on the heat sink with thermal compound. Use star washers for a good electrical connection when bolting rectifiers D_{11} and D_{12} to the board.

Miscellaneous

Use star washers on both sides of the board when mounting C_5. IC sockets may be used for IC1, IC2 and Opto 1.

ALWAYS use an isolation transformer when connecting an oscilloscope to look at waveforms on the primary side of the power supply.

Do not mount Q_3 until after the initial test procedures.

6

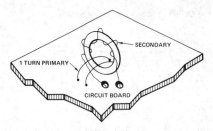

Current Sense Transformer
Figure 8

Power-Up Procedures

Even though this is a proven circuit board, the control circuitry should be checked separately before powering up the complete supply. To do this, connect an isolated + 12 volt supply through a diode between pin 12 of IC1 and ground (Figure 9). Use an oscilloscope to check the drive signals at pins 9 and 11 of IC1. These signals should be in-phase, quasi-square waves at a frequency of 100 kHz (Figure 10a). When these signals check OK, look at the gate waveforms of the MOSPOWER FETs. These waveforms should be out of phase (Figure 10b). Q_7's waveform may have some overshoot. With Q_3 still out of the circuit board, connect a variac or a high voltage DC power supply to the AC input. Slowly increase this power supply voltage (the control circuitry is still running with the floating 12 V supply) while monitoring the output voltage. When the 5 volt output gets somewhere between 4.5 volts and 6.5 volts the supply should begin regulating and further increases in the input voltage

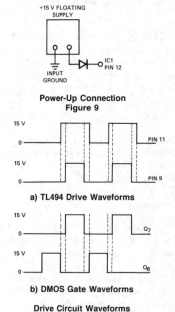

Power-Up Connection
Figure 9

a) TL494 Drive Waveforms

b) DMOS Gate Waveforms

Drive Circuit Waveforms
Figure 10

will not change the output. No significant current should be flowing from the high voltage supply at this time. Check the ± 15 volt outputs for the correct voltage.

While monitoring the supply voltage on pin 12 of IC1 (using a floating voltmeter) connect a load to the 5 volt output to draw about 1 ampere. The supply voltage on pin 12 should increase to about 15-20 volts if the power supply winding # 2 on T_2 is working correctly. If everything works correctly so far, disconnect all power supplies and install Q_3 in the circuit board. Using a variac or DC power supply and a floating voltmeter (or an isolation transformer and a non-floating meter), increase the input voltage to the line cord to about 20 VDC or 40 Vrms while monitoring the supply voltage on pin 12 of IC1. This voltage should level off around 10-12 volts. Connect a minimum load to the power supply (5 Ω, 5 W) and increase the line voltage to full voltage. IC1's supply voltage should increase to about 15-20 volts.

The power supply is now ready for use. Adjust the output voltage to 5 volts using R_{10}. Adjust the current limit R_{24} with the ± 15 volts fully loaded and the + 5 volt output delivering about 25 amperes. A minimum current of about 5 amperes must be drawn from the 5 volt output for the ± 15 volt outputs to be able to deliver 1 ampere each.

The power supply may now be plugged in directly to the power line for operation. The only requirement necessary is to have a minimum load of about 1 ampere at all times on the + 5 volt output.

Power Supply Features and Specifications

5 Volt Output
- 20 amperes output current
- 0.2% line regulation (± 20% line variation)
- 0.4% load regulation (no load to full load)
- < 100 mVp-p ripple and noise at full load
- Output over-current protection
- ≤ 0.5 mS transient response time (no load to full load)
- Over-current protected

15 Volt Outputs
- 1 ampere output current each
- 0.2% line regulation
- 1.0% load regulation
- < 10 mV ripple
- Short circuit current limiting

VN4000A Features
- 4000 volt BV_{DSS}
- < 1 ohm on-resistance
- < 100 nS switching times
- Rugged safe-operating area
- No secondary breakdown

Parts List

Part #	Quantity	Description	Recommended Mfg.
Resistors			
R_1	1	5.1 KΩ ±5% 1/4 W Resistor	Allen Bradley
R_2, R_6, R_9, R_{12}, R_{13}, R_{15}, R_{19}	7	1 KΩ ±5% 1/4 W Resistor	
R_3, R_7	2	330 Ω ±10% 1/2 W	
R_4	1	33 KΩ ±10% 1 W	
R_5, R_8	2	50 Ω ±5% 1/4 W	
R_{10}	1	500 Ω±10% Trimpot	
R_{11}	1	1.2 KΩ ±5% 1/4 W	
R_{14}, R_{17}	2	100 KΩ ±5% 1/4 W	
R_{16}, R_{23}	2	470 Ω ±5% 1/4W	
R_{18}	1	20 KΩ ±5% 1/4 W	
R_{20}	1	18 Ω ±10% 1 W	
R_{21}	1	100 Ω ±5% 1/4 W	
R_{22}	1	180 Ω ±5% 1/4 W	
R_{24}	1	100 Ω ±10% Trimpot	
Capacitors			
C_1	1	0.001 μf Ceramic Disc	
C_2	1	0.05 μf Ceramic Disc	
C_3, C_{14}, C_{16}, C_{17}	4	0.1 μf, 25 V Ceramic Disc	
C_4	1	1.0 μf, 25 V Electrolytic	
C_5	1	710 μf, 200 V Electrolytic (32D)	Sprague
C_6, C_8, C_9	3	47 μf, 25 V Electrolytic	
C_7	1	1.0 μf, 400 V TRW-35	TRW
C_{10}, C_{11}	2	2 μf, 25 V Tantalum	
C_{12}	1	200 μf Mica	
C_{13}, C_{19}	4	220 μf, 10 V Tantalum	Mallory 227K010P1G
C_{15}	1	5 pf Mica or Ceramic	
C_{18}, C_{21}, C_{22}	3	0.01 μf, 5 K V Ceramic	
C_{19}	1	10 μf Electrolytic	
C_{20}	1	6.8 μf Electrolytic	
Integrated Circuits			
IC1	1	TL494 PWM IC	Texas Instruments
IC2	1	CA3130 Op-Amp	RCA
IC3	1	TL430 Voltage Reference	Texas Instruments

6

Parts List (continued)

Part #	Quantity	Description	Recommended Mfg.
Integrated Circuits (continued)			
IC4	1	μA7815UC+15 V Regulator (μA7812UC+12 V)	Fairchild
IC5	1	μA7915UC−15 V Regulator (μA7912−12 V)	Fairchild
Diodes/Rectifiers			
D_1, D_3 – D_6, D_{14}, D_{15}	6	1N4148 Diode	Motorola
D_2	1	1N4107 Zener	Motorola
D_7 – D_{10}	4	1N5804 Fast Recovery	Unitrode
D_{11}, D_{12}	2	1N5831 Schottky	Unitrode
D_{13}	4	1N5406 Rectifier	Motorola
D_{14}	1	1N5360 Zener	Motorola
Transistors			
Q_1, Q_4	2	2N4400	Motorola
Q_2, Q_5	2	2N4403	Motorola
Q_3	1	MPSU10	Motorola
Q_6, Q_7	2	VN4000A MOSPOWER FET	Siliconix
Ferrites & Accessories			
T_1	1	F1146-1-06 Pot Core	Indiana General
T_1 Bobbin	1	B475-1	Indiana General
T_2	1	IR8031-1	Indiana General
T_2 Bobbin	1	B680-2	Indiana General
T_3	1	BBR7727-1 Toroid	Indiana General
L_1, L_2	2	F1146-1-TC9-315	Indiana General
L_3	1	F2037-1-TC9	Indiana General
Miscellaneous			
Opto 1	1	H11A1 Opto-Isolator	G.E.
TH_1	1	3D304 Thermistor	Midwest Components, Inc.
F_1	1	2 A Fast Blow Fuse	Buss
TO-3 Heat Sink	2	LAT03B5CB	IERC
TO-220 Heat Sink	2	LAD66A4CB	IERC
TO-3 P.C. Sockets	4	LSG-3DG2-1	Augat
D_{11}, D_{12} Heat Sink	1	E240-001	IERC
Output Banana Jacks	5		
3-Wire Line Cord	1		
Fuse Block for F_1	1		
Transformer Tape		Type 10 Epoxy Film	3M
Copper Foil		Type 1194	3M

6.2 Solid State RF Generators for Induction Heating Applications

Abstract

Radio frequency power for induction heating has traditionally been generated by vacuum tube oscillators. These generators use high voltage tubes and generally operate in the frequency range of 100–500 kHz, 1-400 kW power output at approximately 50–60% efficiency. Although thyristor SCRs and bipolar transistors are marketed at this time, they are unsuitable due to either slow switching or low power performance. Until just recently, solid state devices with both high-frequency and high-power capabilities were not available.

A new power semiconductor — the power MOSFET — a highspeed field effect transistor, can now meet these criteria. The power MOSFET switching time is in the order of 50–100 nanoseconds and can generate many kilowatts of power at frequencies to 500 kHz.

This paper describes a solid state RF generator using MOSFET transistors for the power semiconductors. The RF generator is a load tracking, resonant inverter capable of full power output over a frequency range of 100 kHz to 500 kHz. Output power levels are in the kilowatt range with induction coil KVAs to 200 KVA (400 V and 500 A) and more. This solid state RF generator has similar characteristics to the more familiar low frequency power converters presently used in the industry; namely, all solid state, high efficiency (approaching 90%), small and compact, and no vacuum tubes or moving parts.

Introduction

The art of induction heating is to uniformly heat metallic workpieces at specified temperature and times. For surface heating and non-magnetic parts, high frequency power is required. Therefore, the power source must be capable of supplying a minimum of several kilowatts of power in the radio frequency range. In addition, the power supply must have adequate control, protection and monitoring circuits.

Over the past 50 years, the high voltage vacuum tube was the only device available for radio frequency power generation. Semiconductors, such as SCRs and bipolar and Darlington transistors, have appeared in the last decade. These components work well in the frequency range up to 10 kHz, but they do not have the characteristics that are required for a radio frequency power source necessary for induction heating applications. In addition, these solid state devices are limited by slow switching speeds and/or low power ratings.

A new switching device, the power field effect transistor, is capable of both higher power outputs and significantly higher switching speeds. Some various market names for these transistors are VMOS, MOSFET, HEXFET and DMOS (these are the manufacturer's distinct trade names). The availability of such components have been disclosed and described in increasing detail since 1979 or so [1, 2, 3, 4]. As a result, Westinghouse Electric Corporation has developed and manufactured a solid state, radio frequency power supply using the power MOSFET in the inverter circuit design of the RF generator.

Power MOSFET Transistor

The power MOSFET is a field effect transistor with ratings presently ranging from 50 V/60 A to 1000 V/

3 A among the various manufacturers. Figure 1 depicts an N channel and a P channel MOSFET transistor showing polarity of the terminal potentials and current flow for normal transistor action. The MOSFET contains an internal, reverse P-N junction diode (shown dotted in Figure 1) which has the same current rating as the transistor.[5] This diode is a viable circuit element that can be used, just as any externally connected discrete diode, for a reverse current path around the transistor.

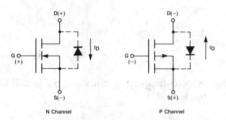

Power MOSFET Symbol Shown with Internal Diode & Polarities of Voltage & Current
Figure 1

The MOSFET transistor is a voltage controlled device with an insulated gate (G) which controls current flow between the drain (D) and source (S) terminals. A positive gate-source potential applied to the N channel MOSFET causes current to flow from the drain to the source. For the P channel MOSFET, the opposite is true: a negative gate-source potential causes current flow from the source to the drain. The drain current can be controlled linearly by the gate-source voltage or can be ''switched on'' by gate overdrive to a level determined by the impedance of the external circuit. With sufficient gate overdrive, the drain-to-source characteristic appears resistive with ''ON'' resistances for high current MOSFETs being less than an ohm. This ''ON'' resistance has a positive temperature coefficient which forces current sharing among paralleled devices, eliminating the need for external current balancing components and making paralleling relatively easy.

Drain-source switching speeds can be very fast, as low as 20–100 nanoseconds, depending on drive circuit source impedance and gate capacitance. MOSFET transistors do not exhibit minority carrier storage effects as do bipolar transistors and switching times are determined primarily by how fast the gate capacitance can be charged and discharged. The fast switching speeds result in a low and thermally manageable switching energy loss per cycle which is requisite for RF operation.

The steady-state gate-source input impedance is very high, being essentially capacitive. But dynamically the gate must be sourced and sinked by a high current source to produce fast switching speeds. For

example, a MOSFET having an input capacitance $C_{iss} = 1000$ pf, passes a gate current of 1 A when driven with a gate-source voltage rise time of 20 V/20 ns. At RF frequencies, these drive requirements can become formidable.

A limiting factor with the MOSFET is the recovery time of the reverse diode. This internal diode is relatively slow, having a recovery time which can be in the order of 200 ns or more compared to FET transistor switching times of 20–100 ns. These differences in speeds must be recognized and dealt with in certain switching circuits.

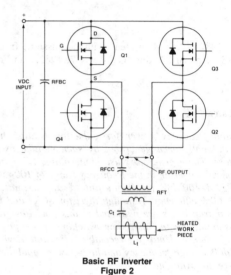

Basic RF Inverter
Figure 2

Solid State RF Generator

A. Basic RF Inverter

Figure 2 shows the basic circuit of a 1 kW resonant inverter developed for induction heating. MOSFET transistors Q1–Q4 are configured as a full bridge, voltage fed inverter with the internal diode of each providing the return path to the DC source for reactive currents. For higher powers, paralleled MOSFET transistors are used; for example, four paralleled 6 A/450 V devices in each leg for a 3 kW rating. The DC terminals of the inverter are tightly coupled to RF bypass capacitor (RFBC) whose capacitance is sufficient to pass the AC component of inverter input without substantially changing its DC potential. The AC terminals of the inverter drive the RF load circuit which is essentially a high Q series resonant circuit formed by tuning capacitor, C_t, and induction load coil, L_t. Transformer RFT matches the load impedance to the VA capability of the inverter, while coupling capacitor RFCC prevents any DC current from flowing in the primary winding and saturating the core. These two components are sized for minimal effect on the resonant circuit parameters, namely:

$$RFCC >> \left(\frac{Ns}{Np}\right)^2 C_t \qquad (1)$$

$$\text{Leakage Inductance of RFT} << \left(\frac{Np}{Ns}\right)^2 L_t \qquad (2)$$

$$\text{Primary Inductance of RFT} >> \frac{1}{w_{min}} \times \frac{E_o}{I_o} \qquad (3)$$

Where:

$\dfrac{Np}{Ns}$ Primary–secondary turns ratio of RFT

E_o Rated RMS output voltage of inverter

I_o Rated RMS output current of inverter

w_{min} Minimum output frequency of inverter

In operation, the MOSFET transistors are switched as diagonal pairs: Q1 & Q2 alternating each half cycle with Q3 & Q4 to provide a square wave voltage output at the AC terminals of the inverter. The waveform of the output current depends on the inverter output frequency which is the switching rate of the MOSFET transistors. Driving the series resonant load off resonance — i.e. at a MOSFET switching frequency differing from the natural resonant frequency of C_t and L_t — results in low output current, while driving it at resonance results in maximum power to the load coil. In fact the output current is controlled in a closed loop by varying the driving frequency.

The differences in switching speeds between the MOSFET transistor and its relatively slow reverse diode imposes an operational limitation on the inverter. Had MOSFET diodes 3 & 4 been conducting at the time MOSFET transistors Q1 and Q2 were switched on, they would have failed to recover in time to prevent short circuit currents from flowing between the positive bus and negative bus through the oncoming transistors and offgoing diodes. Commutation of current from a diode to a FET in adjacent legs of the inverter occurs whenever the output current leads the output voltage. One simple solution to the problem is to avoid leading currents and assure that the output current always lags the output voltage (lagging power factor). This is accomplished by always gating the MOSFETs at a frequency greater than the resonant frequency of C_t and L_t. With such a lagging power factor, current is always commutated from "its own diode" to the MOSFET transistor being gated on.

Thus, MOSFET diodes 1 & 2 are conducting when MOSFET transistors 1 & 2 are turned on eliminating short circuit recovery problems.

Figure 3 shows "light load" inverter waveforms resulting from driving of the series load at a frequency

$$f >> \frac{1}{2\pi \sqrt{L_t C_t}}$$

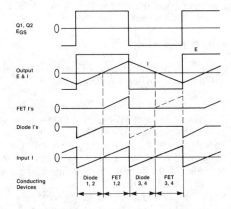

Inverter Waveforms for Light Load
Figure 3

The output current takes the form of an "inductive ramp" with positive current from the DC source flowing through MOSFET transistors and a nearly equal negative current being returned to the source by the MOSFET diodes. The net DC current is nearly zero and very little power is delivered to the highly reactive load (PF ≈ 0). Figure 4 shows the waveforms

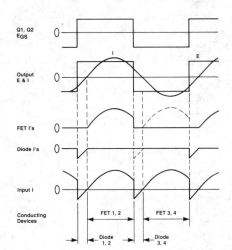

Inverter Waveforms for Heavy Load
Figure 4

6

for the same series load being driven at nearly resonance by decreasing the inverter output frequency. The output current is a sinusoid with a high amplitude. In this situation a mostly positive current flows from the DC source through the MOSFET transistors with very little negative current being returned by the MOSFET diodes. The net DC current is large and "heavy power" is delivered to the highly resistive load (PF $\approx$ 1). In both extremes, the commutation of current from a MOSFET diode to "its own MOSFET transistor" is illustrated.

B. Gate Drive Circuits

The gate drive circuitry shown in Figure 5 is designed to perform the following functions:

1. Charge and discharge the input capacitances (C_{iss}) of the inverter MOSFETs for drain-to-source switching times of 100 ns or less.
2. Provide a dead time between turn on and turn off of alternate firing MOSFETs serially connected in one pole of the inverter to prevent short circuit "shoot through" currents.
3. Provide a low impedance to the gate and source of each MOSFET during steady state inverting operation to prevent possible misfiring by the switching of the adjacent device. [6]
4. Prevent misfiring of the MOSFETs during shutdown.

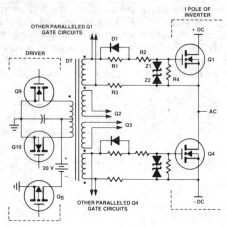

Basic Gate Drive Circuits
Figure 5

MOSFETs Q9 and Q10 and transformer DT form a push-pull driver with four isolated secondaries for gating the inverter power devices. The turns ratio from ½ primary to each secondary is 1:1 being specially wound for low leakage inductance in the order of 0.4 μH. This unavoidable transformer leakage inductance plus stray inductance in the same order of magnitude as the leakage forms a RLC circuit to charge the input capacitance of the MOSFET. Resistors R1, R2 & R3

are in the MOSFET "turn on" charging path being dimensioned for a circuit Q $\approx$ 1. The "turn off" discharge path includes only R2 & R3 (D1 shorts R1) increasing the circuit Q $\approx$ 2 resulting in a faster discharge than charge of the input capacitance. This provides an asymmetrical gate drive signal with relatively slower turn on time and relatively faster turn off time. The result as shown in Figure 6 is that the "off going MOSFET" reaches its gate threshold voltage before the "on coming MOSFET" attains its threshold, providing a gating dead time of about 50 ns for the pair. Figure 7 shows G-S turn on tracking waveforms of diagonally fired MOSFETs.

Since the gate drive is an alternating source, a significant negative potential (see Figure 6) is on the gate of the "off-going MOSFET" when the "on coming MOSFET" begins to switch. Thus, transient D-G currents caused by switching of the adjacent MOSFET should not cause the "off going" MOSFET to turn back on spuriously.

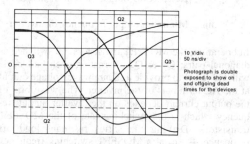

MOSFET G-S Turn On–Turn Off Waveforms
for Alternate Firing Devices
Figure 6

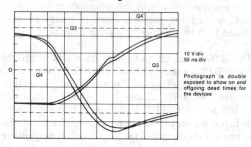

MOSFET G-S Turn On–Turn Off Waveforms
for Series Fired Devices
Figure 7

If the driver were stopped by removing pulses from driver MOSFET transistors Q9 and Q10, the current established in the primary of DT would freewheel through the other ½ primary and reverse diode of the device that was off. This would result in possible unscheduled gating of the inverter MOSFETs until the energy in DT transformer was dissipated and then result in an open circuited G-S drive connection thereafter. If the tuned load on the inverter was still ringing

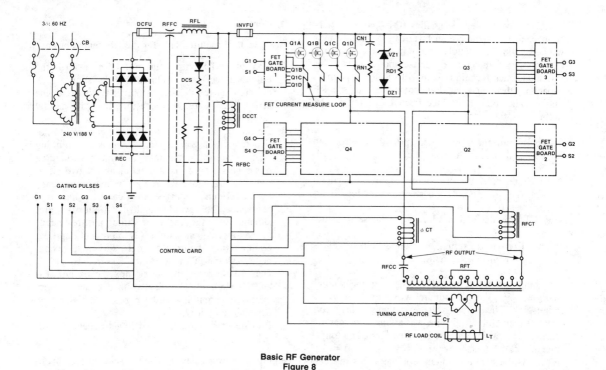

Basic RF Generator
Figure 8

down through the inverter MOSFET diodes to RF bypass capacitor, the switching of the diodes — i.e., rapid charge of D-S voltage — might now gate on the open circuited adjacent MOSFET causing a "shoot through" and failure. To alleviate this condition, driver MOSFETs Q9 and Q10 are both turned on at the STOP command while series MOSFET Q_S is turned off. Thus, the driver transformer is shorted at turn off and secondary drive voltage is reduced to nearly zero immediately. This presents a low impedance to the G-S drive circuits diverting any transient D-G currents from the gate of inverter MOSFET and preventing abnormal turn on.

C. KW RF Generator Circuits

The basic schematic of the overall 3 KW RF generator is shown on Figure 8. A 3 phase full wave diode rectifier is used to effectively utilize the voltage rating of the MOSFET transistors in the inverter. This is so because a single phase full wave rectifier would stress the inverter transistors with a 1.57/1 peak to average voltage (about 393 V peak for a 250 V DC supply) unless an elaborate DC filtering and lightload bleeder were used. The peak to average stress with the 250 V DC 3 phase supply is 1.049/1 or only 263 V peak. Since induction heating loads rarely have waveform modulation restrictions (especially when 3 phase diode rectifiers are used) filtering is quite unnecessary. Thus, about 50% more power can be handled by an

inverter powered from 3 phase than from single phase given the same voltage stress on the transistors.

Radio frequency filtering by RFL choke and RFFC feed through capacitor minimizes RF feedback (at twice the output frequency) to the AC power lines. The DC surge suppressor holds down any surge voltage across bypass capacitor RFBC during the following situations:

1. Turn on of the main circuit breaker and LC resonant charging of RFBC.
2. Turn off of the inverter gating under heavy load and transfer of energy from the RFL choke to RFBC.
3. Turn off of the inverter gating under light load and transfer of energy from the high Q output circuit to RFBC.

The inverter bridge has four paralleled MOSFET transistors per leg mounted on a common water cooled heat sink. The gate circuits of each transistor is as shown in Figure 5. Four such circuits are mounted on a FET drive board and assembled to the heat sink for each leg. Paralleling of the four gate drive circuits is done by one set of leads from the appropriate driver transformer secondary. Each inverter leg assembly contains a snubber network (RN, CN), tranzorb voltage suppressor network (VZ, DZ) and DC equalizing resistor (RD).

6

The four inverter leg assemblies (Q1–Q4) are interconnected in a symmetrical layout by multiple sandwiched, flat bus, such that equal and opposite currents are flowing to minimize inductances. This low inductance technique is used from RFBC bypass capacitor, through each drain-source tie, to the RF output. The total inductance from DC input to RF output is in the order of 0.3 μH.

Three RF current transformers provide control and protection signals to the controller:

- RFCT provides a closed loop feedback signal for control of the RF output current and for load overcurrent trip.
- ϕ CT provides a signal for output phase measurement and limiting.
- DCCT provides a signal for instantaneous trip of the gating for an inverter fault.

The output rating of the 3 KW RF generator is nominally a 250 V peak square wave at 18.7 Arms (sinusoid). The output voltage remains nearly constant, except for 60 Hz line variations, whereas the current will vary in amplitude as required. The fundamental of the square wave ($E_1 = .9 \times 250 = 225$ Vrms) is the exciting voltage for the series resonant load circuit. RF induction load coils exhibit Q's in the order of 20–60. Thus quite high potentials (4500 V–13,500 V ideally) would be developed across a coil and tuning capacitor connected directly to the RF output terminals when driven at the resonant frequency for maximum power. In general, RF load coils at 3 kW or so, can have much less impedance than 4500 V/18.7 A = 240 ohm and generally require impedance matching.

The usual RF matching transformer is air core with considerable exciting current and poor primary to secondary coupling. Such a transformer is not suitable for use with an inverter because of the reactive current and voltage requirements.

A universal, core type transformer (RFT) was developed to suit a 54:1 load impedance matching range.[7] Split primary and secondary windings on a 2 mil strip wound core are seriesed or paralleled to provide output voltages and currents from 5.1 V/823 A to 37.5/112 A in approximately 15 steps. Thus induction coils with impedances ranging from 102 V/823 A to 2250 V/112 A ($20 \leq Q \leq 60$) can be matched at the 3 kW level. The transformer is designed to operate from 100 kHz to 500 kHz with a maximum primary exciting current of 1 amp and maximum loss of 94 watts.

D. Control

Basic control requirements for the RF generator are:

1. Control the RF output current (and necessarily the output power) by a closed loop system which effects changes in output frequency to excite the series load nearer to or further from its natural resonance.
2. Detect the phase angle (ϕ) of output current relative to output voltage and "switch in" limiting phase control to maintain a worst case lagging phase nearly zero. This is requisite for two situations: (1) light load where the "asked for current" by the reference is higher than can be achieved at resonance. This would result in an out-of-control situation with the frequency sliding past resonance to the low frequency limit. (2) to prevent improper commutation of current from the slow reverse connected MOSFET diodes to the fast transistors.
3. Detect voltage on the series tuning capacitor (C_t) and "switch in" limiting voltage control where excessive potential would occur as with higher than expected circuit Q.
4. Provide high frequency and low frequency limiting.

Figure 9 shows the basic control scheme which implements the aforementioned requirements. It is a straightforward closed loop system with the operator's control potentiometer PT1 providing the current reference I_c^* and RFCT/REC1 providing the current feedback signal I_{fb} to OA1 error integrator. The output of OA1 decreases the frequency of pulses from the VCO (which is twice the output frequency) until zero error is achieved; i.e. the magnitude of output current is satisfied. Pulses from the VCO clock the timing flip-flop whose Q and $\bar{Q}$ outputs drive OR gates IC12A,B and IC12C,D alternately at ½ the VCO frequency. The OR gates drive emitter follower EFT1 and EFT2 which, in turn, provide low impedance gating signals to the driver N channel MOSFETs Q9 and Q10.

Rise time and fall time at the gates of Q9 and Q10 are about 10 ns with drain-source switching times in the order of 20 ns. The IC12A,B and IC12C,D paralleled pairs of OR gates are paralleled within one quad IC to increase current sinking and sourcing to enhance their switching times. All logic elements are CMOS operated at 15 volts. This allows direct interfacing with bipolar and FET transistors in the controller without need for level conversion.

A high STOP or FAULT signal from the sequencing and protective logic circuits simultaneously make the

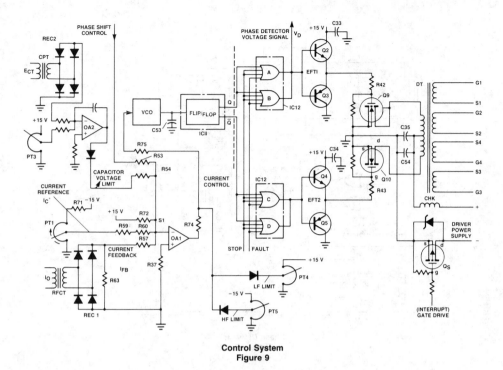

Control System
Figure 9

outputs of OR gates IC12A,B and IC12C,D a high. Driver MOSFETs Q9 and Q10 are both turned on shorting the primary of the driver transformer (DT) for low impedance stopping of the inverter gate drive pulses. At the same time the -15 V gate drive signal to the P channel series clamp MOSFET Q_S is removed, disconnecting the drive power supply from the shorted driver.

Series capacitor voltage limiting occurs when the voltage sensed by transformer CPT and rectified by REC2 tends to exceed the maximum level desired. Capacitor voltage limit pot PT3 sets the maximum level at V_c^*. When rectified voltage signal V_c is less than V_c^*, integrator OA2 has a negative output back biasing D15 and disconnecting OA2 from summing node S1 to OA1. Should V_c tend to exceed V_c^*, the output of OA2 goes positive and is connected to OA1 summing node. This constitutes additional negative feedback to OA1 which "makes up" the existing difference between I_c^* and 1_{fb} limiting the VCO frequency from further reduction and holding the capacitor voltage at the limit level.

The phase detector circuitry is shown in Figure 10. Basic phase detection occurs with the EXCLUSIVE NOR gate IC3C. The inputs are square waves representative of the RF output current (V_I) and RF output voltage (V_{dl}). The truth table for EXCLUSIVE NOR gate IC3C is as follows.

Input		Output
V_I	V_{dl}	V_ϕ
0	0	1
1	0	0
0	1	0
1	1	1

Thus, an output occurs from the EXCLUSIVE NOR gate whenever two 1's or two 0's overlap, resulting in an average output voltage versus phase of the two square waves as shown in Figure 11. This "phase output" after processing by OA5 integration, OA6 inverter and diode D16 provides negative feedback to S1 summing node of OA1 whenever it exceeds the minimum desired phase determined by the setting of PT2 potentiometer. The VCO frequency is limited from further reduction holding the phase angle between voltage and current at the limit level.

The interesting part of the phase detector lies in the development of the V_I and V_{dl} input signals to the EXCLUSIVE NOR gate IC3C.

First it should be recognized that the phase characteristic shown in Figure 11 is really not suitable for stable closed loop control. This is because the phase

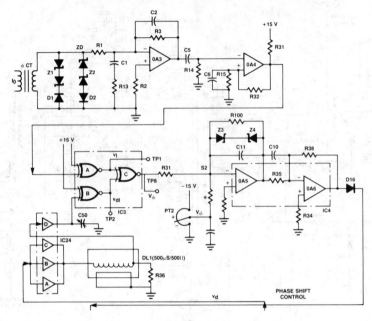

Phase Detector
Figure 10

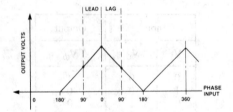

Average Output of EXCLUSIVE NOR Phase Detector
Figure 11

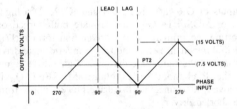

Average Output of EXCLUSIVE NOR Phase Detector
with 90° Lag
Figure 12

detector average output voltage does not monotonically increase with variation of ϕ from 90° lag to 90° lead. At $\phi = 0°$, the slope changes and the average output decreases for further increasing leading phase. Therefore, detection at $\phi = 0°$ is hazardous since the phase control loop would become unstable with each incremental increase in ϕ incrementally decreasing the error signal to 0A5. This results in increasing error to OA1 which necessarily drives the frequency of the VCO to its lower limit giving nil current output at $\phi = 90°$ lead. The solution to this problem is to phase shift the current 90° lagging and change the characteristic to that shown in Figure 12. Thus, at $\phi = 0°$, should the phase tend to go leading, the phase error would tend to increase resulting in more (not less) feedback to hold the phase at the desired level. This is accomplished by wideband amplifier OA3 connected as an integrator. OA3 integrates the current signal from ϕCT which is squared by ZD

clamp network into a ramp (see Figure 13). The integrated current waveform from OA3 is AC coupled to comparator OA4 which generates a square wave at 90° lagging (almost) from the original current signal. The lagging square wave is further "squared up" by EXCLUSIVE NOR gate IC3A into V_I.

The second problem is in the "almost" 90° lag generated by OA4 comparator. It happens that there is a constant, added delay time (t) of about 200 ns between the actual RF current and output of OA4 — mostly caused by OA4 (an LM211 high speed comparator). This fixed time delay is formidable since it represents a variable phase angle delay of 7.2°–36° when the frequency is varied from 100 kHz to 500 kHz. The problem is further compounded by a constant lead time (t') of V_d voltage signal from IC12A,B which is a digital representation of the RF output voltage. V_d leads the actual RF output voltage by about 300 ns.

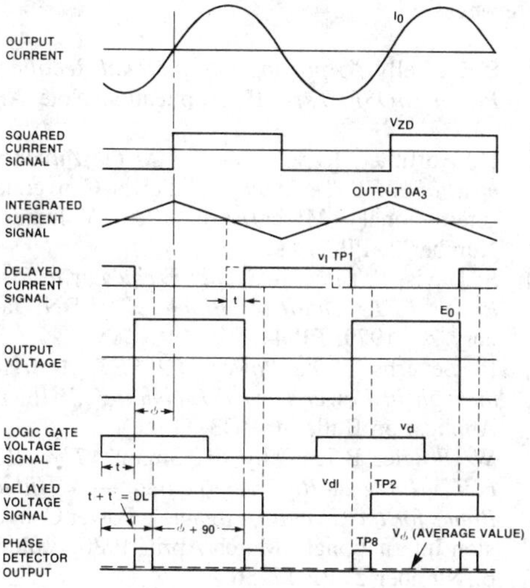
OUTPUT CURRENT
I0
SQUARED CURRENT SIGNAL
VZD
INTEGRATED CURRENT SIGNAL
OUTPUT 0A3
DELAYED CURRENT SIGNAL
VI TP1
E0
OUTPUT VOLTAGE
LOGIC GATE VOLTAGE SIGNAL
Vd
DELAYED VOLTAGE SIGNAL
t + t' = DL
Vdl
TP2
PHASE DETECTOR OUTPUT
TP8
Vφ (AVERAGE VALUE)

Phase Detector Waveforms
Figure 13

The solution to this problem is to make the voltage signal V_{dl} lag the actual RF output voltage by the same fixed time as the current signal V_I lags the output current. Since the EXCLUSIVE NOR phase detector measures the relative phase of V_I and V_{dl}, the absolute phase of each is unimportant as long as they are equal. To accomplish this, delay line DL1 and adjustable vernier capacitor C50 on the output of buffer gate IC24D provide most of the required time delay ($t' + t$). The remaining is provided by the propagation times of the squaring gate IC3B and paralleled delay line buffers IC24A,B,C.

E. Test Results

The RF generator was tested from a 1 kW output level using 4 HPWR-6501 MOSFETs to over 3 kW using 16 MOSFETs, 4 paralleled per leg. Some pertinent data and statements about those results follow:

1. Waveforms of voltage and current (see Figures 14 and 15) were as expected, except that MHz ringing at current commutations were significant. Worst case D-S ringing occurred off resonance with maximum E_{DS} = 370 V at 250 V DC.
2. Current sharing among the paralleled MOSFETs was excellent with a worst case unbalance of 13.5% from the mean and a typical unbalance of less than 10% measured. The only precaution taken for paralleling was symmetry in circuit layouts and matching of gate threshold voltages (V_t) as follows:

   V_t within ½ V for devices in one inverter leg.
   V_t within ½ V for groups in alternate firing legs.
   V_t within 1 V for groups in diagonal firing legs.

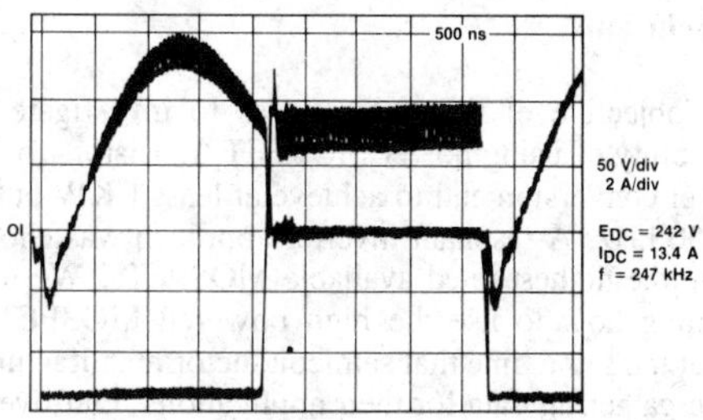
500 ns
0I
50 V/div
2 A/div
EDC = 242 V
IDC = 13.4 A
f = 247 kHz

MOSFET Voltage and Current Waveforms
Figure 14

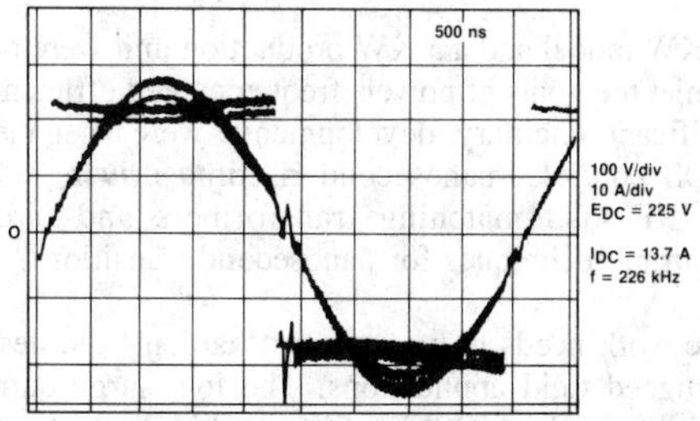
500 ns
0
100 V/div
10 A/div
EDC = 225 V
IDC = 13.7 A
f = 226 kHz

RF Output Voltage and Current Waveforms
Figure 15

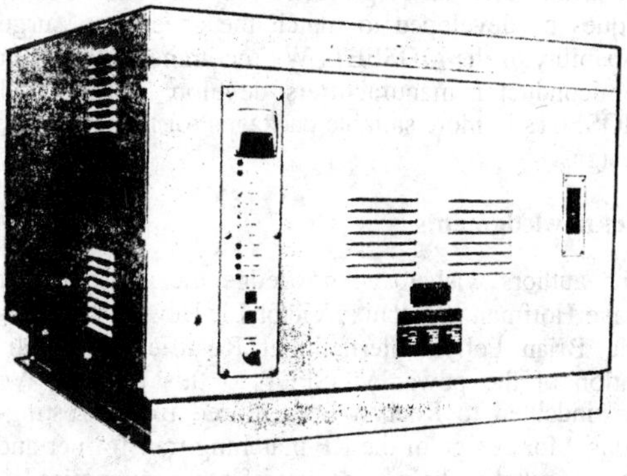

Solid State 3 KW, 100 kHz–450 kHz RF Generator
Figure 16

3. Over 3 KW of output power was achieved with a typical set of recorded data tabulated below:

Edc Volts DC	Idc Amps DC	E_{01} Volts rms	I_{01} Amps rms	f kHz	ϕ Degrees	Po Watts (Calculated)	DC-RF Efficiency % (Calculated)
236	16.6	198	18.74	387.2	20.9	3466	88.5

(1) Above data for RLC load — no transformer

L_t = 30 μH C_t = 0.0055 uf R = 9.87 ohm

(2) MOSFET TO3 case temperature rise above inlet water <5°C

6

Conclusions

The objective of this project was to investigate the feasibility of using power MOSFET transistors for RF power conversion and to achieve at least 1 KW output at 450 kHz. A resonant inverter approach was chosen using the highest rated available MOSFETs. We were learning how to use the high powered MOSFET at about the same time that semiconductor manufacturers were gathering data for their applications. Our overall objective was to develop basic techniques to eventually achieve tens of kilowatts as higher rated devices became available.

A 1 KW model and a 3 KW production unit were built that met the goals of power, frequency and efficiency. Significant ancillary developments were megahertz CMOS controls, nanosecond risetime drivers, iron-core RF load matching transformers and circuit measuring techniques for nanosecond transitions.

More work needs to be done to "harden" the design for rugged field applications. The low surge current capability of the MOSFET (like the bipolar transistor, but unlike the SCR) makes it difficult to protect even with high speed gate suppression. We would recommend that compatible protective devices and/or techniques be developed to match the speed and surge capability of the MOSFET. We recommend also, that semiconductor manufacturers develop higher rated MOSFETs in more suitable packages for higher power designs.

Acknowledgments

The authors wish to acknowledge the assistance of Dave Hoffman, Siliconix; Victor Li, Hewlett-Packard; and Brian Pelly, International Rectifier, in application of the power MOSFET to this project. We are indebted to Reuben Lee, retired from Westinghouse, for design of the RF matching transformer and for consultative help in design of other magnetics.

References

[1] B. R. Pelly, *"Applying International Rectifier's Power MOSFETs,"* IR Application Note AN-930.

[2] D. Hoffman, *"VMOS — Key to the Advancement of SMPS Technology,"* Power Conversion International - March/April, 1980, Volume 6, Number 2, PP 37–42.

[3] S. Davis, *"Switching-Supply Frequency to Rise; Power FETs Challenge Bipolars,"* EDN, January 20, 1979, PP 44–50.

[4] R. Severns, *"The Power MOSFET, A Breakthrough in Power Device Technology,"* Intersil Application Bulletin A033.

[5] W. Fragle, B.R. Pelly, B. Smith; *"The HEXFET's Integral Reverse Rectifier — A Hidden Bonus for the Circuit Designer,"* Power Conversion International - March/April, 1980, Volume 6, Number 2, PP 17–36.

[6] S. Clemente, *"Gate Drive Characteristics and Requirements for Power HEXFET's,"* IR Application Note AN-937.

[7] W.E. Frank, Reuben Lee; *"New Induction Heating Transformers,"* IEEE Transactions on Magnetics presented at the 3rd Joint Intermag-Magnetism and Magnetic Materials Conference; Montreal, Canada, July 20–23, 1982.

6.3 Using Power MOSFET Transistors to Interface from IC Logic to High Power Loads (AN79-6)

Power MOSFETS are an ideal element to interface power loads to integrated circuit logic. Although circuit design is simple, there are a few rules and precautions to observe in order to minimize power dissipation and to have reliable operation, which are not obvious at first glance.

Topics considered are (1) the nature of the load, (2) general driving requirements of MOSFETs, and (3) the output characteristics of the logic element.

Load Considerations

Freedom from second breakdown limitations makes driving highly inductive or capacitive loads a natural application for MOSFETs. Inductive loads include transformers, solenoids or relays. High current inrush loads such as incandescent lamps, pulse forming networks, and motors also are generally handled easily. Some attention must be given to the load characteristics, however.

In common with bipolar semiconductor devices, Power MOSFETS can be damaged if their voltage ratings are exceeded. Although their avalanche energy capability is much better than that of bipolar transistors, it is not good design practice to have the MOSFET absorb inductive energy unless the part is rated for this type of service. The spikes generated from inductive loads may have tremendous energy content and usually some means of limiting their amplitude must be provided.

In addition the transient power generated during the turn-on and turn-off intervals must be determined in order to check for excessive channel temperatures. Highly inductive loads may generate significant power

on turn-off, whereas capacitive-like loads cause power surges on turn-on. The power waveform should be obtained, a suitable rectangular model derived, and peak junction temperature computed using techniques discussed elsewhere [1,2]. This calculation is especially important for incandescent lamp and motor loads as their current surges last from tens to thousands of milliseconds which causes a significant surge in the temperature of the MOSFET power driver.

Inductive Loads

Usually with inductive loads the peak voltage spike should be limited to a value below the breakdown rating of the transistor. Three techniques are commonly employed: free-wheeling diodes, peak clipping and snubbing. Typical circuits are shown in Figure 1.

The spikes caused by most electromechanical inductive loads such as solenoids or relays are effectively handled by the free-wheeling diode in part (a). The low impedance of the diode usually causes the current to have a long decay time, however, which may be intolerable in some applications. Speed may be traded for overshoot voltage, by using a resistance, R, in series with the diode [3].

The free-wheeling diode may be an inexpensive rectifier such as the 1N4002. However, junction rectifiers do exhibit a turn-on transient which may allow excessive overshoot if the MOSFET is being driven off rapidly. For high speed switching, a Schottky or low voltage ion-implant rectifier is required; the ordinary fast recovery rectifier does not have a fast turn-on time.

6

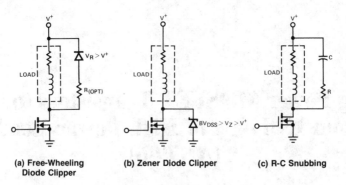

| (a) Free-Wheeling Diode Clipper | (b) Zener Diode Clipper | (c) R-C Snubbing |

Methods of Limiting Inductive Spikes
Figure 1

Often the safest and least expensive limiting technique is to use a zener diode as shown in part (b). The zener responds in picoseconds and can protect the MOSFET from supply transients as well as the inductive spike; consequently, zener limiting is particularly attractive on raw power buses. In a manner similar to using a resistor in series with a free-wheeling diode, faster decay of load current is achieved by clipping at a level above the supply voltage.

The R-C snubber is commonly used in power conversion circuits to limit spikes caused by transformer leakage inductance and wiring inductance. It also reduces power dissipation by shaping the load line to appear more resistive. Resistor R, in series with the capacitor, is required to limit the current surge on turn-on (a good idea even when MOSFETs are used) and to ensure that the circuit is adequately damped. Since the circuit is basically a resonant tank, it will exhibit a damped oscillation unless the circuit Q is 1/2 or less. Values are usually empirically determined. The peak voltage across the network will not exceed that calculated using the energy relationship: $1/2\ LI^2 = 1/2\ CV^2$. Solving for the voltage, it is found that

$$V = I\ \sqrt{L/C}$$

The resonant frequency can be calculated from the usual relationship and R selected so that $Q \simeq 1/2$ by using

$$R = 4\pi fL$$

The equations and experience indicate that larger values of C lower the peak voltage and resonant frequency and consequently the resistor R also must be reduced. An optimum value exists for a given L-C combination which results in minimum overshoot. Another consideration is to minimize the power dissipation in the transistor; various techniques are discussed elsewhere [4,5].

Capacitive and High Inrush Loads

Usually no auxiliary circuitry is required with capacitive-like loads. Although the MOSFET has no failure mode akin to secondary breakdown, it is necessary to observe the safe area curves of the MOSFET in order to avoid excessive temperature excursion during current inrush. When inrush power is excessive, usually increasing the gate drive will reduce it and may hold it within bounds. A lamp circuit and waveforms are shown in Figure 2.

Figure 2.1 illustrates the typical transient current, voltage and transistor dissipation of an incandescent lamp load being driven by a MOSFET circuit that is not drive limited — it can supply all the current that the load demands while maintaining operation in the ohmic region. Under these conditions, the output voltage quickly swings from its off state (V_1) to on state in a time dictated by the transient gate drive current and the MOSFET's capacitances. The peak current is a function of the lamp's cold resistance and decays quasi-exponentially as the lamp filament heats and resistance increases.

The transistor dissipation waveform is similar to the power waveform of Figure 2.1(c); this can be equated to the rectangular power pulse of 2.1(d) to simplify peak and average power calculations.

The case where the power transistor is drive limited (where it cannot supply all of the peak current that the load demands) usually — but not always — results in greater transistor dissipation. This is illustrated in the transient waveforms of Figure 2.2 where the peak current is much less than in the previous example, but also where transistor operation does not enter the ohmic region initially during the switching transition. The resulting power dissipation pulse is greater and may be destructive.

MOSFETs make ideal drivers for incandescent lamps because they can handle high current surges without

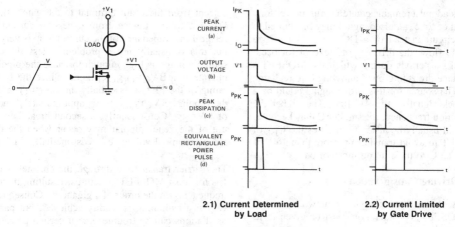

2.1) Current Determined by Load

2.2) Current Limited by Gate Drive

Waveforms When Driving an Incandescent Lamp, High Inrush
Loads Develop Similar Waveforms
Figure 2

failure caused by secondary breakdown. Lamp drivers may need to handle two types of surges, cold resistance inrush and flashover.

Cold resistance inrush occurs on all lamps during turn-on; the peak is between 12 and 18 times the steady state current. Furthermore, the inrush current may be from 2 to 5 times rated current 5 milliseconds after power application. The inrush depends somewhat upon the lamp's design and the cold temperature of the bulb, which in turn is dependent upon the lamp's operating duty cycle and ambient temperature. Since the turn-on surge is a repetitive transient, for maximum MOSFET reliability, it is usually desirable to have sufficient gate drive to place the operating point in the ohmic region during the surge in order to minimize power dissipation.

Driving Grounded Loads

In many cases, the load is connected to ground and cannot be arranged as shown in Figures 1 and 2. Driving a grounded load forces the MOSFET to be used in a source follower circuit. The difficulty with a follower is that, to keep the MOSFET in the ohmic region with a large drain current flow, the gate must be about 10 volts above the source potential, which is only slightly lower than the supply. Therefore, the gate drive voltage must come from a voltage source which is about 10 volts above the supply voltage. If such a voltage is available, no significant problem exists other than ensuring that the driver circuit has a sufficient voltage rating. When no fixed voltage source is available, it may be generated using the bootstrap technique.

A bootstrap circuit is shown in Figure 3. Operation is as follows: when the driver bipolar is on, the gate potential is near ground and the MOSFET is off. Capacitor C is charged to V_{DD} through the load and diode D1. When the driver goes off, the gate voltage rises, turning on the MOSFET, which raises the source potential. If C is many times larger than the input capacitance of the MOSFET, it acts as a voltage source in series with the MOSFET source terminal potential, thereby providing a gate-source voltage close in value to V_{DD}. The capacitor C will lose charge with time through the reverse resistance of diode D1, so that it becomes impractically large if the load must be held on more than a few seconds, unless a strobing technique is used as discussed in the following paragraph. However, many loads, such as hammer drivers in high speed printers, are actuated for under a millisecond so that capacitors in the range of 0.1 μF are adequate. In cases when the load is a solenoid, it may be permissable for the MOSFET to come out of the ohmic region as a result of a partial loss of gate drive after the solenoid has pulled in, because the hold-in current is very low. However, the power dissipation of the MOSFET may increase significantly unless the drop in gate drive reduces drain current to a rather low value.

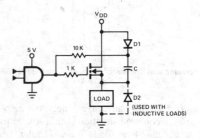

Driving Loads Connected to Ground by Using Bootstrapping
Figure 3

6

6-83

When the load must remain actuated at full power for a long period of time, a pulsed signal may be applied to the drive circuit instead of a DC level. When load current is to flow, the driver is off, except for occasional brief on periods during which the MOSFET's source voltage drops to ground allowing C to recharge to V_{DD}. The source for this strobe signal could be a system clock signal or the AC line. The higher the pulse repetition frequency, the smaller C may become, but C should usually be at least ten times the C_{iss} of the MOSFET to avoid transferring more than 10% of the charge on C to the gate during turn-on.

General Driving Considerations

Regardless of the type of logic or network used to drive a MOSFET, consideration must be given to its properties such as:

1. The input protection zener diode
2. The high frequency response
3. The capacitive input impedance

The input protection zener diode integral with some devices places restrictions on the drive levels. The positive voltage on the gate with respect to the source should not exceed the maximum voltage rating of the zener diode nor should the zener become forward biased by allowing the gate-to-source voltage to become negative.

The reasons for these restrictions are evident from the circuit of Figure 4 where it is seen that the zener is actually the base-emitter junction of a bipolar transistor whose collector is connected to the drain. Consequently, the zener exhibits a negative resistance characteristic similar to the BV_{CEO} of a transistor. Above a few hundred microamperes, the voltage may switch from, for example, 20 volts to 8 volts. The lower voltage will limit the amount of current available from the MOSFET and excessive dissipation in the zener and drive circuitry may occur.

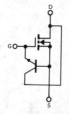

A Parasitic npn Transistor in Zener Protected MOSFETs
Figure 4

If the drive circuit can cause a negative gate-source voltage — a common situation with a source follower driving a capacitive load — the bipolar becomes turned on in the forward direction causing a flow of current from the drain terminal to the gate terminal. Damage to the device from excessive current may result if the impedance of the current loop is very low, but this is usually not a problem unless the drain-source voltage is high enough to cause the bipolar to operate in a $BV_{CEO(SUS)}$ mode. Since the BV_{DSS} rating of the FET is essentially the BV_{CBO} rating of the bipolar, $BV_{CEO(SUS)}$ is approximately one half of BV_{DSS}. Consequently, a second breakdown failure of the zener bipolar may occur when the drain-source terminal voltage exceeds one half of BV_{DSS}.

The carrier transit time through the channel is under 1ns for most MOSFET structures, resulting in cut-off frequencies on the order of a gigahertz. Consequently, very fast switching is readily achieved, but parasitic oscillations can be troublesome if certain precautions are not observed.

Usually oscillations are prevented by observing one or more of the following guidelines:

1. Keep lead and trace lengths short.
2. Place ferrite beads on the gate lead close to the gate terminal or use a resistor of 10 to 1000 ohms in series with the gate.
3. Avoid a layout which may couple output signal to the input.
4. Surround the MOSFET with a ground plane and shield output from input.

Since the MOSFET gate input resistance is essentially infinite, many can be driven from a CMOS or TTL output. However the input impedance of a MOSFET is capacitive and the drain current essentially follows the voltage on the gate. Although switching speed, per se, is not important when driving a lamp or electromechanical load, the limited transient current available from the logic element may result in switching slow enough to cause significant transient power dissipation, particularly when a number of MOSFET stages are being driven in parallel. Accordingly, a transient analysis of some sort is usually required.

A fairly simple, yet quite accurate, analysis is to use a charge control approach as described by Evans and Hoffman [6]. For any particular time interval,

$$\Delta t = \frac{(\Delta V_{GS})(C_{in})}{I_G}$$

where

ΔV_{GS} is the gate-source voltage change

C_{in} is the effective input (gate-source) capacitance

I_G is the average gate current during switching

(all values must be determined for the time interval of interest).

Table 1
Pertinent Switching Relationships

Interval	Symbol	Gate Voltage Change	Capacitance
Turn-On Delay	$t_{d(on)}$	$V_{G(TH)} - V_{G(off)}$	C_{iss}
Rise Time	t_r	$V_G @ I_{D1(on)} - V_{G(TH)}$	$C_{iss} + \dfrac{\Delta V_{DS}}{\Delta V_{GS}} C_{rss}$
Turn-Off Delay	$t_{d(off)}$	$V_{G(on)} - V_G @ I_{D2(on)}$	C_{iss}
Fall Time	t_f	$V_G @ I_{D2(on)} - V_{G(TH)}$	$C_{iss} + \dfrac{\Delta V_{DS}}{\Delta V_{GS}} C_{rss}$

Table 1 shows the appropriate quantities to use in the equation. For completeness, turn-on and turn-off delay relations are included, but these intervals are rarely of interest in power circuits. However, dissipation may be a problem during rise and fall time.

Capacitance values used are the average values as V_G varies over the ranges shown for the time interval of interest. Appropriate V_{DS} values must also be used to determine the capacitance. Key V_{GS} points are:

$V_{G(off)}$ = Off state gate voltage prior to turn-on

$V_{G(TH)}$ = Threshold gate voltage

$V_G @ I_{D1(on)}$ = V_G corresponding to the peak value of drain current for capacitive or resistive loads or the value of drain current when the drain voltage enters the ohmic region for inductive loads

$V_{G(on)}$ = On-state gate voltage prior to turn-off

$V_G @ I_{D2(on)}$ = V_G corresponding to the value of drain current flowing prior to turn-off

Driving from CMOS Logic

The widely used CMOS logic elements are ideal for direct coupling to power MOSFETs because CMOS can operate with supplies up to 15 volts, a level which provides ample drive for the MOSFET. Since switching speed and transient load handling capability is related to the output impedance of the drive source, a brief examination of the CMOS circuit follows.

All CMOS circuits usually have an output configuration as shown by the inverter of Figure 5. The inverter

consists of a p-channel MOSFET connected in series with an n-channel MOSFET (drain-to-drain) with the gates tied together and driven from a common

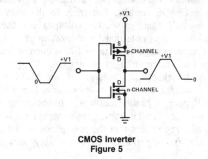

CMOS Inverter
Figure 5

signal — hence, the name CMOS (complementary MOS). When the input signal goes positive (+ V), the p-channel MOSFET is essentially off and conducts only I_{DSS} (picoamperes). The n-channel unit is forward biased but since only I_{DSS} is available from the p-channel, V_{DS} is very low. Conversely, when the input goes low (zero), the p-channel device is turned full on, the n-channel device is off, and the output will be very near + V. Since the current (without a load) is extremely small, the inverter dissipates almost no power in either stable state; the only dissipated power of consequence occurs during the switching transitions as capacitances are charged. Due to the extremely high input impedance of a MOSFET, the CMOS gate has the capability of interfacing with many MOSFETs when only static conditions are considered.

The DC resistance between drain and source when the device is turned on is generally labeled "ON resistance" R_{ON} or $R_{DS(ON)}$. However, the CMOS gate has a limited output capability determined by the gain of the n- and p-channel devices. Equivalent circuits of the CMOS output are shown in Figure 6.

6

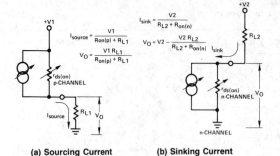

$$I_{source} = \frac{V1}{R_{on(p)} + R_{L1}}$$

$$V_O = \frac{V1\, R_{L1}}{R_{on(p)} + R_{L1}}$$

$$I_{sink} = \frac{V2}{R_{L2} + R_{on(n)}}$$

$$V_O = V2 - \frac{V2\, R_{L2}}{R_{L2} + R_{on(n)}}$$

(a) Sourcing Current **(b) Sinking Current**

CMOS Source/Sink Capabilities
Figure 6

A look at the characteristic curves of CMOS transistors will provide insight into the dynamic driving source impedance presented to the MOSFET gate. Figure 7 shows the characteristic curves of n-channel and p-channel enhancement mode transistors typically found in CMOS circuits. Referring to the curve of $V_{GS} = 15V$ (gate-to-source voltage) for the n-channel transistor, note that for a constant drive voltage V_{GS}, the transistor behaves like a current source for drain-to-source voltages greater than $V_{GS} - V_T$ (V_T is the threshold voltage of an MOS transistor — on the order of a volt or two). For a V_{DS} below $V_{GS} - V_T$, the transistor behaves essentially like a resistor. Similar curves are obtained for lower values of V_{GS} except that the magnitude of the current is significantly smaller and, in fact, I_{DS} increases

approximately as the square of increasing V_{GS}. The p-channel transistor exhibits similar, but complemented, characteristics with less gain and a more gradual transition from a current source to a resistor.

When driving a capacitive load the initial voltage change across the load will be a ramp due to the current source characteristic, followed by a rounding off due to the resistive characteristic dominating as V_{DS} approaches zero. For fastest turn-on and therefore lower dissipation in the MOSFET, the peak output current should be achieved while the CMOS inverter is operating in the constant current mode. To accomplish this, the maximum current from the MOSFET is that which corresponds to a V_{GS} that is a few volts below the CMOS supply voltage, V_1. From Figure 7, note that operating the CMOS driver at higher V_{CC} will have a profound effect on switching speed because the CMOS output current increases roughly as the square of V_{GS} and the voltage where rounding occurs has been pushed to a higher level.

Therefore, the optimum interface from CMOS to a MOSFET is shown in Figure 8. In this configuration, the turn-on current is supplied from the p-channel FET which has the poorest characteristics of the CMOS pair, but when operating at 15 volts, serious rounding of the MOSFET's gate waveform can usually be made to occur at a level above that required to handle the load current. The turn-off current is supplied from the n-channel FET; it maintains good drive capability down to the threshold voltage of the MOSFET's gate which minimizes tailing of the drain current.

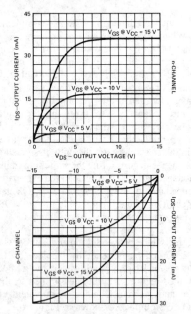

Transistor Output Characteristics of a CMOS Inverter
Figure 7

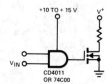

Driving VMOS with a CMOS Gate
Figure 8

Driving from TTL Logic

The lower logic levels used in TTL make it much less satisfactory than CMOS for direct coupling to power MOSFETs; however, TTL can be directly coupled when lower output currents are required, or some additional circuitry can be added to make the match more universally applicable. The coupling problems are readily appreciated by analyzing the output circuit of TTL.

Figure 9 shows the totem-pole output circuit configurations commonly used in TTL. When the driver is off, the output is high, however, the output is slightly over two diode drops below the supply voltage for either configuration. Since the nominal supply is 5 volts, the output is approximately 3.5 volts, a level too low to fully utilize a power MOSFET.

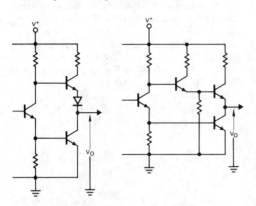

(a) Medium-Speed TTL Output Configuration (5400/7400 Families)   **(b) High-Speed TTL Output Configuration**

Basic TTL Output Configurations
Figure 9

When the driver is on, the lower output transistor is also on; the output is low, on the order of a few tenths of a volt. The low level is satisfactory to ensure that the power MOSFET is cutoff, in most cases, but the high level needs to be raised.

A method of boosting the TTL output level is shown in Figure 10. The external resistor allows the full-supply voltage to be applied to the MOSFET gate, but in doing so, the TTL output transistor becomes cut-off as the level increases above 3.5 volts. Consequently, the high drive capability and low output impedance of TTL is not effectively utilized, as the drive to the final value of gate voltage must come solely from the pull-up resistor. To maintain a reasonably fast rise time, it is necessary to limit the peak drain current to the value obtained for a particular MOSFET type when the gate voltage is at 3.5 volts; the additional 1.5 volts are used as overdrive to place the operating point in the ohmic region.

Full utilization of power MOSFETs is achieved with open collector TTL as shown in Figure 11. The open collector circuits do not have the top transistors and the lower transistors are designed to be used with supplies up to 15 volts. The MOSFET rise time is now mainly dependent upon the external resistor used. For fast rise time, the lower resistance values required may cause objectionable dissipation when the TTL output is low; use of the circuit of Figure 12 will provide high speed and low dissipation. It essentially restores the totem pole output to the TTL circuit using an external high voltage transistor. Since the bipolar transistor does not saturate, a general purpose transistor with a high f_T will provide fast drive signal to the gate.

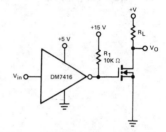

Open Collector TTL is Used to Provide Greater Enhancement Voltage
Figure 11

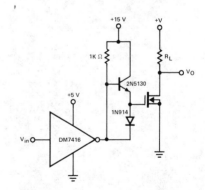

A "Totem Pole" Driver Increases Switching Speed and Reduces Dissipation
Figure 12

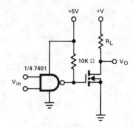

Driving the VMOS with Standard TTL
Figure 10

6

Summary

MOSPOWER FETs can easily interface power loads to integrated circuit logic. Spikes from inductive loads usually must be limited to a level below the breakdown voltage of the MOSFET. High inrush, capacitive-like loads usually require high gate drive to place operation in the ohmic region during inrush.

To avoid deleterious operation and ensure fast switching, consideration must be given to the input protection zener diode — if present — and capacitive input impedance. Direct coupling to CMOS is usually very satisfactory if the IC supply is 10 to 15 volts. With T^2L, open collector elements are usually required to obtain sufficient gate voltage for the power MOSFET.

References

[1] Bill Roehr and Bryce Shiner, "Transient Thermal Resistance — General Data and Its Use"; Motorola Application Note, AN569, Motorola Semiconductor Products Inc., Phoenix, AZ.

[2] Bill Roehr, et al. "Silicon Rectifier Handbook," 2nd Edition, Chapter 2, pp. 30–39, Motorola, Inc., Phoenix, AZ.

[3] loc. cit. "Silicon Rectifier Handbook," Chapter 8, pp. 117–121.

[4] E.T. Calkin and B.H. Hamilton, "Circuit Techniques for Improving the Switching Loci of Transistor Switches in Switching Regulators," IEEE Conference Record, 1972 IAS Annual Meeting, pp. 477–484.

[5] Rollie J. Walker, "Circuit Techniques for Optimizing High Power Transistor Switching Efficiency," Proceedings of Powercon 5, May 1978, Power Concepts, Inc., Oxnard, CA.

[6] Arthur Evans and Dave Hoffman, "Dynamic Input Characteristics of VMOS Power Switch," Siliconix Application Note, AN79-3.

6.4 MPP500: The First Single Package Complementary MOSPOWER Device

Introduction

The new MPP500 series of complementary power MOSFET pairs provide increased convenience for designers of small motor drives, inverters, and similar devices with load ratings of up to 150 W. Mechanical design is simplified because only half as many packages are needed to construct either a single-phase or a three-phase bridge. Electrical design is simplified because device matching by the designer is not required and because complicated isolated drives for the upper switches in the bridge are no longer necessary. This last simplification alone can result in a dramatic savings in system cost.

MPP500 Device Characteristics

An MPP500 power MOSFET pair consists of one n-channel and one p-channel MOSFET which are matched for die size, breakdown voltage, and power handling capability. Die size matching means that the p-channel member of the pair has higher ON-resistance ($R_{DS}(on)$) than the n-channel member and that both FETs have equal input capacitance.

The reason for the difference in $R_{DS}(on)$ is that holes (the majority carriers in p-channel devices) have less mobility in a semiconductor structure than do electrons (the majority carriers in n-channel devices). To construct devices with equal $R_{DS}(on)$, the p-channel device would require approximately 2.4 times the active area of the equivalent n-channel device and would, consequently, have 2.4 times as much input capacitance. This would mean that for equal gate drive impedance the p-channel device would switch more slowly than the n-channel device, and in certain applications, this would complicate system design considerably.

In the MPP500, both FETs have equal die size and are mounted on a common header. Thus, the power handling capability of the two devices is, essentially, equal; for most applications, the difference in $R_{DS}(on)$ can be ignored. Three different drain-source voltage ratings are available: 100 V (MPP500), 80 V (MPP501), and 60 V (MPP502); all three devices will handle 2.3 A continuous or 7.7 A pulsed.

Input Drive Requirements

The real advantage of the MPP500 lies in the ways it may be driven. Because the upper FET is a p-channel, it is only necessary to lower its gate terminal 10 V below the positive power rail to fully turn ON the device. This can be accomplished with any open-collector or open-drain logic output. It is not even necessary for the drive transistor to withstand the full power supply voltage. Because, if the power is stable, a suitable zener may be interposed between the current sink and the p-channel gate. Shut off can easily be accomplished with one pull-up resistor. Furthermore, the same current sink that turns on the p-channel FET ON can be used to simultaneously turn the complementary n-channel FET OFF. This is one of the benefits of using FETs with equal input capacitance. In fact, in Figure 1a the CMOS gate used to drive the MPP500 does not even have an open drain output. Further, because the PLUS 40 process with a 40 V gate breakdown is used in the MPP500, no zener is required for power input voltages less than 40 V. Some overlap conduction will occur under these circumstances during the switching transition as it does in CMOS logic, and for the same reason. A zener between the gates minimizes this.

Reprinted courtesy of PCI July, August 1984

6

The MPP500 as a Power CMOS Driver

The circuits shown in Figure 1 are useful for a variety of solenoid, lamp, and relay driver applications. Inverse diodes are inherent in all power MOSFETs, so additional diodes for inductive loads are not required unless very fast switching speeds are necessary [1]. Such fast switching speeds are seldom required for mechanical loads like motors and relays.

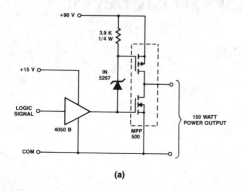

(a)

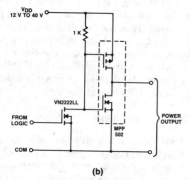

(b)

a) Very Large CMOS Inverter
b) Power CMOS Buffer
Figure 1

Switching speed is a matter to which some attention must be paid. For resistive loads or situations where no zener is desired between the n and p channel gates, it makes sense to switch as rapidly as possible. Rapid switching minimizes overlap currents and maximizes efficiency. As with all other power MOSFETs, the limit to how fast you may switch depends on how much time and money can be put into the gate drive circuit.

For reactive loads, the situation changes and a slower gate drive circuit may be a better choice. This situation exists because the reverse recovery time of the MOSFET inverse diode is typically between 200 and 400 ns. When switching speeds much faster than the diode reverse recovery time are desired, additional components will be needed. Otherwise, reverse conduction through the intrinsic diode or dV/dt effects during switching will damage the MOSFETs. At input voltages below 40 V, Schottky diodes in parallel with the MOSFET diodes are sufficient, since the forward drop of the Schottkys is less than the MOSFET diode drop. They, thus, will prevent the MOSFET diodes from conducting. At higher voltages, more complex arrays, using blocking diodes and conducting diodes or more complex drive schemes involving active commutation are required. For most inductive load circuits, a slower drive which trades switching losses for conduction losses in the extra diodes is simpler, less expensive, and dissipates less power. Generally, all that is required to switch at less than super-fast rates is a less expensive drive circuit. A CMOS logic gate is generally sufficient. A 4000 series CMOS gate, for example, will drive the MPP500 (both gates) with a current rise/voltage fall time of 300 to 700 ns. A series gate resistor can slow this further if desired.

Single-Phase Bridge Applications

Cross-coupled driving for the upper FETs of a single-phase bridge, as is shown in Figures 2, 3, and 6 is an economical shortcut made possible by the use of p-

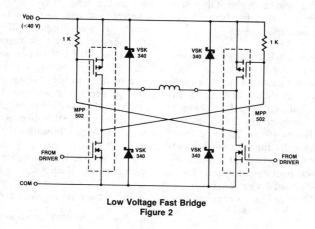

Low Voltage Fast Bridge
Figure 2

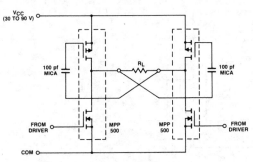

High Frequency Bridge for Resistive Loads
Figure 3

channel FETs in the upper half of the bridge. Each upper FET is driven from the drain of the opposite lower FET, and one or two terminals, referenced to common, control an entire bridge. Dead time between the two drives is still required to prevent possible conduction overlap. Figure 6 shows one way this may be accomplished.

Capacitive coupling for the gates of the upper FETs, as shown in Figures 2 and 3, works surprisingly well if low leakage capacitors are used [2]. It functions by using the input capacitance of the upper FET as half of a capacitive voltage divider which divides the voltage change on the drain of the lower FET to a level appropriate for gate drive on the upper FET.

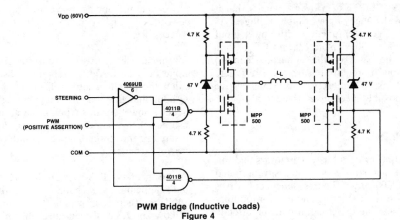

PWM Bridge (Inductive Loads)
Figure 4

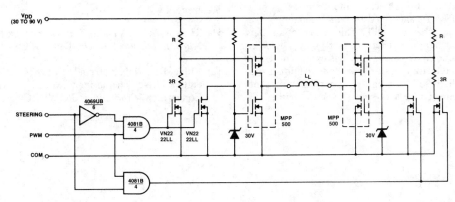

Wide Input Range Bridge with Active Commutation
Figure 5

6

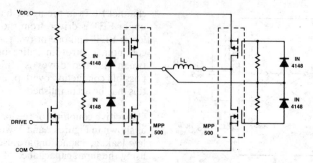

**Unilateral Drive Bridge with Resistor-Diode
Overlap Prevention
Figure 6**

The advantage of this system over zeners is that it works over a wide range of power input voltages. As shown, both bridges will function with power input voltages from 30 to 90 V. Note that no DC path is provided for the gates of the upper (*p*-channel) MOSFETs. None is required. The MOSFET gate itself has no DC path. Similar systems using resistive dividers, and possibly speed-up capacitors, also work well.

Other Possibilities

Controlled output bridges for servo motors and microstepping of stepper motors are easily designed using the MPP500 series. Four independent drives are required—two for steering and two for modulation—but the steering drives can usually be directly coupled to the logic. For pulse width modulation, it is more efficient to modulate the upper FETs rather than the lower FETs because the upper FETs have higher ON-resistance, and system dissipation is reduced when they are OFF. This is one of the few instances where the difference in $R_{DS(on)}$ between the *p* channel and *n* channel FETs is significant. Using an open-collector output PWM, and techniques previously discussed, driving the upper FETs presents no problem. Diode arrays, as shown in Figure 2, or active commutation as shown in Figures 4 and 5 will be required if high speed switching of inductive loads is necessary.

For linear (dissipative) output current control, as might be used for microstepping a disc head drive, either the upper or lower FETs may be modulated without changing the efficiency, but, as always, efficiency is lower for this type of operation. Care must also be taken when operating in the linear mode to ensure that the power dissipation limits of the package are not exceeded.

Complementary MOSPOWER arrays are finally here, and it appears that *p*-channel power MOSFETs can fulfill their theoretical promises. The limits now stand at 100 V and 2.3 A. Expansion both to higher currents and higher voltages may be expected.

References

[1] Severns, R., "dV$_{DS}$/dt Turn-on in MOSFETs," *Siliconix Technical Article,* TA 84-4, 1984.

[2] Ruble, R. and W. S. Treitel, "A Low Cost Regenerative Drive Scheme for Power FET Switching Converters," *Proceedings,* POWERCON 8, 1981.

6.5 Applying 240 Volt MOSPOWER Transistors and Current Limiting Diodes to Electronic Pulse Dialer Circuits (AN83-2)

Introduction

Siliconix MOSPOWER FETs and current-limiting diodes offer simple circuit solutions for operating in the transient-filled telephone environment.

Because of their high voltage rating and low ON resistance, MOSPOWER FETs are excellent devices for use as dial-pulse and muting switches in a dialer circuit. High-voltage current-limiting diodes with less than 1 mA forward current and high impedance work well in dialer circuits to protect low-voltage dialer chips and minimize circuit current requirements.

This application note discusses design considerations when using these devices and also describes device features, characteristics and advantages in dialer applications. An example application is included.

Design Requirements

The major design effort for dialer circuits using current-limiting diodes and MOSPOWER FETs focuses on meeting protection and operational requirements. Protection is needed for low-voltage semiconductor devices (such as dialer chips) that are subjected to high-voltage transients (surges) on the phone line. For example, Bell System publication PUB 48005, paragraph 5.2[1] specifies surge tests for telephone equipment. A portion of Table 5.1 of that document is reproduced here showing the surges applied between tip and ring.

Type	Peak Amplitude V	Peak Available Current A	Maximum Rise Time, μSec	Minimum Delay Time, μSec	Number of Surges of Each Polarity
M1 Normal	600	100	10	1000	4
M2 Normal (Part 68)	800	100	10	560	2
M3 Abnormal	600	100	10	2500	1
M4 Abnormal	1000	200	10	1000	1
L1 Normal	600	200	10	1000	4
L2 Normal	1000	200	10	360	2
L3 Normal (Part 68)	1500	200	20	160	2
L4 Abnormal	600	200	10	2500	1
L5 Abnormal	1500	400	10	1000	1

Figure 1

One source of transients is lightning surges. However, these transients rarely cause damage to the dialing circuitry since it is exposed only when the phone is off-hook (handset lifted from the cradle). For most phones, the off-hook condition occurs on average, a small percentage of the time. Of course, a direct or nearly direct hit may cause damage even if the phone is on-hook and regardless of the type of protective circuitry used.

Voltage transients also appear in the dialer circuit when the phone is answered while it is ringing. The PBX or central office switch can take 100ms or more to trip (disconnect) the ringing voltage after the phone is taken off-hook. During this time, the ring voltage is applied to the dialer circuit. This condition occurs about one-third of the time on incoming calls, with the normal 2-sec on, 4-sec off ringing cycle.

Paragraph 2.6.2 of PUB 48005 specifies ring voltages up to 130Vrms superimposed on a dc voltage of 0 to +105 V for bridged ringing. Connection of the ringer between tip and ring conductors is called bridged ringing. This ringer connection is normally used on single party lines. The magnitude of the voltage across the dialer circuit may be high or low, depending on the switch ring voltage, loop length, ringer impedance and number of ringers attached to the loop.

However, the major source of transients applied to the dialer circuitry is inductive spikes generated by dial pulsing into electromechanical switching equipment. The interruption of loop current flow during breaks in dialing causes large transient voltages produced by the collapsing magnetic field in the line-circuit relays of the telephone switching equipment. Since current flow cannot change instantaneously in an inductor, the voltage rises as high as necessary to maintain the current flow either by circuit components breaking down, contact arcing, or charging of circuit capacitances.

If this capacitance is low, the voltage rise can be substantial until breakdown or arcing occurs. For example, a capacitance of 0.005μF will allow transients of greater than 1500V. Each dial pulse causes a transient, and telephone dialing generates many such transients in rapid succession. On short low-resistance, high-current loops, high-energy pulses can be delivered to the dialer circuitry.

A standard approach for providing protection from inductive spikes is the use of a zener diode with high surge capability to limit the voltage impressed on the dialer circuit. Figure 2 shows allowable percent break and dial-pulse rate for dialers pulsing through a speech network with zener voltages below 300V. As the zener voltage is reduced, the percent break and dial pulse rate specs are tightened. It is desirable, therefore, to use the highest possible zener voltage consistent with protecting circuit components. Maximum zener voltage can now approach 240V for a corresponding

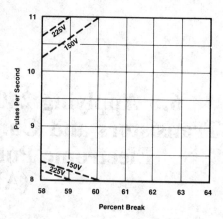

Percent break and pps limits as a function of zener voltage for devices which generate pulses. This diagram is used to determine the allowed area of operation of a dial for a given zener diode in the protective network. Thus, for the zener voltage of 225V the parameters of the dial pulses must lie in the hexagon bounded by the two sloping lines marked 225V, the 58 percent break line, the 8 pps line, the 11 pps line and the 64 percent break line.

Figure 2

increase in allowable operating area using high-voltage MOSPOWER FETs and current-limiting diodes.

Other important design considerations are loop current during breaks in dialing, dialer circuit current requirement and speech attenuation. Loop current must be below 1mA during the break period to maintain an acceptable break condition. Additionally the current requirement of the dialer circuit during talking periods must be low to prevent loss of efficiency of the carbon transmitter (if one is used) on long loops. A low supply current dialer circuit design is important for these reasons.

Speech attenuation is also important, especially for long loops. A dialer circuit operating in parallel with the speech circuitry should present a high shunt impedance to minimize speech level attenuation. A series dial-pulse or muting switch should exhibit low ON resistance for the same reason.

240-Volt MOSPOWER Device

Important attributes of MOSPOWER devices in dialer applications include high gate-input impedance, low ON resistance, high breakdown voltage, high current capability and low cost. The Siliconix VN2410M, for example, was designed for applications such as dialer circuits. As in all MOS devices, the gate requires only minute continuous drive current to hold it ON, unlike bipolar transistors. Gate current flows when the gate input capacitance charges during a change of state. The only other gate-current components is a few nanoamperes of input leakage current.

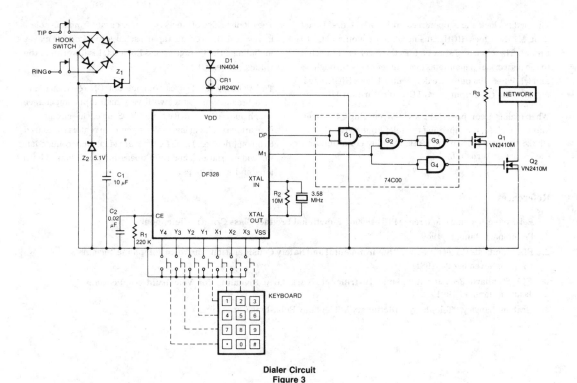

Dialer Circuit
Figure 3

Drain-to-source ON resistance for the VN2410M is specified as 10 ohms maximum. If the device is placed in series with a speech network, this low ON resistance contributes negligible speech-signal attenuation on long loops. The device breakdown voltage rating of 240V allows the use of protective clamping devices with voltage ratings approaching 240V. High voltage clamping allows a relaxation in percent break and dial-pulse rate limits relative to circuits using lower voltage clamping.

Other devices in the series have breakdown voltages of 170V (VN1710M) and 120V (VN1210M) for lower voltage requirements. The 0.25A continuous drain-current rating can handle full loop current on short loops. These devices are packaged in a TO-237 plastic package which is equipped with a power tab. The TO-92 package is available for lower power applications.

240-Volt Current-Limiting Diode

The JR240V current-limiting diode has a breakdown voltage rating of 240V. Other available breakdown voltage ratings are 220V (JR220V), 200V (JR200V), 170V (JR170V) and 135V (JR135V). These devices which are supplied in a 2-lead TO-92 plastic package feature a forward-current range of 200 to 770μA. Thus when used in series with a dialer circuit, they limit current to less than the required 1 mA during breaks. Since dynamic impedance is typically 2 megohms, there is no speech-signal attenuation.

Dialer Circuit Example

Figure 3 shows a dialer circuit using MOSPOWER devices for the dial-pulse switch and muting functions plus a current-limiting diode for dialer-circuit current regulation and protection. Dial-pulsing is performed through a resistor connected in shunt with the speech network. This circuit features a minimum number of components and low-current operation. The dialer chip is the Siliconix DF328, but other IC's can be used in similar designs.

Lifting the handset applies power through the diode bridge to the dialer circuit. Current flows through D1, CR1 and Z2, establishing Vdd for the DF328 and charging C1. When the minimum operating voltage (Vdd) is reached, power-on reset occurs via the CE network of C2 and R1. Initially, both DP and M1 are LOW, providing a LOW at the gate of Q1 to hold it off. A HIGH at the gate of Q2 turns it on, connecting the telepone network.

The current-limiting diode CR1 serves two purposes. First, it limits the total dialer-circuit current drawn from the loop to less than 1 mA, and second, it maintains a high dialer-circuit shunting impedance across the telephone set network. Z1 is a high-voltage zener diode with high surge capability that protects against loop transients and office inductance spikes. The device should limit all transients to less than the breakdown voltages of CR1, Q1 and Q2 (240V).

6

The DF328 clock starts upon recognition of the first keyed digit. M1 then goes HIGH, causing Q2 to turn off and Q1 to turn on. This change of states mutes the receiver connected to the network and maintains loop current flow through R3 and Q1. When dial pulse breaks occur, DP goes HIGH, causing the gate G3 output to go LOW and turn off Q1.

When dialing is complete, M1 goes LOW, causing Q1 to open and Q2 to close, reconnecting the telephone network. The DF328 then returns to the static standby condition and the oscillator turns off.

The diode bridge protects the dialer circuit from line polarity reversal. D1 prevents rapid discharge of C1 during makes in dialing if the voltage across R3 and Q1 is lower than the voltage across C1.

The MOSPOWER devices combine low ON resistance and high breakdown voltage with very high input impedance which allows direct drive from CMOS logic. This circuit configuration results in low dialer-circuit current as compared to bipolar devices. These features are ideal for accommodating wide variations in circuit operating conditions of both long and short loops.

References

1. Bell System Technical Reference PUB 48005 **"Functional Product Class Criteria - Telephones"** Preliminary, January 1980.
2. EIA Standards Project PN-1361 **"Environmental and Safety Considerations for Voice Telephone Terminals,"** Draft 5, November 24, 1981.
3. EIA Standard RS-470 **"Telephone Instruments with Loop Signaling for Voiceband Applications,"** Issue 1, January 1981.
4. Graham Langley, **"Telephony's Dictionary"**, Telephony Publishing Corp., 1982.

6.6 Audio Amplifiers

6.6.1 The Autobias Amplifier:
A New (AC Coupled) Topology for Automatically Biased Audio Amplifiers Using Power MOSFETs (TA82-1)

Abstract

An obstacle blocking wide acceptance of power MOSFETs in audio amplifiers is the lack of an automatic bias technique. A unique circuit topology senses and maintains quiescent current despite the half-wave pulses inherent in class AB operation. Performance of the circuit, consisting of little more than a differential amplifier driving a totem pole output, rivals that of more complex circuits.

Introduction

MOSPOWER FETs offer a number of significant improvements in the characteristics of interest for amplifiers as compared to their bipolar counterparts, namely:

1. MOSPOWER transistors are comparatively immune to second breakdown because no transverse current flow exists, as in a bipolar, to cause current concentrations. Therefore, no complex power limiting or protection circuitry is needed.

2. The temperature coefficient of transconductance is negative. Consequently, design of thermally stable circuits is easy and devices may be readily used in parallel for increased output current.

3. The transfer curve (I_D vs. V_{GS}) is linear over most of the operating current range of the transistor which produces a low distortion output. Furthermore, the curvature which is present adheres approximately to a squarelaw, so that the resulting even-order products may be cancelled by a push-pull connection.

4. The capacitances are low which allows high frequency open-loop response. Also, FETs do not exhibit minority carrier storage time since conduction is solely by majority carriers. Consequently it is simple to build amplifiers which do not exhibit the various forms of transient distortions.

The chief problem in using MOSPOWER is providing the proper biasing. The average gate voltage must be held a few volts positive with respect to the source, depending upon current desired and the characteristics of the transistor, especially the threshold voltage. Unfortunately, threshold voltage not only shows a dependence upon temperature (approximately $6\,mV/\,°C$), but is subject to lot-to-lot variations. Consequently, some kind of feedback scheme must be used to maintain the idle current despite variations on the order of 2 volts in the required gate voltage. The unresolved problem to date is how to sense the output idle current in the presence of the high current half-wave pulses which occur under class AB operation. The solution of this problem is the subject of this article.

The Auto Bias Scheme

The method discovered which accomplishes biasing is shown in Figure 1. In this scheme, the bias idle current is maintained by comparing the voltage obtained from the current sense resistor to a reference level. When the signal is large enough to cause cut-off of the current on negative half-cycles, thereby producing an asymmetrical waveform, the peak clipper and filter produce a voltage which is substantially the same as the zero signal level, regardless of the amplitude of the peak current. For proper operation, the DC level caused by the idle current must be set to one-half the level of the peak output from the clipper as shown in Figure 2. Note that the current waveform of Part A has an average level proportional to the peak current while the output from the clipper remains essentially constant as shown on B. Proof that this scheme really works will be given with the data obtained from the practical designs.

Reprinted by permission of the Audio Engineering Society, April 1982.

6

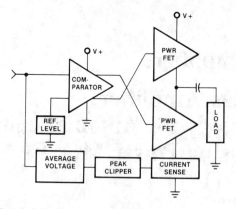

Automatically Biasing (Autobias) Scheme for Single Supply Class AB Amplifiers
Figure 1

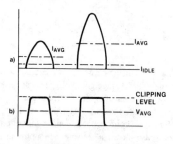

Waveform of a) One Output Transistor Current and b) the Clipper Output. The Varying Average Level of the Current Wave Causes Negligible Shift in the Average Level of the Clipper.
Figure 2

General Description of Amplifier Topology

The objective of this project was to produce a simple circuit which would fully utilize the advantages of MOSFETs and not require any adjustments. Consequently, a direct coupled scheme using complementary devices, so much in vogue with bipolar amplifiers, was rejected as being too complex, especially when a second differential amplifier to control the quiescent current would have to be added. In addition, p-channel FETs suffer from the severe disadvantage that holes are lower mobility carriers than electrons. The practical result is that to produce a p-channel FET with gain equivalent to a given n-channel FET, the die area must be about twice as large, causing a corresponding increase in capacitances and higher cost. The topology chosen is a direct implementation of the block diagram shown in Figure 1. As shown in Figure 3, the circuit using npn transistors and n-channel FETs consists of a differential amplifier driving a push-pull totem pole output configuration. The diff-amp controls the bias current of Q_4 which acts as a current source for Q_3. This arrangement necessitates AC coupling to the load even if a split supply were used because the voltage at the drain of Q_4 is affected by several component values.

Bias is accomplished by comparing the voltage drop across R_{21} to the voltage from the reference diode, D_1. To achieve accuracy in setting the bias, the DC resistance in the base circuits of Q_1 and Q_2 must be equal (i.e., R_{10} in parallel with $R_{11} = R_2 + R_{16} + R_{17}$ and the h_{FE} of Q_1 and Q_2 should be matched). Matching of V_{BE} is not important because of the drops across R_3 and R_4 and the base circuit resistance. Should the current in Q_4 tend to increase, the resulting drop across R_{21} coupled to Q_1 lowers its collector voltage (V_{C1}) which is coupled through R_6 to the gate of Q_4, thereby holding the current of Q_4 nearly constant.

Under large signal conditions, the high current peaks must be clipped and the waveform filtered. The best place to clip is right across R_{21}. Diode D_3 performs this function and also provides dynamic bypassing. In this location, power output is maximized, as the total voltage loss across the diode is only about a volt. Total harmonic distortion also is less with the diode in this position; apparently the decreasing diode incremental resistance with increasing current compensates for a nonlinearity in the MOSPOWER FET. The diode introduces a generous number of low level high order harmonics, but a considerable reduction of the second and third harmonics occurs. R_{18} and R_{20} often can be chosen to achieve a minimum in the total harmonic distortion; their value determines the composition of harmonic content. R_{18}, R_{19} and D_2 balance both halves of the push-pull stage. At low signal levels, particularly when Q_3 and Q_4 are not well matched, R_{19} and R_{21} cause a reduction in even-order distortion products. The resistor R_{17} and diode D_4 form a second stage clipper which keeps the peak output quite constant regardless of the audio signal level across R_{21}. R_{16} and C_5 form a low pass filter to prevent the clipped waveform from being mixed with the input audio which is also applied to the base of Q_1. Distortion increases considerably if C_5 is eliminated.

To provide maximum power output from a given supply, the do level at the drain of Q_4 (V_{D4}) must be one-half of the power supply voltage (V_A) at full signal output. This level is controlled by the divider composed of R_{13}, R_8 and R_9. The level V_{D4} is thus determined by the fixed level placed on the gate of Q_3 minus the V_{GS} drop of Q_3, which usually is consistent within two volts from a given MOSPOWER production line.

The differential amplifier acts as a voltage comparator to maintain the DC bias level and as a phase splitter to drive the MOSFET output transistors. The interstage circuits are somewhat unusual. The load for Q_2 is R_7, but the signal from Q_2 must be referenced to the source of Q_3 since it is driven as a common source amplifier. This is accomplished by the bootstrap circuit R_{13} and C_4. To maximize R_7 for higher gain, R_{13} must be small, on the order of 1 KΩ; this requires C_4 to be about 400μF for satisfactory low frequency response.

Note that the divider composed of R_8 and R_9 is also boot-strapped by C_4 and R_{13} to provide benefits which are not generally obvious. The distortion is reduced slightly, but of more importance is the reduction of the turn-on current surge caused by charging the output coupling capacitor C_8. The time constant of R_{13} and C_4 determines the surge magnitude and duration.

The load resistors for Q_1 and Q_2 are R_5 and R_7, which should be matched for the lowest distortion. The DC level at the collector of Q_1 is dropped by R_6 to the appropriate level at the Q_4 gate. Since current through R_6 is supplied by a low valued constant current source, negligible loss in DC gain occurs through the coupling network. However, capacitor C_2 is needed to prevent high frequency rolloff caused by the time constant composed of the input capacitance of the MOSPOWER FET and R_6.

A feedback signal is developed by the R_{23}-R_{24} divider and applied to the base of Q_2 via C_7. Resistor R_{22} works with R_{23} and R_{24} to set the voltage V_{D4} to half of V_A during troubleshooting and these resistors also perform the function of a bleeder for the power supply filter capacitor. Resistor R_{25} insures that the output coupling electrolytic (C_8) has a charging current path should the amplifier be turned on with speakers disconnected. Feedback could be taken from a tap on R_{25}; this lowers distortion at low frequencies but overload recovery is poorer.

Since MOSPOWER is not plagued by second breakdown, involved overload protection circuits are not required. However, gain does not decrease with operating current, so enormous output current could be developed into a short circuit load. To prevent this, the zener diodes Z_1 and Z_2 are used to prevent excessive gate drive. Although one diode per gate will limit the peak current, in some cases the circuit can cope better with large input signal overloads if two diodes are used in series. They should be chosen to allow equal positive and negative swings about the nominal idle gate voltage.

Since most MOSPOWER devices have a cut-off frequency in the giga-cycle range, parasitic RF oscillations can be troublesome. Most assemblies require an RF bypass C_6, on the order of 0.22 μF, physically connected as shown. In addition, R_{14} and R_{15} (100Ω or so) are included as parasitic suppressors. Ferrite beads could be used in place of the resistors. In either case, the suppressors should be mounted closely to the gate terminal. In addition, leads running to the gate terminal should be either shielded or used as a twisted pair with a ground lead.

Design Criteria

In the simple circuit of Figure 3, primary tradeoffs occur between output slew rate, input impedance, matching requirements and bias point accuracy. For example, with a given pair

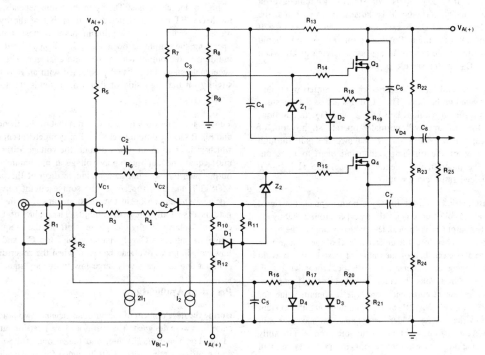

Basic Autobias Circuit
Figure 3

of output transistors an increase of slew rate requires that the current in the diff-amp increases. The higher currents through the resistors in the base circuits of Q_1 and Q_2 produce larger voltage drops with a greater likelihood of an error in the output idle current unless the h_{FE} matching between Q_1 and Q_2 is tightened or the values of the resistors in the base circuit are proportionally reduced. If the resistors are reduced, then to preserve the low frequency response, capacitors, C_1, C_5 and C_7 must increase at an added cost which becomes significant if values over 1 μF are necessary, since it is desirable to avoid electrolytics.

The values of the components in the bias feedback loop from R_{21} to the base of Q_1 should be proportioned in a particular manner, since both bias and audio signals are present at the Q_1 base. A large signal which overloads the diff-amp disturbs the idle current setting. Best recovery was empirically determined to occur when the capacitor values for C_1, C_7 and C_5 are equal and R_2 is at least three times R_{16}. If C_5 is made proportionally larger, the point where low frequency distortion starts to increase can be made lower; however, the idle current recovery waveform begins to assume the character of a damped oscillation.

In order to prevent a significant portion of the DC level on C_5 from biasing the diode D_4, R_{16} should be at least twenty times R_{17}. R_{17} is chosen so that the clipped output voltage from D_3 sends a current through D_4 such that the peak level from D_4 equals twice the open circuit level at the base of Q_2. R_{17} should be large enough such that the peak level across D_4 is essentially constant regardless of the amount of current flowing through R_{21} and should produce the same current in D_4 as is flowing in D_1 when D_1 and D_4 are the same type diode.

Resistors R_3 and R_4 used in series with the emitters of Q_1 and Q_2 provide two benefits. The DC drops across them lessens the V_{BE} match required between Q_1 and Q_2. In addition, since the diff-amp will normally operate slightly out of balance, these resistors tend to balance the gain through both sides of the diff-amp. Choosing them to be about twice the junction incremental resistance ($25 \ \Omega/I_E(mA)$) has yielded satisfactory results.

Because the load resistor for Q_2 is bootstrapped, the signal swing at the collector of Q_2 with respect to ground slightly exceeds the output peak-to-peak level which is close to the power rail voltage. Therefore, the quiescent level of the voltage V_{C2} must be above one half of the rail by at least 3 volts to avoid signal clipping or nonlinearity of the peak negative signal. Another 2 or 3 volts should be allowed for diff-amp unbalance. These requirements force the maximum value for $R_{13} + R_7$ to be $((V_A/2) - 6)/I_1$, assuming that the current drawn by the R_8, R_9 divider is negligible. To maximize gain and prevent diff-amp current fluctuations from significantly influencing the voltage at the gate of Q_3, R_{13} should be small compared to R_7; however, the smaller R_{13} becomes, the larger C_4 must become in order to have adequate low frequency response and satisfactorily limit the turn-on current surge as C_8 charges.

R_5 must equal R_7 to keep the diff-amp in balance electrically. For thermal balance V_{C1} should equal V_{C2} by satisfying the equation $(I_1 + I_2)R_5 = I_1(R_7 + R_{13})$. R_6 is chosen to place the proper gate voltage on Q_4 by consulting typical data for the particular MOSFET being used.

The correct value for V_{G4} depends on the idle current required to minimize crossover distortion from the MOSFET. This is empirically determined and must be about 300 mA for the transistors used in the designs shown later. Using ordinary silicon diodes for the clippers yields a clipping level of 0.6 to 0.7 volts depending upon the diode characteristics. This dictates that the reference level on the base of Q_2 should be from .3 to .35 volts. If $R_{10} = R_{11}$ and D_1 and D_4 are the same type and operated at the same current, these requirements are met and a 1Ω resistor for R_{21} will set the idle level between 300 and 350 mA. To preserve the balance of the push-pull output, R_{19} must equal R_{21} and correspondingly, $R_{18} = R_{20}$ and $D_2 = D_3$.

The value for R_{20} and R_{18} is empirically determined. If R_{20} is greater than twenty times R_{21}, its effect on distortion and power output is negligible; the output spectrum will be relatively free of high order harmonics but maximum power output will not be achieved. As R_{20} approaches the value of R_{21}, the "dynamic bypass" action becomes noticeable, power output increases and a marked reduction in 2nd and 3rd harmonics occurs. This results in a decrease of total harmonic distortion, but the output spectrum will have minute amounts of high order harmonics. The optimum value depends upon the MOSFET characteristics as well as that of the bypass diode. A search for the optimum point is best done by operating the amplifier open loop, i.e. R_{24} shorted, and monitoring the output with a spectrum analyzer. The most acceptable spectrum is usually obtained with an R_{20} value slightly on the high side of that which yields the lowest THD.

The remaining network to be designed is that of the divider R_8 and R_9. It sets the output level V_{D4}. For symmetrical clipping, which yields maximum power output, the voltage at the gate must equal one-half the supply voltage at maximum power output plus the nominal gate to source drop of the power MOSFET and the drop across the source circuit network. The resistances can be in the megohm range, as the FET gate is essentially an open circuit. The value of C_3 is chosen to have negligible phase shift in the frequency range of interest. Overly large values for C_2 and C_3, however, improve overload behavior when the zener diodes conduct due to excessively large low frequency signals.

Practical Applications

During the development of the autobias scheme, two practical amplifiers were designed. The first is a 25 watt version intended for home high-fidelity use, and the second is a 50 watt design of lesser power bandwidth intended for public address use.

A 25 Watt Design

Figure 4 shows a practical 25 watt amplifier. Transistors are used for the current sources shown in Figure 3. Base drive for these transistors is derived from the main power supply V_A, so that their collector current is proportional to the rail voltage. This feature holds the voltages on the diff-amp collectors close to $V_A/2$. The sensitivity of I_Q to V_A is about 3.4 mA/volt when V_B is held constant; the sensitivity of I_Q to V_B is -15 mA/volt when V_A is held constant. In a practical amplifier with a non-regulated supply, variations in power output will cause fluctuations in V_A, but will not affect V_B; therefore, having I_Q increase slightly with power output as discussed later will tend to compensate for the 3.4 mA/volt I_Q/V_A sensitivity. In the case of line voltage variations, since V_A is about five times V_B, the sensitivities tend to cancel, leaving a net sensitivity of about 2 mA/volt.

The circuit arrangement causes ripple cancellation to take place. Ripple from V_A is applied via R_5 and R_6 to the gate of Q_4. An out of phase ripple component is also applied to the gate via the current source Q_6. By properly proportioning the filter components of the supplies V_A and V_B, output ripple can be reduced to a level acceptable for most applications without the use of unduly large capacitors or an extra filter section connected between R_5 and V_A. The bootstrap capacitor (C_4) effectively filters the ripple from the gate of Q_3. The filter for V_A should be at least 5000 μF.

The sensitivity of the MOSFET quiescent current, I_Q, to threshold voltage ($V_{GS(TH)}$) depends upon the loop gain of the bias stabilization portion of the circuit. With the components used in Figure 4, the sensitivity is about 16 mA/volt, which is felt to be satisfactory to accommodate the typical 2 volt tolerance of a MOSPOWER production line.

It is not possible to measure the change in idle current level as the power output from the amplifier is varied using the usual sine wave test signal because the idle level is obscured by the output current. However, waveforms similar to the ones shown on Figure 5 can be fed into the amplifier; by viewing the voltage across R_{21} during the time of the zero level step, the idle level is visible. It is found that a fairly precise setting of the voltage across R_{21} is necessary if the idle level is to remain invariant with power output changes.

An increase of idle level with power output is a result of the average voltage output from the clipper and filter decreasing with signal level. The decrease results in the diff-amp raising the gate voltage to compensate for this apparent decrease in idle level. A small amount of this behavior is desirable in that it compensates for the decrease in idle level experienced as the supply voltage, V_A, drops due to increased audio power output. Should the clipper output increase with power output, the circuit will reduce the idle current level. This situation yields improved power efficiency, but the output may show evidence of cross over distortion at the higher power levels. For optimum performance, it is therefore necessary that the clipping level be maintained at slightly below twice the zero signal idle voltage from the clipper, regardless of the power output from the amplifier. The two stage clipper used is necessary to reasonably achieve this goal.

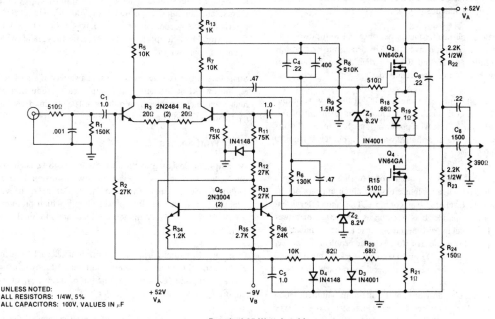

Practical 25 Watt Autobias
Figure 4

6

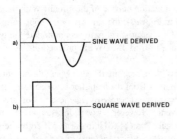

a) ——— SINE WAVE DERIVED

b) ——— SQUARE WAVE DERIVED

**Test Waveforms Used to Observe Idle Level
Under Large Signal Conditions
Figure 5**

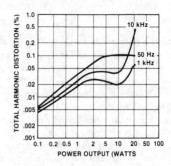

**Distortion vs Power Output for 25 Watt Amplifier
Figure 6**

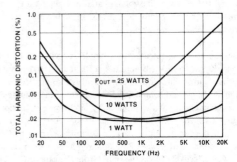

**Distortion vs Frequency for 25 Watt Amplifier
Figure 7**

Identical tests were run on an amplifier at ambient temperatures of 25 °C and 45 °C. The current at all power levels dropped about 22 mA at 45 °C, a predictable result because the reference diode, D_1, has a temperature coefficient of 2 mV/°C, which is divided in two by the resistors R_{10} and R_{11}. Distortion as a function of power output remained essentially unchanged at the test frequencies of 75 Hz, 1 kHz, and 5 kHz.

Figure 6 shows total harmonic distortion at selected frequencies as a function of power output and Figure 7 shows distortion at 1.0, 10 and 25 watts as a function of frequency. In the important mid-range, THD is under 0.1%. The increase in THD at high frequencies is caused by the extra drive required of the diff-amp to handle the MOSFET input capacitance, while the increase at low frequencies is caused somewhat by the coupling capacitors, primarily the output capacitor (C_8), but mostly is attributed to the bootstrap capacitor (C_4), and the bias filter capacitor (C_5).

The usually published frequency response at 1 watt power output is not shown as it conveys little information. Response is down 1/2 dB at 19 Hz and 100 kHz when driven from a 1 KΩ source with the input filter removed.

The power efficiency calculates at 53% at the 25 watt output level. The power dissipation is essentially independent of frequency and varies little with power output. It is about 17 watts at idle and increases to about 22 watts at output levels from 10 to 25 watts.

Most assemblies show no evidence of slew rate distortion until at least 30 kHz when a slight crossover glitch appears on the waveform at a level of 15 watts or more. The glitch is level sensitive due to imperfect bias tracking as a function of power output. Bias tracking also becomes worse as the frequency increases, probably because of stored charge problems in the rectifier diodes used for D_2 and D_3. This tracking error would normally not be encountered on speech or musical signals.

Choice of suitable transistors for the diff-amp is limited. Good results have been obtained by using matched pairs of type 2N2484. The gain typically is about 400 at 2 mA and although the 60 volt V_{CEO} rating seems marginal, no problems have been experienced.

The output devices are Siliconix type VN64GA, which typically have a transconductance of 3 mhos, an $r_{ds(on)}$ of 0.3 Ohms, and an input capacitance of 700 pF. These characteristics are needed in order to have sufficient gain in the current bias loop, good power efficiency, and wide open loop bandwidth.

Some listening tests have been run using a variety of associated equipment. Most listeners notice improved reproduction of high frequency transients and have difficulty in detecting overloads and clipping unless excessive.

A 50 Watt Design

An amplifier designed to produce 50 W into 8 Ω is shown on Figure 8. For this application low distortion is only required over a range from 50 Hz to 10 kHz. It was adapted to use an available power supply which produces + 78 V and − 52 V under no load and nominal line conditions.

Figure 9 shows total harmonic distortion at selected frequencies as a function of power output and Figure 10 shows distortion at 1, 10, and 50 watts as a function of frequency. In the important mid range, THD is under 0.1%.

The narrower power bandwidth of the 50 watt amplifier as compared to the 25 watt amplifier is a direct result of the higher input capacitance of the higher power output devices.

To avoid serious high frequency distortion the diff-amp current was increased from 2 mA to 4 mA; the higher current necessitated a 2:1 reduction in base circuit resistance in order to avoid an excessively tight match on the current gain of the diff-amp pair with the result that the bypassing action of C5, kept at 1 μF, is less effective at low frequencies. The narrower power bandwidth of the 50 watt amplifier does not reflect a significant difference in the frequency response at 1 watt as compared to the 25 watt design.

Although not shown, data at 5 watts output is similar to that at 10 watts. The low frequency distortion is higher at these levels than at 50 watts because of the imperfect filtering action of C5. Because of the clipper circuit, the voltage on C5 is a larger percentage of the input signal as the output drops from the 50 watt level.

The harmonic distortion spectrum is shown on Figure 11. The dynamic range of the instrument used is 90 dB. Note that the third harmonic is prominent at all power levels. The even harmonics quickly disappear, but at the 50 watt level, odd harmonics up to the 13th were detectable. Probably none of these harmonics is discernable by ear.

Power data is shown on Figure 12. The power efficiency calculates at 63% at the 50 watt output level. The power dissipation is essentially independent of frequency. The heat sink must handle about 40 watts of power.

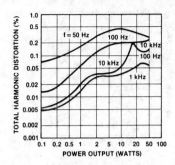

Distortion vs Power Output for 50 Watt Amplifier
Figure 9

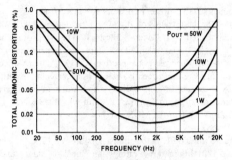

Distortion vs Frequency for 50 Watt Amplifier
Figure 10

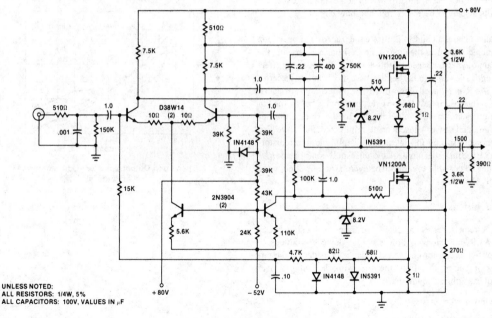

UNLESS NOTED:
ALL RESISTORS: 1/4W, 5%
ALL CAPACITORS: 100V, VALUES IN μF

Practical 50 Watt Autobias
Figure 8

6

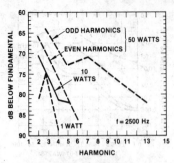

Locus of Harmonic Spectrum
Figure 11

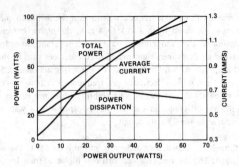

Power Requirements
Figure 12

Finding suitable driver transistors for the diff-amp was not easy. The best transistor discovered to date is a D38W14 manufactured by General Electric. The transistor has $BV_{CEO} > 100V$, which allows ample voltage margin, and $h_{FE} > 400$ which places base current $< 10~\mu A$ at 4 mA of collector current. Consequently, base current matching to within $1~\mu A$ is not too difficult and this match will produce a maximum error of 20 mV in the diff-amp which translates into a 20 mA error in the MOSFET idle current.

The MOSPOWER output transistors are VN1200As. They have 120 V breakdown ratings, an "on" resistance under 0.2 ohms, transconductance of 5 Siemens, and an input capacitance of 1200 pF. This excellent combination of characteristics plays a major part in achieving the excellent results displayed by the simple autobias circuit.

Finally, Figures 13 and 14 illustrate the tradeoffs in open loop harmonic content as the diode series resistance is varied in the dynamic bypassing scheme. With $R = 47\Omega$, the bypassing effect is negligible, and no harmonics past the 11th are discernible. As R is reduced, high order harmonics are introduced which increase in level. However, harmonics below the 6th minimize at various values of R. In the final amplifier design, $2/3\Omega$ was chosen as it minimized the large 3rd harmonic resulting in lowest overall total harmonic distortion. By better matching of the push-pull configuration, it should be possible to reduce the even harmonics below the levels shown.

Conclusions

The circuit scheme presented illustrates that using MOS-POWER transistors as power output devices produces an amplifier of extraordinary performance considering the circuit simplicity. It offers the following advantages over bipolar amplifier counterparts:

1. Only one driver stage

2. Simple overload protection

3. Stable bias point

4. Power efficiency independent of frequency.

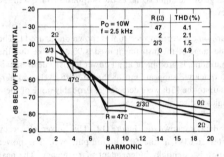

Open Loop Output Spectrum Even Harmonics
Figure 13

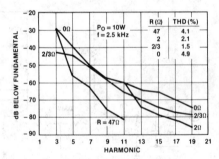

Open Loop Output Spectrum Odd Harmonics
Figure 14

6.6.2 A Simple Direct-Coupled Power MOSFET Audio Amplifier Topology Featuring Bias Stabilization

Abstract

Utilizing the high gain and high input impedance of short channel power MOSFETs, a simple circuit has been devised to provide sufficient drive for amplifiers up to 200 watts output. The circuit described features automatic control of the quiescent bias level and offers performance which meets criteria for high quality audio reproduction.

Introduction

Short-channel power MOSFETs offer several characteristics attractive for audio power amplifier applications namely:

1. Low distortion at all frequencies because of a linear transfer characteristic and a very high cut-off frequency.
2. A low drive power requirement because of a nearly infinite input resistance, relatively low capacitance, and high transconductance.
3. Failure immunity because second breakdown occurs well above the power rating.
4. Freedom from minority carrier storage delay time following signal overloads because FETs are strictly majority carrier devices.
5. High power efficiency because of low "on" resistance.

One of the major design problems has been providing a means of stabilizing the idle current level because short-channel FETs exhibit an increase of drain current with temperature at the desired idle level. Furthermore, as a result of production line variations a threshold voltage range on the order of 2 volts must be accommodated. Consequently, these FETs must be used within a closed loop bias system. The puzzler has been to provide a means of sensing the idle level despite the presence of the high current half-wave pulses which occur under Class–AB operation. A successful solution to the bias problem was the subject of an earlier paper by the author [1], however, the simple circuit in that paper used AC coupling of the audio signal. The circuit described in this paper applies this bias technique to a direct-coupled design. The resulting circuit is simple and offers outstanding performance.

The Autobias Principle

The scheme previously reported extracts the idle current from the total current by means of a clipper circuit and filter which produces a voltage related to the idle level. A small resistor in series with the source of one of the power FETs is used to sample output current. The voltage at idle is one-half of the clipping level so that when the signal is large enough to cause current cut-off on negative half cycles (thereby producing an asymmetrical waveform) the peak clipper and filter produce an average voltage which is substantially the same as the zero signal level, regardless of the amplitude of the peak current, as illustrated by the waveforms of Figure 1. The filtered voltage is used

6

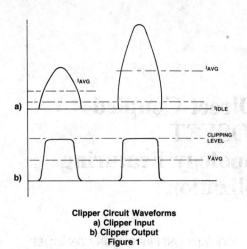

Clipper Circuit Waveforms
a) Clipper Input
b) Clipper Output
Figure 1

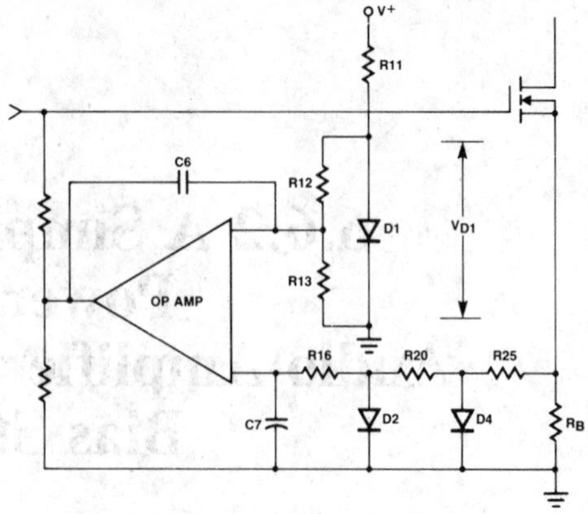

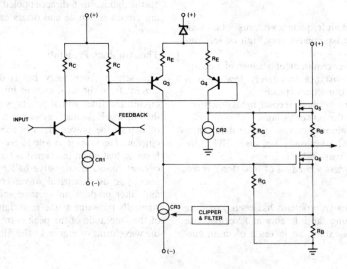

Op-Amp Bias Control for Class AB
Figure 2

in a feedback loop to control the FET gate voltage and thereby maintain the bias current within close limits.

A simple technique for achieving a stable automatic bias current is shown in Figure 2. The op-amp serves as a differential amplifier in a voltage comparator loop which serves to maintain the levels at its inputs equal. Thus the idle current is simply $V_{D1}/2R_B$, since R_{12} and R_{13} are equal. The loop gain is so high that variations of FET threshold voltage or other circuit constants have no measurable effect upon the idle current level. It is important to use a two stage clipper and also match the diodes D_1 and D_2 to avoid shifts in idle level under large signal conditions; a dual matched signal diode is an ideal choice. Circuit resistors should be chosen to force equal currents through

D_1 and D_2 under large signal conditions. In addition, R_{16} should be over twenty times R_{20} to prevent a significant portion of the DC level on C_7 from biasing D_2. Capacitors C_6 and C_7 filter the clipped audio signal.

Circuit Topology

The general circuit topology is shown in Figure 3. It consists of an input differential amplifier using n-p-n transistors directly coupled to a p-n-p push-pull stage which drives the output power FETs in a totem pole arrangement. A similar topology has been previously used by Sampei, et al [2], and Harvey [3], however, the implementation is different and neither circuit has automatic bias.

Amplifier Topology for Power MOSFETs
Figure 3

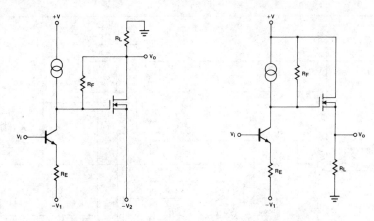

A Real Source Follower and a Quasi Source Follower
Figure 4

Sampei used complementary output devices whose gates are driven from a common node. The capacitive loading at the gate node is about three times higher than would be the case if n-channel FETs were driven as shown in Figure 1, because a p-channel device has about twice the input capacitance of an n-channel transistor with the same gain and on-resistance. This unfortunate situation is a result of having to construct a larger p-channel die to overcome the lower mobility of the p-type carriers. The circuit used by Harvey uses a complicated arrangement of current sources and emitter followers to achieve high gain from the input diff-amp while the p-n-p drivers act primarily as level shifters and current drivers operating at low gain.

Both designs provide excellent performance, but for bias stability, FETs having a relatively long channel should be used so that the temperature coefficient of drain current is zero at the proper bias current for class AB operation. This type of FET design has poorer linearity, lower gain, and higher "on" resistance than the short channel devices now being produced in volume for general purpose applications. The design in this paper accommodates any type of FET and provides an automatic closely regulated bias control.

Totem pole output designs have not been used extensively in high quality amplifiers because of the inherent unbalance in input resistance which occurs. This is not a problem with FETs; however, an unbalance in driver frequency response occurs because the signal swing at Q4 is several times larger than that at Q3, which results in differing Miller capacitances. This effect was masked in bipolar designs by the poor frequency response of the output transistors but shows up when FETs are used as a small phase shift which can cause a crossover glitch at high frequencies. The

problem is easily corrected by using Miller compensation capacitors of a proper ratio on Q3 and Q4. In the design shown, crossover distortion is barely evident at 100 kHz.

Circuit Implementation

Figure 5 shows a complete circuit with values suitable for a 50 watt output into an 8 Ω load. The bias control loop consists of the op-amp, CR3, Q6, and associated components. The op-amp serves as a high gain differential amplifier and integrator which further filters the output of the clipper diodes, D4 and D2, and the filter composed of R16 and C7. The op-amp controls the current source CR3 which maintains the required gate voltage for the idle current (IQ) desired. The best idle level is found by measuring distortion at a low level where class AB operation begins to occur. For the power transistors used, the effect of idle level upon low level distortion is shown in Figure 6. When making this test, it is important to select a power level where operation is beginning to move from class A to class AB.

The topology has the interesting property of allowing the output stages to operate either as common source or source follower amplifiers by simply changing the return point of the gate resistors. Figure 4 shows the follower configuration. In an amplifier with a fairly large amount of feedback, a recent paper [4] shows that the follower connection offers no advantage over a gain stage, however the follower may be useful if a high gain amplifier with a low output impedance is desired without using an overall feedback loop. Closed loop distortion tests using either configuration yielded essentially the same results with the amplifier in this paper.

6

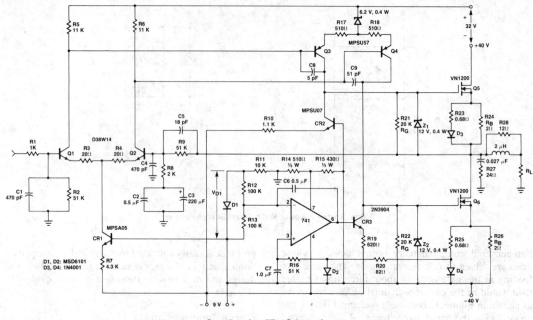

Complete Amplifier Schematic
Figure 5

The loads for the p-n-p drivers are provided by current sources (CR2 & CR3) and high-valued load resistors (RG) R21, R22. This technique permits high gain to be achieved while allowing all the driver current to charge the input capacitance of the output devices. The drivers may work class AB when the input is a fast rising signal thereby providing a high peak current in phase with the drain current to provide a high slew rate.

The drivers and the diff-amp stages are designed so that they cannot be driven into saturation. To do this requires that the stages load resistances and quiescent currents be chosen to limit the available voltage swing; to achieve reasonable gain from the diff-amp requires a driver supply voltage which at first seems unnecessarily high (32 V) but is readily available from a standard 24 volt transformer. The current drain from this supply is fairly low, however, requiring only 8 mA for each driver plus 2 mA for the diff-amp.

In addition, the driver supply is non-critical with regard to hum or absolute value. It is possible to obtain this supply by bootstrapping from the amplifier output, but a very large capacitor would be required to avoid a serious increase in low frequency distortion. The result is that bootstrapping works out to be less satisfactory and more expensive than the approach shown. The supply need not ride on the +40 V rail. It is just as satisfactory to have a +72 V supply with respect to ground; it could serve both channels of a stereo pair.

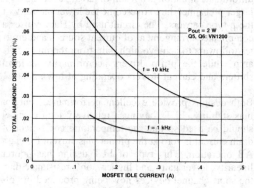

Effect of Idle Level on Harmonic Distortion
Figure 6

The −9 V line which rides on the −40 V rail also supplies about 20 mA and must be well filtered to prevent AC ripple from being amplified by CR3 and injected into the gate of Q6. A zener diode is an effective filter and also provides a reference level for the current source transistors CR1 and CR2. Neither the absolute value nor the regulation of the other supplies is critical. Any desired power output may be obtained from the circuit by simply altering the value of the main rails. Automatic switching between two supplies could also be used to improve power efficiency. No circuit changes are necessary other than altering the divider to the supply terminal, pin 7, of the op-amp to avoid exceeding its voltage rating (30 V) and using transistors with sufficient voltage breakdown.

The diode D4 serves not only as part of the bias clipper network but also provides dynamic bypassing of the 2 ohm sense resistor. The diode reduces third harmonic distortion and increases power output at the expense of introducing a number of low-level high order harmonics into the output. This action is illustrated in detail in reference [1]. A similar network must be used in the source circuit Q5 to balance both halves of the push-pull output stage.

The zener diodes Z_1 and Z_2 provide overload protection with a shorted output. They are chosen to limit the gate drive to a level such that $I_D < P_{D(rated)}/V+$. In addition, the zeners prevent large voltages from breaking down the FET gate-oxide layer. The zeners also conduct in the normal diode forward direction when the gate signal would normally swing the gate signal below the source. This action not only protects the gate but prevents the signal from saturating CR3 which would render the bias control loop inoperative.

Since most power MOS devices have a cut-off frequency in the gigacycle range, parasitic RF oscillations can be troublesome. Most assemblies require bypasses on the order of 0.22 μF, connected from supply lines at the FET location. In addition, leads running to the gate terminal should be either shielded or used as a twisted pair with a ground lead.

Performance

Curves of distortion as a function of frequency and power output are shown in Figure 7. In summary, total harmonic distortion is generally under 0.1% throughout the audio range. The unusual rise in 20 Hz distortion at ½ watt is caused by the imperfect filtering of the output of the clipper circuit in the bias control loop. Because of the clipper circuit, the voltage on C5 becomes a larger percentage of the input signal as the output drops from the 50 watt level.

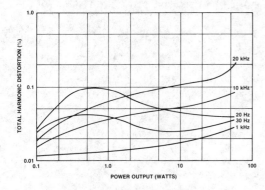

Distortion Characteristics
Figure 7

Power dissipation and efficiency as a function of power output are shown on Figures 8 and 9. Power efficiency is about 60% at full output and is essentially independent of frequency.

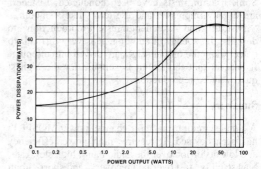

Power Dissipation in Both Output Transistors versus
Audio Power Output
Figure 8

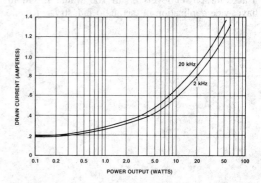

Current from Main Supplies versus Audio Power Output
Figure 9

Figure 10 shows a 100 kHz square wave output when the input filter R_1 C_1 is removed. With the filter in place, the slight 1 MHz ringing disappears. Rise time and frequency response can be varied by altering the value of C5.

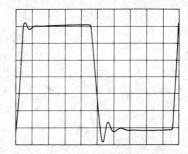

Output Voltage with 100 kHz Square Wave Input. Input Filter
Removed (5 V/Division)
Figure 10

6

The output coupling filter permits capacitive loads to be driven without introducing oscillation. With a 1 μF load, distortion rises slightly at high frequencies.

Some listening tests have been conducted. Listeners are usually impressed and use terms like "smooth," "musical," and "excellent transient reproduction" to describe the sound. In addition, it is very difficult to detect when the output stages are driven into clipping. It appears that the circuit is suitable for the highest quality audio reproduction.

Summary

A simple circuit using short-channel power MOSFETs has been devised. It features high power efficiency and low distortion over the audio band, high slew rate, bias stability, fast overload recovery, and short-circuit protection. Listening tests have rated it very high in audio accuracy. Since it is potentially low cost and usable for power levels up to 200 watts, it should have wide application in the audio industry.

References

[1] Bill Roehr, *"The Autobias Amplifier, A New Topology for Automatically Biased Audio Amplifiers Using Power MOSFETs,"* Journal of the Audio Engineering Society, Vol. 30, No. 4, April 1982, pp. 208–216.

[2] Tohru Sampei, Shin-ichi Ohashi, and Shikayaki Ochi, *"100 Watt Super Audio Amplifier Using New MOS Devices,"* IEEE Transactions on Consumer Electronics, Vol. CE-23, No. 3, August 1977, pp. 409–416.

[3] Barry Harvey, *"Power-FET Amplifier Designs Boost Fidelity, Reduce Complexity,"* EDN, Vol. 25, No. 17, Sept. 20, 1980, pp. 137–142.

[4] Edward M. Cherry and Gregory K. Cambrell, *"Output Resistance and Intermodulation Distortion of Feedback Amplifiers,"* Journal of the Audio Engineering Society, Vol. 30, No. 4, April 1982, pp. 178–191.

6.6.3 A MOSFET Power Amplifier with Error Correction

Abstract

Power MOSFETs are emerging as the device of choice for high-quality power amplifiers because of their speed, reduced need for protection and falling cost. A low-distortion power amplifier design is illustrated which includes output stage error correction to reduce the effect of transconductance droop in the crossover region and thus allow operation at more efficient bias levels.

Introduction

The rapid evolution of power MOSFETs during the last few years has brought them to the point where they are now very attractive for use in audio amplifier power output stages. Important improvements which have been made include increased voltage, current and dissipation ratings, reduced "on" resistance, availability of complementary pairs and greatly reduced cost. Although a 75-watt MOSFET is still more expensive that a 150-watt bipolar transistor, the premium is small when considered relative to total amplifier cost and improved performance.

The purpose of this paper is to demonstrate the level of performance achievable with current technology and illustrate practical circuit techniques for achieving this performance.

Power MOSFETs have several fundamental advantages over bipolar power transistors, most notably speed and freedom from secondary breakdown. The latter provides higher "usable" power dissipation, improved reliability, and freedom from safe-area limiter circuits which can misbehave and cause audible degradation. MOSFETs also have some disadvantages in comparison with bipolar transistors. These include higher turn-on voltage drive requirements and smaller transconductance at low current levels. The former tends to contradict generalizations which have been made to the effect that drive circuits for power MOSFETs are less expensive, at least for the reliable source-follower configuration. The latter results in transconductance droop in the crossover region if bias currents are not fairly high. Such transconductance droop can result in crossover distortion.

In this paper we will illustrate a high-performance amplifier design which utilizes the advantages of the power MOSFET while dealing with the drawbacks of the device. Although not taken to an extreme, the underlying philosophy of the design is that small-signal silicon is inexpensive, i.e., that the overwhelming portion of expense in a power amplifier is in items like the power transformer, filter capacitors, power transistors, heat sinks, chassis and related hardware. Thus, in order to take full advantage of the performance achievable with the MOSFET output stage, a very high quality front-end and driver are provided. The driver, operating from regulated boosted supplies, is capable of providing high voltage and current swings to the power MOSFETs with good headroom. Output stage transconductance droop is dealt with by employing a simple but very effective output stage error correction technique proposed by Hawksford. [1] The resulting design achieves a 20-kHz THD figure of less than 0.0015 percent at an idle bias of only 150 mA.

MOSFET vs. Bipolar Output Stages

Design of MOSFET power amplifiers is quite straightforward and conventional as long as differ-

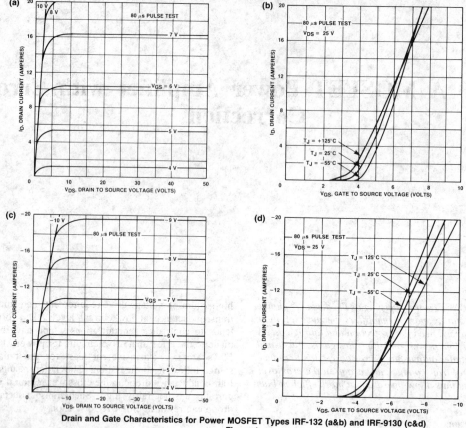

Drain and Gate Characteristics for Power MOSFET Types IRF-132 (a&b) and IRF-9130 (c&d)
Figure 1
(Drawings courtesy of International Rectifier, Inc.)

ences between MOSFETs and bipolar transistors are understood. Figure 1 shows the drain and gate transfer characteristics typical of the vertical DMOS devices used in this project (International Rectifier types IRF-132 and IRF-9130). [2] The important point to see here is that these enhancement devices require about 3 volts of forward gate bias to begin to turn on (e.g., gate threshold voltage, V_t) and may require as much as 10 volts to conduct high currents (12 amperes). While the required bias voltage is thus higher than for bipolar transistors, it can still be generated by the traditional V_{be} multiplier circuit. Thermal bias stability for the power MOSFETs is much better than that for bipolar transistors, even though the vertical DMOS devices have a Vgs temperature coefficient of about −5.0 mV/°C at a typical bias current of 150 mA. This is partly due to the MOSFET's lower transconductance at the bias point. The MOSFETs negative temperature coefficient of transconductance also tends to improve thermal bias stability. As a result, in some cases, a V_{be} multiplier transistor or associated reference diode need not be mounted on the heat sink.

If the popular source-follower output stage configuration is used, the substantial gate drive voltage required for high currents means that the driver stage should be provided with a "boosted" power supply voltage greater than that of the main high-current supply in order to take full advantage of the voltage swing available from the latter. The current requirements for the boosted supply are small, and it can be regulated at little additional expense, thus reducing hum, crosstalk and modulation distortion. Several high-quality bipolar power amplifiers also use boosted driver supplies, some regulated.

While bipolar transistors are regularly placed in parallel with small individual emitter ballast resistors, the paralleling issue is not as straightforward for power MOSFETs, at least in linear applications. It has been said that the negative temperature coefficients of transconductance and "on" resistance of MOSFETs act to suppress current hogging by one transistor, thus permitting easy paralleling of MOSFETs without ballast resistors. This appears to be true for hard-switching applications where the paralleled devices are all fully turned-on together (i.e., channels fully enhanced by forward gate voltage) so that current and dissipation imbalances are only a result of mismatched "on" resistance.

However, the issue is more complex for linear, and especially low-distortion, applications because the operating region of interest is not the fully turned-on region, but rather the linear region wherein drain current at a specified gate voltage is important. Specifically, recognizing that the gate threshold voltage specification for these devices is 2–4 volts, an examination of the gate transfer characteristics of Figure 1 indicates that a very serious current imbalance can exist unless gate threshold voltages among paralleled devices are reasonably matched. It is also apparent that reasonable temperature differentials will not adequately reduce the imbalance. Because of the size of the worst-case threshold voltage differentials possible, source ballast resistors are not a reasonable approach to achieving balance. It thus appears that for high quality audio applications where paralleled devices are necessary, both threshold voltage and transconductance of paralleled devices should be matched. Threshold matching that guarantees that all devices are "on" to some extent in the quiescent bias state, and transconductance matching to within 20% is probably adequate.

Modern complementary MOSFETs, with maximum "on" resistances of only about 0.3 ohms, are just about as efficient as bipolar transistors in terms of voltage dropped from supply rail to load in output stages. However, they typically require a higher operating current to achieve a given transconductance. This characteristic is illustrated in Figure 2. The device transconductance in a source-follower or emitter-follower output stage is important because it determines the small signal voltage drop through the stage as a function of current. This is especially important in Class-AB stages where it is desirable that the sum of the effective transconductances of both halves be high and be constant with current so as to avoid crossover distortion. It can be seen from Figure 2 that approaching this condition with power MOS-FETs requires fairly substantial bias current (as a rough starting point, that current where transconductance is one-half its high-current asymptotic value), on the order of a few hundred milliamperes.

In contrast, bipolar power transistors are typically biased at a much lower current, but this is not entirely advantageous. A typical bipolar output stage will often be biased approximately where the dynamic emitter resistance of the output devices ($1/g_m$) at crossover is equal to the associated ballast resistance as a compromise in achieving approximately constant total output stage transconductance as a function of current.

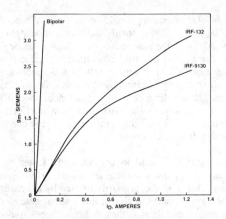

Power MOSFET Transconductance versus Drain Current. For comparison, note much greater ratio of transconductance to operating current typical of bipolar transistors
Figure 2

This is done because both halves are "on" and contribute transconductance in the crossover region while only one half contributes transconductance at currents well outside the crossover region. This usually results in bias currents of less than 100 mA per output transistor, sometimes as low as 20 mA. This small amount of bias current compared to several amperes of signal current being handled can sometimes result in unexpected temporary bias inadequacy, resulting in crossover distortion, because a small change in circuit parameters (about 50 mV) can cause the bias current to vary considerably. This can happen as a result of the time delay in the intentional thermal feedback loop created by placing the bias generator sensing junction on the heat sink; the temperature of this junction will differ from that of the power transistors as a result of thermal delay, low-pass filtering and attenuation. After a high dissipation interval ends, the amplifier may find itself temporarily underbiased because the power transistor junctions cooled down faster than the bias transistor.

Bipolar output stages can be operated in an overbiased mode, but the penalty can be dangerously reduced thermal stability if larger heat sinks are not used, or increased crossover distortion if larger ballast resistors are used. Much less thermal feedback is necessary for thermal stability of MOSFET output stages and their higher bias is less likely to disappear under transient thermal conditions. Compared to bipolar designs, Class-AB MOSFET power output stages also tend to have a wider Class-A region of operation and a smoother transition to the Class-B region of operation.

Although power MOSFETs require virtually no drive current at low frequencies, their substantial input capacitance means that drive circuits with forward and reverse drive capabilities similar to those employed with bipolar transistors should be used for wideband, high slew-rate circuits for demanding audio applications. Although the gate-source capacitance can be on the order of 700 pF, this capacitance is effectively "bootstrapped" in a source follower output stage, typically reducing its effect by about an order of magnitude. The smaller gate-drain capacitance, about 100 pF, is also present. A 100 $V/\mu s$ slope with an effective capacitance of 170 pF thus requires a 17 mA current capability from each driver.

Power MOSFETs tend to be inherently faster than bipolars, partly because there are no minority carrier effects. Their speed is primarily limited by the ability of the drive circuitry to charge the internal gate electrode capacitance through the effective gate resistance. The wider bandwidth, reduced excess phase and reduced variation of device speed with voltage and current tend to allow greater high-frequency negative feedback with greater stability. The higher switching speed also tends to reduce dynamic crossover distortion. Furthermore, the MOSFET's higher switching speed greatly reduces the flow of Class-AB common-mode current at high frequencies which poses such a destructive threat to many bipolar designs.

However, power MOSFETs do have a tendency to very high frequency parasitic oscillations in real-world circuits. This appears to be a result of the natural high speed of the device combined with the substantial drain-source capacitance (300 pF) typical of these devices, making formation of a Colpitts oscillator easy if inductance is present in the gate circuit. This often necessitates the use of a small resistor in series with the gate. In combination with the device input capacitance, this resistor (typically about 100 ohms) can create an additional pole which tends to reduce the high-frequency improvement over bipolar transistors.

Perhaps one of the most important advantages of power MOSFETs for audio use is their freedom from secondary breakdown and large safe operating area (SOA). A highly simplified explanation of secondary breakdown in bipolar transistors is that it results from localized current hogging which in turn results from localized thermal "hot spots." Transistor current at a given base-emitter voltage has a very strong positive temperature coefficient. Thus, a "hot spot" carries more current and gets even hotter as a result. This regenerative process, once started, can be very rapid and unforgiving. It can persist even after the external voltage and current conditions re-enter the safe operating area, leading to destruction. The relationships in a power MOSFET are in contrast *degenerative* in nature because hotter regions exhibit reduced transconductance and thus tend to conduct less of the total current. This tends to equalize the temperature across the chip. The safe area of a MOSFET is thus primarily governed by simple thermal considerations of how

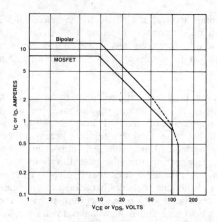

Safe Operating Area (SOA) Comparison of Bipolar Ring-Emitter Transistor (2SA-1072) and a Power MOSFET (IRF-9130). Rated power dissipations are 120 W and 75 W, respectively
Figure 3

much energy (product of power and time) is required to raise the temperature of the hottest point on the chip (which will not be much different than the average temperature of the whole chip) to a dangerous point.

Figure 3 shows a comparison of safe operating area for a power MOSFET and a typical bipolar power transistor. Notice that there are no steep secondary breakdown SOA slopes at high voltages for the MOSFET — it is essentially limited by simple power dissipation over its full voltage range. This is also true for short-term dissipation well in excess of rated continuous dissipation, where thermal time constants govern the allowable excess dissipation. For example, a 25-ampere peak with 100 volts across the MOSFET can be handled for 10 microseconds. Figure 3 illustrates that "usable" dissipation (SOA at higher voltages) for a MOSFET may be equal to that of a bipolar power transistor of substantially higher rated power dissipation. Safe operating area at high voltages is particularly important when difficult reactive loads are being driven. In many power amplifiers the use of multiple paralleled output devices is for reasons of increased safe operating area rather than simple thermal considerations. Finally, freedom from secondary breakdown means freedom from complex safe-area limiter circuits, some of which are notorious for their misbehavior. [3]

Figure 4 illustrates a simple 50-watt MOSFET power amplifier design. It is notably similar to what a simple bipolar power amplifier design would look like. Transistors Q1 and Q2 comprise the input differential amplifier whose output is converted to a single-ended current by current mirror Q3, 4. This current feeds the common-emitter pre-driver Q5 which is provided with a constant-current load. Capacitor C1 provides Miller-effect feedback compensation and establishes a

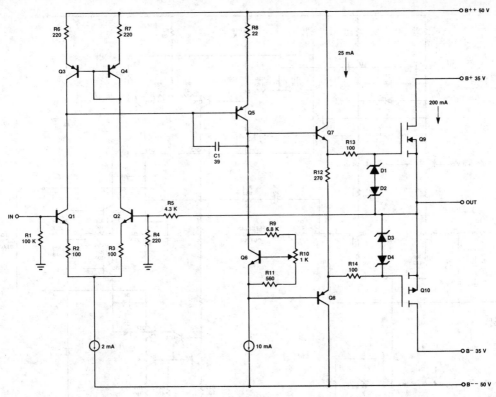

A Simple MOSFET Power Amplifier. Note use of boosted supplies for driver circuitry to satisfy power MOSFET gate drive requirements
Figure 4

stable gain-crossover frequency of approximately 2 MHz. Transistor Q6 is connected in a conventional V_{be} multiplier circuit to provide adjustable bias (nominally about 8 V) for the output stage. If thermal feedback is required, a sensing diode placed directly in series with the emitter of Q6 (with appropriate modification of associated resistor values) and mounted on the heat sink will provide approximately the correct degree of compensation. Emitter-follower drivers Q7 and Q8 provide a low-impedance drive for the gates of power MOSFETs Q9 and Q10. The drivers isolate the high-impedance pre-driver collector circuit from the nonlinear input capacitance of the MOSFETs and provide adequate charge and discharge current for the MOSFET gate circuits. The boosted supplies for all circuits prior to the output stage enable the drive circuitry to provide adequate gate voltage to fully turn-on the MOSFETs while maintaining margin against saturation. Zener diodes D1–D4 protect the MOSFETs from excessive gate-source voltages of either polarity.

High-Performance Input and Driver Circuits

As would be the case with a bipolar design as well, many improvements can be made to the "front-end" of the simplified amplifier of Figure 4 in order to provide higher performance and take full advantage of the capability of the MOSFET output stage. Although substantially adding to the complexity of the schematic in appearance, such improvements primarily involve only small-signal, low-voltage transistors and inexpensive passive components, and thus contribute only a small percentage increase to total amplifier cost.

The front-end for the amplifier to be discussed here is shown in Figure 5. The input stage is a differential JFET-bipolar cascode with a constant current bias supply. The cascode allows the use of a low-noise dual JFET, achieving a referred input noise of less than 6 $nV/\sqrt{Hz}$. It also provides good common-mode and power supply rejection, necessary because negative feedback is not very effective in reducing power supply and common-mode impairments introduced at the input stage. The degenerated JFET input stage can handle fairly large open-loop input signals with relatively low distortion, making the amplifier relatively immune to transient intermodulation distortion (TIM) and RFI effects.

The input stage is loaded by current sources (Q6, Q7) to provide high open-loop gain at low frequencies.

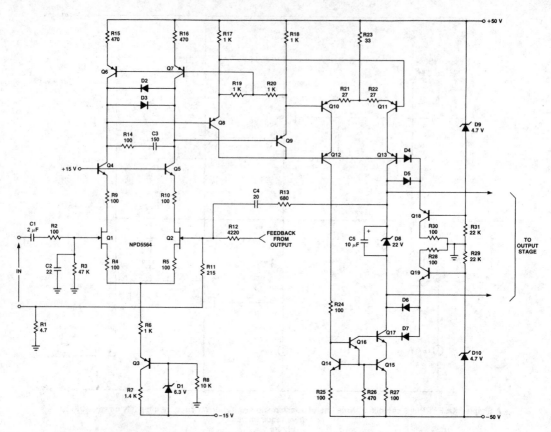

MOSFET Power Amplifier Front End. Differential cascode circuitry minimizes distortion
Figure 5

Emitter followers Q8 and Q9 isolate the input stage from second-stage (pre-driver) loading effects and produce a combined common-mode feedback to properly bias Q6 and Q7. This also provides additional common-mode rejection by reducing the common-mode impedance seen by the collectors of the input stage. Limiter diodes D2 and D3 prevent excessive signal swings at the collectors of Q4 and Q5 when the amplifier is clipping.

The complementary pre-driver stage consists of a differential cascode (Q10–13) loaded with a Darlington-cascode current mirror (Q14–17) to provide a single-ended drive for the output stage. The cascode achieves high speed by eliminating Miller effect and allowing the use of fast low voltage transistors in the common-emitter differential amplifier. Elimination of Miller effect is also important in reducing high-frequency distortion resulting from nonlinear collector-base junction capacitance. [4] The cascode configuration also improves low-frequency linearity and power supply rejection by reducing Early effect. The complementary pre-driver structure, made possible by the current mirror, greatly reduces second-order distortion.

Transistors Q18 and Q19 provide regulated bias for the cascode bases and emitter-follower collectors. Adequate current is available so that these voltages remain stable even under clipping conditions. Diodes D4–D7 prevent the cascodes from saturating when the amplifier clips. Zener diode D8 provides for two identical drive signals offset by 22 volts to allow for biasing and error correction in the output stage.

Overall negative feedback connections and frequency compensation are also shown in Figure 5. R11 and R12 set the closed-loop gain at approximately 20. The resistance of this divider was chosen to be fairly low to avoid noise and maintain good high-frequency characteristics. As a result, current flow and dissipation is not insignificant (100 mW in R12 at 50-watt operating level). To avoid thermally-induced distortion at low frequencies, these resistors should be over-sized metal-film types, 1-watt and 2-watt respectively.

Feedback compensation is provided by C4 and R13 which implement rolloff feedback from the output of the pre-driver to the inverting amplifier input, establishing a stable gain crossover frequency of about 2 MHz. Providing compensation by feedback to the input stage tends to allow improved slew rate and reduced power supply coupling; the latter because both ends of the network are ground-referenced (in contrast to the Miller-effect compensation of Figure 4). Elements C3 and R14 act to stabilize the loop formed by C4 and R13. This front-end design enables the amplifier to achieve a slew rate in excess of 300 $V/\mu s$.

Output Stage and Error Correction

In virtually any well-designed power amplifier the output stage ultimately limits performance. It is here where both high voltages and large current swings are present, necessitating larger, more rugged devices which tend to be slower and less linear over their required operating range. The performance-limiting nature of the output stage is especially true in Class-B and -AB designs, where the signals being handled by each "half" have highly nonlinear "half-wave-rectified" waveforms and where crossover distortion is easily generated. In contrast, it is not difficult or prohibitively expensive to design front-end circuitry of exceptional linearity.

Overall negative feedback greatly improves amplifier performance (including dynamic distortions such as transient intermodulation distortion [4]), but it becomes progressively less effective as the frequency or speed of the errors being corrected increases. High-frequency crossover notch distortion is a good example. For this reason, several high-performance amplifier designs now employ feedforward error correction in addition to conventional negative feedback. However, some of these designs can be complex and expensive. The philosophy of this design is based on the observation that only the output stage needs extra error correction and that such local error correction can be less complex and more effective.

While the power MOSFET has many advantages, it was pointed out that the lower transconductance of the MOSFET will result in considerable crossover distortion unless rather high bias currents are chosen. Figure 6 illustrates this effect by showing the individual and summed transconductances of both halves of a Class-AB MOSFET output stage as a function of net output current. At a bias current of 150 mA and a load of 8 ohms, this transconductance variation can result in open-loop output stage harmonic distortion on the order of one percent, as pictured in Figures 7a and 7b. Mismatches in the transconductance characteristics of the top and bottom output devices also contribute to the distortion of Figure 7. Again, while bipolar transistor transconductance is high enough and consistent enough that it is relatively unimportant in an emitter-follower stage, MOSFET transconductance is smaller

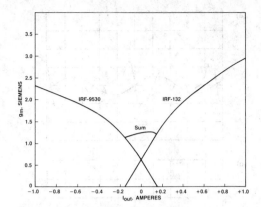

Output Stage Transconductance versus Output Current (I_{bias} = 150mA). Reduced Total Transconductance in Central Region can Cause Crossover Distortion.
Figure 6

and less consistent, making it a significant parameter in source-follower stages.

Figure 8 illustrates an error correction technique described by Hawksford which is well-suited to this application. [1] Here the output stage, being a source follower, is modeled as having exactly unity gain with an error voltage e(x) added. This error represents any departure from unity gain, whether it is a linear departure due to less than unity gain, a distortion due to transconductance nonlinearity, or injected errors like power supply ripple. A differential amplifier, represented by summer S2, merely subtracts the output from the input of the power stage to arrive at e(x). This error signal is then added to the input of the power stage by summer S1 to provide that distorted input which is required for an undistorted output. Note that this is an error cancellation technique like feedforward as opposed to an error reduction technique like negative feedback. This technique is in a sense like the dual of feedforward. It is less expensive because the point of summation is a low-power internal amplifier node. It is less critical of component tolerances and frequency response matching because less circuitry is enclosed and that circuitry is simple. Feedforward tends to become less effective at very high frequencies because the required phase and amplitude matching for error cancellation becomes progressively more difficult to maintain. The technique of Figure 8 also tends to become less effective at very high frequencies because, being a feedback loop (albeit not a traditional negative feedback loop), it requires some amount of compensation for stability, detracting from the phase and amplitude matching.

A schematic of the MOSFET power amplifier's output stage and error correction circuit is shown in Figure 9. The error correction circuit is a slightly modified ver-

6

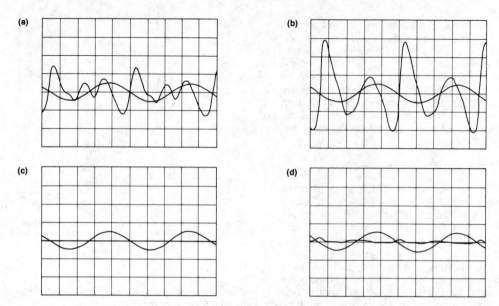

Output Stage Open-Loop distortion (THD); a) 1 kHz, No Error Correction; b) 20 kHz, No Error Correction; c) 1 kHz, with Error Correction; d) 20 kHz, with Error Correction. Vertical Distortion Scale 0.5 percent/div. All Measurements at Full Power (50 W).
Figure 7

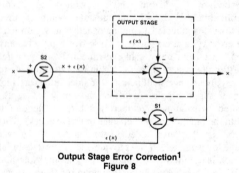

Output Stage Error Correction[1]
Figure 8

sion of one illustrated in reference 1. Emitter followers Q20 and Q21 isolate the high-impedance predriver output nodes from the output stage and provide a low-impedance signal for the error correction summation process. Double emitter followers Q24, Q26 and Q25, Q27 provide a high-current drive capability for the MOSFET gates and isolate the error correction summing nodes from the MOSFET gate loads. Note that Q24 and Q25 can be fast, inexpensive small-signal transistors. Transistors Q22 and Q23 and resistors R38–R45 comprise the differential amplifier for summer S2 of Figure 8. The output current of these transistors is summed with the input signal (summer S1 of Figure 8) by means of R34 and R35. For error correction, the top and bottom halves (Q22, Q23) work independently to produce identical correction signals. Input signals offset by ±11 volts, as supplied by the pre-driver circuit, provide DC operating voltage for these circuits. These offset voltages must be

adequate to allow for the maximum bias plus V_{GS} signal swing required by the MOSFETs. Transistors Q22 and Q23, in conjunction with R32 and R33, control the DC voltage drop across R34 and R35. They thus set the bias for the MOSFETs by means of a V_{be}-referenced feedback loop which also includes Q24, Q25, Q26 and Q27. Resistors R38 and R39 control the loop gain of the bias loops and improve stability. Overall frequency compensation of the error correction and bias loops is provided by R36, R37, C6, C7 and C10.

Figures 7c and 7d show open-loop distortion of the output stage with error correction to be less than 0.1%, illustrating an improvement of better than an order of magnitude, even at 20 kHz. This was achieved with 5% tolerance resistors. While use of closer-tolerance resistors would improve the correction at lower frequencies, where it is unnecessary, their use would make a smaller improvement at 20 kHz because performance there is beginning to be limited by the speed of the error correction loop. Sensitivity of 20-kHz THD to tolerance in the error correction circuit has been measured to be approximately 0.0002% per percent in the closed-loop amplifier. For ultimate performance, a pot can be placed between the junctions of R38, R39 and R44, R45.

The output stage is completed by C8, C9 and R50–53 for control of parasitic oscillations and D11–D14 for protection of the MOSFET gates from excessive drive voltages. As mentioned in Section 2, power MOSFETs are considerably more prone to high-frequency

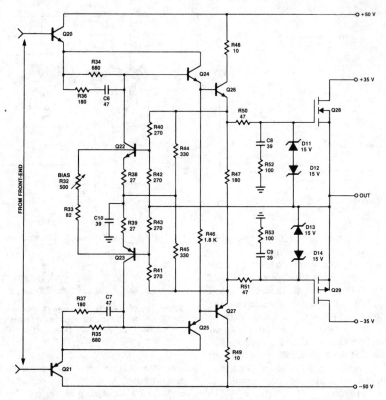

MOSFET Power Amplifier Output Stage. Q22 and Q23 Provide Error Correction Signals
Figure 9

parasitic oscillations than bipolar power transistors because of their inherent high-speed nature and because of their substantial drain-source capacitance, making it easy to form an efficient Colpitts oscillator structure with inductance in the gate circuit. The amount of series gate resistance required for suppression of parasitic oscillations grows in proportion to the amount of inductance in the gate circuit. For high-speed output stage operation it is therefore important to minimize this inductance. Although not employed in Figure 9, this can be done especially well by shielding the gate leads back to the driver transistors, grounding the shield to the local bypass ground at each end. Then only a 10-ohm series resistor at the driver end and a ferrite bead at the gate end are necessary.

Performance

This amplifier employs substantial amounts of negative feedback (40 dB at 20 kHz), and 20-kHz total harmonic distortion (THD) was the primary performance metric used in the design process. In recent years several new forms of distortion have been described, sometimes in the belief that they were caused by large amounts of negative feedback and that traditional measures of distortion (e.g., harmonic and inter-

modulation) would be ineffective in detecting them. Some of these beliefs have been shown to be unfounded. 4–10 Nevertheless, it was decided to include some of these newer measures of distortion in the performance evaluation.

In spite of the error correction, which improves performance by more than an order of magnitude, transconductance variation in the output stage is still the dominant source of distortion in this amplifier. For this reason, output stage bias current continues to influence performance and tradeoffs can be made. The measurements presented here were made at a bias current of 150 mA, resulting in a quiescent output stage power dissipation of 11 watts for the 50-watt amplifier. It should also be noted that the transconductance characteristics of the N- and P-channel output devices were not matched.

A word about measurement technique is in order. In many cases the distortions being measured were below those levels measurable by conventional equipment and techniques. In order to add dynamic range to that provided by the equipment employed, a distortion magnifier circuit was utilized. This circuit scales the output level of the noninverting amplifier under test

6

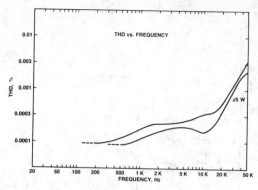

Total Harmonic Distortion (THD) as a Function of Frequency
Figure 10

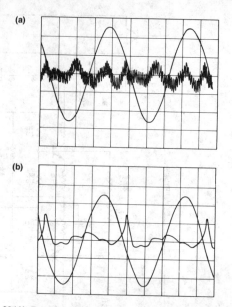

20 kHz Total Harmonic Distortion (THD) Products at Full Power (50 W) a) without Error Correction (THD Analyzer Reads .02%); b) with Error Correction (THD Analyzer Reads .0006%).
Figure 12

down to that of the input, subtracts the two, re-introduces 11 percent of the scaled-down amplifier output signal, and finally multiplies the result by 9 for presentation to the measuring equipment. The net effect is to provide unity gain for the fundamental and a gain of ten to distortion products generated by the amplifier under test. Amplitude and phase balance adjustments were incorporated into the output signal path prior to the subtraction to achieve a fundamental null of greater than 60 dB to frequencies beyond 20 kHz. The excellent noise and distortion performance of the 5534-type operational amplifiers employed make this approach effective.

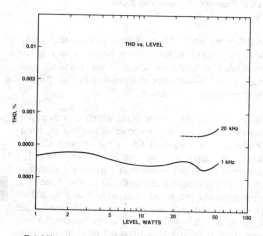

Total Harmonic Distortion (THD) as a Function of Level
Figure 11

To measure harmonic distortion, for example, a sensitive THD analyzer [11] with a 20-kHz measurement floor of about 0.001 percent was employed in combination with this distortion magnifier to achieve a residual of about 0.0003 percent at 20 kHz, primarily limited by noise of the power amplifier under test. The distortion output of the analyzer was then observed with both an oscilloscope and a spectrum analyzer.

The latter further improves the measurement floor in most cases. Most of the other distortion tests employed a similar arrangement. Due to an oscilloscope calibration error, all vertical deflections in the photographs are 6.4 percent low.

Figures 10 and 11 show total harmonic distortion as a function of frequency and power. Dashed portions of the curves indicate that distortion is below the residual of the measuring system. Figure 12 shows the appearance of the 20-kHz full-power harmonic distortion products without and with output stage error correction. Figure 13 illustrates a virtually unmeasurable level of SMPTE* intermodulation distortion (60 & 6000 Hz, 4:1).

Dynamic intermodulation distortion (DIM), a test for measurement of transient intermodulation distortion (TIM), is shown in Figure 14. [12] In this test a 3.18-kHz square wave and a 15-kHz sinewave are mixed 4:1 and passed through the amplifier. A spectrum analyzer is used to measure the in-band inter-modulation components. Performance is shown for both 30-kHz and 100-kHz first-order low-pass filtering of the square wave source (DIM-30 and DIM-100). As predicted by the good 20-kHz THD performance and high slew rate of this amplifier, both DIM-30 and DIM-100 distortion levels are very low; in fact, the former is unmeasurable.

Interface intermodulation distorion (IIM) [9] is measured by applying 1000 Hz to the amplifier under test

---

* Society of Motion Picture and Television Engineers.

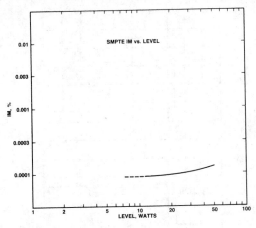

SMPTE Intermodulation Distortion as a Function of Level
Figure 13

and 60 Hz to a test amplifier, each of which drives opposite ends of an 8-ohm load resistor. A spectrum analyzer is then used to measure distortion products at the output of the amplifier under test. Both amplifiers are operated at half the rated power of the amplifier under test and distortion products are referred to the 1-kHz level at the output of the amplifier under test. For this test the spectrum analyzer was preceded by a modified version of the distortion magnifier to produce a magnification of 100. IIM was unmeasurable, at less than 0.0001 percent.

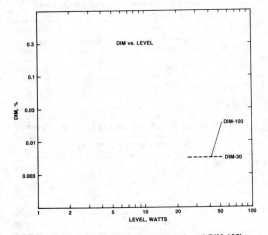

Dynamic Intermodulation Distortion (DIM-30 and DIM-100) as a Function of Level
Figure 14

Phase intermodulation distortion (PIM) is shown in Figure 15. [10] PIM is measured in the same way as SMPTE-IM, except that phase modulation of the carrier is measured instead of amplitude modulation. The phase modulation is then expressed in time (e.g., rms nanoseconds).

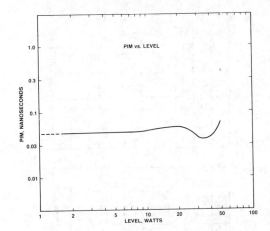

Phase Intermodulation Distortion (PIM) as a Function of Level. Note that Phase Modulation is Expressed in rms nanoseconds.
Figure 15

Damping factor (DF) as a function of frequency is shown in Figure 16, and is extremely high. It is high for three reasons: 1) the power MOSFETS present very light loading to the drivers, producing a low open-loop output impedance essentially equal to the inverse of their transconductance; 2) the error correction circuit tends to drive this open-loop output impedance to zero; 3) substantial overall negative feedback further reduces the output impedance by an amount approximating the feedback factor (40 dB at 20 kHz). Inclusion of a parallel R-L network (0.5Ω, $0.5 \mu H$) at the output in series with the load for complete capacitive load stability will reduce the high-frequency damping factor to 125 at 20 kHz.

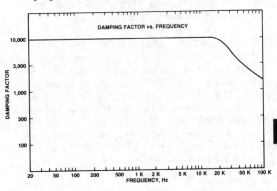

Damping Factor as a Function of Frequency
Figure 16

Although the need for this much damping factor is doubtful, the importance of DF on frequency response and coloration has sometimes been underestimated. This is explained by the fact that most speaker systems are designed assuming they will be driven by a pure voltage source (sometimes, it seems, with limitless

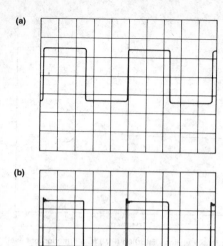

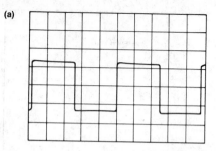

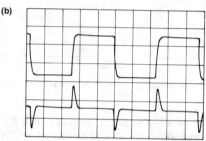

20 kHz Square Wave into an 8 Ohm Load;
a) Small-Signal (1 V/div); b) Full Power (20 V/div)
Figure 17

20 kHz Square Wave into a 1-Ohm Resistor in Series with a
1-Microfarad Capacitor; a) Small-Signal (1 V/div); b) Full-Power
(Input Bandlimited to 200 kHz). Top Trace 20 V/div, Bottom
Trace Is Output Current at 20 A/div. Timebase is 10 μs/div.
Figure 18

current capability as well!). For example, the impedance of a nominal 4-ohm system may dip to 2.5 ohms and rise to over 50 ohms at various points across the frequency band due to driver and crossover resonances. A typical bipolar amplifier may have a damping factor of 100 (perhaps less at high frequencies), resulting in frequency response deviations on the order of 0.3 dB with such a load. Coloration due to low DF may also partly explain audible differences among vacuum-tube and low-feedback designs.

Figure 17 illustrates small-signal and full-power 20-kHz square waves into an 8-ohm load. Figure 18 shows small-signal and full-power 20-kHz square waves into a reactive load consisting of 1-ohm and 1 μF in series. In the full-power case the square wave has been band-limited to 200 kHz by a first-order low-pass filter. Figure 19 shows 500 kHz small-signal square waves into an 8-ohm resistive load and a reactive load consisting of 1-ohm and 1-μF in series. It also shows a full-power 500 kHz square wave into an 8-ohm load. Few bipolar amplifiers would survive this test.

Table 1 summarizes overall performance of the amplifier.

Table 1
Summary of MOSFET Power Amplifier Performance

POWER OUTPUT ($R_1 = 8\Omega$)	50 W
TOTAL HARMONIC DISTORTION (20–20 kHz)	<0.001%
SMPTE IM DISTORTION	0.00013%
DYNAMIC INTERMODULATION DISTORTION (DIM-30) (DIM-100)	<0.006%* 0.014%
INTERFACE INTERMODULATION DISTORTION (IIM)	0.0001%
PHASE INTERMODULATION DISTORTION (PIM)	<0.1 ns
SLEW RATE	>300 V/μs
RISE TIME	100 ns
DAMPING FACTOR (20–20 kHz)	>5000
S/N ("A" WTD, re 1 WATT)	108 dB

*Below Measurement Floor

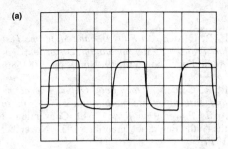 (a)

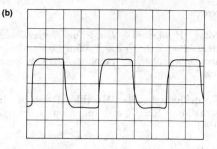 (b)

(c)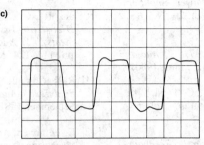

500 kHz Square Wave Response; a) Small-Signal, 8-Ohm Load (1 V/div); b) Small-Signal 1-Ohm and 1-μF Series Load (1 V/div); c) Full-Power, 8-Ohm Load (20 V/div.). Timebase is 0.5 μs/div.
Figure 19

Conclusion

Power MOSFETs are capable of exceptional performance when used in combination with good drive circuitry and simple error correction circuitry. Their ability to operate without complex and unreliable safe-area limiting circuitry makes them especially useful for demanding audio applications. Compared with bipolar transistors, the major disadvantage of MOSFETs (and source of distortion) seems to be the lower transconductance, but this can be dealt with effectively by means of a simple error correction circuit. Although a MOSFET power amplifier can still be expected to cost a little more, the improved characteristics seem to justify the small premium in applications where performance is important.

Editors Note:

This paper is intended to illustrate design techniques and performance achievable, and not as a construction project.

6

References

[1] M. J. Hawksford, *"Distortion Correction in Audio Power Amplifiers,"* 65th Convention of the Audio Eng. Soc., preprint #1574, February 1980.

[2] International Rectifier HEXFET Databook, International Rectifier Corporation, Copyright 1981.

[3] T. Holman, *"New Factors in Power Amplifier Design,"* J. Audio Eng. Soc., Vol. 29, No. 7/8, pp. 517–522, July/August 1981.

[4] R. R. Cordell, *"Another View of TIM,"* Audio, Vol. 64, Nos. 2 & 3, February and March 1980.

[5] W. G. Jung, M. L. Stephens, C. C. Todd, *"An Overview of SID and TIM,"* Audio, Vol. 63, Nos. 6–8, June–August 1979.

[6] P. Garde, *"Transient Distortion in Feedback Amplifiers,"* J. Audio Eng. Soc., Vol. 26, No. 5, pp. 314–321, May 1978.

[7] E. M. Cherry, *"Transient Intermodulation Distortion; Part I–Hard Nonlinearity,"* IEEE Trans. Aconst., Speech, Sig. Proc., Vol. ASSP-29, pp. 137–146, April 1981.

[8] R. R. Cordell, *"A Fully In-band Multitone Test for Transient Intermodulation Distortion,"* J. Audio Eng. Soc., Vol. 29, No. 9, pp. 578–586.

[9] R. R. Cordell, *"Open-Loop Output Impedance and Interface Intermodulation Distortion in Audio Power Amplifiers,"* 64th Convention of the Audio Eng. Soc., preprint #1537, November 1979.

[10] R. R. Cordell, *"Phase Intermodulation Distortion-Instrumentation and Measurements,"* J. Audio Eng. Soc., Vol. 31, No. 3, pp. 114–124, March 1983.

[11] R. R. Cordell, *"Build a High-Performance THD Analyzer,"* Audio, Vol. 65, Nos. 7–9, July–September 1981.

[12] E. Leinonen, M. Otala, J. Curl, *"A Method for Measuring Transient Intermodulation Distortion (TIM),"* J. Audio Eng. Soc., Vol. 25, No. 4, pp. 170–177, April 1977.

6.6.4 Boost OP-AMP Output Power With Complementary Power MOSFETs (AN83-5)

Introduction

Many high-quality, monolithic op-amps are available today at low cost. They offer low noise and distortion, high slew-rate and wide bandwidth. However, due to the limited chip-size, their small output transistors cannot deliver any appreciable power to a load.

Higher power monolithic and hybrid op-amps are available, but their cost is inevitably much higher than their low-power counterparts. Also, their performance tends to be inferior in the categories of noise, distortion and slew-rate.

Clearly, a circuit topology that combines a low-power, high quality op-amp, a pair of inexpensive complementary power MOSFETs and a few additional components, is desirable. This combination should be expected to have very good performance, since power MOSFETs are inherently linear devices.

This Applications Note presents a power-boosted op-amp circuit using Siliconix N and P channel power MOSFETs. It is configured as a current-boosted DC amplifier operating from standard op-amp power supply voltages. The design takes advantage of a new thermally-stable biasing technique for the MOSFETs. Emphasis is placed on the ease of interfacing complementary power MOSFETs to an op-amp, and the high performance obtainable.

A Current-Boosted DC Power Amplifier

Circuit Topology

The schematic for this simple circuit is shown in Figure 1. It is configured as a non-inverting DC amplifier with a closed-loop gain of 11. Drive for the power MOSFET output stage is taken from the op-amp's power supply pins, rather than its output pin. This may seem somewhat unconventional, but there are very good reasons for doing this. The output pin could be used to drive a pair of MOSFETs configured as a complementary source-follower. However, the source-follower output-stage has two major disadvantages: a voltage gain of less than unity, and a substantial DC threshold offset-voltage. This results in peak gate-drive voltages, from the op-amp, several volts greater in magnitude than the output at the MOSFET sources. Also, the output of most op-amps can only swing to within about 2 volts of the supply rails. Thus, the peak voltage swing at the output of the source-follower is restricted to significantly less than the power-supply rail voltages.

The amplifier shown in Figure 1 does not have these limitations. Both the op-amp's output stage and the MOSFET current-booster stage provide voltage gain. The op-amp's output stage is used as an inverting common-emitter phase-splitter, while the MOSFETs are used as an inverting common-source output-stage. Since the collectors

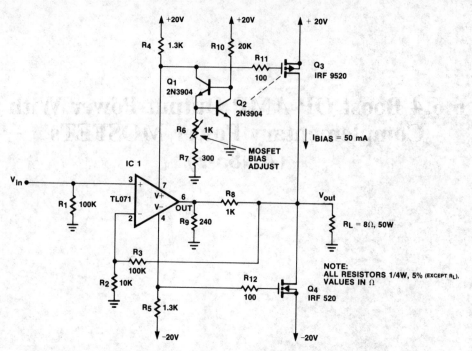

A Current Boosted DC Power Amplifier Using an Op-Amp, a Pair of Complementary Power MOSFETs
and a Few Additional Components
Figure 1

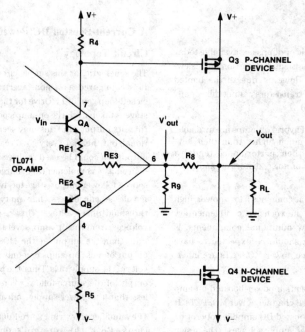

A Simplified View of the Current Boosted Amplifier, Showing the Secondary Feedback Loop. This is Formed
by the Op-Amp's Output Stage, the MOSFETs, and the Feedback Network R8 and R9.
Figure 2

of the output transistors in almost all monolithic op-amps are connected to their power supply pins, they can be used to drive the MOSFETs.

Load resistors R_4 and R_5 are connected from the op-amp's supply pins to the power supply rails. They generate the necessary bias and drive voltages that the MOSFETs require. Since these voltages are referenced to the power supply rails rather than the output voltage, a significantly larger peak output swing is possible. This circuit easily generates peak gate-to-source enhancement voltages for each MOSFET in excess of 10 volts. Thus, the output voltage swing is limited only by the peak output current multiplied by the $R_{DS(on)}$ of the MOSFETs.

Quiescent bias current for the op-amp also flows through resistors R_4 and R_5. Since this current is regulated internally by the op-amp, it does not affect circuit performance. In fact, it proves beneficial because some DC bias voltage for the MOSFETs is generated across R_4 and R_5. More will be said about DC biasing for the MOSFETs later.

It should be noted that small resistors (100Ω) are included in series with the gates of each MOSFET. These are to suppress any potential parasitic oscillation that might occur in the circuit. The resistors should be placed as close to the MOSFET packages as possible, for maximum suppression. Care should be taken in the layout of the circuit to avoid excessively long lead lengths as well as any ground-loops.

Looking at Figure 2, it can be seen that a secondary feedback loop is formed by the op-amp's output stage, the power MOSFETs and the feedback network R_8 and R_9. The effect of this feedback loop is to stabilize the voltage gain seen, going from V_{in} to V_{out}. It can be shown through linear circuit analysis, that at low frequencies:

$$\frac{V_{out}}{V_{in}} = \frac{g_{mA}g_{m3}R_4R_L}{1 + \beta g_{mA}g_{m3}R_4R_L}\left(\frac{r_{eA}}{r_{eA} + R_{E1} + R_{E3}}\right) \quad (1)$$

and

$$\beta = \frac{V'_{out}}{V_{out}} = \frac{R_9}{R_8 + R_9} + \frac{R_8R_9}{R_8 + R_9}\left(\frac{1}{g_{m3}R_4R_L}\right) \quad (2)$$

where:

$g_{mA}(Q_A)$		$\simeq$	40	mA/V
r_{eA}	$= 1/g_{mA}$	$=$	26	Ω
R_{E1}		$=$	64	Ω
R_{E2}		$=$	128	Ω

and

$g_{m3}(Q_3)$	$\simeq$	2.5	A/V
R_4	$=$	1.3	KΩ
R_8	$=$	1.0	KΩ
R_9	$=$	240	Ω
R_L	$=$	8	Ω

If, as in this case;

$$\frac{R_8R_9}{R_8 + R_9} \ll g_{m3}R_4R_L$$

then,

$$\beta \simeq \frac{R_9}{R_8 + R_9} \quad (3)$$

Thus:

$$\frac{V_{out}}{V_{in}} = \frac{124.0}{1 + \beta(124.0)}$$

and so,

$$\frac{V_{out}}{V_{in}} \simeq 1/\beta = \frac{R_8 + R_9}{R_9} \quad (4)$$

This analysis has been done for the upper half of the output stage only. Since the upper and lower halves are reasonably well-matched, this analysis will suffice for both.

Equation (4) shows that the low-frequency voltage-gain of the combined output-stage is set independently of the op-amp. This additional gain reduces the voltage swing at the output of the op-amp, and improves the bandwidth, slew-rate and distortion of the amplifier. Theoretically this improvement should be equal to $1/\beta$ or about a factor of 5 in this case. However, this value is only observed in the reduction of distortion over the source-follower output-stage. The rather large input capacitance of the MOSFETs (several hundred pF), degrades the frequency response somewhat and slows the amplifier down. The actual improvement in slew-rate was observed to be closer to a factor of 3, which is still more than acceptable. The full power bandwidth of this amplifier was well over 100KHz.

A detailed frequency response analysis of this amplifier has not been presented here, because it is somewhat lengthy. Suffice to say, however, that the additional gain provided by the combined output-stage increases the overall gain-bandwidth-product of the circuit. As well, the MOSFET stage adds an extra pole to the open-loop transfer function. Thus the amplifier is unstable for closed loop gains less than about $1/\beta$, or 5 in this case. If unity or very low values of gain are desired, an externally compensated op-amp such as the TL080 should be used in place of the TL071. An external compensation capacitor can then be chosen so that the circuit will not oscillate at the particular closed-loop gain used.

Biasing the MOSFET Pair — A New Approach

As mentioned previously, the op-amp's DC operating current generates some bias voltage for the MOSFETs, across R_4 and R_5. This voltage is intentionally made less than either MOSFET's threshold voltage. Consequently the biasing circuitry comprised of Q_1 and Q_2, has full control over the MOSFET idle current. The MOSFET threshold voltages are typically 3 volts for the N-channel device and 4 volts for the P-channel.

Texas Instruments' TL071 BIFET op-amp was chosen for its low supply current of 2.5mA (max) as well as its high speed and low noise.

The low supply current ensures that the voltage across R_4 and R_5 can be reduced to less than 3 volts. With the bias adjust trimpot (R_6) set for minimum bias, the MOSFETs should not conduct any current. In some cases though, the values of R_4 and R_5 may have to be changed to ensure that the MOSFETs do turn off with R_6 set for minimum bias.

Q_1 and Q_2 form a variable, temperature-compensated current-sink, connected to the positive supply pin of the op-amp. As trimpot R_6 is reduced from its maximum value, the additional current drawn by Q_1, through R_4, increases. At some point, the voltage across R_4 will reach the threshold voltage of the P-channel MOSFET and it will begin to conduct current.

This current will tend to force the output of the amplifier positive, if the N-channel device is conducting a lesser current. But the output of the amplifier is DC coupled back to the inverting terminal of the op-amp. Thus, the op-amp adjusts the voltage across R_5 to force the N-channel device to conduct the same current as the P-channel device. As trimpot R_6 is varied, the voltage at the op-amp's output pin can be seen to go more negative as the bias is increased. This makes sense because more current is then flowing out of the op-amp's negative supply pin, which increases the gate-to-source voltage of the N-channel MOSFET.

It can be seen that the op-amp is used not only as the main gain element in this amplifier, but also as the bias controller for the N-channel MOSFET. Any variations in the idle current of the P-channel MOSFET are mirrored, under control of the op-amp, by the N-channel device. This keeps the output centered at zero.

Table 1
Current Boosted DC Amplifier Performance

Conditions: T_A = 25°C, SUPPLIES	= ±20V, R_L = 8Ω
MOSFET BIAS CURRENT	= 50mA, CLOSED-LOOP GAIN (A_{VCL}) = 11

Output Voltage Swing:	+18.5V, −19.0V
Output Sink/Source Capability:	2.5A DC
Rise Time:	1μS
Slew Rate:	35 V/μS
Overshoot:	+ve: 15%, −ve: 20%
Output Offset: (Equal to $A_{VCL} \times V_{IO}$ (op-amp))	75mV
Noise Floor (at 10Hz bandwith):	−90dB at 100mW into 8Ω
THD:	0.0075% at 100Hz and 20W
	0.0162% at 1KHz and 20W
	0.1350% at 10KHz and 20W
Bias Stability: After 20W output for ½ hour the bias decreased to 43mA.	

Some provision must be made to compensate for the variation in MOSFET idle-current with temperature, at fixed gate-to-source voltages. The temperature coefficient of drain-current at low bias levels is positive, which is due to the decrease in threshold voltage with increasing temperature. This variation in threshold voltage is approximately −5mV/°C. If the voltage across R_4 is changed by this amount as the output devices heat up, then the idle current will be thermally stable.

This is easily achieved by thermally bonding (i.e. glueing) Q_2 of the bias circuitry to the heatsink on which the MOSFETs are mounted. As Q_2 heats up its base-emitter voltage decreases by about 2.2mV/°C. The effective change in voltage across R_4 with temperature is thus:

$$\frac{dV(R_4)}{dT} = \frac{R_4}{R_6 + R_7} \times (-2.2mV/°C)$$

In this amplifier, the sum of $R_6 + R_7$ was set to about 420Ω for a MOSFET bias current of 50mA. From equation 5, this gives an effective decrease in voltage across R_4 of −6.9mV/°C. This decrease is somewhat larger than that of the threshold voltage with temperature. Consequently the MOSFET idle current will decrease slightly as the temperature increases. This is desirable because the output stage power dissipation at idle will then decrease as the MOSFETs heat up. In this amplifier, the idle current decreased by approximately 15% for a 60°C temperature rise at the heatsink surface.

Amplifier Performance

Table I lists the amplifier performance measured on a breadboard version. Of particular note is the large output voltage-swing possible, with a low impedance load. This makes it suitable for use in applications such as Winchester head-actuator drive amplifiers. Here, the larger the output voltage swing, the faster the head accelerates across the surface of the disc.

The excellent noise, distortion and slew-rate specifications make this amplifier suitable for audio use as well. Supplied by ±20V generated from a 12V battery, through a DC-to-DC converter, two of these amplifiers would be more than adequate as a car-stero booster-amplifier.

Small robots too, could benefit from servo power-amplifiers based on the design presented here.

6.7 Bipolar-FET Combinational Devices

Introduction

Because of the fundamental difference in operating mechanisms, FETs and BJTs have different operating characteristics. From the user's point of view, FETs have the advantages of drive simplicity, fast switching speed, and second breakdown ruggedness, but have the disadvantage of higher conduction resistance. To reduce FET conduction resistance and minimize the power loss, a larger chip area is required. For a typical 400 V device with the same conduction drop, FETs require between 2.5 and 3 times the chip area of an equivalent BJT.

As chip size increases, yield decreases exponentially. This drives the cost per useful chip up dramatically. This is the fundamental reason why FETs are not cost competitive in high voltage applications. By combining FETs and BJTs in a single device, one may produce a device with the advantages of both.

In this paper, four combinational configurations will be described, and the application characteristics of each will be discussed.

Combinational Devices and Characteristics

In general, FETs are used in combinational devices to simplify drive and to increase switching speed, while BJTs are used for current carrying capability. Four configurations are considered in this paper. Shown in Figure 1a is a FET-Darlington configuration. Figure 1b is a cascode configuration, Figure 1c shows a parallel configuration, and Figure 1d is a FET-Gated Configuration (FGT). The application characteristics of each of the four configurations are given below.

Darlington Configuration [4, 6]

In this configuration, a FET and a BJT are connected in Darlington fashion with the FET as driver transistor and the BJT as output transistor. When the gate voltage of the FET is high, drain current, I_D, is conducted into the base of Q_1 providing base current

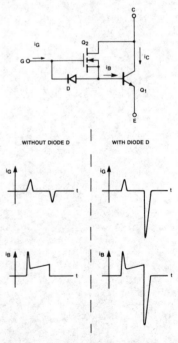

WITHOUT DIODE D | WITH DIODE D

FET – Darlington (or Cascade) Configuration and Waveforms
Figure 1a

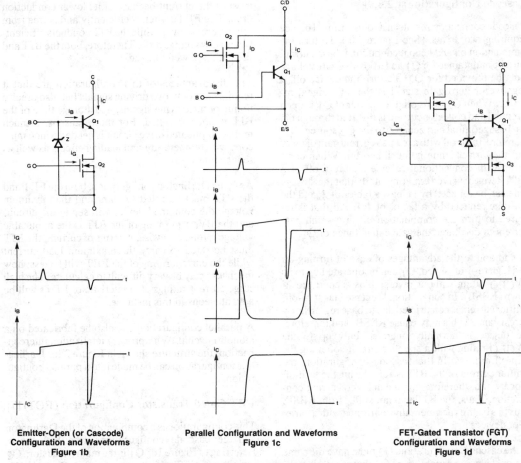

**Emitter-Open (or Cascode)
Configuration and Waveforms
Figure 1b**

**Parallel Configuration and Waveforms
Figure 1c**

**FET-Gated Transistor (FGT)
Configuration and Waveforms
Figure 1d**

for the BJT which then conducts the load current. When the gate voltage is lowered, the FET turns off immediately, but the BJT does not turn off until the stored charge in its base is exhausted. To speed up turn-off of the BJT, a diode, D, may be added to provide a path for withdrawal of excess carriers in the base of Q_1. Without the diode, excess carriers in the base of Q_1 decay by recombination only, and the BJT turns off slowly. With the diode, Q_1 can be turned off rapidly, but the gate drive must provide a large reverse current spike. At the instant of turn-on, when Q_2 is conducting and Q_1 is turning on, most of the load current passes through Q_2 into the base of Q_1. This provides an inherent base current "kick" to Q_1 as shown in the waveforms in Figure 1a. The turn-on speed of this combination device is usually somewhat faster than the speed of Q_1 alone.

In the Darlington configuration, both the FET and the BJT must be rated to withstand the maximum voltage that the combined device will see in the circuit. The applicable voltage rating for the BJT is its BV_{CEO} rating, not any of the other, higher, ratings

such as BV_{CBO} or BV_{CEX}. Only the BJT, however, must be rated to carry the maximum load current. The drain current rating of the FET need only be equal to the maximum base current for the BJT, which will be the maximum load current of the BJT divided by its forward gain (β) at that current. Both parts of this combination may be fabricated monolithically as suggested by Figure 2 [6] from a common epi layer. The FET and BJT portions are then interconnected in Darlington configuration.

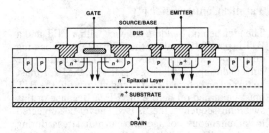

**Device Structure for Monolithic Q_1 and Q_2
Figure 2**

Cascode Configuration [1,2,3,9]

Cascode connection results in a four-terminal device requiring two drives. (See Figure 1b.) To turn the combination on and keep it conducting, base current must be maintained for Q_1 and a positive gate voltage must be provided for Q_2. To turn the device off, a negative or zero gate signal is the only signal required. When Q_2 turns off, the emitter of Q_1 is open circuited and collector current is diverted through the Q_1 base terminal and zener Z to the Q_2 source. Q_1, therefore, turns off with a reverse current gain of unity and the turn-off time is much shorter. When compared with conventional reverse-bias turn-off of a BJT, it was observed experimentally that the storage time can be reduced by at least a factor of 2, and the fall time reduced by a factor of 1.4. All that is required to turn the combination off is a small gate current to discharge the gate capacitance of Q_2.

In addition to the advantages of ease of turning-off and high turn-off speed, Q_1 can be operated up to its BV_{CBO} rating without reverse bias second breakdown (RBSB). In conventional reverse bias turn-off, emitter current is constricted due to base region lateral resistance. This may cause RBSB, and therefore, the device is normally operated only up to the BV_{CEO} rating with a snubber circuit softening the turn-off process. In the cascode configuration, the emitter current of the BJT is zero during the turn-off process and, therefore, there can be no current constriction. Thus, the BJT will not suffer from RBSB. This has been experimentally confirmed with a variety of devices [1].

In the cascode configuration, Q_2 must have the same current carrying capability as Q_1, but the drain voltage requirement need be only 10 V. By using a low voltage FET, the voltage rating of Q_1 can be enhanced from BV_{CEO} to BV_{CBO}. This is often a 50 to 100 percent improvement. Furthermore, as the FET required is a very low voltage part, it will be much smaller than a high voltage part of similar current rating. The disadvantages, from the user's point of view, are higher conduction voltage drop and more complicated drive signals.

Parallel Configuration [4]

The idea behind building a device with a FET and a BJT in parallel is to allow the FET to take the beating during switching and allow the BJT to conduct the current during conduction. At turn-on, the FET must be turned on before the BJT, and at turn-off, the BJT must be turned off before the FET. Because of this switching sequence, the BJT will never be subjected to simultaneous voltage and current stress during switching. The FET is, in effect, an active snubber for the BJT and, therefore, no additional snubber is needed. When the device conducts, the BJT carries most of the current because of its lower conduction drop. The FET switches efficiently and is free from second breakdown, while the BJT conducts efficiently with low voltage drop. Therefore, both the BJT and the FET are well utilized.

The disadvantages of this configuration are that it requires not only two driving signals but also precise timing of the driving signals, and storage time of the BJT is not controlled. For these reasons, a much more complicated drive circuit is needed which must cope with temperature and loading effects as well as sequencing.

As in the Darlington configuration, both the FET and the BJT must be rated to withstand the maximum voltage the combined device will see in the circuit, and the BV_{CEO} rating of the BJT is the applicable voltage rating. Likewise, in terms of current, the BJT must be rated to carry the maximum load current while the current rating of the FET will be less; how much less may be very difficult to calculate. The high surge current ratings characteristic of FETs will be advantageous in this instance.

A parallel configuration can also be fabricated on a monolithic chip. With proper metalization interconnection, the structure shown in Figure 2 for Darlington configuration can be made into a parallel configuration.

FET-Gated Transistor Configuration (FGT) [7]

This configuration is a combination of the Darlington and the cascode configurations described above. Referring to Figure 1d, Q_1 is the main transistor; Q_2 and Q_3 are FETs that are used to turn Q_1 on or off. By connecting the gates of the FETs, the configuration becomes a three-terminal device. When the gate voltage is high, both Q_2 and Q_3 conduct. This provides base current to Q_1 and the entire assembly turns on. When the gate voltage is low, both Q_2 and Q_3 turn off, and Q_1 is turned off by the opening of its emitter as in the cascode combination. The zener diode Z now serves two purposes. The first is to provide the path for Q_1 base current when the emitter of Q_1 is opened during turn-off. The second is to provide a path for Q_2 gate discharge current at turn-off. As can be seen from the Q_1 base current waveform, there is a base current spike at turn-on, a large reverse base current spike at turn-off, and a current approximately proportional to the load during conduction. This is an ideal base current waveform for a BJT and is inherent in the configuration. It is accomplished with very little drive. Q_1 base current is derived from the load instead of a base power supply.

In this configuration, both the BV_{DSS} rating of Q_2 and the BV_{CBO} rating of Q_1 must be greater than the maximum voltage that the combination will see in the circuit, while Q3 remains a low voltage device. On the other hand, Q3 must have a drain current capability equal to Q1, while Q2 can be rated for the base current of Q1. That is, the maximum load current of Q1 divided by its forward gain at that current. Both Q2 and Q3 are, therefore, much smaller in chip size than a comparable single FET which would control the same amount of power. The voltage rating of Q2 and Q1 and the current rating of Q3 and Q1 become the rating for the FGT configuration.

Besides the advantage of a simple drive, there are other advantages to this configuration. Smaller gate capacitance, due to small total chip area of Q2 and Q3, is one advantage. From the user's point of view, small gate capacitance translates into smaller gate current requirement and higher dv/dt capability. Another advantage is that the total chip size of this configuration is much smaller than the size of a comparable single FET [7]. The disadvantage of this configuration is the high conduction voltage drop which is equal to the sum of the V_{CE} of Q1, and V_{DS} of Q3.

Figures 3 through 5 show oscillogram waveforms for the FGT configuration. Figure 3 shows the base current (top) and collector current (bottom) waveforms for the BJT. Both turn-on and turn-off current spikes are evident. Figure 4 shows device voltage (top), device current (center), and BJT current (bottom) during turn-off. It can be seen that the device voltage rises long after emitter current is cut off. In other words, the emitter current crowding phenomenon which commonly occurs in conventional turn-off does not occur in this configuration, and thus, the device is free from reverse-bias second breakdown. Also from Figure 4, it can be seen that the device storage time is about 100 nsec and fall time is about

30 nsec. This is very fast for a BJT of this rating. Figure 5 shows device voltage (top), BJT base current (center), and device current (bottom) during turn-off.

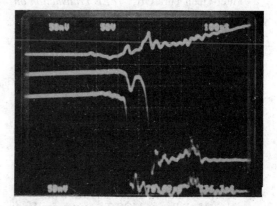

Figure 4

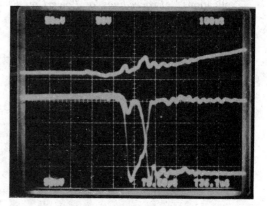

Figure 5

Conclusions

BJT-FET combinational devices extend the capabilities of bipolar transistors. A Darlington configuration provides an easy-to-drive BJT. A cascode configuration enhances the voltage capability of a BJT. A parallel configuration is essentially an actively snubbered BJT. And the FGT configuration extends the power range of a BJT. Each of these configurations loses some useful characteristics, and none of these devices can solve all problems. However, each configuration does have applications. Of the four configurations, only cascode and Parallel have been reportedly fabricated in monolithic form. The cascode configuration has been fabricated as a power hybrid circuit. The FGT has been demonstrated as a discrete circuit. Further research in this area is needed.

6

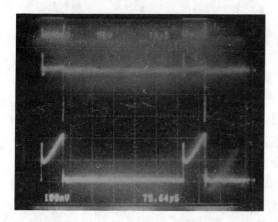

Figure 3

References

[1] D. Chen and B. Jackson, "Turn-off Characterisitcs of Power Transistors Using Emitter-Open Turn Off," *IEEE Transaction on Aerospace and Electronic Systems,* Vol. AES-17, No. 3, May 1981, pp. 386–391.

[2] R. Skanadore, "A New Bipolar High Frequency Power Switching Technology Eliminates Load-Line Shaping," *Proceedings of PowerCon 7,* 1980, pp. D2-1–D2-14.

[3] D. Chen and J.P. Walden, "Application of Emitter-Open Turn-Off Technique for High Voltage Inverter Applications," *IEEE Power Electronics Specialists Conference,* June 1981, pp. 252–257.

[4] R. Blanchard, R. Baker and K. White, "A New High-Power MOS Transistor for Very High Current, High Voltage Switching Applications," *Proceedings of PowerCon 8,* April 1981, pp. N1-1.

[5] M.S. Adler, "A Comparison Between Bimos Device Types," *IEEE Power Electronics Specialists Conference,* June 1982, pp. 371–374.

[6] N. Zommer, "The Monolithic HV BIPMOS," *IEEE International Electron Device and Material Conference,* December 1981, pp. 263–266.

[7] D. Chen and S. Chandrasekran, "An FET-Gated High Voltage Power Transistor," *IEEE Industrial Application Society,* 1982, pp. 713–720.

[8] D. Chen and S. Chin, "Design Considerations for FET-Gated High Voltage Transistors," *IEEE Power Electronics Specialists Conference,* 1983, pp. 144–149.

[9] S. Clements, B. Pelly, R. Ruttonsha, and B. Taylor, "High Voltage High Frequency Switching Using a Cascode Connection of Bipolar Transistor and Power MOSFET," *IEEE Industrial Applications Society,* 1982, pp.1395–1405.

6.8 Power FET Base Drive Circuit for High Power Darlington Transistors

The increasing availability of very large power bipolar junction transistors has opened up many new avenues for power electronics applications. With present device technology, power transistors having a V_{CBO} of 600 V and an I_C of 300 A are being manufactured. This development makes it possible to extend the application of transistors to integral horsepower DC and AC motor drives and similar high power and high frequency applications.

The power MOSFET with its high input impedance and surge current capability is an ideal device for driving large BJTs. It allows the designer to couple low level logic control circuits directly to high power transistors without other intermediate elements.

Described in this paper are:

1. A comparison of drive methods for high power Darlington transistors.

2. A base drive circuit design using two power FETs.

A discrete power Darlington, using a Westinghouse D7ST as the output stage and D60T as the driver stage, is used as an example in the following discussion. The composite device is rated at 125 amperes continuous and 450 volts. It is connected as a four-terminal device with the base leads of both the driver and output transistors available.

Comparison of Drive Methods

In a power Darlington the primary drive problem occurs at turn-off. There are several ways in which the negative drive at turn-off can be applied. The first involves the use of negative bias on both base-emitter junctions (see Figure 1a). Another method utilizes a diode connected between the bases of the two transistors (see Figure 1b) to provide a path for discharge current from the base of the output transistor. This path is necessary because the driver transistor (Q_2) cuts off before all base charge has been removed from the output transistor (Q_1). Without a diode or a resistance between the bases (R_1 in Figure 1), the only method of charge removal from the power device would be internal recombination which is a gradual process. The presence of the diode insures that reverse potential is applied to the output device after the driver turns off.

Both methods of turn-off were compared on the D60T-D7ST transistor pair. The first two photographs (Figures 2a and 2b) show the collector current, reverse base current and collector-emitter voltage for a single reverse drive and dual reverse drive. Figure 2b shows that the device has a storage time of 5.4 microseconds at a collector current of 80 amperes. Collector-emitter voltage was 140 volts. A forward drive current of 4 amperes into the base of the driver transistor was employed so that upon application of reverse drive the D60T device was heavily saturated. The reverse bias potential (7 V) applied to the two devices resulted in a reverse base current of 5 amperes. Seven volts reverse bias was used because it was the maximum voltage which the emitter-base junction of the power device could withstand without avalanching.

There are two distinct reverse current levels during the storage time. The primary level which lasts for

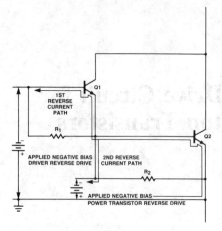

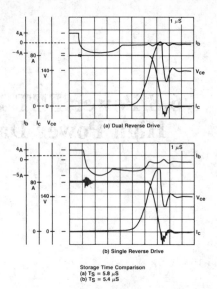

Storage Time Comparison
(a) $T_S = 5.8\ \mu S$
(b) $T_S = 5.4\ \mu S$

Single versus Dual Drive T_S Comparison (80 Amperes I_C) Figure 2

Dual Reverse Drive Figure 1a

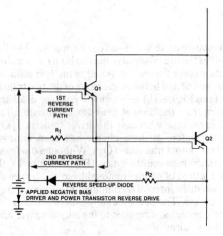

Single Reverse Drive (with Speed-Up Diode) Figure 1b

2.4 microseconds corresponds to the interval when charge is removed from both the driver device and the power device. The smaller secondary reverse current level, which occurs after the driver turns off, corresponds to the removal of charge from the power device through the reverse speed-up diode. Note that the smaller current level lasted approximately 3 microseconds, indicating that contributions to the overall storage time were roughly the same for both devices.

Figure 2a, a dual reverse drive, exhibits a 6 microsecond storage time which is slightly greater than the single reverse storage time. All the conditions were identical to those employed in the single reverse drive

circuit. In this case, however, the 7 volt negative bias voltage was applied to both base-emitter junctions at turn-off. The magnitude of the reverse current measured at the base of the driver transistor was slightly smaller than it was when using a single reverse drive. This happened because both reverse currents were provided by the same negative bias source. It is conceivable that with a stronger negative bias, or a system using two independent negative biases, the dual drive scheme could be more effective than the single drive scheme.

From the above experimentation, the single drive scheme seems to provide a good compromise between drive complexity and fast device turn off. Additionally, in the single drive Darlington configuration, the designer has some flexibility in selecting the number of speed-up diodes employed. The speed-up diodes turn on as the driver device turns off and provides a negative bias of one or more diode drops to the driver's base-emitter junction. The internal construction of the large transistor makes it a somewhat distributed device so that reverse bias voltages across emitter-base junction may also prevent dv/dt turn on of the transistor. Some empirical measurements with two and three speed-up diodes are presented (Figures 3a and 3b). The results of these comparisons indicate a choice of two anti-parallel diodes would minimize the Darlington devices' storage time effects. The storage time comparison for 1, 2, and 3 series speed-up diodes using a collector current of 80 amperes was 5.4, 4, and 4.3 microseconds, respectively.

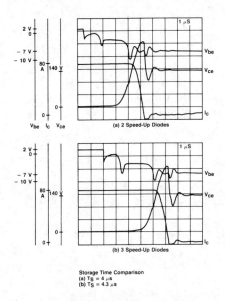

Storage Time Comparison
(a) $T_S = 4\ \mu s$
(b) $T_S = 4.3\ \mu s$

**Single Drive, Series Speed-Up Diodes Storage Time Comparison (80 Amperes)
Figure 3**

The Base Drive Circuit

The base drive circuitry for the Darlington power switch is shown in Figure 4 and Table 1. At the extreme left of Figure 4, the optical coupler HP2601

provides isolation between the control logic and the base drive which was DC coupled to the power circuit. The HP2601 utilizes a Schottky transistor output. The open collector output of the Schottky transistor is coupled into the bases of identical *npn* switching transistors whose collectors were connected to a positive 15 volt supply through resistors R5 and R4. In turn, the outputs of transistors Q2 and Q3 are low pass filtered using the networks R6-C6 and R7-C7. The diodes D1 and D2 increase the noise margin of the transistors Q2 and Q3.

The transistor pairs (Q7, Q6) and (Q5, Q4) are complementary emitter followers. The resistor-capacitor network at the output of (Q6, Q7) coupled with the internal capacitance of the FET IRF531 provides a slower voltage rise at the gate of the FET at turn on. In this case, a compromise had to be reached between the requirements of fast turn on for the Darlington pair and oscillations caused by stray inductance in the circuit. The zener diodes (ZD4, ZD5) and (ZD2, ZD3) provide transient protection to the gates of the FETs. R9, with a value of 1/4 ohm, damps oscillations excited by stray reactances at turn-off, while R10 and the internal resistance of the FET provide a means of controlling the forward current. It should be noted that the forward and reverse drive are arranged in a symmetrical push-pull arrangement. Either the *n*-channel FET will be ON providing forward drive, or the *p*-channel FET will be ON providing reverse drive to the power Dar-

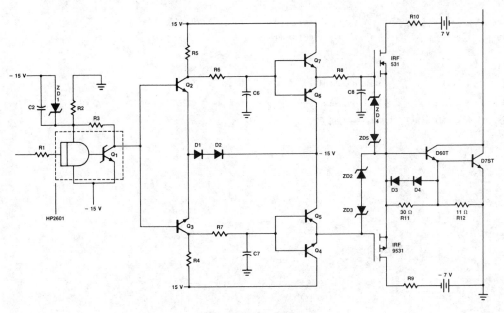

**Darlington Base Drive Circuit
Figure 4**

6

lington. The switching waveforms shown in the oscillograms were made using the base drive circuit described.

Table 1
Component Values for the
Darlington Base Drive Circuit

(All resistances are rated 1/4 Watt, 5% unless otherwise specified)

C_2	22 uF	25 V tantalum
C_6, C_7	87.6 pF	
C_8	130 pF	
D_1, D_2		Rectifier Diodes
D_3, D_4	A 114 F	Fast Recovery Diodes
Q_1	HP 2601	Optocoupler
Q_2, Q_3	TCG 161	*npn* switching transistor
Q_4, Q_7	mm 5262	*npn* switching transistor
Q_5, Q_6	mm 3726	*pnp* switching transistor
R_1	360	
R_2	180	
R_3	510	
R_4, R_5	1.8 k	1/2 Watt
R_6, R_7	1.6 k	
R_8	680	
R_9	1/2	1 Watt
R_{10}	11/6	24 Watts (6-11 4 Watt noninductive resistors)
R_{11}	30	1 Watt
R_{12}	11	4 Watts noninductive
ZD_1		4.7 V zener diode
$ZD_2, ZD_3,$ ZD_4, ZD_5	1N4746A	18 V zener Diode

6.9 A High Power, High Frequency FET Inverter for a Low Frequency Transmitter

Introduction

This paper describes a 3 kW, 150 kHz FET inverter module that is part of a low frequency transmitter. The module is a single phase bridge inverter providing a quasi-square wave output voltage. A total of 16 such modules are cascaded to provide a variable frequency, variable voltage DC-to-sinusoidal power source for feeding a low-frequency transmitter antenna. A Read Only Memory (ROM) is used to coordinate the switching of the 64 FETs in the system and to achieve sinusoidal output with controllable voltage and frequency.

A brief description of the system will be given first. Then attention will be focused on power inverter operation. Breadboard construction and experimental waveforms obtained will also be presented.

DC-to-Sinusoidal Inverter

Figure 1 shows the DC-to-Sinusoidal inverter using 16 inverter power modules. The secondary windings of the transformers were connected in series. Through ROM-controlled timing of inverter switching, the upper 8 modules synthesize a stepped sinusoidal voltage, and the lower 8 modules synthesize a second stepped sinusoidal voltage. The resultant output is a vector summation of the two stepped waveforms. The amplitude of the output voltage is controlled by adjusting the phase angle between the two stepped waveforms, and the frequency is controlled by varying the frequency of the clock feeding the ROM.

Figure 2 shows the timing relationship between one voltage waveform and the individual secondary winding voltages that make it up. Since the load can be inductive, capacitive, or resistive, the load current may be leading, lagging, or in phase with the output voltage. Therefore, the primary current of each inverter may also have any phase relationship to the primary voltage. Current commutation in the inverter takes place from transistor to diode or vice versa, and problems associated with power MOSFET parasitic diode recovery could occur in the inverter. Schottky diodes are used to remedy this situation as will be described in the next section.

FET Bridge Inverter

Figure 3 shows a diagram of one inverter power bridge. The source voltage is 300 V, and the maximum switched current is approximately 15 A. A low voltage Schottky diode, in series with each FET, is included. The reason for using the Schottky diodes is to prevent momentary "shoot through" caused by slow recovery of the FET parasitic diodes. In case of reverse current flow, the Schottky diodes divert the current from the parasitic diode to the discrete fast recovery diode. This avoids the problems associated with parasitic diode recovery. Because of the speed and power requirements in this application, this is necessary to prevent Mode 3 dv/dt problems with the power MOSFETs.

6

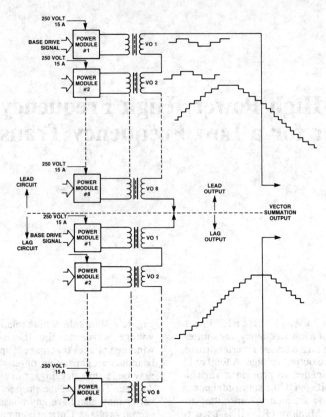

Block Diagram for Vector Summation Scheme
Figure 1

Gate Drive Circuit

Figure 4 is a diagram of a typical gate drive circuit. The dotted portion of the diagram provides an auto-protection mechanism which prevents the FET drain current from exceeding a preset value. This feature is essential in the present application because of the number of FETs involved in the system.

In the drive circuit, both positive and a negative 12 V power supplies are used. A negative voltage supply is used to ensure the cut-off of FETs. This was done in this system because switching noise generated by other FETs in the inverter may induce spurious gate voltages. A negative gate voltage supply minimizes this risk.

Clamping Mode of Operation

To have a quasi-square output voltage waveform, the inverter FETs must be gated according to the wave-forms shown in Figure 3. A dead time of 1 μs is provided in the gating waveforms between A and D and also between B and C to prevent shoot through. When switches A and B conduct, the output voltage is negative. When one of the four pairs (Q_B, D_D), (Q_D, D_B), (Q_C, D_A) and (Q_A, D_C) conducts, the output voltage is clamped to approximately 0 V. Which of the above four pairs conducts depends on the timing of the base drive waveform and the nature of the load. In order to use the vector summation method of obtaining an output voltage, this mode of operation is necessary. The clamping mode allows the output voltage of each inverter module to be independently controlled regardless of the direction of the system load current.

Due to such operation, however, the electrical stress on the four FET switches may be different. Two switches may turn off at a higher level of current than the other two.

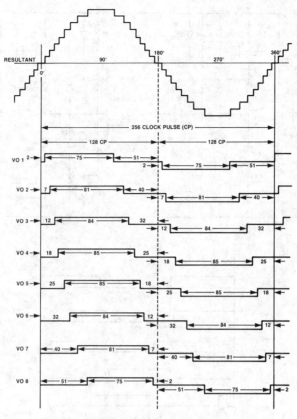

**Timing Waveform for Synthesizing
Sinusoidal Output Voltage
Figure 2**

Component Selection and Design

Semiconductor Devices: Two power FETs (IRF450) were paralleled to handle the required current level (15 A). With available MOSFETs, paralleling was necessary. 600 V fast recovery diodes (200 nsec recovery time) were used for reverse conduction around each FET pair and a Schottky diode was used to block reverse current through each FET pair.

Transformer: At 150 kHz operation, Litz wire was used to reduce skin effect losses, and bifilar winding was used to minimize leakage inductance. A Ferrite core was used. The transformer was designed such that the maximum operating flux density was 500 Gauss. A toroidal core was chosen to minimize flux leakage.

Capacitors: Two types of capacitors were used in the circuit. C_L was used to block DC voltage and prevent the transformer core from saturating. A polypropylene AC capacitor was selected for its low ESR. The other capacitors were connected directly across the DC bus and physically very close to the inverter leg. These capacitors were used to decouple line inductance and minimize voltage spiking when a FET turned off. A low inductance type DC capacitor was chosen for this service.

Snubber Circuit: The voltage spiking caused by FET turn-off can be controlled by tight power circuit layout and putting capacitors across the input lines. Normally, snubber circuits not only add to the parts count but also cause undesirable circuit interaction in a bridge inverter. Nonetheless, a simple RCD snubber with a small capacitor (1000 pF) was used

6

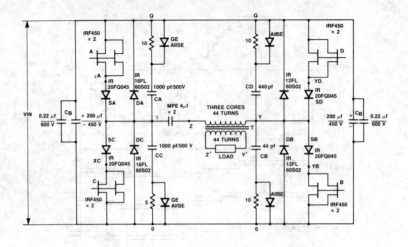

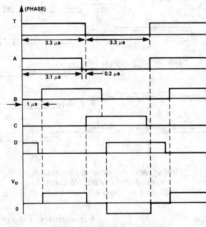

**FET Power Inverter Circuit
and Gate Drive Timing Waveforms
Figure 3**

around each FET pair in the breadboard. This was found to be necessary to reduce turn-off speed and noise. Since the snubber capacitor is very small, the power loss associated with the snubber remains small, and the problems associated with snubber circuit interaction are minimized.

Breadboard Construction: Figure 5 shows a photo of the actual breadboard. All the FETs and diodes are mounted on the same heatsink with insulating material between them. This reduces the length of the power path. Gate drive boards are on the left and the transformer is on the right.

Experimental Results

Figures 6 through 9 show oscillograms from the operating breadboard. Figure 6 shows the gate voltage and current waveforms. A reverse gate voltage is applied to the FETs during the off period. Figure 7 shows the output voltage waveform, and Figure 8 shows the load current waveform. The ringing in the voltage waveform is caused by the resonance of the transformer parasitic capacitance and circuit inductance. Figure 9 shows the turn-off switching waveforms of a FET.

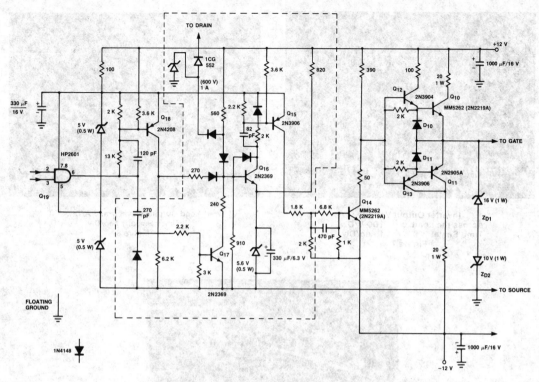

Gate Drive Circuit with Auto Protection Circuit
Figure 4

Inverter Breadboard
Figure 5

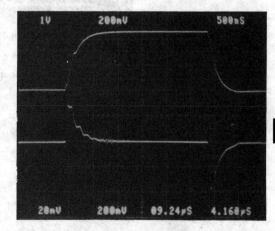

Gate Voltage:	10 V/Div.
Gate Current:	200 mA/Div.
Time Scale:	500 nsec/Div.

Figure 6

6

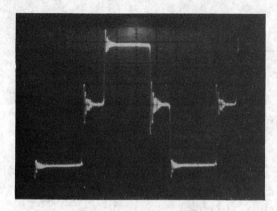

Inverter Output Voltage
Across the Load: 100 V/Div.
Time Scale: 1 sec/Div.
Figure 7

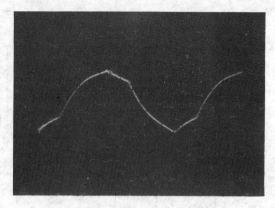

Load Current Waveform: 8 A/Div.
Times Scale: 1 sec/Div.
Figure 8

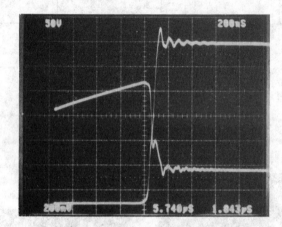

Turn-Off Voltage and Current Waveforms
Drain Current: 4 A/Div.
Drain Voltage: 50 V/Div.
Time Scale: 200 nsec/Div.
Figure 9

6.10 Linear Voltage and Current Regulators Using Power MOSFETs

Introduction

A wide variety of linear integrated circuit voltage regulators are presently available in monolithic and hybrid form. These devices accept input voltages in the range of 0 to 120 V, provide regulated output voltages between 0 and 30 V, and handle load currents up to 10 A. A small number of monolithic current regulators are also available, but these devices are primarily limited to low current and low voltage applications.

When the voltage and/or current requirements of a particular application exceed the capabilities of the available IC regulators, one must turn to a discrete design. Power MOSFETs when operated in their saturation region [1] can be used to replace bipolar transistors as series pass elements in linear regulators. There are two major advantages in using MOSFETs over bipolar transistors:

1. The average gate drive current for a MOSFET is considerably lower than the average base current of a comparably rated bipolar transistor.

2. A MOSFET can be operated at rated breakdown voltage and maximum drain current under certain pulsed conditions. Consequently, MOSFETs have a full rectangular SOA curve whereas bipolar transistors do not. [2]

Additionally, MOSFETs are easy to parallel if the proper precautions are taken, and, unlike bipolar transistors, they do not require current equalizing resistors. Reference 3 comprehensively discusses the correct techniques for successful parallel operation of these devices.

Low power gate control circuitry can be designed using CMOS or BIFET op-amps that draw very little standby current. A MOSFET linear regulator that delivers 10 A or more can be built with relatively few components. A separate driver transistor, necessary to boost the output current of an op-amp or other amplifier to drive a large series pass bipolar transistor, is unnecessary with a power MOSFET.

This section describes four selected voltage and current regulator circuits that make use of both open-loop control (no feedback) and closed-loop control (with negative feedback) to regulate the output. Each circuit has its own inherent benefits and drawbacks that will be discussed in detail.

Voltage Regulator Circuits

Linear voltage regulators generally take the form of a series-pass circuit in which the pass transistor sustains the difference between the input and output voltages. The simplest series-pass MOSFET regulator is shown in Figure 1. This is an open-loop circuit in which the IRF130 is operated as a DC source follower. The output voltage is equal to the MOSFET gate voltage (V_G) less the value of gate-to-source voltage (V_{GS})

necessary to sustain the load current. Since the gate voltage is equal to the voltage generated across R_1 by current I_1 the output voltage can be shown to be:

$$V_{OUT} = I_1R_1 - V_{GS} \qquad (1)$$

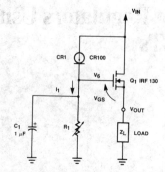

This Simple Adjustable Output Voltage Regulator Uses an IRF130 MOSFET and a Small Number of Additional Components
Figure 1

To ensure that the power MOSFET always operates in its saturation region,

$$V_{DS} \geq V_{GS} - V_T \qquad (2)$$

where V_T is the MOSFET threshold voltage.

In the circuit of Figure 1, the MOSFET will always have a value of V_{DS} that is greater than its V_{GS}. This is because the current-regulator diode requires a certain minimum voltage drop across itself ($\cong 1.0$ V) to function properly. Consequently, the MOSFET always operates in its saturation region. The maximum output voltage the circuit can generate is

$$V_{OUT}(max) = V_{IN} - [V_F(CR1) + V_{GS}] \qquad (3)$$

while the minimum output voltage is zero.

Additionally, the value of drain-to-source voltage for the MOSFET should never exceed its rated BV_{DSS}. Thus,

$$|V_{IN} - V_{OUT}|max = BV_{DSS} \qquad (4)$$

Despite its simplicity, the circuit shown in Figure 1 does have a major disadvantage which may limit its usefulness. This is quite evident when the transfer characteristics of the IRF130 MOSFET, shown in Figure 2, are examined. Because the output voltage is dependent on V_{GS}, as V_{GS} varies with load current (I_{OUT}) and junction temperature (T_j), V_{OUT} varies too.

When, for example, the output voltage of the regulator is adjusted to 5 V at a load current of 1 A, as I_{OUT} is increased to 10 A, V_{OUT} drops by 2.3 V to 2.7 V. This drop in output voltage with increasing load current is undesirable, but is an inherent characteristic of the circuit. Consequently, this type of regulator is not suited to low voltage applications because the percentage change in output voltage can be as high as 50%. Rather, it is suited to the higher voltage applications (20 V and above) where the percentage change in V_{OUT} will be much lower (10% or less).

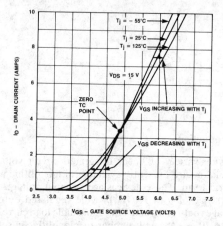

The Transfer Characteristics of the IRF130 MOSFET Show that for a Constant I_D, Below the Zero TC Point V_{GS} Decreases with Temperature While Above It V_{GS} Increases with Temperature
Figure 2

Additionally, the output voltage varies with junction temperature, and this variation tends to aggravate the situation described above. At low values of I_{OUT}, the temperature variation of output voltage is dominated by the threshold voltage drift with temperature. This change in threshold voltage is typically in the range of -3 to -6 mV/°C and causes the output voltage to increase with temperature in equal magnitude. For a constant value of I_{OUT}, an increase in T_j of approximately 100°C can cause V_{GS} (and, hence, V_{OUT}) to change by as much as 600 mV.

When the load current is increased, a point is reached where the output voltage ceases to change as the junction temperature varies. This point is known as the zero temperature coefficient (or zero TC) point and is characteristic of all power MOSFETs. Above the zero TC point, the output voltage decreases with increasing junction temperature for a constant value of load current. This occurs because the MOSFET drain current is proportional to both the channel carrier mobility (μ), as well as the square of (V_{GS}–

V_T). Since μ decreases with increasing T_j, (V_{GS}–V_T) must increase to sustain a given value of load current. Above the zero TC point, even the decrease in V_T with temperature is insufficient to compensate for the decrease in μ with temperature and, thus, V_{GS} increases. At the zero TC point, however, the competing effects of $\frac{\partial \mu}{\partial T}$ and $\frac{\partial V_T}{\partial T}$ completely cancel each other and V_{OUT} remains constant.

When the performance of the open-loop regulator is unacceptable for a particular application, a closed-loop method of controlling the output voltage becomes necessary. A slightly more complex closed-loop voltage regulator using an IRF130 MOSFET and a TL071 BIFET op-amp is shown in Figure 3. The advantage of this circuit is that the TL071 functions as a DC error amplifier and keeps the output voltage virtually equal to the reference voltage (V_{REF}) at its non-inverting input. The output voltage is almost completely independent of the MOSFET gate-to-source voltage as shown by the following equation:

$$V_{OUT} = \frac{(V_{REF} + V_{IO})A_{VOL}(DC) - V_{GS}}{1 + A_{VOL}(DC)} \quad (5)$$

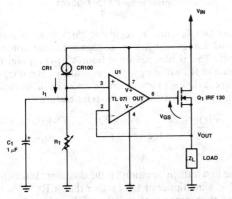

This Simple Adjustable-Output Closed-Loop Voltage Regulator Provides a V_{OUT} that is Independent of the IRF130 MOSFET's Gate to Source Voltage
Figure 3

Here, $A_{VOL}(DC)$ is the DC open-loop voltage gain of the op-amp, V_{IO} is the input-offset voltage of the op-amp, and V_{REF} is equal to I_1R_1. Since $A_{VOL}(DC)$ is large for most op-amps (100 dB or more is typical), the variation of V_O with V_{GS} is insignificant. This regulator will have a very low DC output resistance because negative feedback is used to maintain a constant output voltage. The closed-loop output resis-

tance can be determined using the approximate relation:

$$R_{OUT} \cong \frac{1/g_m}{1 + A_{VOL}(s)} \quad (6)$$

where g_m is the transconductance of the MOSFET used in the output stage and $A_{VOL}(s)$ is the open-loop gain of the op-amp as a function of frequency. Since $A_{VOL}(s)$ for most op-amps decreases linearly with frequency, the value of R_{OUT} will rise linearly with frequency. This is somewhat undesirable, particularly if the regulator must power circuits whose supply currents pulsate at high frequencies (i.e., audio and video amplifiers). However, a suitable capacitor connected between V_{OUT} and ground in the circuit of Figure 3 will overcome the increase in R_{OUT} with frequency.

The temperature variation of V_{OUT} in this regulator can be seen to be virtually independent of any temperature variation in V_{GS}. This is obvious if equation 5 is differentiated with respect to temperature:

$$\frac{\partial V_O}{\partial T} = \frac{\partial V_{REF}}{\partial T} + \frac{\partial V_{IO}}{\partial T} \quad (7)$$

Here we have assumed negligible variation in $A_{VOL}(DC)$ with temperature which is valid in most cases. Thus, the drift in V_{OUT} with temperature is mainly dependent upon the input offset voltage drift with temperature and the reference output voltage drift with temperature. Both of these drift components can be minimized by using precision components for the reference voltage generator and the DC error amplifier. In some cases, the temperature coefficients of V_{REF} and V_{IO} may partially cancel each other if their polarities are opposite. This would then result in a temperature coefficient for V_{OUT} that is less than either that of V_{REF} or V_{IO}.

The designer should be aware that since this regulator uses a negative feedback loop, oscillation can occur if the circuit has an insufficient phase margin (ϕ_M).[4] Most modern internally-compensated op-amps have a single low-frequency dominant pole (usually 10 Hz to 100 Hz) as well as several higher order poles and zeroes. The frequency of the dominant pole is normally chosen so that at the unity-gain crossover frequency, the op-amp has a phase margin of approximately 50° to 80° and is stable.

The addition of a MOSFET output stage to the op-amp in Figure 3 adds two poles and a zero to the loop transfer function[5]. This addition can result in an overall phase margin that is near zero or even negative. For an overall phase margin of 30° or less, instability in the regulator is very likely. The HP-41 program described in Reference 5 allows one to calculate the gain, poles, and zero of a MOSFET source follower. The program can be used as an aid in plotting the Bode diagrams of loop gain and loop phase for the circuit shown in Figure 3.

It is difficult to provide typical Bode diagrams of loop-gain and loop-phase versus frequency for the circuit of Figure 3 because the poles of the MOSFET source-follower vary widely with different load impedances. For each load condition, a different set of bode plots would have to be generated. This can be rather time consuming. Alternatively, the regulator itself can be built and then tested with different load impedances in an attempt to make it oscillate. This, in fact, was done with the circuit of Figure 3.

It was found the regulator was unconditionally stable with no load, a purely resistive load, and a parallel resistive/capacitive load. The values of capacitance used varied from 0.1 μF to 1000 μF. With a purely capacitive load of more than 10 μF, however, oscillation in the gate drive voltage was noticed at the output of the op-amp. The frequency of oscillation was rather low (100 to 200 Hz), and the waveform was that of a distorted sinewave. The amplitude of oscillation was quite large, being equal to the limits of the op-amp's output swing (approximately 20 Vpp), but the DC output voltage showed no signs of any perturbation. Presumably, this was due to the filtering action of the purely capacitive load. The DC output voltage remained constant at a value almost exactly equal to V_{REF}. Consequently, this type of instability could be tolerated in certain designs since the regulator appears to operate normally as far as the capacitive load is concerned.

Current Regulator Circuits

Discrete current regulators generally take the form of a series pass circuit like their voltage regulator counterparts. In these circuits as well, the pass transistor sustains the difference between the power supply voltage and the voltage across its load. The pass transistor is forced to conduct a constant current by suitable biasing circuitry. This circuitry prevents the output current from fluctuating due to variations in power supply voltage and load impedance.

The simplest current regulator, an open-loop circuit, is shown in Figure 4. The regulator uses the base-emitter junction of a bipolar transistor (Q_1) as a reference voltage generator across the MOSFET source resistor R_S. This forces the MOSFET to conduct a drain current, independent of its drain-to-source voltage, that is equal to

$$I_{OUT} = \frac{V_{BE(Q1)}}{R_S} \tag{8}$$

The 1 mA constant-current diode used in the circuit provides operating bias for the bipolar transistor and ensures that its emitter current, and hence, V_{BE} does not vary substantially with power supply voltage. Consequently, the circuit will have a very high degree of rejection to power supply voltage fluctuations.

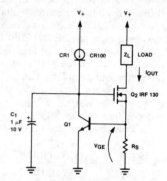

This Simple Open-Loop Current Regulator Uses a Simple V_{BE} Voltage Reference to Establish a Constant Drain Current in the IRF130 MOSFET
Figure 4

This simple current regulator does have a major disadvantage just like its voltage regulator counterpart. This is obvious since both $V_{BE(Q1)}$ and the value of R_S will vary with temperature. The temperature variation in I_{OUT} can be found by differentiating equation 8 with respect to temperature:

$$\frac{\partial I_{OUT}}{\partial T} = \frac{1}{R_S} \frac{\partial V_{BE(Q1)}}{\partial T} - \frac{V_{BE(Q1)}}{R_S^2} \frac{\partial R_S}{\partial T} \tag{9}$$

The first term in equation 9 is the dominant term since it has a multiplier of $1/R_S$ rather than $1/R_S^2$. Therefore, the temperature variation of $V_{BE(Q1)}$ is mainly responsible for the drift in I_{OUT}. Since equation (9) does contain two terms of opposite sign, it is possible to make the output current independent of temperature if the type of resistor used for R_S is chosen according to the relation

$$\frac{1}{R_S} \frac{\partial R_S}{\partial T} = \frac{1}{V_{BE}} \frac{\partial V_{BE}}{\partial T} \tag{10}$$

which basically states that the fractional change in R_S per °C must be equal to the fractional change in $V_{BE(Q1)}$ per °C (about -0.34% per °C).

A modified open-loop current regulator that is similar to the previous one is shown in Figure 5. This circuit employs a slightly more complex Widlar band-gap, reference[6], instead of a simple temperature sensitive V_{BE} reference. The advantages of using the band-gap voltage reference in a current regulator are improved output current accuracy and temperature stability over the V_{BE} referenced circuit. These advantages outweigh the need for the additional components.

The band-gap reference has a characteristic zero temperature-coefficient point that can be made to occur at any desired ambient temperature by simply choosing the circuit values appropriately. For the circuit in Figure 5, the zero TC point occurs at an ambient temperature of approximately 25°C, and the reference output voltage is equal to the band-gap voltage of silicon (1.205 V) plus a small constant (typically less than 0.1 V) that is circuit dependent. In this case, the reference output voltage is $\cong 1.220$ V, and the output current is therefore equal to

$$I_{OUT} = \frac{1.220 \text{ V}}{R_S} \qquad (11)$$

It should be noted that in order to achieve relative insensitivity in I_{OUT} to changes in temperature, the resistors used in the band-gap voltage reference must be low temperature-coefficient types. This necessitates the use of high-quality metal-film or wire-wound resistors and not the less expensive carbon-composition type. The reference resistor R_S, of course, should also be of high stability.

We now turn our attention to Figure 6 where the circuit diagram for a two terminal closed-loop current regulator is shown. This circuit uses a CA3140 BI-MOS op-amp to drive an IRF130 power MOSFET that is used as the series pass element. A low-value,

high-wattage resistor is used as the current sensing element, and the voltage generated across this resistor is fed back to the inverting terminal of the op-amp. The op-amp thus forces the current flowing through the MOSFET to be equal to

$$I_{OUT} = \frac{V_{REF} + V_{IO}}{R_{SENSE}} \qquad (12)$$

where V_{REF} is equal to $I_1 R_1$ and V_{IO} is the input offset voltage of the op-amp.

The variation of I_{OUT} with temperature will be rather complicated due to the fact that all four terms in equation 9 (I_1, R_1, V_{IO}, and R_{SENSE}) vary with temperature. For the CA3140 op-amp, the variation of V_{IO} with temperature is only about 5 V/°C and does not constitute a major component of drift in the output current. If we also assume that R_1 is a low TC potentiometer, then

$$\frac{\partial I_{OUT}}{\partial T} \cong \frac{R_1}{R_{SENSE}} \times \frac{\partial I_1}{\partial T} - \frac{(I_1 R_1 + V_{IO})}{(R_{SENSE})^2}$$
$$\times \frac{\partial R_{SENSE}}{\partial T} \qquad (13)$$

Here, notice that the second term in equation (10) will be dominant since it has the term $(R_{SENSE})^2$ as its denominator. Thus, the variation in I_{OUT} will be primarily dependent upon the stability of resistor R_{SENSE} as it heats up.

Because this circuit is configured as a two terminal regulator, it must operate from a minimum supply voltage of about 12 V. This ensures proper operation of the voltage regulator circuitry that supplies 10 V to the op-amp. If this minimum operating voltage is

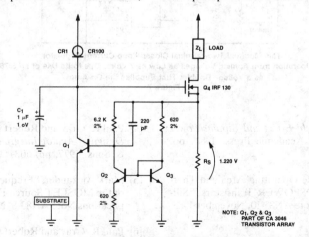

This Modified Open-Loop Current Regulator Uses a Widlar Band-Gap Voltage Reference for Improved Accuracy and Temperature Stability
Figure 5

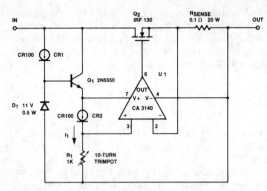

A Closed Loop Two-Terminal Regulator Can Be Built with a CA3140 B1MOS Op-Amp, a Low-Value High-Wattage Current Sensing Resistor, and a Few Additional Components
Figure 6

can thus regulate currents up to 10 A from a source voltage as low as 7 V. The maximum voltage that either of the circuits in Figure 6 or 7 can sustain is equal to the breakdown voltage of the IRF130 MOS-FET, which is around 100 V.

Since this current regulator uses a negative feedback loop like the voltage regulator in Figure 3, a sufficient phase margin for stable operation is necessary. Since the MOSFET is configured as a source-follower with a resistive load of 0.1Ω, using typical values from the IRF130 data sheet as input data for the HP-41 source-follower pole/zero finding program [5], we obtain

gain at DC	$= +0.2857$
zero frequency	$= -979.4\,\text{MHz}$
pole 1 frequency	$= -4.707\,\text{MHz}$
pole 2 frequency	$= -2.753\,\text{GHz}$

These pole and zero frequencies of the MOSFET source-follower combined with the 30 Hz dominant pole of the CA3140 and its higher order poles and zeroes, allow for stable operation with a feedback factor of unity as used in the circuits of Figures 6 and 7.

higher than desired, the op-amp and voltage reference can be powered from a voltage doubler using the Si7661 CMOS voltage converter IC as shown in Figure 7. This IC doubles the 5 V output from the voltage regulator to 10 V. With this modification, the circuit

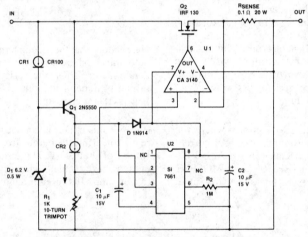

This Modified Two-Terminal Closed-Loop Current Regulator Permits Operation from Applied Voltages as Low as 7 Volts, Due to Its Use of an Si7661 as a Voltage Doubler That Supplies the Op-Amp
Figure 7

References

[1] Adolph Blicher, *Field Effect and Bipolar Power Transistor Physics,* Academic Press, 1981, pp. 238–243.

[2] Rudy Severns, "Safe Operating Area and Thermal Design for MOSPOWER Transistors," Siliconix Application Note AN83-10, November 1983, pp. 13–15.

[3] Rudy Severns, "Parallel Operation of Power MOSFETs," Siliconix Applications Note TA84-5, June 1984.

[4] Paul R. Gray and Robert G. Meyer, *Analysis and Design of Analog Integrated Circuits,* John Wiley & Sons, 1977, pp. 503–505.

[5] Mark Alexander, "Frequency Response Analysis of the MOSFET Source-Follower," Siliconix Applications Note AN83-8, November 1983.

[6] Paul R. Gray and Robert G. Meyer, *Analysis and Design of Analog Integrated Circuits,* John Wiley & Sons, 1977, pp. 258–260.

6.11 Frequency Response Analysis of the MOSFET Source-Follower (AN83-8)

Introduction

Power MOSFETs are becoming increasingly popular as output stages of linear power amplifiers. This trend will continue as the cost of these devices relative to bipolar transistors continues to decrease.

Perhaps the most widely used MOSFET output-stage is the source-follower configuration. It offers certain advantages over other types of output-stages:

- When used in a complementary arrangement with N and P-channel devices (see Figure 1), only single-node drive is required.

- Since the voltage-gain is positive and close to unity, the effective gate-to-source capacitance is *reduced*.

- The output impedance is only of the order of a few ohms or less.

These characteristics make the source-follower suitable for use in wide-band, high-speed amplifiers, particularly for audio applications. Most designers of audio power amplifiers with MOSFET output-stages, however, use only the single-capacitance MOSFET model (see Figure 2) in their frequency response calculations. Experiments have shown that the actual -3dB frequency of a source-follower can be less than half that calculated using the single-capacitance model.

Clearly, the single-capacitance MOSFET model is inadequate for frequency response calculations. What is needed is a more comprehensive model that will enable a designer to accurately predict how a MOSFET output-stage will behave with frequency. The three-capacitance model satisfies this requirement.

Analysis of the Three-Capacitance MOSFET Model

Figure 3 shows the schematic diagram of the three-capacitance MOSFET model. R_{gen} is the generator output impedance used in the test setup although the output impedance of a driver amplifier stage could be substituted instead. R_{ge} is an external gate resistor usually included to suppress any potential parasitic oscillation that might occur in the circuit. R_{gi} is the intrinsic gate-electrode resistance of the MOSFET, and R_L is the load resistance.

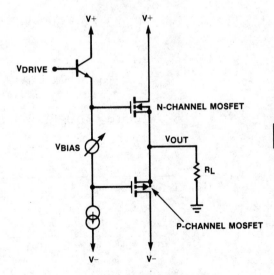

A Complementary MOSFET Output Stage that Needs Only Single-Node Drive
Figure 1

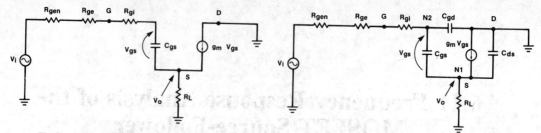

**The Single Capacitance AC-Model of the MOSFET
Source-Follower
Figure 2**

**The Three Capacitance AC-Model of the MOSFET
Source-Follower
Figure 3**

The analysis is as follows (refer to Figure 3)

1.

$$\text{KCL at N1 } (\Sigma i_{N1} = 0)$$

$$\frac{V_o - 0}{R_L} + (V_o - 0)sC_{ds} - g_m V_{gs} - V_{gs}sC_{gs} = 0$$

$$V_o(1/R_L + sC_{ds}) - V_{gs}(g_m + sC_{gs}) = 0$$

$$V_o = \frac{g_m R_L + sR_L\,C_{ds}}{1 + sR_L C_{ds}} V_{gs} \tag{1}$$

2.

$$\text{KCL at N2 } (\Sigma i_{N2} = 0)$$

$$((V_{gs} + V_o) - 0)sC_{gd} + ((V_{gs} + V_o) - V_o)sC_{gs} + \frac{((V_{gs} + V_o) - V_i)}{R_{gi} + R_{ge} + R_{gen}} = 0$$

$$(V_{gs} + V_o)sC_{gd} + V_{gs}sC_{gs} + \frac{((V_{gs} + V_o) - V_i)}{R_{gi} + R_{ge} + R_{gen}} = 0$$

$$V_{gs}(sC_{gd} + sC_{gs} + 1/(R_{gi} + R_{ge} + R_{gen})) + V_o (sC_{gd} + 1/(R_{gi} + R_{ge} + R_{gen})) - V_i/(R_{gi} + R_{ge} + R_{gen}) = 0$$

$$V_{gs} = \frac{1}{s(C_{gd} + C_{gs}) + 1/(R_{gi} + R_{ge} + R_{gen})} \left[\frac{V_i}{R_{gi} + R_{ge} + R_{gen}} - V_o(sC_{gd} + 1/(R_{gi} + R_{ge} + R_{gen})) \right]$$

$$V_{gs} = \frac{1}{1 + s(R_{gi} + R_{ge} + R_{gen})(C_{gd} + C_{gs})} \left[V_i - V_o (1 + s(R_{gi} + R_{ge} + R_{gen})C_{gd}) \right] \tag{2}$$

3. Substituting (2) into (1) and simplifying,

$$\frac{V_o}{V_i} = \frac{g_m R_L}{1 + g_m R_L} \times \frac{1 + sC_{gs}/g_m}{1 + s\left[\dfrac{R_L((C_{gs} + C_{ds}) + (R_{gi} + R_{ge} + R_{gen})(C_{gs} + (1 + g_m R_L)C_{gd}))}{1 + g_m R_L}\right]} \dots$$

$$\dots \frac{}{+ s^2\left[\dfrac{R_L(R_{gi} + R_{ge} + R_{gen})(C_{ds}(C_{gd} + C_{gs}) + C_{gs}C_{gd})}{1 + g_m R_L}\right]} \tag{3}$$

Since this transfer function is of the form:

$$\frac{V_o}{V_i} = DC\ gain\ \frac{(1 + s\tau_3)}{(1 + s\tau_1)(1 + s\tau_2)}$$

where

$$f_{zero} = -1/(2\pi\ \tau_3)$$

$$f_{pole1} = -1/(2\pi\ \tau_1)$$

$$f_{pole2} = -1/(2\pi\ \tau_2)$$

by setting the denominator of (3) equal to zero, it is then of the form $ax^2 + bx + c = 0$. The quadratic formula can then be used to find the two pole frequencies.

HP-41 Program "SFANL"

An easy to use program for the HP-41CV programmable calculator was written to perform the somewhat tedious calculations involved. It features prompting for the eight input variables (R_{gen}, R_{ge}, R_{gi}, R_L, C_{gs}, C_{gd}, C_{ds}, and g_m) as well as the four output variables (DC gain, f_{zero}, f_{pole1}, and f_{pole2}).

The program has two entry points, the first being the initialization section where input data is prompted for and put into registers R_0 through R_7. The second is the entry point into the calculation section of the program, which assumes that input data is already present in R_0 through R_7. This second entry point is useful if only one or two input variables must be changed before a subsequent run of the program.

The program listing is as follows:

a) Main program SFINT/SFANL

01	LBL ᵀSFINT	entry point for initialization
02	ᵀENTER GM	begin initialization
03	PROMPT	
04	STO 00	
05	ᵀENTER CGS	
06	PROMPT	
07	STO 01	
08	ᵀENTER CDS	
09	PROMPT	
10	STO 02	
11	ᵀENTER CDG	
12	PROMPT	
13	STO 03	
14	ᵀENTER RGI	
15	PROMPT	
16	STO 04	
17	ᵀENTER RGE	
18	PROMPT	
19	STO 05	
20	ᵀENTER RGN	
21	PROMPT	
22	STO 06	
23	ᵀENTER RL	
24	PROMPT	
25	STO 07	Initialization done
26	LBL ᵀSFANL	Entry point for calculations
27	RCL 00	Begin calculations
28	RCL 07	
29	*	
30	ENTER↑	
31	ENTER↑	
32	1	
33	+	
34	STO 08	
35	/	
36	STO 09	DC gain calculated, save in R9
37	RCL 00	
38	RCL 01	
39	PI	
40	*	
41	2	

6

42	*	
43	/	
44	CHS	
45	STO 10	f_{zero} calculated, save in R_{10}
46	RCL 08	
47	RCL 03	
48	*	
49	RCL 01	
50	+	
51	RCL 04	
52	RCL 05	
53	+	
54	RCL 06	
55	+	
56	*	
57	RCL 01	
58	RCL 02	
59	+	
60	RCL 07	
61	*	
62	+	
63	RCL 08	
64	/	
65	STO 11	Denominator s coefficient calculated, save in R_{11}
66	RCL 01	
67	RCL 03	
68	+	
69	RCL 02	
70	*	
71	RCL 01	
72	RCL 03	
73	*	
74	+	
75	RCL 04	
76	RCL 05	
77	+	
78	RCL 06	
79	+	
80	*	
81	RCL 07	
82	*	
83	RCL 08	
84	/	
85	STO 12	Denominator s^2 coefficient calculated, save in R_{12}
86	1	
87	RCL 11	
88	RCL 12	
89	XEQ ^{T}QROOT	Find roots of denominator
90	FS? 09	Complex roots?
91	GTO ^{T}CMPLX	Go if yes
92	PI	

93	/	
94	2	
95	/	f_{pole1} calculated, save in X
96	X< >Y	
97	PI	
98	/	
99	2	
100	/	
101	X< >Y	
102	^{T}GAIN =	
103	ARCL 09	
104	PROMPT	Display DC gain
105	^{T}FZER =	
106	ARCL 10	
107	PROMPT	Display f_{zero}
108	^{T}FPL1 =	
109	ARCL X	
110	PROMPT	Display f_{pole1}
111	^{T}FPL2 =	
112	ARCL Y	
113	PROMPT	Display f_{pole2}
114	GTO ^{T}SFANL	Loopback
115	LBL ^{T}CMPLX	
116	^{T}CPX POLES	
117	PROMPT	Prompt if poles are complex
118	END	

b) Subroutine QROOT (finds roots of $ax^2+bx+c=0$)

01	LBL ^{T}QROOT	Entry point for QROOT
02	CF 09	
03	STO 19	Save a in R_{19}
04	RDN	
05	STO 18	Save b in R_{18}
06	RDN	
07	RCL 19	
08	*	
09	4	
10	*	
11	RCL 18	
12	X↑2	
13	X< >Y	
14	−	$b^2 - 4ac$ calculated
15	X<0?	Complex roots?
16	GTO ^{T}REAL	Go if not
17	SF 09	Set flag 9
18	CHS	
19	TONE 9	
20	LBL ^{T}REAL	
21	SQRT	
22	RCL 19	
23	2	
24	*	

25	STO 19		
26	/		
27	RCL 18		
28	RCL 19		
29	/		
30	CHS		
31	FS? 09		
32	RTN	Complex roots calculated, return	
33	STO 18		
34	X<>Y		
35	STO 19		
36	−		
37	RCL 18		
38	RCL 19		
39	+		
40	RTN	Real roots calculated, return	
41	END		

Program Use and Verification

Note: SIZE on the HP-41CV should be set to 20 or greater, and the ENG 3 display-mode should be selected before using this program.

Running the program is a very simple procedure. When starting from label SFINT, after each prompt the appropriate input value should be entered into the X register and the R/S key pressed.

The input register map and associated prompts are listed below:

VARIABLE	REGISTER	PROMPT
g_m	R_0	"ENTER GM"
C_{gs}	R_1	"ENTER CGS"
C_{ds}	R_2	"ENTER CDS"
C_{gd}	R_3	"ENTER CGD"
R_{gi}	R_4	"ENTER RGI"
R_{ge}	R_5	"ENTER RGE"
R_{gen}	R_6	"ENTER RGN"
R_L	R_7	"ENTER RL"

After the eighth input value has been entered and the R/S key pressed, the HP-41CV will calculate and then display the four variables. After each variable is displayed, the R/S key should be pressed to see the next one. When the last variable is displayed, a subsequent press of the R/S key will restart the program at label SFANL.

The output-variable sequence and sample display is shown as follows:

VARIABLE	SAMPLE DISPLAY
DC gain	"GAIN = 888.9E-3"
f_{zero}	"FZER = −265.3E6"
f_{pole1}	"FPL1 = −1.964E6"
f_{pole2}	"FPL2 = −61.21E6"

A test was run to verify the program does indeed calculate values that match those taken from a test setup. In this case, a Siliconix VN1200A N-channel power MOSFET was connected as a class A source-follower with 24.0V between its drain and source. Also, the DC gate bias was adjusted so that the DC drain current was equal to 750 mA.

Under these conditions, the following data was recorded:

VARIABLE	VALUE	UNITS
g_m	1.5	A/V
C_{gs}	770	pF
C_{ds}	150	pF
C_{gd}	45	pF
R_{gi}	5	Ω
R_{ge}	4.3	KΩ
R_{gen}	50	Ω
R_L	8	Ω
f_{-3dB}	345.6	kHz
DC gain	0.930	—

The non-dominant pole and zero frequencies were not measured since they were outside the measurement range of the equipment used.

Using the test data as input to the HP-41 program, the following results were calculated:

VARIABLE	VALUE	UNITS
DC gain	0.923	—
f_{zero}	−310.0	MHz
f_{pole1}	−350.7	kHz
f_{pole2}	−171.7	MHz

Comparing the calculated dominant pole (f_{pole1}) with the measured value, it can be seen that the program does indeed work very well. In this case, the error in the calculated dominant pole was approximately 1.5%.

Clearly, the three capacitance MOSFET model is quite adequate for frequency response analysis in the range of DC to a few MHz. At higher frequencies, though, package inductance and capacitance must be included in the MOSFET model for an accurate analysis.

6

6.12 Stepper Motor Interface

"Commands, controls and data are more and more in digital form since digital information is readily manipulated, inexpensively, with integrated circuits. Thus, when an output motion is desired, a digital device (the stepping motor) provides a logical link between digital information and mechanical translation."

That's the opening statement in the introduction to one stepper motor manufacturer's handbook. The book then continues for the next 60 pages explaining this "Digital Device" in terms of slew rates, angular accuracy, ramping rates, inertia, load, torque, etc., none of which is digital terminology as encountered in everyday conversation.

Introduction

As long as physical laws are enforced, motion control systems will never be fully digital. However, if the stepper motor selected satisfies the load requirements, the motor/controller subsystem can be made to appear as a digital appendage to the logic system. In the past, attempts to achieve this relationship between stepper motor, controller, and logic system resulted in difficult and clumsy designs. But with the recent availability of new integrated circuits, power switches, and software specifically designed for stepper motor control, system design has been greatly simplified.

Between the logic system's commands and the motor lie two distinctly different problems. One problem is applying power to the motor's windings with sufficient voltage and current, at the precise time, with minimal power losses in the drive circuitry. The second problem is to create the necessary logic for motor control

Flexible logic, not fixed, that can interpret instructions and issue control commands to the motor may be designed to take full advantage of the motor's strengths while avoiding its weaknesses.

Driver Circuits

The simplest motor drive scheme is called unipolar drive, (illustrated in Figure 1). There are two windings per motor phase, one for each polarity of magnetic field to be generated. Thus for one polarity of field, we close SW1; for the opposite polarity in the same motor phase, close SW2. This is a common arrangement which keeps drive circuitry simple, but does not make optimal use of the motor windings. Only half of the copper is in use at any given time.

We can make better use of the motor windings by passing current in either direction, as appropriate, through each winding. Many high performance motors have only one winding per motor phase and require that "bipolar" drive be used. With other

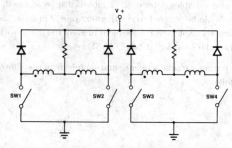

Simplified Diagram of a Unipolar, Resistance-Limited Drive
Figure 1

motors, the two windings for a given phase can be connected in series. Since twice as many turns are energized, only half the current is needed for a given motor torque. In this article, we will concentrate on ways of implementing this "bipolar" drive.

The term "bipolar" in this context refers to the two directions of current carried by each winding–not to bipolar transistors. In fact, we will use power MOSFETs exclusively as power switches since they are an advance which eases our job.

Bipolar drive can be obtained by two methods. The half-bridge design, Figure 2, maintains one end of the coil at ground and switches the other between V + and V −, thereby reversing the current through the coil. The second method powers the coil with a single supply in an H-bridge configuration, Figure 3. In this case, either SW1 and SW4 can be on with SW2 and SW3 OFF, or vice versa, causing the polarity at the ends of the coil to reverse.

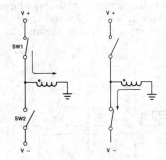

Simplified Diagram of a Bipolar Drive
Figure 2

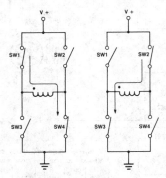

Simplified Diagram of an H-Bridge Circuit Providing
Bipolar Drive
Figure 3

Although the H-bridge drive reduces the number of power supplies required, this occurs at the expense of logic complexity. However, the use of p-channel devices for switches to the V + supply and n-channel

devices for switches to ground allows the use of the same logic sequence used for a unipolar coil drive.

The H-bridge also requires twice as many power devices per coil as the two-supply bipolar method. Nevertheless, cost does not increase proportionately because the power-switch's voltage rating has been reduced by half while maintaining the same voltage drive across the coil.

The Power Switches

Power MOSFETs have been improving steadily and are now ready to take over the job of H-bridge switches. Power MOSFETs are now commonly available up to 400 V at 20 A which provides a range capable of driving most available stepper motors to their performance limitations.

Several direct advantages are gained with power MOSFETs when compared to power transistors. One distinct advantage is their speed. In power switching applications such as these, device heating is always a major concern. The faster any power device can be switched through its linear region to a state of minimum drop across the device, the more its internal heating will be minimized. Speed in the power MOSFETs, combined with the fact that they are non-saturating, also allows balancing their turn-on and turn-off times. This eliminates the necessity of adding excessive "dead time" to avoid simultaneous device conduction when reversing the direction of current flow through the coil.

With all this speed, at the power we're talking about, comes a good-news/bad-news story. The good news is that even very large power MOSFETs are extremely fast when driven as hard as the Si7250 drives them. In one lab setup, an IRF9530 switching three amperes turned off in seven nanoseconds, for a rate of change of current near 450 amperes/microsecond! One would expect it to switch even larger currents with comparable switching times, for still larger dI/dt. Switching-type power regulation (covered later in this article) can be done very efficiently with negligible power loss.

The bad-news part of the story is that these rapid current changes make stray wiring inductance very important. That aforementioned test setup kept losing power devices when operating at higher currents, well within the device ratings. The problem was a piece of number 16 wire three inches long connecting the p-channel drain (which was switching) to the n channel drain (and internal diode). Its 70 nanohenry inductance developed a 30 volt spike when the p-channel device turned off. At full current, that spike rose to 90 volts, which, combined with the V + supply, exceeded the device's drain voltage rating.

The cure for the problem is easy. With the TO-220 drain tabs mounted side by side and strapped together, the inductance, and resulting voltage spike, are greatly reduced. As a similar precaution, the entire power path from the p-channel source through the supply bypass to the n-channel source should have a comparably low inductance. Wide foil conductors and switcher-rated capacitors are strongly recommended. In cases where inductance is unavoidable and extreme switching speed is not essential, one can simply reduce the switching speed. In the lab setup mentioned previously, a 100 ohm resistor in series with the device gate was sufficient to obtain safe operation even before fixing the offending wire. The switching times were still negligible.

Another concern when using power MOSFETs is the nature of the load they present to their drivers. This load is purely capacitive and, under certain circumstances (during transitions), can be sizeable. This capacitance need not be of particular concern if thoroughly understood.*

*Oxner, Edwin S., "Power FETS and Their Applications," Prentice-Hall, Inc., Englewood Cliffs, N.J., 1982.

Switch Drivers

Another advantage of power MOSFETs is that once they are turned ON, they need no drive power to remain ON. If a driver was then defined that would take advantage of these properties, it would have sufficient current capability to drive highly capacitive loads during transitions. But upon reaching a static condition, it would consume minimal power. One such integrated circuit designed especially to drive power MOSFETs is the Si7250.

The Si7250 provides four sets of complementary outputs specifically designed to drive two H-bridges. Internal logic locks out any "illegal" inputs which could result in turning ON both devices in one leg of either H-bridge. Another feature is that the timing offset (skew) between each set of complementary outputs is specified, which eliminates some possible timing concerns.

Combining MOSFET H-bridges with the Si7250 driver is a simple, efficient way of obtaining a bipolar drive. Figure 4 shows how the Si7250 and a pair of VQ3001s combine to drive a stepper motor. The VQ3001

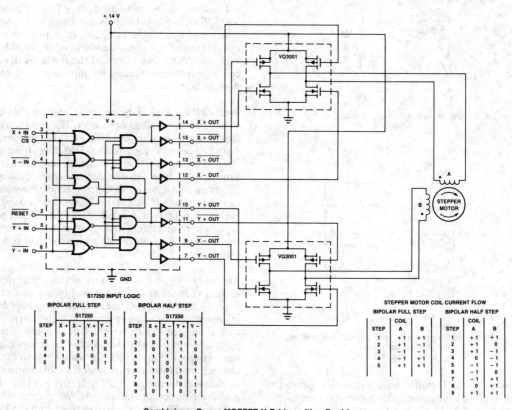

S17250 INPUT LOGIC

BIPOLAR FULL STEP

S17250

STEP	X +	X −	Y +	Y −
1	0	1	0	1
2	0	1	1	0
3	1	0	1	0
4	1	0	0	1
5	0	1	0	1

BIPOLAR HALF STEP

S17250

STEP	X +	X −	Y +	Y −
1	0	1	0	1
2	0	1	1	1
3	0	1	1	0
4	1	1	1	0
5	1	0	1	0
6	1	0	1	1
7	1	0	0	1
8	1	1	0	1
9	0	1	0	1

STEPPER MOTOR COIL CURRENT FLOW

BIPOLAR FULL STEP

	COIL	
STEP	A	B
1	+1	+1
2	+1	−1
3	−1	−1
4	−1	+1
5	+1	+1

BIPOLAR HALF STEP

	COIL	
STEP	A	B
1	+1	+1
2	+1	0
3	+1	−1
4	0	−1
5	−1	−1
6	−1	0
7	−1	+1
8	0	+1
9	+1	+1

Combining a Power MOSFET H-Bridge with a Predriver
Bipolar Drive
Figure 4

provides an H-bridge, two *p*-channel and two *n*-channel devices, in each fourteen-pin package. This compact three-package set results in an efficient driver for stepper motors up to a maximum drive voltage of 14V, the VDD limit of the Si7250.

For stepper motors requiring higher voltages, the Si7250 controlling the *p*-channel devices in the V+ end of the H-bridges must be isolated from the rest of the circuit and referenced to V+, as in Figure 5. A separate Si7250 is then used to drive the *n*-channels.

A Watt Saved is a Watt Earned

All the circuits discussed thus far depend on series resistance to limit the coil current. With high performance motors, this can waste considerable power. To achieve high step rates, large supply voltages must be used to cause the rapid build-up of current in the windings. The time required to cause a current change (ΔI) in an inductance (L) is just $\Delta I = L/V$, where V is the voltage applied. Once the current is up

to the desired level, we must prevent it from rising further. Smaller motor applications commonly use a series resistor (perhaps winding resistance) to limit the current. This also slows down the current rise somewhat. Worse yet, the resistors necessary to limit the current in a large motor could easily dissipate a kilowatt of wasted power. They can also dissipate budgets and space.

Another method of limiting coil current is to remove the applied voltage once the current reaches the required level. In Figure 6a, for example, SW1 and SW4 are turned ON to cause forward current through the motor's coil. Soon the current has built up enough to cause the current sensor in series to open SW1, the coil rings back against diode D3, and current flows through the path shown in Figure 6b. Since there is little voltage drop around this new path, the current decreases only slowly, but eventually drops below the current sensor's lower threshold. This causes SW1 to close again bringing the current back up to its nominal value.

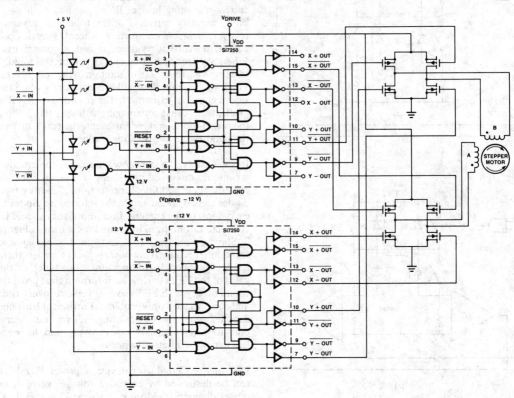

Variation of the Circuit in Figure 4 for Higher-Voltage Stepper Motors
Figure 5

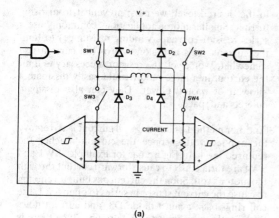

(a)

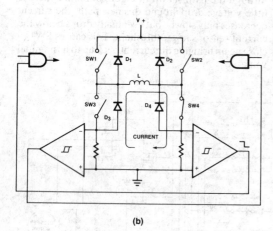

(b)

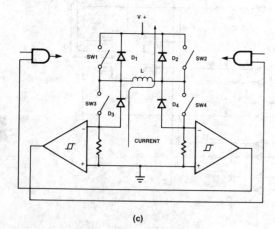

(c)

Simplified diagram of a current-limiting coil driver. Connecting the power supply to the coil causes coil current to rise, (a). When current reaches the required level, SW1 opens and current starts to decay, (b). Diode D3 retards the rate of current decay, its anode may be connected to ground for improved efficiency. With all switches open, the coil returns energy to the power supply, (c).
Figure 6

This cycle resembles the action of a switching-type (pulse width modulated) power supply. The average current drawn from the power supply is a small fraction of the coil current. Not only is power saved, but the *p*-channel switches can be sized for their smaller average current. Since *p*-type silicon has a lower mobility than *n*-type, *p*-FETs occupy more real estate for a given current rating, and silicon real estate is expensive. The potential savings from requiring less of the *p*-channels can be considerable.

Another advantage of the H-bridge is illustrated in Figure 6c. Coil current is brought to zero simply by opening all the switches. The current path is through the diodes D2 and D3. The voltage across the coil is large and reversed, so the current decays as quickly as it originally built up. The energy stored in the coil is returned to the supply capacitor.

Although the source-to-drain diodes inherent in power MOSFETs can be used for D1 through D4 in Figure 6, they are not well suited for use in a high frequency current-regulating bridge. Being *pn* diodes, they exhibit minority-carrier storage time and reverse-recovery delay. Consequently, when D3 carries coil current in a current-regulating bridge, it stores charge. When SW1 recloses, that charge is removed rapidly, creating a current spike in addition to the coil current. This extra current flow will cease abruptly when the stored charge is exhausted. The resulting large value of dI/dt can cause a damaging voltage spike. While the origin of this spike is different, the result may be the same as previously: bridge failure.

In some cases, the same low inductance construction techniques discussed previously will be an effective remedy. For additional protection, Schottky-type diodes can be added across the internal *pn* diodes to carry most of the current. Preferably, they should be connected from drain to ground, as shown in Figure 7. Standard-recovery (not fast) *pn* diodes may also work here because their slow recovery time lets them act as "snubbers". Adding series resistance to the gates of the FETs to slow their turn-on time provides compatibility with FET diode reverse recovery time and can also solve this problem. Further, at least one supplier appears to be adjusting FET process parameters to soften internal diode recovery, and the problem may soon quietly disappear.

Like any semiconductor device, a power MOSFET can be destroyed by excessive voltage, current, or power dissipation. Moreover, its extreme speed capability makes over-voltage (spiking) of particular concern if circuit inductance is not carefully controlled. But these are the extent of their failure mechanisms. Current crowding and secondary breakdown phenomenon associated with bipolar transistors do not exist with power MOSFETs.

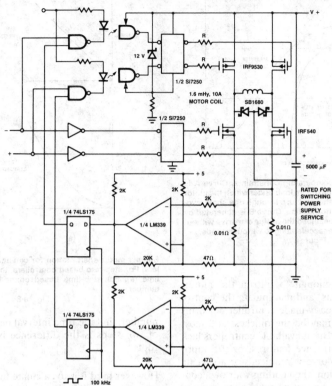

Diagram of a practical current-limiting coil driver. The _p_-channel devices are switched off by current sensors when coil currents reaches 10 A. Operation is similar to that of a switching type power supply. The Schottky diodes and resistors R are anti-spike-measures discussed in the text.
Figure 7

These devices excel in handling high peak current so long as average current and dissipation ratings are observed. Current gain does not sag at high current levels, nor does leakage to the gate endanger stability at high temperatures. Finally, MOSFETs parallel easily; the positive temperature coefficient of saturation voltage tends to make the devices share current evenly.

A Circuit in Practice

A practical current-regulating H-bridge is shown in Figure 7. The current-sensing comparators have thresholds of 95 mV and 105 mV for a 10% hysteresis. Using low inductance current-sensing resistors, filter capacitors, and power distribution conductors is essential, but even with all reasonable precautions taken, the comparators are likely to do peculiar things for a few microseconds after the bridge switches. (A 10 A, 100 ns, 60 V switch is, after all, a fearsome neighbor!!!) The clocked D flip-flop is included to let

the comparator collect its wits before being asked its opinion of the current level. This precaution can avoid a spectacular oscillation.

Generating Driver Signals

Fixed logic may suffice in stepper motor applications not requiring long or fast traverses. But to obtain high slew rates, the use of fixed logic is ruled out by the stepper motor characteristics demonstrated in Figure 8. To avoid losing synchronization during start up, a stepper motor must accelerate from a "rest" position by some tapered acceleration profile, up to its maximum step rate. The same dynamics, of course, apply to decelerating the motor and load to a stop. Both problems can easily be solved by controlling the motor with algorithms in a microcontroller.

Companies offering "motor controller ICs" have developed motor control software and stored it in the permanent memory of standard microcontrollers.

6

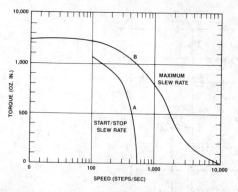

Determining stepper motor operating constraints. Curve A is generated by plotting the maximum load under which the motor can start at various step rates and turn through one shaft revolution without losing synchronization. Curve B is generated by plotting the greatest frictional load the motor can drive without losing synchronization after accelerating to various speeds.
Figure 8

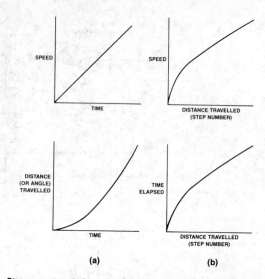

Stepper-motor shaft motion for constant acceleration controllers. For step-rate based controllers, the relevant plots are μs time (a). For step-time based controllers, they are μs step number.
Figure 9

Any processor fast enough to outrun the motor, execute the algorithms, and manipulate the control lines will work. If a ready-made controller performs satisfactorily then it may be the quickest and most economical solution. But if available controllers lack some required performance features, or if a more specific acceleration curve is required for a specific motor/load combination, then coding your own controller may be the answer. Setting out to do so requires going back into the analog world of motion.

Available torque for a typical stepper motor is nearly constant over a wide range of speeds, (Figure 8). Consequently, with a proper drive sequence, the motor can maintain constant acceleration. If acceleration does not exceed torque capability, the motor will reach its maximum step rate without skipping steps, an essential requirement for open loop motion control systems.

Hard-wired acceleration controllers usually generate a pulse train with a linearly increasing frequency since linear voltage-controlled oscillators are easy to make in analog circuitry. This approach was convenient with hard wired controllers; however, microprocessor-based controllers require a different approach.

Several microprocessors are available with built-in timers fast enough for motor step timing, but all of them are programmed in terms of time interval, not frequency. A microprocessor must be told how long to wait before issuing the next motor step. This information can be easily derived from a curve showing speed and elapsed time versus distance (shaft angle or step number), Figure 9. Both speed and elapsed time are square root functions of the step

number, and the time interval between the initiation of two steps is the difference between two square roots.

However hard it is to calculate the length of the next step, we need only do it once for each step and save the results in the form of a look-up table. For applications where the system's properties are known when the code is written, calculate the step lengths directly in microseconds or in multiples of whatever units your timer counts in. For a more general controller, all we need is a table of relative step durations to be multiplied by a constant for any particular controlled system. The results are easy to use (Attachment, "Accelerating the Stepper").

Once the motor accelerates to its maximum step rate, the step intervals are held constant. To decelerate the motor, the microprocessor reads backward through the look-up table, commanding increasingly longer step intervals until the motor comes to a stop. The number of steps needed for deceleration will be equal to the number required for acceleration. If the required traverse is small, there may not be enough steps available to accelerate to the maximum step rate within the motor's torque limits. In this case, half the traverse will be spent accelerating and the other half decelerating with the maximum step rate never being reached. If the number of steps is large enough, the motor will accelerate to its maximum step rate with more than half the steps remaining. It will command full-speed steps until just enough steps remain for deceleration, then decelerate back to stop.

Available torque may decrease as motor speed rises due to friction and other effects. In such cases, acceleration must be tapered down as speed increases. One convenient, approximate way to do this is by adding a constant to the step interval. For the first few steps where the intervals are long, the addend has little effect. As the speed increases and the step intervals shorten, the addend becomes increasingly larger in proportion to step length and prevents the motor speed from accelerating as fast as it otherwise would.

A good choice for the constant addend is one-half of the full-speed step interval. This value increases the duration of the shortest step by 50%, and, to reach the same final speed, the acceleration sequence must include 9/4 times as many steps. The time required to reach final speed will increase by a factor of 15/8. This tapered sequence can drive the motor to greater final speeds than the linear sequence for any given initial single-step time interval.

The look-up table approach for step interval control can be used to establish the sequence in which motor currents are applied. Stepper motors are commonly available in a rich variety of types with 2, 3, 4, or 5 coils, "halfstep" or "fullstep" operating modes, and either clockwise or counterclockwise shaft rotation. Regardless of the motor type used, the sequence of coil-current commands can be listed in a pair of tables–one for clockwise and one for counterclockwise steps. The last command to the motor drive can be saved as an index for moving to the next command.

Figure 10 shows how this technique operates for a two-coil stepper motor in a half-step mode. The command bits are simply the required inputs for the Si7250 driver. Each 4-bit word is used both as a command sent to an output port and as an index for looking up the next command. Changing from one motor or mode to another only requires changing the look-up tables.

This strategy results in efficient and compact microprocessor code. The microprocessor used should have an appropriate timer and a fast multiply instruction, such as found in the MC68701 single-chip microcomputer. Using this processor, the required code and tables occupy less than one third of the internal 2K EPROM and provide motor speeds up to 7000 steps/second. Driving a pair of H-bridges controlled with a pair of Si7250s as shown in Figure 7, the controller can solve a variety of industrial motion-control problems.

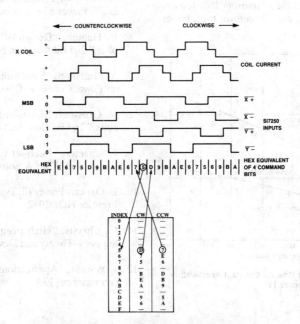

10. Look-up table with hexadecimal commands for clockwise and counterclockwise rotation. If the coil command is a 5, then D steps clockwise and 7 steps counterclockwise for a 2-coil, 4-phase motor operating in half-step mode. Table entries marked (−) denote illegal indexes and may be replaced by the nearest legal state.

Figure 10

Attachment

Accelerating the Stepper

Single-step time T is the first step interval when starting the motor from rest. This value, determined by available motor drive currents and load inertia, is obtained either by calculation or by experiment. Knowing T and the step size S in degrees of shaft rotation or units of linear load motion, acceleration is equal to $2S/T^2$.

If the first step interval of an acceleration sequence is of duration T, then step number n occurs at time $T_n = T(\sqrt{n+1} - \sqrt{n})$. As n increases, step rate increases until the motor reaches full speed at step number N. This speed is approximately $2\sqrt{N}/T$ steps per second, and total acceleration time is exactly $T\sqrt{N}$.

The plot of n versus single-step speed multiples can be used to estimate the time required to complete a given motion. For example, suppose a motor must accelerate from rest to 1000 steps/second and then stop after moving a load through 200 steps. If T is determined to be 10 ms then motor speed must increase by a factor of 10. From the curves in the graph, a $10\times$ speed increase is reached at N = 25, and total acceleration time is 5T or 50 ms. Since acceleration and deceleration both require 50 ms and 25 steps, the load moves the remaining 150 steps at 1000 steps/second in 150 ms. Therefore, total travel time is 250 ms.

Since only moderate values of N are required for many applications, it is often practical to precalculate step lengths ($\sqrt{n+1} - \sqrt{n}$) and multiply by T (in units of 1/4 msec.) to get step duration in microseconds. When using an MC68701 microprocessor, the multiplication is 8×8, giving a 16-bit result directly usable by the microprocessor's 16-bit timer.

Suggested Reading List

1. R. Middlebrook and S. Cuk, "Advances in Switched-Mode Power Conversion," Volumes I-III, Teslaco Inc., 490 S Rosemead Blvd., Suite 6, Pasadena, CA 91107, 2nd Edition, 1983

2. W. T. McLyman, "Transformer and Inductor Design Handbook," Marcel Dekker, Inc., NY 1978

3. W. T. McLyman, "Magnetic Core Selection for Transformers and Inductors," Marcel Dekker, Inc., NY 1982

4. Proceedings of Powercon 1 through 11, Power Concepts, Inc., PO Box 5226, Ventura, CA 93003

5. Proceedings of the IEEE Power Electronics Specialists Conferences 1970-1984, sponsored by the IEEE Power Electronics Council

6. E. Hnatek, "Design of Solid State Power Supplies," 2nd Edition, Van Nostrand-Reinhold, 1982

7. A. Pressman, "Switching and Linear Power Supply Power Converter Design," 1977, Hayden Books

8. N. Grossner, "Transformers for Electronic Circuits," 2nd Edition, McGraw-Hill, 1983

9. J. Schaefer, "Rectifier Circuits–Theory and Design," John Wiley & Sons, 1965

10. E. Oxner, "Power FETs and Their Applications," Prentice Hall, 1982

11. G. Chrysis, "High Frequency Switching Power Supplies–Theory and Design," McGraw-Hill, 1984

12. J. Watson, "Applications of Magnetism" Wiley-Interscience, 1980

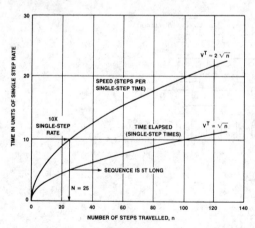

Motor Speed and Elapsed Time for Constant Acceleration
Figure 11

6.13 Design Tips

6.13.1 Solving the Stepper Motor Interface Problems (AN82-1)

Introduction

Stepper motor drive techniques are evolving rapidly with their logic signals now very often being generated by microcontrollers. Off the shelf controllers such as those named here are serving many needs. As well, many users are choosing to develop their own software and apply standard microcontrollers to more efficiently suit their needs. In either case, the outputs of the controller will be TTL levels. The problem then is one of interface, applying those signals efficiently across the motor.

To resolve the interface problem, Siliconix now offers the Si7250 H-Bridge predriver and controller, and the MOS-POWER devices necessary to complete the system.

The Si7250 was designed specifically as a predriver to take full advantage of MOSPOWER's characteristics. MOSPOWER devices present a significant capacitive load to the driver during transitions. The Si7250 is capable of driving a 500pf load in less than 25ns. The optimum signal range for MOSPOWER gates is consistent with the 10V output swing of the Si7250. MOSPOWER devices once switched, require no drive to sustain their "on" or "off" state, the Si7250, in either static condition, draws exceptionally low power. When unselected, it powers down to an even lower current, resulting in a total motor drive circuit, both predriver and H-Bridges, which consumes less than two milliamps.

The Si7250 has TTL compatible inputs and four sets of complementary outputs, capable of driving MOSPOWER devices large enough to exercise even the largest stepper motors available, to their performance limitations.

Another distinct advantage of using the Si7250 as an H-Bridge controller is the safety provided by its internal logic. This eliminates the possibility of ever shorting two power devices directly across the power supply to ground.

Si7250 TRUTH TABLE

INPUT PINS						OUTPUT PINS							
1 CS	2 R	3 X+.	4 X-.	5 Y+.	6 Y-.	7 Y-.	9 Y-.	10 Y+.	11 Y+.	12 X-.	13 X-.	14 X+.	15 X+.
1	X	X	X	X	X	0	1	0	1	0	1	0	1
X	0	X	X	X	X	0	1	0	1	0	1	0	1
8 INPUT STATES DECODED													
0	1	0	1	0	1	0	1	1	0	0	1	1	0
0	1	0	1	1	0	1	0	0	1	0	1	1	0
0	1	0	1	1	1	0	1	1	0	1	0	0	1
0	1	1	0	0	1	0	1	1	0	1	0	0	1
0	1	1	0	1	0	1	0	0	1	1	0	0	1
0	1	1	0	1	1	0	1	1	0	1	0	0	1
0	1	1	1	0	1	0	1	1	0	0	1	0	1
0	1	1	1	1	0	1	0	0	1	0	1	0	1

The 8 remaining input states are **not** decoded resulting in a reset output condition. This is to prevent the inadvertent shorting of any power drivers directly across the supplies in a standard H-Bridge configuration.

INPUT PINS						OUTPUT PINS							
0	1	0	0	0	0	0	1	0	1	0	1	0	1
0	1	0	0	0	1	0	1	0	1	0	1	0	1
0	1	0	0	1	0	0	1	0	1	0	1	0	1
0	1	0	0	1	1	0	1	0	1	0	1	0	1
0	1	0	1	0	0	0	1	0	1	0	1	0	1
0	1	1	0	0	0	0	1	0	1	0	1	0	1
0	1	1	1	0	0	0	1	0	1	0	1	0	1
0	1	1	1	1	1	0	1	0	1	0	1	0	1

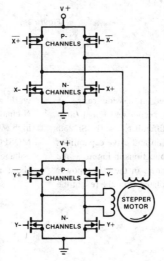

Figure 1

In this H-Bridge configuration P-channel devices are used to switch the coil to the (V+) power supply and the N-channel devices are used to switch the coil to GND.

NOTE: An N-channel device driven by a logic "1" is turned on while a P-channel device driven by a logic "1" is turned off.

6

Figure 2 shows the Si7250 and MOSPOWER H-Bridges driving a stepper motor using the VQ3001 which consists of two P-channels and two N-channels in one 14 pin package.

The limitation of this circuit is that the motor drive voltage cannot exceed 14V, the maximum operating voltage of the Si7250.

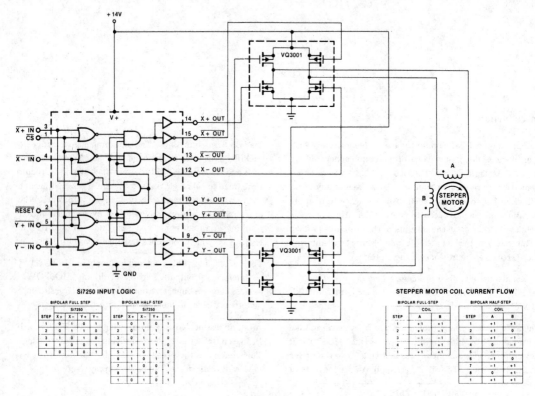

Figure 2

Stepper motors may be operated in a wide range of voltages to achieve the desired responses. For higher voltage applications, it's still not difficult to use the Si7250 and take advantage of the protection and drive capabilities it provides. For higher voltage applications in Figure 3, the Si7250 which controls the P-channels in the (V+) power supply end of the bridge is opto-isolated from the rest of the world and referenced to V+. A separate Si7250 drives the N-channel end. Now the only components which care how high V+ operates are the devices in the H-Bridge. Standard MOSPOWER devices are now available that extend this voltage range up to 100V (@ 20 AMPs) which should meet about any stepper motor's needs.

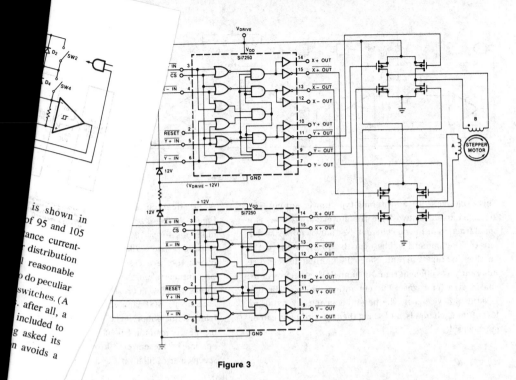

Figure 3

...circuits shown so far depend on their ...nent series resistance to limit the coil ...igh performance motors, added series ...waste a lot of power. At high step rates, ...oltages must be used to make the current in ... build up quickly. The time taken to cause a ...nge I in a winding of inductance L is just ...here V is the voltage across the coil (the voltage ...ninus the back EMF of the motor). Once the ...is up to the desired value, we must somehow ...it from rising even higher. Small motors commonly ...series resistor (perhaps winding resistance) to limit ...urrent, and this slows down the current rise somewhat. ...se yet the necessary resistors for a large motor can ...ily dissipate a kilowatt of wasted power.

A better way of limiting coil current is to remove the applied voltage once the current is large enough. Figure 4 illustrates the basic idea. To send a forward current through a motor coil, SW1 and SW4 turn on as before in (Fig. 4(a)). After a time IL/V+, the current has built up to a value I, and the current sensor in series with SW4 opens SW1. The coil rings back against diode D3, and for awhile the current flows through the path shown in Fig. 4(b). There is little voltage drop around the path, so the current decreases slowly. Eventually it will drop below the current sensor's lower threshold, and SW1 will close again for a short time to bring the current back up to its nominal value.

6

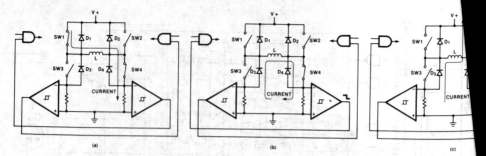

Figure 4

This cycle acts like a switching-type power supply. The average current drawn from the power supply (V+) is a small fraction of the coil current. Not only is power saved, but the P-channel switches can be sized to carry the smaller average current. Since P-channel devices are inherently less efficient than N-channels they occupy more silicon area for a given current rating, which is directly related to device cost. The potential savings by requiring less of the P-channels, and sizing them accordingly, can be considerable.

A practical current-regulating H-Bridge
Figure 5. The comparators have thresholds
millivolts for a 10% hysteresis. Low-induc
sensing resistors, filter capacitors and powe
conductors are essential, but even with a
precautions taken, the comparators are likely t
things for a few microseconds after the bridge
ten ampere, 100 nanosecond, 60-volt switch is
fearsome neighbor!) The clocked D flipflop is
let the comparator collect its wits before bein
opinion of the current flowing. This precautic
spectacular oscillation.

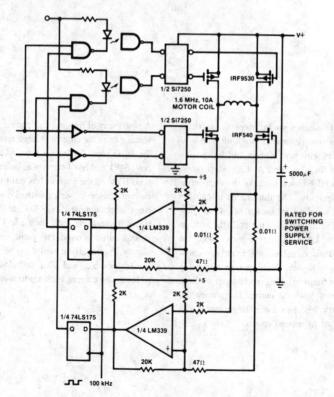

Figure 5

6.13.2 The D469:
An Optimized CMOS Quad Driver for
MOSPOWER FET Switches (AN83-11)

Introduction

Many applications of power MOSFETs call for the devices to be used as on-off switches. While they can be directly driven by TTL or CMOS logic if moderate switching times are acceptable, to take full advantage of their high speed switching capabilities, a more robust driver is required. This driver should be able to provide at least 10 to 12 volts of gate drive at peak current of several hundred milliamps.

Until now, the only devices capable of driving MOSPOWER FETs in high speed switching applications were discrete bipolar devices or IC's such as National's DS0026 MOS clock driver. While the DS0026 can drive a 1000pF load with typically 20nS rise and fall times, it does suffer from some serious drawbacks that restrict its use. Being a bipolar device, it draws a fair amount of current quiescently (30mA) and also requires a large logical "1" input current (10mA). Because the DS0026 is configured internally as a dual inverting driver, two additional TTL inverters are required to make the part usable when non-inverting drive is required.

The DS0026 is unsuitable for use in many applications, particularly where the control logic is low-power CMOS. Its configuration as an inverting-only driver is also an inconvenience to the designer. Clearly a device that retains the high speed drive capabilities of the DS0026 but has none of its disadvantages is desirable.

Enter the D469 IC, which was designed as an optimized driver for MOSPOWER FET switches from the start. Internally it contains four independent drive channels, and each one can be configured as either a logically inverting or non-inverting driver (see Figure 1). Since it is a CMOS device, the D469 is compatible with low-power CMOS logic and microprocessors and draws very little current quiescently (2.5mA). Each of the D469's outputs can drive a 500pF load with typically 20nS rise and fall times and can be connected in parallel to drive larger capacitive loads if desired.

LOGIC DIAGRAM

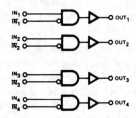

PIN CONFIGURATION

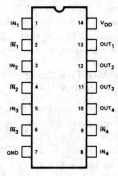

The Block Diagram of the D469 Reveals that Each Channel can be Configured as Either an Inverting or Non-Inverting Driver
Figure 1

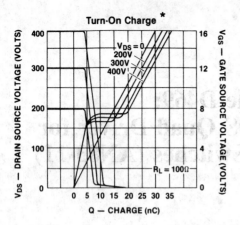

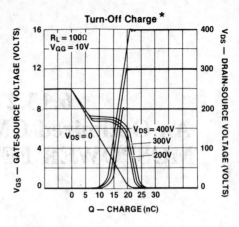

Switching Time Equations

$$t_{d(on)} = \frac{Q_{g1}}{V_{g1}} R_{gen} \ln\left(\frac{V_{GG}}{V_{GG} - V_{g1}}\right)$$

$$t_{d(off)} = \frac{Q_{g3} - Q_{g2}}{V_{GG} - V_{g2}} R_{gen} \ln\left(\frac{V_{GG}}{V_{g2}}\right)$$

$$t_r = \frac{Q_{g2} - Q_{g1}}{V_{g2} - V_{g1}} R_{gen} \ln\left(\frac{V_{GG} - V_{g1}}{V_{GG} - V_{g2}}\right)$$

$$t_f = \frac{Q_{g2} - Q_{g1}}{V_{g2} - V_{g1}} R_{gen} \ln\left(\frac{V_{g2}}{V_{g1}}\right)$$

The Charge Transfer Characteristics of a High-Voltage Power MOSFET Show the Occurence of Maximum Input-Capacitance
(C = dQ/dV) at a Gate-to-Source Voltage of Approximately 7V
Figure 2

This applications note introduces the charge transfer characteristics of a power MOSFET and shows that the D469 is very well suited to gate drive applications. A brief discussion of additional driver power dissipation due to gate current then follows. To prove that the D469 is a better choice than the DS0026 as a MOSFET gate driver, a comparative analysis between the two devices in switching time tests is performed. Finally, some applications of the D469 in DC motor drive circuitry are described.

The D469 and MOSFET Charge Transfer Characteristics

Each of the four output drivers in the D469 has an output resistance of approximately 8 ohms and is capable of sinking or sourcing a peak current of 500mA. The output voltage swing at peak output current is then limited to 4 volts above and below the ground and V_{DD} rails respectively.

At a cursory glance, the limited output voltage swing at peak current might seem to be a disadvantage. However, if one examines the charge-transfer characteristics of some of the higher current power MOSFETs, it becomes apparent that the maximum input capacitance occurs at gate-to-source voltages in the region of 7V. These charge transfer characteristics are now included in all Siliconix data sheets. Figure 2 shows pertinent data for the VNDA type geometry.

The D469 is ideally suited to driving MOSPOWER FETs because its peak source and sink current capability occurs in the output voltage ranges of 0 to 8 volts and 12 to 4 volts respectively, with a 12 volt power supply. When used to drive common-source connected N or P channel MOSFETs, maximum drive current is available at the point of maximum input capacitance.

The currents that flow in a D469's output stage during charge/discharge cycles of the effective gate capacitance

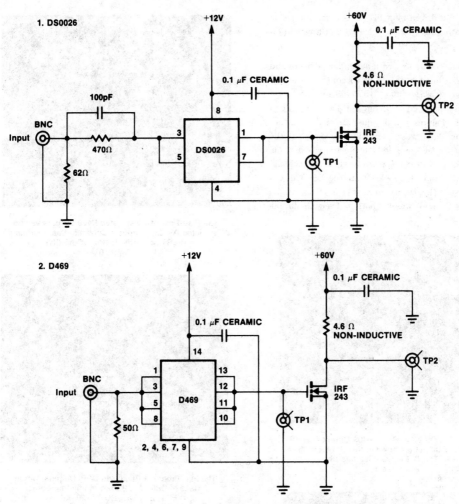

**The Schematic of the Fixture Used to Perform the
D469 versus DS0026 Switching Time Tests
Figure 3**

cause additional power to be dissipated by the driver. Equation (6) of reference 2 shows that this power is proportional to the gate-drive voltage as well as the switching frequency. For switching frequencies below 100 kHz, the additional power dissipated by the driver is so low it can be neglected. However, when the switching frequency is above 500kHz and the gate-drive voltage is appreciable, the additional driver power dissipation cannot be neglected. It should be calculated and added to the DC power dissipation of the driver. This value can then be used to determine if the operating junction temperature of the driver is within the maximum data sheet limits.

A Comparative Analysis
The D469 vs. The DS0026 in Switching Time Tests

D469 and DS0026 IC's were incorporated into a test fixture, driving identical power MOSFETs with identical loads. Figure 3 shows the schematic diagram of this fixture. The power MOSFETs used were Siliconix IRF243s with typical gate-to-source capacitance of 1600pF and drain-to-gate capacitance of 300pF. The load resistance for both MOSFETs was approximately 4.6 ohms (non-inductive), and the power-supply voltage was 60V. Both the DS0026 and D469 were powered from a separate 12 volt

6

power supply and driven by a TTL level pulse train. The pulse train had a pulse width of 150nS and a repetition rate of 1KHz.

Since the DS0026 contains two independent drivers and the D469 four, a fair test would be to compare the two devices as dual drivers. Thus, two D469 outputs were paralleled and compared to a single DS0026 output in drive capability. Figures 4 and 5 indicate that on a "per chip" basis the D469 was essentially the equal of the DS0026. To be precise, the D469 waveform's rise and fall times were 36nS and 44nS respectively while the DS0026's were 26nS and 42nS. It appears that the DS0026 was somewhat faster (by 10nS) on the rising edge of its waveform than the D469 was, but both had very similar fall times.

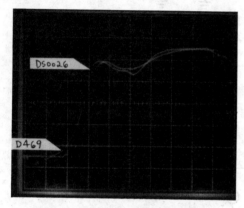

The Rise Time of 2 Paralleled D469 Outputs versus 1 DS0026 Output is Shown in this Photograph
(VERT: 2.4V/Div,HORIZ: 20 ns/Div)
Figure 4

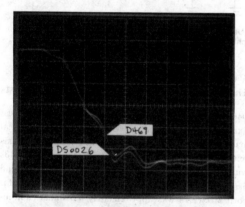

The Fall Time of 2 Paralleled D469 Outputs versus 1 DS0026 Output is Shown in this Photograph
(VERT: 2.4V/Div, HORIZ: 20 nS/Div)
Figure 5

Some further tests were conducted to see how the D469 with two or more outputs connected in parallel compared to the DS0026 with both of its outputs connected in parallel.

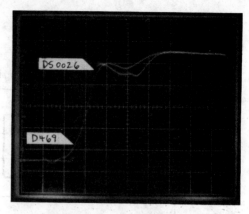

The Rise Time of 4 Paralleled D469 Outputs versus 2 Paralleled DS0026 Output is Shown in this Photograph
(VERT: 2.4V/Div, HORIZ: 20 nS/Div)
Figure 6

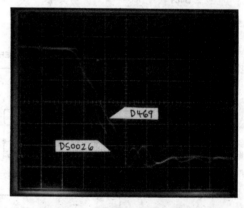

The Fall Time of 4 Paralleled D469 Outputs versus 2 DS0026 Output is Shown in this Photograph
(VERT: 2.4V/Div, HORIZ: 20 nS/Div)
Figure 7

Figures 6 and 7 show the rise and fall times of the D469 with four outputs in parallel versus the DS0026 with two outputs in parallel. Here it can be seen that the DS0026 with rise and fall times of 22nS and 30nS respectively was still the faster of the two — but not by much. The D469 came in a very close second with rise and fall times of 32nS and 42nS respectively. Interestingly enough, the rise and fall times of the D469 with four outputs connected in parallel were virtually the same as those with two outputs in parallel. The rise and fall times for the DS0026, on the other hand, showed some improvement.

The main difference observed between the D469 and the DS0026 was the propagation delay-time each device exhibited. For the D469 this parameter was approximately 50nS while that of the DS0026 was in the range of 5 to 10nS. Figure 8 clearly shows this difference. It should be noted that Figure 8 is the only one in which the difference in propagation delay is shown. Figures 4 to 7 do not show

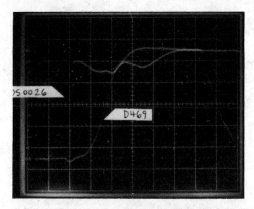

The 50 nS Difference in Propagation Delay Between the DS0026 and the D469 is Clearly Seen in this Photograph (VERT: 2.4V/Div, HORIZ: 20 nS/Div)
Figure 8

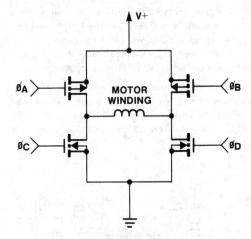

A Bipolar H-Bridge Motor Drive Circuit Using N- and P-Channel MOSFETs
Figure 9

this because it is easier to compare switching waveforms when the oscilloscope traces are overlaid with one another.

Summary

In summarizing the results of this comparative analysis, we can say that the D469 is really a DS0026 in a "CMOS disguise". Although the DS0026 was faster than the D469 in both cases, its speed advantage was no more than about 10nS in any of the rise and fall time tests. As a 2-channel driver the D469 is the obvious choice, especially when one considers its power consumption compared to the DS0026. In the test fixture shown in Figure 2, the DS0026 drew an average supply current of 30mA and thus consumed about 360 mW of power while the D469 drew slightly more than 2 mA and consumed about 25mW of power. Needless to say, the DS0026 ran quite hot to the touch while the D469 was cold. The 50nS propagation delay of the DS469 should not pose any problem if all the MOSFETs in a particular circuit are driven by D469s.

Applications of the D469 Quad Driver

The D469 quad driver is well suited to applications such as DC motor control. Motors ranging in size from fractional up to several horsepower can be directly driven with power MOSFET devices, if they are provided with suitable gate drive. The D469 fulfills the gate-drive requirements and provides the interface between the MOSFETs and low-power CMOS or TTL control logic.

Figure 9 shows a typical "H-bridge" motor drive circuit in which the winding current flow is bidirectional. N-channel MOSFETs are invariably used in the lower legs of the bridge since they can be directly driven by the D469. In the upper legs of the bridge, both N- and P-channel devices can be used. In each case, driving the upper MOSFETs presents a

problem that is common to a number of applications other than motor drives. The problem is that the upper MOSFET switches which control the positive power supply to the load cannot have their gate-drive voltage referenced to ground.

When the P-channel MOSFETs are used as the upper switches, their source terminals are connected to the positive power rail. The gates of these MOSFETs are then driven 10 to 12 volts below the positive rail to turn them on. Because the gate drive is always referenced to the positive rail, the gates of the N and P-channel MOSFETs in each half of the bridge can be driven directly by the D469 if the power supply voltage is 14V or less. Operation at higher supply voltages, however, requires that the P-channel devices be isolated in some fashion from the D469 that is ground referenced.

The isolation problem crops up again if N-channel MOSFETs are used as the upper switches in the bridge. Since their source terminals are connected to the load, one can see that the gate waveforms must swing to at least 10 or 12 volts above the positive rail if they are to turn on completely. Also, when either of the lower N-channel MOSFETs is conducting, the respective upper N-channel device must be off or cross-conduction will occur. To prevent this, the gate-to-source voltage of the upper MOSFET must be zero when the respective lower MOSFET is on and vice versa. This means that the gate-voltage of the upper N-channel devices must swing from the $R_{DS}(on)$ voltage drop of the lower MOSFETs all the way to 10 or 12 volts above the positive rail. Even with a 12 volt power supply, the gate voltage swing is almost 24 volts, and thus the D469 cannot be used to directly drive the upper MOSFETs. Again, some form of driver isolation is required.

6

There are many methods of providing isolation between the D469 and the gates of the upper MOSFETs when using elevated supply voltages. As may be expected, certain methods will work better with N-channel MOSFETs, and others will work better with P-channel devices.

Three methods of providing gate-drive isolation will be considered. Two of them provide DC isolation between the D469 and the gates of upper MOSFETs while the third provides isolation between a floating D469 and the control-logic inputs.

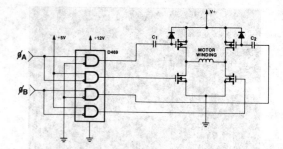

An H-Bridge Drive Using Capacitive Isolation for the P-Channel MOSFETs
Figure 11

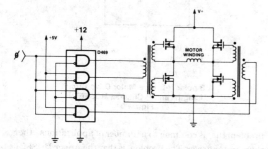

An All N-Channel H-Bridge Drive Using Transformer Isolation
Figure 10

a. Transformer Isolated Gate Drive

Figure 10 shows an all N-channel H-bridge motor drive using transformer isolation between the D469 and the gates of both the upper and lower MOSFETs. Each transformer is driven by an AC waveform generated by two D469 driver channels

In many cases the transformer will prove to be one of the most compact, least expensive and simplest methods of isolation to implement. It also works well with either N- or P-channel MOSFETs in the upper half of the bridge.

However, there is a drawback with this method. The interwinding capacitance of the isolation transformers allow noise spikes appearing at the MOSFET gates to be coupled back to the D469 and cause possible damage. Care must be taken to ensure that any transients coupled back to the driver are of sufficiently low magnitude such that no damage is done. They can be minimized by the inclusion of a Faraday shield between the primary and secondary windings of the isolation transformer.

b. Capacitor Isolated Gate Drive

A capacitor isolated N- and P-channel H-bridge motor drive is shown in Figure 11. In this circuit, only the P-channel MOSFETs are isolated from the D469; the N-channel devices are driven directly. The diodes are included to clamp positive gate-to-source transitions on the P-channel MOSFETs to +0.7V.

This method of isolation is even more compact and less expensive than that using transformers. Although a coupling capacitor several times the effective input capacitance of the P-channel MOSFETs is required, it is still small and inexpensive. This method does have the drawback of coupling noise spikes appearing at the MOSFET gates back to the outputs of the D469. Also capacitor isolation does not work well with N-channel MOSFETs in the upper half of the bridge. This is because the sources of the upper N-channel MOSFETs are not connected to an AC ground (such as the positive supply rail). Thus, there is no fixed reference for the AC coupled signal on the MOSFET side of the capacitor.

c. Floating D469 With Opto-Isolated Logic Inputs

Figure 12 shows a different method of providing isolation, namely referencing a D469 driver to the positive power rail and using an opto-coupler to provide the DC isolation between it and the low-voltage control logic. A -12V power supply, referenced to the positive rail, is generated for the D469's ground pin with a zener-diode, capacitor and resistor. The D469 thus provides gate-drive waveforms to the P-channel MOSFETs that swing from the positive rail to 12 volts below it. If the reservoir capacitor C is made large enough, it will be able to supply enough charge to the D469 for transfer to the gates of the MOSFETs. This ensures that the upper MOSFETs will have fast switching times.

Although this circuit is more complex than the previous ones, DC operation of the motor coils is provided if so required. Since the upper D469 floats with respect to the positive power rail, any noise transients on this rail will not affect the operation of the driver or be coupled back to the logic inputs. This method has the disadvantage of being difficult to implement with an all N-channel bridge. The large voltage swing required at the gates of the upper N-channel devices precludes the use of the circuit in this case.

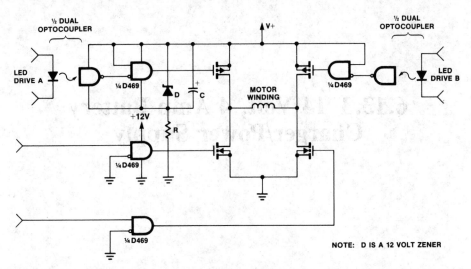

A Different Method of Isolation: Floating a D469 Driver and Optically Coupling the Logic Drive for the P-Channel MOSFETs
Figure 12

In summarizing this look at motor drive circuits using the D469 as a gate driver, one can see that if the correct methods of isolation are used then the D469 certainly simplifies the task of interfacing logic circuitry to a DC motor.

1. Ed Oxner, Siliconix, inc., "Correlating the Charge Transfer Characteristics of Power MOSFETs with Switching Speed". *Proceedings of Powercon 9*, H-4 pages 1 to 5, 1982.

2. Rudy Severns, INTERNATIONAL RECTIFIER, Inc., "Simplified HEXFET Power Dissipation and Junction Temperature Calculation Speeds Heatsink Design". *IR HEXFET Catalog*, 1980. Pages A-80 to A-86.

3. Siliconix, inc., *D469 Data Sheet*, May 1983.

4. NATIONAL SEMICONDUCTOR CORP. DS0026 *Interface Data Book 1980*, Pages 6-10 to 6-16.

6

6.13.3 14 Volt, 4 Amp Battery Charger/Power Supply

Very low drive power makes MOSPOWER FETs more desirable than bipolar transistors in this simple, minimum-parts–count battery charger. A low-power operational amplifier is used to control the VN64GA. A power bipolar transistor in the same circuit would require over 100 mA of base current and a much more complicated drive circuit thus reducing overall efficiency and increasing circuit complexity.

Operational amplifier A_1 directly drives the VN64GA with the error signal to control the output voltage. Peak rectifier D_1, C_1 supplies error amplifier A_1 and the reference zener. This extra drive voltage is necessary because the MOSFET gate voltage must exceed its source voltage by several volts for the VN64GA to pass full load current.

The output voltage is pulsating DC which is quite satisfactory for battery charging. To convert the system to a regulated DC supply, capacitor C_2 is increased and another electrolytic capacitor is added

across the load. The response time is very fast, being determined by the op-amp.

The 2N4400 current limiter circuit prevents the output current from exceeding 4.5 A. However, maintaining a shorted condition for more than a second will cause the VN64GA to exceed its temperature ratings. A generous heat sink, on the order of 1°C/W, must be used.

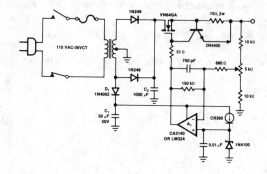

6.13.4 400 Volt, 60 Watt
Push-Pull Power Supply

An increase in DC/DC-converter efficiency is achieved by using MOSPOWER FETs in the power stage. Because the FET's drive requirements are low, a standard switching-regulator IC can provide both control and drive functions.

The design shown delivers a regulated 400 V, 60 W output. The TL494 switching regulator governs the operating frequency and regulates output voltage. R_1 and C_1 determine switching frequency, which is approximately $0.5\,RC$ — 100 kHz for the values shown.

The TL494 directly drives the FET's gates with a voltage-controlled, pulse-width-modulated signal. Each FET requires only 10 mW; the remaining drive power is dissipated by the turn-off resistors, R_2 and R_3. Operating in a push-pull configuration, the FETs switch primary current on in about 150 ns and then off in 300 ns. These short switching times — about a tenth of equivalent bipolar devices — and the resulting high operating frequency permit using physically small reactive components.

After full-wave rectification, the output waveform is filtered by a choke-input arrangement. The 1 μH, 75 μF filter accomplishes the job nicely at 100 kHz. A feedback scheme using R_4, R_5 and R_6 provides for output-voltage regulation adjustment, with loop compensation handled by C_2. Diodes D_1 and D_2 provide isolation and steering for the 33 V zener transient clamp, D_3.

Output regulation is typically 1.25% from no-load to the full 60 W design rating. Regulation is essentially determined by the TL494. Output noise and ripple consists mainly of positive and negative 0.8 V spikes occurring when the output stage switches.

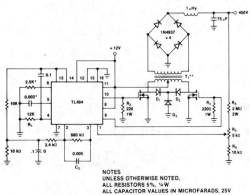

NOTES
UNLESS OTHERWISE NOTED,
ALL RESISTORS 5%, ¼W
ALL CAPACITOR VALUES IN MICROFARADS, 25V
Q_1 & Q_2: VN64GA ON HEAT SINK
D_1 & D_2: 1N4934
D_3: 33V, 3W ZENER
T_1: PRI: 12T, CT, NO 18 AWG
      SEC: 275T, NO 24 AWG
      CORE: IND GEN 8231-1

6

6.13.5 Self Oscillating Flyback Converter

A low-power converter suitable for deriving a higher voltage from a main system rail in an on-board application is shown below. It uses the core characteristics to determine frequency. With the transformer shown, operating frequency is 250 kHz. Diode D_1 prevents negative spikes from occurring at the MOSFET gate, the 100 Ω resistor is a parasitic suppressor, and Z_1 serves as a dissipative voltage regulator for the output and also clips the drain voltage to a level below the rated power FET breakdown voltage.

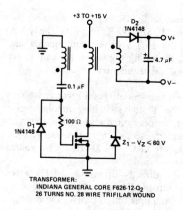

TRANSFORMER:
INDIANA GENERAL CORE F626-12-Q2
26 TURNS NO. 28 WIRE TRIFILAR WOUND

6.13.6 Positive Input/Negative Output Charge Pump

A charge pump is a simple means of generating a low-power voltage supply of opposite polarity from the main supply. The 74C14 IC is a self oscillating driver for the MOSFET power switch. It produces a pulse width of 6.5 μs at a repetition frequency of 100 kHz. When the MOSFET device is off, capacitor C is charged to the positive supply. When the power MOSFET switches on, C delivers a negative voltage through the series diode to the output. The zener serves as a dissipative regulator. Because the MOS-FET switches fast, operation at high frequencies allows the capacitors in the system to be small.

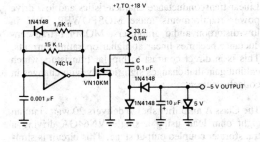

6.13.7 20 Watt, Class A Audio Amplifier

Linear transconductance characteristics and low drive power requirements make MOSPOWER ideal in low-distortion audio amplifiers. MOSFET transconductance becomes linear at higher operating currents. This is in direct contrast to bipolar transistors which exhibit gain that changes significantly with changes in collector current.

The Class A amplifier shown delivers 20 watts into an eight ohm load using a single VN64GA driving a transformer coupled output stage. This circuit is similar to the audio output stage used in many inexpensive radios and phonographs. Distortion is less than 5 percent at 10 watts using very little feedback (3%) with the VN64GA biased at 3 amperes.

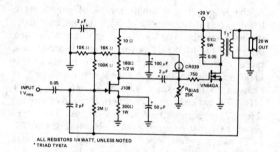

ALL RESISTORS 1/4 WATT, UNLESS NOTED
* TRIAD TY67A

6.13.8 Astable Flip-Flop with Starter

A pair of non-zenered MOSPOWER transistors, a pair of LEDs and a simple RC circuit make an easy sequential flasher with almost unlimited sequencing time — from momentary to several seconds.

The infinite input resistance of the MOSFET gate allows for very long sequencing times that are impossible when using bipolars. One precaution, though, don't wire your circuit using phenolic or printed circuit boards when you're looking for slow sequencing (they exhibit too much leakage!).

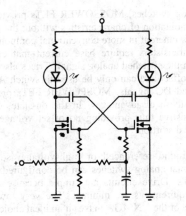

6.13.9 High Current Analog Switches

For analog switches, MOSPOWER FETs provide an ideal combination of characteristics without the usual trade-offs inherent in more conventional components. Bipolar transistors require base current that causes offsets in the switched analog signal. Triacs also produce an offset and can only be used to switch AC or interrupted DC signals. MOSPOWER FETs provide none of these disadvantages. In the on-state, MOS looks resistive thus providing no offset voltage and very low distortion.

Either ground referenced or high-voltage isolated bidirectional analog switches can be configured with MOSFETs. Drive circuits are simple because drive power requirements are minimal. The very low on-resistance of the VN64GA makes it an ideal choice for analog switch circuits.

currents up to 12.5 amperes. Two MOSPOWER devices are necessary to enable current control in both directions. This is not possible with a single device because of the drain-source diode which is inherent in the MOSFET. Input signal range for this switch is limited by the maximum allowable gate-source enhancement voltage of ±30 volts. Since 12 volts enhancement is required for 12.5 amperes output, then ±18 volts is the maximum signal range. Higher voltage inputs begin to turn off the VN64GA at the positive signal peaks thus limiting the MOSFET to pass less current. Turn-on is accomplished by driving the gates high (+30 V) and turn-off by driving the gates low (−30 V). Off isolation is enhanced by biasing the

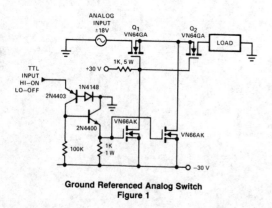

Ground Referenced Analog Switch
Figure 1

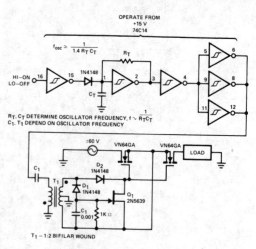

High Voltage Isolated Switch
Figure 2

The circuit above shows a ground referenced, bidirectional switch with a signal range of ±18 volts for

common source node at -30 volts. This has the effect of shunting to ground any signal that feeds through the drain-source capacitance.

Isolated gate drive is required to operate the analog switch over its full ±60 volt operating range. One way to accomplish this is to use a transformer isolated drive in a DC-to-DC converter as shown in Figure 2. The transformer can supply high voltage isolation, thus allowing the analog switch to operate as a true isolated switch up to thousands of volts above ground.

The oscillator, buffer and control functions are all performed by a single hex CMOS Schmitt trigger. The oscillator is gated on or off from a CMOS compatible input signal controlling the input Schmitt trigger. The VN64GA's input capacitance acts as both filter and storage capacitor for the half-wave rectified oscillator control voltage. Turn-on time of the analog switch depends on the oscillator frequency and is typically 10 microseconds for a 1 MHz oscillator.

When the switch is on, JFET Q_1 is biased off by the negative supply D_1, C_1 and therefore has no effect on operation. When the oscillator is turned off, Q_1 loses its bias and turns on, thus shorting the MOSFET gate to source for fast run-off.

6

6.13.10 Laser Diode Pulsers

The VN64GA can switch 10 amperes in 10 to 20 ns when driven by a low impedance source. This extremely fast switching speed is excellent for driving high-current laser diodes and to modulate these diodes at very high frequencies. A typical 2 watt peak power laser diode requires a maximum 200 ns, 10 ampere pulse at a 0.1% maximum duty cycle.

The Laser Diode Pulser is a simple drive circuit capable of driving the laser diode with 10 ampere, 20 ns pulses. For a 0.1% duty cycle, the repetition rate will be 50 kHz. A complementary emitter-follower is used as a driver. Switching speed is determined by the f_T of the bipolar transistors used and the impedance of the drive source.

A faster driver circuit is shown in Figure 2. It can supply higher peak gate current to switch the VN64GA very quickly. This circuit uses a MOS-POWER totem-pole stage to drive the high power switch.

The upper MOSFET is driven by a bootstrap circuit. Typical switching times for this circuit are about 10 ns for both turn-on and turn-off.

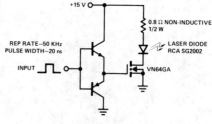

Laser Diode Pulser
Figure 1

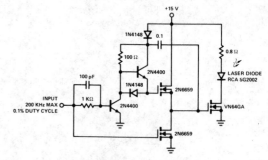

High-Speed Laser Diode Driver
Figure 2

6.13.11 A One-MOSPOWER FET Analog Switch

Using four diodes in an array as shown allows using only one MOSPOWER transistor for analog switching. Current flow is controlled keeping the source-base connection of the MOSFET towards the load [the importance of this is stressed in AN77-2]. Be sure to use diodes capable of handling the load current and be sure to use a transistor whose breakdown voltage specification exceeds the peak analog voltage anticipated.

Operationally, by increasing the gate-to-source bias voltage the MOSFET turns on. For applications other than either full on or full off, care must be taken not to exceed the dissipation of the MOSPOWER transistor.

A suitable *heat sink* cannot be overstressed in such applications.

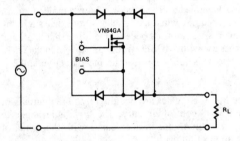

6

6.13.12 Stepping Motor Driver

Stepping motors find wide use in disk drives and machine control. MOSPOWER transistors are ideal motor drivers because of their freedom from second breakdown. The circuit shows how simple a motor drive becomes when using them. Note that snubbing networks are not used because load line shaping is not necessary with MOSPOWER and the inductance of the motor is fairly low so that the inductive spike is small.

The MOSFET gates are tied directly to the outputs of the CMOS control circuitry. The logic is arranged to sequence the motor in accordance with the needs of the application.

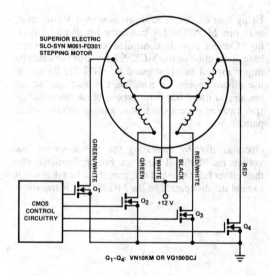

6.13.13 Constant Speed Motor Controller

DC motors enjoy wide usage in applications varying from electric trains to electric drills. In many applications a constant speed characteristic is desirable. This circuit uses MOSFETs and a single IC to achieve this speed control. One feature of this circuit is that no tachometer is required for monitoring the speed. Speed is determined by sampling the reverse EMF generated by the motor. Motor speed is controlled by pulse width modulation. Be sure only to use this circuit for DC supply rails because MOSFET devices can only block voltage in one direction.

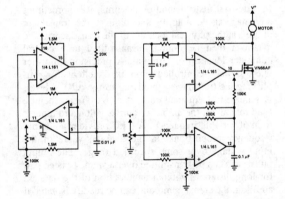

6.13.14 Voltage-to-Frequency Converter with Digital Line Driver

In industrial instrumentation systems, it is sometimes advantageous to amplify and digitize the transducer outputs at the physical location of the transducers. The data can then be digitally transmitted to a central computer location or terminal. A precision amplifier with digitally-controlled gain is excellent for amplifying the transducer signals up to the ±10 V level. A voltage-to-frequency converter (VFC) can then be used to convert the amplifier output voltage to a pulse train of variable frequency. A VFC provides an output frequency proportional to the average value of the input voltage over some specified range. The output pulse train can be transmitted over a single twisted pair for long distances without degradation of the data. In addition, the signal commons can be readily isolated if desired by coupling the output pulse train through an electro-optical coupler or pulse transformer.

Low-cost integrated VFC circuits are adequate for some applications, but are generally limited in output drive capability. Some also require numerous external components. The VFC circuit shown can provide good accuracy and can drive very long cables or low-impedance loads.

A two-channel FET switch driver, the Siliconix D169, is used as a comparator. The D169 comparison threshold is internally 1.2 V with V_R connected to ground and the complementary outputs both swing ±14 V when connected as shown. The ±14 V complementary output drives two MOSPOWER transistors connected in a totem-pole configuration. The MOSFETs ON-resistance will be approximately 2 to 3 ohms with ±14 V of gate drive, so the output voltage V_O will swing between +5V and zero even with a low-impedance load.

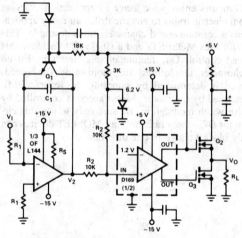

NOTE: ALL RESISTORS ARE IN OHMS UNLESS OTHERWISE NOTED.
$f_o = V_1 / R_1C_1\Delta V - 10 < V_1 < 0$

Voltage-to-Frequency Converter

To visualize the circuit operation, assume that the integrator output V_2 is ramping positive in response to a negative input V_1. This circuit configuration requires that the input voltage be negative. The D169 output (OUT) is negative and transistor Q_1 is OFF. Since OUT is approximately −14 V, the 6.2 V zener will be conducting and the input to the comparator will be determined by the R_2 voltage divider. The voltage at IN, the D169 input, will be half of the integrator output V_2 minus half of the 6.2 V zener voltage.

When the integrator voltage ramps positive up to +8.6 V, the voltage at the D169 input will be +1.2 V and

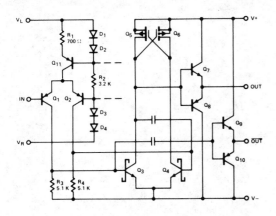

Logic	OUT	$\overline{\text{OUT}}$
0	V −	V+
1	V+	V−

D169 Switch Driver

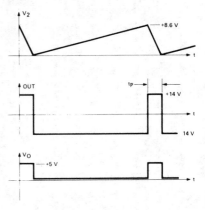

VFC Waveforms

the comparator will trip. The OUT pin will suddenly go positive and thereby gate Q_1 ON which will reset the integrator. With the OUT pin positive, the zener-diode will conduct as a forward biased diode and the voltage applied to the input of the D169 comparator (IN pin) will be half of V_2 plus half of the diode drop (0.6 V). When the integrator output drops down to approximately zero during the reset cycle, the comparator will again trip and OUT will again go negative. Since the saturation voltage of Q_2 is much less than 1.8 V, we can be sure that the integrator will always ramp down sufficiently to trip the comparator. After resetting, the integrator will again ramp positive at a rate set by the input voltage V_1. With OUT at −14 V, the zener-diode will again conduct as a zener and apply − 6.2 V through R_2 to the D169 input. The VFC waveforms are shown for one cycle. The width of the positive 5 V output pulse is determined by the integrator reset time, which will be constant. The pulse repetition rate is directly proportional to the negative input voltage; higher input voltage causes the integrator to ramp positive faster and to thereby increase the pulse repetition rate.

For a DC input of $-V_1$ and integrator output change ΔV, the output frequency will be determined by the integration time interval Δt. From $i = C dV/dt$, we have

$$\frac{V_1}{R_1} = C_1 \frac{\Delta V}{\Delta t}$$

Neglecting the short reset time, output frequency will be $1/\Delta t$ which is equal to $V_1/R_1 C_1 \Delta V$. The non-linearity error caused by finite reset time, tp, is approximately $tp/\Delta t$ multiplied by f_0.

This basic VFC design can be easily tailored to specific design requirements by judicious choice of integrator op amp and component values. For example, using 1/3 of a Siliconix L144 triple op amp with a 120 K current-setting resistor provides a low current drain circuit capable of about 1.3 μsec reset time. A capacitor C_1 of 0.001 μF was used in one application along with an R_1 of 100 K. This made the full-scale frequency for a −10 V input equal to:

$$f_{MAX} = \frac{10V}{100 \text{ K} \times 0.001 \text{ } \mu\text{F} \times 8.6 \text{ V}}$$
$$= 11.628 \text{ Hz}$$

The other component values are not critical; an 18 K resistor to the base of Q_1 will be adequate and 3 K to the 6.2 V zener with R_2 of 10 K. For very wide dynamic range, a DG211 quad switch can be used to switch in different values of input resistor R_1.

The output can be taken directly from the complementary outputs of the D169 comparator/driver or from the totem-pole MOSPOWER transistors. The VN35AK power transistors can readily drive a 50 ohm load with 5 V, one microsecond pulses.

Although this circuit requires two IC packages rather than one, it has many advantages in flexibility. It can be designed for high-frequency operation, minimal power drain, or maximum output pulse power.

6

7.1 Understanding MOSPOWER Transistor Characteristics Minimizes Incoming Testing Requirements (TA84-1)

Introduction

As MOSPOWER transistors grow in popularity, more test engineering groups are given the task of developing incoming screening procedures for these devices. One approach that might be taken is to faithfully measure all of the parameters specified on the manufacturer's data sheet. This approach leads to an unnecessary investment of engineering time and talent. An effective alternate approach is to first understand the testing and characterization procedure that led to the MOSPOWER transistor parameters specified on the manufacturer's data sheet. The screening procedures for the transistors at incoming electrical test may then be matched to the specific application. This approach considerably simplifies the testing necessary to gain assurance that the transistors will work as required in the application.

This article begins by looking closely at a typical MOSPOWER transistor data sheet. Next, the device characterization procedures that are followed to obtain the data sheet are discussed. The specific tests used to obtain the data sheet values are covered in detail. With this information, it is possible to determine the measurements needed to verify the data sheet and the measurement techniques to follow.

The Data Sheet of a Typical MOSPOWER Transistor

The performance of all MOSPOWER transistors is documented in a data sheet. This information is presented in several distinct blocks. These blocks of data are discussed in detail below.

Product Description Block

The performance of the transistor is highlighted for the potential user. This block of information allows a design engineer to decide quickly whether to look more closely at the data sheet. The data given in this section presents the MOSPOWER transistor in the best possible light. Closer investigation is required to understand the full capability of the device.

Absolute Maximum Ratings

This block of information indicates the limits of device performance. The data is divided into two types:

1. Parameters that are 100% tested.

2. Parameters guaranteed by design or manufacture and not verified by 100% testing.

Electrical parameters such as breakdown voltage and maximum drain current fall into the first category. Parameters such as operating and storage temperature fall into the second category. Power dissipation usually falls into the second category. It is often measured on a sample basis before a production lot is released.

Electrical Characteristics and Test Circuits Used to Obtain These Results

This section of the data sheet defines the tests that the transistor undergoes before it is shipped to a customer. A test engineer uses the information in this section to set up an incoming test procedure. It is the values in this table that are discussed in detail in this article.

Typical Performance Curves

The Table of Electrical Characteristics provides information on device performance at extremes of current, voltage, etc. A circuit designer must know device behavior over the entire range to complete a satisfactory design. The typical device performance curves provide this information.

Reprinted with permission from the January 1984 issue of Test & Measurement World.

MOSPOWER Transistor Structure

A MOSPOWER transistor has the four separate device regions shown in Figure 1. The purpose of each region is:

Source: Carriers for conduction are supplied by this region.

Body: This region electrically separates the source from the drain when the transistor is 'off'.

Gate: The conduction of the transistor is controlled by the voltage between this terminal and the source.

Drain: Carriers flow to this region when the transistor is 'on'.

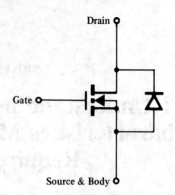

The Symbol for an N-Channel MOSPOWER Transistor
Figure 2

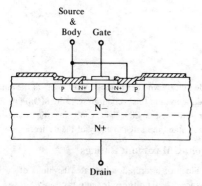

Cross-Section of a MOSPOWER Transistor Showing the Four Regions and the Three Terminals
Figure 1

In conventional MOS transistor fashion, the source and body region are electrically shorted together by metal on the surface of the device. This configuration permits operation of the transistor using only three external leads. The electrical symbol for an N-channel MOSPOWER transistor is shown in Figure 2. The diode from body to drain emphasizes the asymmetrical nature of this type of device. Inclusion of the diode symbol also aids in explaining the electrical behavior of the transistor. The abbreviated terminology used to describe the electrical behavior of MOSPOWER transistors is explained more fully in the data sheet. Table 1 shows a typical set of electrical characteristics for a MOSPOWER transistor, including the definition of the various symbols. It is this table that will receive our attention for the remainder of this article.

Static Device Parameters

These parameters generally determine whether or not a device works in a circuit. They should usually be sampled at incoming inspection. The static device parameters may be measured using most conventional curve tracers as long as the leads are properly biased. Most curve tracers still label the three terminals using bipolar transistor terminology. The MOS transistor leads correspond as follows to the bipolar labels:

Source and Body = Emitter

Gate = Base

Drain = Collector

As long as one remembers that MOS transistors are voltage driven while bipolar transistors are current driven, measurements are easily made. The static device characteristics are first explained.

BV_{DSS} – drain-to-source breakdown voltage with the gate shorted to the source. This voltage is determined by forcing a specified current from drain-to-source and measuring the resulting voltage as shown in Figure 3. The device will allow this current flow as long as the power dissipation of the package is not exceeded. It is wise to limit the device power dissipation to prevent inadvertent device damage.

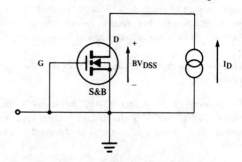

Device Configuration for BV_{DSS} Measurement
Figure 3

ELECTRICAL CHARACTERISTICS (T_C = 25°C unless otherwise noted)

STATIC

	Parameter	Type	Min.	Typ.	Max.	Units	Test Conditions
BV_{DSS}	Drain-Source Breakdown Voltage					V	V_{GS} =
						V	I_D =
$V_{GS(th)}$	Gate-Threshold Voltage					V	$V_{DS} = V_{GS}$, I_D =
I_{GSSF}	Gate-Body Leakage Forward					nA	V_{GS} =
I_{GSSR}	Gate-Body Leakage Reverse					nA	V_{GS} =
I_{DSS}	Zero Gate Voltage Drain Current					mA	V_{DS} = Max. Rating, V_{GS} = 0
						mA	V_{DS} = Max. Rating, V_{GS} = 0 T_C = 125°C
$I_{D(on)}$	On-State Drain Current[1]					A	V_{GS} = , V_{DS} =
						A	V_{GS} = , V_{DS} =
$V_{DS(on)}$	Static Drain-Source On-State Voltage[1]					V	V_{GS} = , I_D =
						V	V_{GS} = , I_D =
$R_{DS(on)}$	Static Drain-Source On-State Resistance[1]					Ω	V_{GS} = , I_D =
						Ω	V_{GS} = , I_D =
$R_{DS(on)}$	Static Drain-Source On-State Resistance[1]					Ω	V_{GS} = , I_D = , T_C =
						Ω	V_{GS} = , I_D = , T_C =

DYNAMIC

	Parameter	Type	Min.	Typ.	Max.	Units	Test Conditions
g_{fs}	Forward Transductance[1]					S ($\mho$)	
C_{iss}	Input Capacitance					pF	V_{GS} = , V_{DS} =
C_{oss}	Output Capacitance					pF	f = 1 MHz
C_{rss}	Reverse Transfer Capacitance					pF	
$t_{d(on)}$	Turn-On Delay Time					ns	V_{DD} = , $I_D \cong$
t_r	Rise Time					ns	R_g = , R_L =
$t_{d(off)}$	Turn-Off Delay Time					ns	(MOSFET switching times are essentially independent of
t_f	Fall Time					ns	operating temperature.)

THERMAL RESISTANCE

	Parameter	Type	Min.	Typ.	Max.	Units	Test Conditions
R_{thJC}	Junction-to-Case					°C/W	
R_{thJA}	Junction-to-Ambient					°C/W	Free Air Operation

BODY-DRAIN DIODE RATINGS AND CHARACTERISTICS

	Parameter	Type	Min.	Typ.	Max.	Units	Test Conditions
I_S	Continuous Source Current (Body Diode)					A	Modified MOSPOWER symbol showing the integral P-N Junction rectifier
						A	
I_{SM}	Source Current[1] (Body Diode)					A	
						A	
V_{SD}	Diode Forward Voltage[1]					V	T_C = , I_S = , V_{GS} =
						V	T_C = , I_S = , V_{GS} =
t_{rr}	Reverse Recovery Time					ns	T_J = , $I_F = I_S$, dI_F/ds = A/μs

1 Pulse Test: Pulse Width $\leqslant$ 300 μsec, Duty Cycle $\leqslant$ 2%

7

$V_{GS(th)}$ - gate threshold voltage. This measurement determines the voltage needed from gate-to-source to produce a given drain current. The drain is normally shorted to the gate as shown in Figure 4.

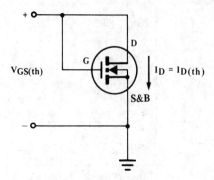

Measurement of Threshold Voltage, $V_{GS(th)}$
Figure 4

I_{GSSF} & I_{GSSR} - gate leakage current (F or R specify 'forward' or 'reverse' leakage). This measurement is made with the source and drain biased to one voltage and the gate held at ground as shown in Figure 5. The capacitance shown in dashed lines indicates that charge must be supplied to the gate to bias it with respect to the source and drain. If an appropriate settling time is not chosen, erroneous readings may result.

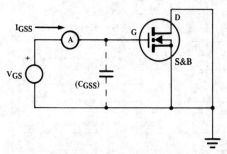

Measurement of the Gate Leakage Current, I_{GSS}
Figure 5

I_{DSS} - zero gate voltage drain current. This 'leakage' current is measured at a voltage typically between 0.8 BV_{DSS} and BV_{DSS}. The measurement circuit configuration is indicated in Figure 6.

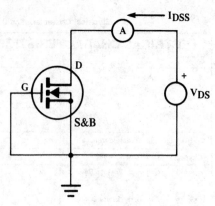

Determination of the Drain-Source Leakage, I_{DSS}
Figure 6

$V_{DS(ON)}$ - static drain-source on-state voltage. This measurement, made as shown in Figure 7, sets a maximum limit on the drain to source voltage for a given gate bias and drain current. Along with $R_{DS(ON)}$, this measurement indicates the power dissipation that will occur when the device is in operation.

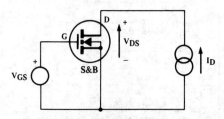

Static Drain-Source On-State Voltage
and Resistance Measurement
Figure 7

$R_{DS(ON)}$ - static drain-source on-state resistance. This measurement is also made using the circuit of Figure 7. In measuring both $V_{DS(ON)}$ and $R_{DS(ON)}$, care must be taken to avoid heating the device. Both of these parameters increase as temperature increases. Low resistance ohmic or Kelvin contacts are required to obtain accurate readings.

Dynamic Device Parameters

These parameters are largely set by device geometry. They need only occasionally be monitored.

g_{fs} – forward transconductance. This parameter is given in units of siemens or mhos and indicates the gain of the device. To understand how gain is measured, we simply divide the change in drain current by the change in gate voltage.

$$*g_{fs} = \frac{\Delta I_D}{\Delta V_{GS}}$$

This parameter is akin to the beta of a bipolar transistor where a change in collector current is divided by a change in base current.

$$h_{fc} = \frac{\Delta I_C}{\Delta I_B}$$

* Measurements are in siemens if ΔV_{GS} is measured in volts and ΔI_D in amperes.

Capacitances – Three capacitance values, C_{iss}, C_{oss}, and C_{rss} are measured in a MOSPOWER transistor. Figure 8 shows the physical origin of these values.

C_{iss} = C_{gs} + C_{dg} (unguarded)
C_{rss} = C_{dg} (guarded)
C_{oss} = C_{ds} + C_{dg} (unguarded)

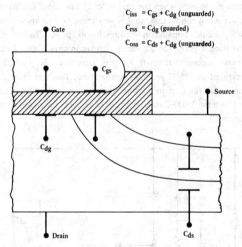

**The Capacitances in a MOSPOWER Transistor
Figure 8**

C_{iss} – input capacitance. Due to the nature of field-effect transistors, the gate-to-source input presents itself as a large capacitor. The value of this capacitor may be several nanofarads on large devices. The magnitude of C_{iss} decreases rapidly within the first 5 to 10 volts of V_{DS}. Typical values are taken at V_{DS} = 25V and V_{GS} = 0. If the quiescent drain voltage is 25 or more volts, the C_{iss} will generally be less than the rated value.

C_{oss} – output capacitance. It is an important factor in determining device switching time. In combination with the resistance of the load, this capacitance determines the turn-off time constant. The device is not fully turned off until this capacitance is charged to V_{DD}. Turn-on time is a function of C_{oss} and $R_{DS(on)}$. This time constant is typically much shorter than the turn-off time constant.

C_{rss} – reverse transfer capacitance. One of the more interesting phenomena in MOSPOWER devices is the negative feedback between the drain and the gate. This feedback is caused by the internal capacitance, C_{rss}. This capacitance limits the maximum operating frequency of the device. It is an inhibiting factor to fast device switching.

$t_{d(on)}$ – turn-on delay time. The ideal device takes zero time to turn on (full-off to full-on), and hence has zero delay between the input and the output signal. All active devices such as vacuum tubes and transistors take some time to react to an input signal. In the case of MOSPOWER transistors in the off state, this finite amount of time is called the turn-on delay time or $t_{d(on)}$. There are two requirements for the test pulse used to switch the device under test. It must switch faster than the device under test (D.U.T.) by a factor of four; hence the rise time of the test pulse must be less than 25% of the rise time of the D.U.T. The driving impedance must be less than 10% of the input impedance of the D.U.T. since it is a capacitive load. $t_{d(on)}$ is measured between the 10% point of the input voltage and the 90% point of the output voltage.

t_r and t_f – the rise and fall times. Measurement of these times also requires a fast pulse generator.

If a pulse source of adequate rise and fall times is not available, the t_r and t_f of the D.U.T. may be calculated knowing the input waveform rise and fall times and the

7

test device rise and fall times by using the equation

$$t_r = [(t_{r2})^2 + (t_{r1})^2]^{1/2}$$

Where:

t_r = actual t_r (or t_f) of the D.U.T.

t_{r1} = input rise (or fall) time (generator)

t_{r2} = output rise (or fall) time (test device)

$t_{d(off)}$ – turn-off delay time. Once a device has reached a stable 'on' state, it requires a finite amount of time to turn off. Intrinsic capacitances such as C_{oss} and C_{rss} have a 'charge storage' that produces a 'turn-off delay time.' This time, plus the fall time, t_f, is called the turn-off time. The $t_{d(off)}$ is measured between the 90% point of the input pulse and the 10% point of the output. The switching test circuit used by Siliconix is shown in Figure 9. Test circuits for these measurements are usually included on the data sheet. A note of caution: inductance of component leads in the drain circuit may cause large errors in the switching data. If the drain *current* waveform differs substantially from the drain *voltage* waveform, the distance of the test circuit components from the D.U.T. should be reconsidered. In order to avoid faulty delay times, be sure the cable lengths measuring input and output are equal.

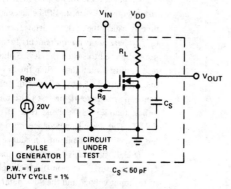

The Circuit Used to Determine Switching Times
Figure 9

Thermal Resistance

These ratings are determined by die size and the thermal characteristics of the package. These values are difficult to measure, but vary little from device to device within a product type assembled on a well controlled production line.

Body-Drain Diode Ratings and Characteristics

MOSPOWER transistors contain an inherent passive device called a 'body-drain' diode. The diode performance need only be monitored if it becomes active in circuit operation. Electrically it is the body-drain junction. The diode conducts when the drain of the device becomes more negative than the source. The diode has the same current capability as the MOS transistor itself. Measurements of this diode are made by shorting the gate to the source.

I_s – the maximum continuous body-drain diode current. This parameter is set by the power dissipation of the package.

I_{sm} – the maximum pulsed body-diode current. The size of the bond wire and its current handling capability generally set this parameter.

V_{SD} – forward-on voltage. As with any diode, the current passing through it causes an IR voltage drop. So it is with the body-drain diode, and this voltage must fall within certain limits at specified currents.

t_{rr} Reverse Recovery Time

When a MOSPOWER transistor is placed in an inductive circuit, the part may commutate or conduct through the body-to-drain diode even when the gate is 'off.' The diode dissipates power when conducting, so it is very important to know how long the diode remains on to compute total power dissipation. The time it takes for the carriers to recombine is called the reverse recovery time (t_{rr}). This parameter is generally measured at 125°C junction temperature. This test must be made with the part in a temperature-controlled environment and with the JEDEC standard reverse recovery tester shown in Figure 10.

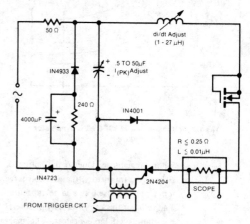

The JEDEC Approved Circuit for Determining t_{rr}
Figure 10

7.2 Accelerated Testing

Imagine being handed a light bulb and then being asked to determine the number of hours of life that the bulb will deliver. Your first reaction might be to screw it into a socket, record the time and date, and then turn the light ON. After several days, weeks, or maybe months you would one day discover the light has burned out. From your observation, you could state with a high degree of accuracy how long the light lasted. Now imagine that you have 100 light bulbs to check. You could repeat the experiment and determine the average life of the light bulbs. You could even describe the failure rate as a function of time or calculate the Mean Time Between Failures (MTBF).

The problem with this approach, of course, is the process of determining the reliability of the light bulb might take months or years to complete. By that time, the information might be too late to be useful to anyone. Or worse, if you were depending on it to help you sell a new light bulb design, you would probably miss the boat by not having the reliability data available in a timely fashion.

How, then, can we accurately predict the reliability of a semiconductor device, which often has an average life of twenty years or longer, without years of testing? Fortunately, there are ways of accelerating the mechanisms that cause these failures. This "accelerated testing" as it is called, is a vital tool that allows us to evaluate new products or processes or to monitor a whole family of products within a reasonable amount of time.

The need for accelerated testing seems obvious, but the actual methods used to accomplish it are often not well understood. The purpose of any accelerated testing program is to cause the devices under test to age at a faster rate than they would under normal usage, while at the same time, not introducing any new failure mechanisms.

Many failure mechanisms associated with semiconductor devices are a result of some kind of chemical reaction occurring on a microscopic scale. To provide an accelerated testing program, we need a way of speeding up these chemical reactions while at the same time determining just how much acceleration we have achieved.

Other failure mechanisms are often due to some kind of mechanical stress on the device. For failures related to mechanical stresses, we need to create some tests that would allow us to establish pass/fail criteria assuring integrity of the various mechanical interfaces of the device (i.e. wire bond, die attach, lid or can seal, etc.).

A Swedish scientist named Svante August Arrhenius first reported the relationship between the rates of chemical reactions and temperature. He described this time temperature relationship by the equation:

$$R = Ce^{-(Ea/kT)}$$

where

R   =   The rate of the chemical reaction.

C   =   A constant.

Ea   =   The activation energy in electron volts (eV) that is associated with the chemical reaction.

Additional Considerations

Guard Banding

Since manufacturers are unable to test devices to exact specifications, it becomes necessary to build in a margin of safety (in the customer's favor) called guard banding. The amount of margin is not an industrial standard; it is an in-house decision made by engineering based on experience and practicality.

Thermal Concerns

To avoid thermal problems, minimize the time the collector voltage is left on to 'viewing time' only. Heat that causes small increases in operating temperature will change the measured device performance leading to erroneous data. Large temperature increases may permanently alter the transistor's characteristics or even destroy the device. It's always advisable to duplicate the test as done by the manufacturer as closely as possible in order to match the results.

Test Equipment and Technique

Test results are only as good as the equipment used and the person making the measurements. The only other variable is how well the equipment is calibrated. If the curve tracer at factory 'A' is as accurate as the curve tracer in 'incoming inspection' at customer 'B' then the parts will test exactly the same for both parties. The common denominator of test method variation is minimized.

Summary and Conclusions

Users of MOSPOWER transistors have many options when deciding on the appropriate incoming inspection. Duplicating the outgoing tests performed by a transistor manufacturer does verify device parameters, but it is a costly procedure. A less costly alternative, use of a curve tracer, has been discussed in this article.

The curve tracer has become an accepted standard as a measuring device in the electronics field. Engineers and technicians the world over rely on these testers daily. When properly calibrated they are extremely accurate and reliable. A curve tracer is highly recommended for the buyer's 'incoming inspection department' lacking automatic test equipment. Most test data published by MOSPOWER vendors can be verified on curve tracers providing these testers are used properly. This curve tracer data is sufficient for most users' purposes.

In some applications, however, the circuit performance depends on device characteristics measured only on special test equipment. These parameters have also been discussed, and circuits are given for these measurements. In many cases these parameters are guaranteed by design, so only sample measurements are required.

Reference

JEDEC Solid State Products Engineering Council. JC-25 Enhancement Type Field Effect Power Transistor for Switching and Amplifier Applications.

7

K = Boltzman's constant (8.62×10^{-5} eV/°K)

T = Temperature in degrees Kelvin

This equation simply states that chemical reactions are accelerated logarithmically as the temperature is increased. The activation energy (Ea) will be different for different chemical reactions. Higher activation energies are used to describe chemical reactions that increase at a faster rate as the temperature is increased.

The actual acceleration factor that describes the time-temperature relationship of a given chemical reaction at two different temperatures is as follows:

$$R1/R2 = e^{-Ea/K(1/T1 - 1/T2)}$$

where

R1 is the reaction rate at the higher temperature.

R2 is the reaction rate at the lower temperature.

T1 is the test temperature in degrees Kelvin.

T2 is the temperature for which the new reaction rate is calculated in degrees Kelvin.

C is the same for both temperatures and has been cancelled out of the equation.

The chemical reactions that most often cause failures in semiconductor devices include metal migration, corrosion, electron trapping, and ionic contamination moving through oxides and silicon. The activation energies associated with these failure mechanisms usually range from 0.4 eV to 1.2 eV.

Knowing the relationship of time and temperature on accelerating failure mechanisms associated with semiconductor devices, it is possible to devise short term evaluation tests that can be used to determine the reliability of a device within a few weeks rather than years. These evaluation tests involve operating the devices under normal or exaggerated conditions (worst case) in a higher than normal temperature environment.

For integrated circuits (ICs), the life test conditions can be either dynamic mode or static mode. In dynamic mode, the device is continuously changing states. For complex circuits, this may be the only way to exercise all parts of the circuit during the evaluation. For other less complex circuits, a static life test mode may prove to be more severe on the device. This can be especially true when stressing the input and output portions of the circuit. For discrete devices such as MOS transistors and JFETs, it is better to do

HTRB (High Temperature Reverse Bias) stress and/or a HTGB (High Temperature Gate Bias) stress since these conditions are usually the worst case conditions for the device. In any case, all the testing described is done at extremely high ambient temperatures (125° C to 175° C or higher). It is the high temperatures combined with the operating voltages that allow us to "age" the devices in an accelerated fashion so the failure rates can be determined.

Having developed the technique of accelerating semiconductor failure mechanisms through high temperature life testing, the next requirement is to understand how to interpret the results and use the life test data to predict actual failure rates of the devices at normal operating temperatures. Several failure distributions that describe the cumulative failure rate versus time have been proposed. The most popular distributions are the Weibull and the Log Normal, however, the Log Normal distribution seems to be emerging as the standard.

To determine the actual failure rate of a device, we need to run a high temperature life test as described earlier and, in addition, we need to understand what activation energies are associated with each failure mechanism found during the life test. There are several ways that can be used to assign an activation energy to a given failure mechanism. One way, which is often used, is to determine the type of failure mechanism through detailed failure analysis. Most major failure mechanisms that commonly affect semiconductor devices have been extensively evaluated and their activation energies reported. For example, ionic contamination has an activation energy of between 1.0 and 1.2 eV. A complete list of common failure mechanisms and the activation energies that are used by Siliconix appears in Table 1.

Table 1
Standard Activation Energies
Used for Common Failure Mechanisms

Failure Mechanism	Activation Energy
Ionic Contamination	1.1 eV
Oxide Defects	0.4 eV
Electron Trapping	1.1 eV
Gold to Aluminum Intermetallics Resulting in Bond Failures.	1.0 eV
Aluminum Electromigration	0.7 eV
Unknown Cause	0.7 eV

7

A second, more accurate method for determining the activation energy of semiconductor failure mechanisms is to determine it experimentally. Using this method, the activation energy of an unknown failure mechanism as well as a known failure mechanism can be determined. The method consists of running two different life tests. If the life tests are run at significantly different temperatures, the failure rates for each failure mechanism can be determined by working backwards through the Arrhenius Equation. This experimental method is very time consuming and is only used when dealing with common failure mechanisms usually associated with semiconductor failures.

Failure Rate Calculations

Calculating the failure rate of a semiconductor device at a specific operating temperature becomes fairly simple once all the parameters are understood. All that is needed is the failure rate of the devices at an accelerated temperature and an understanding of the activation energy associated with the failure mechanism. The following is an example of how this is done:

Example:

A product has undergone extensive reliability evaluation that included high temperature life testing. A total of 600 devices have been placed on life test at 150° C for a cumulation of 593,500 device hours. Thirteen failures have occurred which are all related to leakage problems and have an assigned activation energy of 1.1 eV. We need to determine the failure rate for an operating temperature of 70° C.

Solution:

The first step is to determine the cumulative failure rate at the life test temperature. This is done as follows:

$$\text{Failure Rate} \atop (\%/1000 \text{ hrs.}) = \frac{(\text{no. of failures} \times 100)}{(\text{device hours} / 1000)}$$

$$= \frac{13 \times 100}{593,500 / 1000}$$

$$= 2.19 \%/1000 \text{ hrs.}$$

Using the Arrhenius equation and the activation energy of 1.1 eV, we can determine the acceleration factor between the life test temperature and the use temperature.

$$\text{Acceleration} \atop \text{factor} = e^{-\{Ea/K [1/Ta - 1/T2]\}}$$

$$= e^{-\{(1.1/.0000862)/[1/(150 + 273) - 1/(70 + 273)]\}}$$

$$= 1137$$

To determine the failure rate of the device at the operating temperature of 70° C, simply divide the failure rate found during the high temperature life test by the acceleration factor.

$$\text{Failure Rate at} \atop \text{Use Temperature} = \frac{\text{Life Test Failure Rate}}{\text{Acceleration Factor}}$$

$$= \frac{2.19 \%/1000 \text{ hrs.}}{1137}$$

$$= 0.002 \%/1000 \text{ hrs.}$$

Confidence limits can be assigned if desired. The use of confidence limits is beyond the scope of this discussion. Several good references are listed that can be reviewed if the reader is unfamiliar with confidence limits.

References

[1] Peck, D.S. and Trapp, O.D., *Accelerated Testing Handbook,* Technology Associates and Bell Labs, Portola Valley, CA, 1978.

[2] King, J.R., *Probability Charts for Decision Making,* Industrial Press, Inc., New York, NY, 1971.

[3] Doyal, Ed, Jr., *Microelectronics Failure Analysis Techniques: A Procedural Guide,* RADC Publication.

[4] Peck, D.S., "Semiconductor Reliability Predictions from Life Distribution Data," *Semiconductor Reliability,* Edited by Schwop and Sullivan, Engineering Publishers, Elizabeth, NJ, 1961.

[5] Peck, D.S., "Uses of Semiconductor Life Distributions," *Semiconductor Reliability Volume 2,* Edited by Von Alven, Engineering Publishers, Elizabeth, NJ, 1962.

7.3 Reliability

Central to the concept of all applications is that the products being used are reliable. The wide range of packages and process flows used in the manufacture of MOSPOWER devices demand careful attention to detail—qualification of new designs, both from a process and packaging point of view; complete documentation of all changes in procedures or materials; review of customer qualification and acceptance requirements; dissemination of test reports and other significant data; research and development of testing and evaluation techniques; and a wide range of reliability monitoring programs. The charter for all of these lies with the Director of Reliability. The following summary of the Siliconix Reliability Manual demonstrates the high level of importance this activity is given:

(1.0) The Reliability Department provides Siliconix with a number of programs to define product reliability levels. Among these programs are (1) qualification, (2) monitor, (3) failure analysis, and (4) data collection and presentation.

(2.0) **Qualification Program**

(2.1) Qualification of New Products and Processes.

(2.1.1) Procedures for qualification of new chip designs require reliability participation or approval in design reviews, documentation, characterization, and reliability stress studies.

(2.1.2) New package qualification approval and release for production is granted by Reliability after prescribed environmental tests have been successfully completed.

(2.1.3) New process qualification approval is granted by Reliability after Quality Control and Reliability have completed evaluation of process control engineering studies. Significant modifications to existing processes are treated as new processes for the purpose of qualification.

(2.1.4) Proper documentation of all changes in processes or procedures and any new or improved designs or materials is assured by Reliability approval of all changes and origination of new documentation as required.

(2.2) Qualification of existing products for new applications.

(2.2.1) Customer Qualifications. Reliability is responsible for the review and acceptance of all customer qualification requirements. Where qualification programs or special testing is required, Reliability designs and implements appropriate test plans and coordinates with customer when required.

(2.3) Reporting and publication of data.

7

(2.3.1) Qualification test reports are prepared and distributed by Reliability for all products or processes which are approved by formal qualification. These reports contain a statement of qualification that certifies the process or product for use.

(2.3.2) Where new products require qualification reporting to the field, Reliability is responsible to assist Marketing Services in the preparation to field sales locations and to customers for information or advertising purposes.

(2.4) Research and Development. It is the responsibility of Reliability to investigate new methods and techniques of testing and evaluation of products, as well as methods of improving the reliability of existing products and processes. This includes, but is not necessarily limited to, the following methods.

(2.4.1) Design test plans and programs.

(2.4.2) Accelerated stress testing.

(3.0) **Reliability Monitor Programs**

(3.1) Device and Package Reliability Monitor Programs are effected for all packages using a variety of device types to maximize data usefulness and cost effectiveness of equipment.

Packages are monitored using applicable methods of MIL-STD-883, Class B, and data reported as specified in detailed procedures for each package-chip combination. Package monitor programs include, but are not limited to, the following general tests, using the appropriate conditions specified in MIL-STD-883, Class B, Method 5005.

7.4 Reliability Report Example

Date: February 1983   Report Number: R-0283

SUBJECT OF THIS REPORT: Failure rate calculations for the VNMK/L geometry.

DISCUSSION: This report deals with the new product qualification testing done on the VNMK/L geometry in the plastic TO-237 package.

TEST VEHICLE:

Part Number:	VN10KM
Geometry:	VNMK /L
Package:	TO-237
Assembly Location:	TAIWAN
FAB Facility:	FAB II

QUALIFICATION TEST PLAN: 1910 - 072

OTHER RELEVANT DOCUMENTS:

Assembly Flow No.:	7428
Bond Diagram:	2619
Burn in Diagram:	HTRB/HTGB
Package Outline:	37-3501-2

RESULT OF THIS TEST: The VNMK/L geometry in the plastic TO-237 package meets the qualification requirements for product release to the commercial market. Predicted failure rates of .008%/1000 hours for HTRB and .019%/1000 hours for HTGB conditions have been found.

7

Report prepared by: Approvals:

Stephen Kent   Design Engr. *Robert A. Mulber*

Steve Kent, Reliability Manager   Product Engr. *Timothy Slate*
Siliconix incorporated, QRA
                                             Package Engr. *Suresh G Belani*

VNMK/L FAMILY

60V – 100V   •   5.0 – 7.5Ω

N Channel VMOS Enhancement Mode MOSFET

REPORT CONTENTS

This report contains not only the specifics related to the subject devices, but also includes a discussion of how a user can compare manufacturers' reliability reports to ensure procurement of the most suitable devices.

I. INTRODUCTION

This report presents the reliability data for the VNMK/L device in the TO-237 package. The design makes use of a vertical MOS structure (VMOS). The vertical structure allows for high density and low $R_{DS(on)}$ resistance since the drain contact does not reside on the top of the die, but instead, is the substrate. (See Figure 1 for cross-sectional view.)

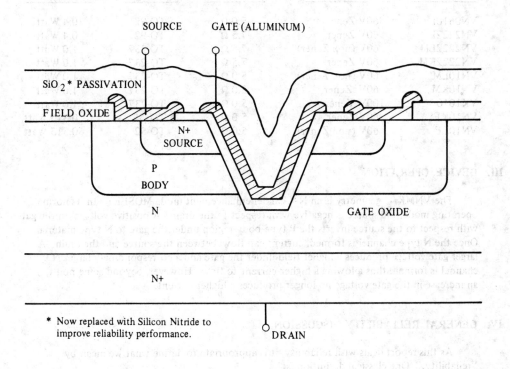

FIGURE 1. VMOS STRUCTURE

The TO-237 package used by Siliconix is similar to a TO-92 package, which is an industry standard for plastic packaged transistors. The major difference between these two packages is due to the heat sink tab that is a part of the TO-237 package. This heat sink tab allows more heat dissipation, thus improving performance (power rating).

The TO-237 package is based on a nickel plated copper frame with spot gold plating in the die attach and lead bond sites. The die attach is gold eutectic and the source and gate connections are made to the frame leads by .002" gold bond wires. The package is encapsulated with epoxy molding compound.

7

II. PRODUCT FAMILY

The VNMK/L geometry is used to manufacture a series of low power small signal MOS transistor devices. These products include the following part numbers.

Part No.	Voltage	Resistance	Package	Power
VN0610L	60V Zener	5.0 Ω	TO-92	0.4 Watt
VN2222L	60V Zener	7.5 Ω	TO-92	0.4 Watt
VN2222LM	60V (non Zener)	7.5 Ω	TO-237	1.0 Watt
VN2222KM	60V Zener	7.5 Ω	TO-237	1.0 Watt
VN10LM	60V (non Zener)	5.0 Ω	TO-237	1.0 Watt
VN10KM	60V Zener	5.0 Ω	TO-237	1.0 Watt
VK1010	100V Zener	5.0 Ω	TO-237	1.0 Watt
VN10KE	60V Zener	5.0 Ω	TO-52	0.315 Watt
VN10LE	60V (non Zener)	5.0 Ω	TO-52	0.315 Watt

III. DEVICE OPERATION

The VNMK/L geometry is an N-Channel enhancement mode MOSFET. In its normal operating mode, the source is negative with respect to the drain. A positive voltage on the gate with respect to the source inverts the P type body region under the gate to N type material. Once the N type channel is formed, current can flow between the source and the drain. A larger gate voltage produces a higher field under the gate and a correspondingly larger N type channel is formed, thus allowing a higher current to flow. However, beyond some point, an increase in the gate voltage no longer produces a higher current.

IV. GENERAL RELIABILITY DISCUSSION

As this report deals with reliability, it is appropriate to define what we mean by "reliability." One classical definition is:

"Reliability is the probability an item will perform as required under stated conditions for a stated period of time."

On the surface, this looks like a relatively simple statement, but it raises some very complex questions. For instance, what are the conditions under which a device must perform? What is a failure? How long can the device be expected to perform?

Because nearly every application has its own unique set of requirements, it is impossible to list a single set of conditions that describes how the devices are to be used. It is also unlikely that a single report will answer all reliability questions for everyone's application. Herein lies the problem that ensues when comparing reliability reports from several different companies to determine which product is most suitable for a given application. Each company will use different test conditions or different tests altogether or, worse, they may each define a failure differently.

How then can the end user read a reliability report, compare it to other reports and make a reasonable decision with regard to which product is going to be used for a given application?

Here are some guidelines that may be helpful:

1. Does the report define what was considered a failure? If not, find out their definition of failure. If it is looser than the data sheet Min/Max spec, you should determine the reason for this inconsistency.

2. Compare test conditions, not just results. A 1% failure rate on thermal shock at −55°C to +150° for 50 cycles may indicate greater reliability than a 0% failure rate for the same test at 0 - 100°C for 5 - 10 cycles.

3. Does the report define the method used to determine the life test failure rates? Choice of activation energy and operating temperature can have a significant impact on the failure rates presented. (Compare apples to apples by making sure you understand the numbers.)

4. Is the report complete? Do you have assurances that secondary reliability considerations such as resistance to corrosion and lead fatigue have been evaluated?

If you are still confused after comparing reliability reports, call the factory and ask to speak to the Reliability Manager. Don't be satisfied until you obtain specific answers to your questions.

V. TEST DESCRIPTIONS

The following descriptions define: the tests used to evaluate this product, the conditions of each test and the criteria used to define failures.

A. PHYSICAL DIMENSIONS

CONDITIONS: MIL-STD-883   METHOD 2016

PURPOSE: This test confirms that the dimensions of the finished product meet the dimensions as stated on the data sheet.

FAILURE CRITERIA:

Any device having one or more dimensions that do not meet the Min/Max requirements specified on the Siliconix data sheet.

7

B. **SALT ATMOSPHERE**

CONDITIONS: MIL-STD-883   METHOD 1009 - 24 Hours Exposure

PURPOSE: This test subjects the devices to a highly corrosive atmosphere of salt and moisture at the elevated temperature of 35°C to simulate long-term exposure to seacoast atmosphere conditions.

FAILURE CRITERIA :

1. Any device that exhibits corrosion over more than 5% of the area of the finish or base metal.
2. Loss of marking legibility.
3. Loss of hermeticity.

(See MIL-STD-883, Method 1009, for a more complete definition of failure criteria.)

C. **LEAD INTEGRITY**

CONDITIONS: MIL-STD-883   METHOD 2004
Condition B2
Three 90° arcs

PURPOSE: This test checks the mechanical strength of the lead for fatigue of the lead and the glass-to-metal interface where the lead is attached to the main body of the package.

FAILURE CRITERIA:

Any device that looses hermeticity, looses a lead, or exhibits indications (at 20X) of breakage, loosening, or cracking of the lead, is a failure.

D. **HIGH TEMPERATURE GATE BIAS**

CONDITIONS: See Figure 2.

V_{GS} = 12.5V

V_{DS} = 0V

T_A = 150°C

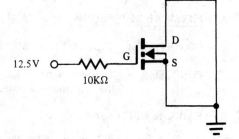

FIGURE 2. HTGB

PURPOSE: This worst-case life test checks the integrity of the gate oxide. Like all MOS devices, MOSPOWER transistors are sensitive to ionic contamination. This test causes failures when the gate oxide is weak (cracked or pitted) and when excessive amounts of sodium or potassium ions are present in the oxide.

FAILURE CRITERIA:

Any device failing to meet the Min/Max limits of the data sheet is a failure.

E. HIGH TEMPERATURE REVERSE BIAS

CONDITIONS: See Schematic Figure 3.

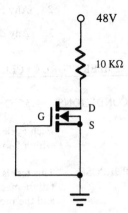

V_{GS} = 0V
V_{DS} = 48V
T_A = 150°C
T_J = 150°C

FIGURE 3. HTRB

PURPOSE: This very important test determines the failure rate of a device. It is a worst-case life test condition that checks the integrity of the field termination and the quality of the body drain junction. This test also detects surface contamination problems and detrimental surface states (especially in the termination area).

FAILURE CRITERIA:

Any device failing to meet the Min/Max specs of the data sheet.

F. THERMAL SHOCK (LIQUID-TO-LIQUID)

CONDITIONS: −55°C to +150°C   150 Cycles

Each cycle consists of a 5-minute exposure at each temperature with a maximum of 5 seconds transfer time between temperatures.

PURPOSE: This test is excellent for evaluating die attach integrity and package hermeticity. In plastic packages, it evaluates the bond integrity as well. Any cracks present in the silicon chip will be propagated by this test, leading to failure.

7

FAILURE CRITERIA:

1. Any device that fails to meet the data sheet Min/Max limits is a failure.

2. Any hermetically packaged device that looses hermeticity is a failure.

3. Any device that shows intermittent behavior is a failure.

G. TEMPERATURE CYCLE (AIR-TO-AIR)

CONDITIONS: $-55°C$ to $+150°C$ 150 Cycles

Each cycle consists of 10-minute exposures at the temperature extremes with a 5-minute dwell at ambient during transfer.

PURPOSE: This test is similar to the Thermal Shock Test, but often activates different failure mechanisms due to the longer exposures to temperature extremes and the more gradual temperature change.

FAILURE CRITERIA :

1. Any device that fails to meet the data sheet Min/Max limits is a failure.

2. Any hermetically packaged device that looses hermeticity is a failure.

3. Any device that shows intermittent behavior is a failure.

H. EXTERNAL VISUAL

CONDITIONS: MIL-STD-883   METHOD 2009

PURPOSE: To verify that the external appearance of the package meets requirements.

FAILURE CRITERIA:

Any device that shows indications of poor molding, damaged leads, or illegible marking is considered a failure.

I. PRESSURE POT

CONDITIONS: 121°C   15 PSIG   100% RH (No Bias)

PURPOSE: To check the performance of the device in humid environments. This is an excellent test to check for passivation defects, poor metal to plastic seal, contamination level during assembly and materials compatability.

FAILURE CRITERIA:

Any device that fails to meet the Min/Max limits of the data sheet, or becomes intermittent is considered a failure.

VI. TEST RESULTS

TEST	CONDITIONS	EVAL. #	DATE	SAMPLE SIZE	RESULTS			
					24 Hrs.	48 Hrs.		
PRESSURE POT	15 PSIG 121°C 100% RH	1–283D	2-25-82	46	0	0		
		1–284B	2-12-82	48	0	0		
		1–285	3-4-82	49	0	0		
		2–111C	6-15-82	50	0	0		
					160 Hrs.	320 Hrs.	660 Hrs.	1160 Hrs.
HTRB	150°C V_{DS}=48V V_{GS}= 0V	1–283J	3-15-82	49	1 Idoff	0	0	0
		1–284G	3-15-82	50	1 Idoff	0	0	0
		1–285G	3-15-82	50	1 Idoff	0	0	0
		2–079	6-3-82	50	0	0	0	0
		2–081	6-3-82	50	1 Idoff	0	0	0
					160 Hrs.	320 Hrs.	660 Hrs.	1160 Hrs.
HTGB	150°C V_{GS}=12.5V V_{DS}= 0V	1–283K	4-13-82	47	0	0	0	0
		1–285A	4-13-82	49	1 Idoff	0	3 Idoff	0
		2–079B	6-8-82	50	2 Idoff	1	0	0
		2–081B	6-8-82	50	2 Idoff	0	0	0
EXTERNAL VISUAL	883 2009	1–283E	1-12-82	5		0 FAIL		
LEAD INTEGRITY	883 M2004 COND. B2	1–283B	1-12-82	15		0 FAIL		
PHYSICAL DIMENSION	883 M2016	1–283A	1-12-82	5		0 FAIL		
SOLDERABILITY	750 M 2026	1–283F	1-21-82	15		0 FAIL		
		1–284C	1-21-82	15		0 FAIL		
		1–285	1-21-82	15		0 FAIL		

(Cont'd.)

7

TEST	CONDITIONS	EVAL. #	DATE	SAMPLE SIZE	RESULTS		
					25	50	150 Cy
SALT ATMOSPHERE	883 M 1009	1–283C	1-15-82	15		0 FAIL	
	24 Hours	1–284A	1-15-82	15		0 FAIL	
		1–285A	1-15-82	15		0 FAIL	
TEMP. CYCLE	–65°C to +150°C	1–283G	2-10-82	48	25	50	0
	10 Min COLD	1–284D	2-10-82	49	0	0	0
	5 Min ROOM TEMP	1–285D	2-10-82	50	0	0	0
	10 Min HOT	2–111B	7-28-82	50	0	0	0
THERMAL SHOCK	–65°C to +150°C	1–283H	3-17-82	48	25	50	1
	5 Min COLD	1–287	3-17-82	49	0	0	1
	7 Sec TRANSFER	1–285E	4-8-82	50	0	0	0
	5 Min HOT	2–111A	6-15-82	50	0	1	0

VII. FAILURE RATE CALCULATIONS:

Failure rate calculations are based on the HTRB and HTGB life test results. These tests simulate worst case conditions at very high operating temperatures. What does high temperature do for the device? It's not so much what it does *for*, as what it does *to* , the device. High temperatures accelerate all known chemical reactions. (This was first discovered by Svante August Arrhenius, a Swedish scientist.) It is important to know that almost all failure mechanisms associated with semiconductor devices are the result of a chemical reaction. Arrhenius described the relationship of time and temperature in chemical reactions by the equation:

$$R = R_o \epsilon^{-E_A/KT}$$

R   = Rate of the chemical reaction.

R_o = A constant.

E_A = Activation energy in electron volts (eV) that is associated with the chemical reaction.

K   = Boltzman's constant ($8.62 \times 10^{-5} eV/°K$).

T   = Absolute temperature in degrees K (°C + 273°).

To calculate the difference in reaction rates (acceleration factor) between two temperatures, the following equation is used:

$$R_1/R_2 = \epsilon^{-E_A/K(1/T_1 - 1/T_2)}$$

R_1 is the reaction rate at the test temperature.

R_2 is the reaction rate at a different temperature.

T_1 is the test temperature in °K.

T_2 is the temperature for which the new reaction rate is being calculated in °K.

R_o is the same for both temperatures and has been cancelled out of the equation.

The activation energy for semiconductor devices usually ranges from .4eV to 1.2eV. If the activation energy is unknown, then .7eV is often used as a conservative estimate. The attached Arrhenius plot will allow the user to compare failure rate using different activation energies.

Using the high temperature operating life test data and the Arrhenius equation, we can calculate the failure rate of the device at normal operating temperatures.

The HTRB and HTGB data shown are from the Test Results shown in Section VI.

7

HTRB SUMMARY

Total Devices on Life Test   249
Total Device Hours   284,840
Total Failures   4*

FAILURE RATES EXPRESSED IN PER CENT
PER 1000 HOURS

	Confidence Limits**		
	60%	80%	90%
Max Failure Rate at T_J = 150°C	1.8%	2.3%	2.7%
Max Failure Rate at T_J = 85°C	.008%	.009%	.011%

HTGB SUMMARY

Total Devices on Life Test   196
Total Device Hours   220,860
Total Failures   9*

Max Failure Rate at T_J = 150°C	4.6%	5.4%	6.2%
Max Failure Rate at T_J = 85°C	.019%	.022%	.026%

VIII. SUMMARY

The purpose of this report was to demonstrate the reliability of our VNMK/L products in plastic TO-237 packages. The results shown in this report represent the reliability of our standard product. No prescreening of devices was allowed.

In addition to presenting this data, it is our hope that we have given the reader a better understanding of reliability testing and reporting methods used in the industry so that this information can be properly applied to device selection. Should you have further questions regarding the reliability of these devices or any other Siliconix product, please contact us.

* Failure are attributed to ionic contamination. An activation energy of 1.1 eV has been used.

** Confidence limits are used to provide a statistical bracket around the actual data to show Min and Max failure rates. The minimum failure rate for these devices would be 0%/1000 hrs. Due to the relatively small sample sizes, the failure rates shown under the confidence limits may appear higher than they actually are.

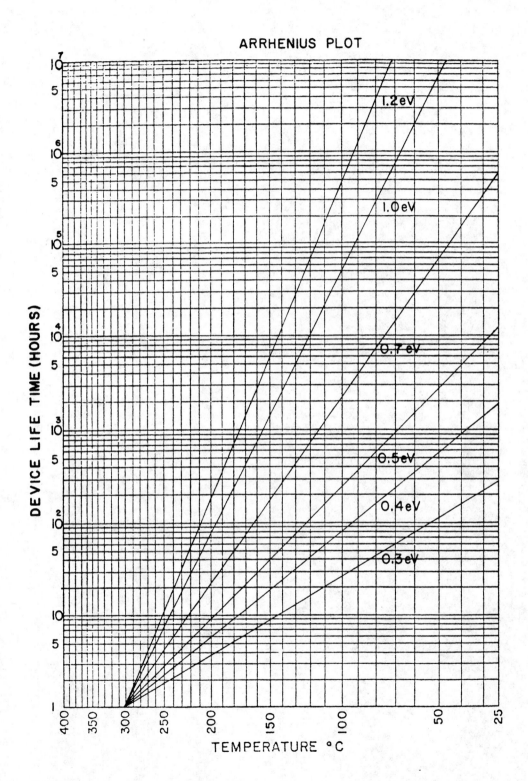

ARRHENIUS PLOT

DEVICE LIFE TIME (HOURS)

TEMPERATURE °C

1.2 eV
1.0 eV
0.7 eV
0.5 eV
0.4 eV
0.3 eV

7

A.1 Calculating the RMS Value of MOSFET Drain Current Using the HP-41CV

Introduction

The conduction loss in a power MOSFET is proportional to the square of the RMS drain current, $I_{D(RMS)}$, flowing through it. To calculate the total power loss and to perform thermal analysis (either graphical or computational), it is necessary to determine $I_{D(RMS)}$. The HP-41CV programmable calculator provides a fast and convenient means of finding $I_{D(RMS)}$, especially if the waveform is complex. This appendix presents the basic equations for calculating the RMS value of a drain current waveform and an HP-41CV program to evaluate them.

The program described here will also be useful for calculating the RMS value of current waveforms in inductor or transformer windings, as well as the RMS value of I_D. To further extend the utility of the program, a calculation for the RMS current in a filter capacitor has been added (case 5).

The RMS Value of a Periodic Waveform

The RMS value of any periodic waveform is defined to be:

$$I_{RMS} = \sqrt{\frac{1}{T}\int_0^T I(t)^2 dt} \qquad (1)$$

where T is the period of the repetitive waveform and I(t) is the function describing the current waveform during the interval T.

Figure 1 shows some common waveforms and the values of I_{RMS} derived from equation (1). The program given in the next section will calculate the RMS value of the waveform in each of the five cases.

Not all waveforms are as simple as those depicted in Figure 1. Very often however, a complex waveform can be broken down into a combination of the waveforms shown in cases one through five, and an approximation is usually sufficiently accurate for most applications. The RMS value of a complex waveform can be determined from the RMS values of its components with the following expression:

$$I_{RMS} = \sqrt{I^2_{RMS(1)} + I^2_{RMS(2)} + .. + I^2_{RMS(N)}}$$

$$(2)$$

If the current waveform cannot be satisfactorily approximated using the above approach, then equation (1) can be evaluated using numerical integration techniques. The HP-41 Math Pac contains programs to evaluate definite integrals using the trapezoidal rule and Simpson's rule. These programs simplify the task of numerically evaluating equation (1) to a few simple keystrokes. For more information on these programs, please consult the Math Pac users handbook.

Program Use

When the RMS program is executed from its entry point, the prompt "ENTER CASE" will appear in the display, and the program will stop. The required case (Figure 1) should be entered into the x-register and the R/S key pressed to resume program execution. The number entered must be in the range of one through five if the program is to proceed to the next step. Otherwise the calculator will repeat the "ENTER CASE" prompt until a valid number has been entered.

8

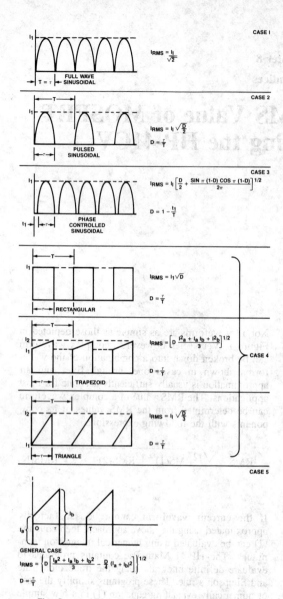

CASE 1 — FULL WAVE SINUSOIDAL — $T = \tau$

$I_{RMS} = \dfrac{I_1}{\sqrt{2}}$

CASE 2 — PULSED SINUSOIDAL

$I_{RMS} = I_1 \sqrt{\dfrac{D}{2}}$, $D = \dfrac{\tau}{T}$

CASE 3 — PHASE CONTROLLED SINUSOIDAL

$I_{RMS} = I_1 \left[\dfrac{D}{2} + \dfrac{\text{SIN } \pi\,(1-D)\text{ COS }\pi\,(1-D)}{2\pi} \right]^{1/2}$, $D = 1 - \dfrac{I_1}{T}$

CASE 4

RECTANGULAR — $I_{RMS} = I_1 \sqrt{D}$, $D = \dfrac{\tau}{T}$

TRAPEZOID — $I_{RMS} = \left[D\,\dfrac{I_a^2 + I_a I_b + I_b^2}{3} \right]^{1/2}$, $D = \dfrac{\tau}{T}$

TRIANGLE — $I_{RMS} = I_1 \sqrt{\dfrac{D}{3}}$, $D = \dfrac{\tau}{T}$

CASE 5

GENERAL CASE

$I_{RMS} = \left\{ D\left[\dfrac{I_a^2 + I_a I_b + I_b^2}{3} - \dfrac{D}{4}(I_a + I_b)^2 \right] \right\}^{1/2}$

$D = \dfrac{\tau}{T}$

Five Cases of Periodic Waveforms and the Derivation of I_{RMS} for Each One
Figure 1

Once the correct case has been established, the calculator will prompt for I_1, D and I_2 (Figure 1) — whichever is required. Only those input variables that are required for each case will be prompted for. As before, after each prompt the appropriate value should be entered into the X register and the R/S key pressed.

After all data has been input, the program will calculate the RMS value of drain current, then display it and stop. A subsequent press of the R/S key will restart the program at the "ENTER CASE" prompt.

The program listing is as follows:

Program Listing

01◆LBL "IRMS"	37 GTO "IRMS6"	73 RCL 02
02 CF 21	38◆LBL "IRMS2"	74 *
03 "PROGAM IRMS"	39 RCL 01	75 RCL 00
04 AVIEW	40 2	76 X ↑ 2
05 PSE	41 /	77 +
06◆LBL "IRMS0"	42 SQRT	78 RCL 02
07 5	43 RCL 00	79 X ↑ 2
08 ENTER ↑	44 *	80 +
09 "ENTER CASE"	45 GTO "IRMS6"	81 RCL 01
10 PROMPT	46◆LBL "IRMS3"	82 *
11 X>Y?	47 RAD	83 3
12 GTO "IRMS0"	48 1	84 /
13 X<=0?	49 RCL 01	85 FS? 05
14 GTO "IRMS0"	50 −	86 GTO "IRMS5"
15 SF IND X	51 PI	87 SQRT
16 "ENTER I1"	52 *	88 GTO "IRMS6"
17 PROMPT	53 ENTER ↑	89◆LBL "IRMS5"
18 STO 00	54 SIN	90 RCL 00
19 FS? 01	55 X<>Y	91 RCL 02
20 GTO "IRMS1"	56 COS	92 +
21 "ENTER D"	57 *	93 X ↑ 2
22 PROMPT	58 PI	94 RCL 01
23 STO 01	59 2	95 X ↑ 2
24 FS? 02	60 *	96 *
25 GTO "IRMS2"	61 /	97 4
26 FS? 03	62 RCL 01	98 /
27 GTO "IRMS3"	63 2	99 −
28 "ENTER I2"	64 /	100 SQRT
29 PROMPT	65 +	101◆LBL "IRMS6"
30 STO 02	66 SQRT	102 "IRMS="
31 GTO "IRMS4"	67 RCL 00	103 ARCL X
32◆LBL "IRMS1"	68 *	104 PROMPT
33 RCL 00	69 DEG	105 GTO "IRMS0"
34 2	70 GTO "IRMS6"	106 END
35 SQRT	71◆LBL "IRMS4"	
36 /	72 RCL 00	

A.2 HP-41CV Transient Thermal Impedance Program

Introduction

This program enables curves of normalized transient thermal impedance versus pulse width to be plotted for repetitive waveforms of various duty cycles. Input data for the program is taken from a graph of normalized transient thermal impedance versus pulse width for a single-shot (i.e., zero duty cycle) waveform.

Theory

Each point that the program calculates for a given duty cycle and pulse width requires three points from the zero duty-cycle curve to be entered as input data. The program calculates each point according to the equation (Reference 1):

$$Z_{JCN} = D + (1-D) Z_1 + Z_2 - Z_3 \qquad (1)$$

Where:

τ = pulse width of repetitive waveform (TAU)
T = period of repetitive waveform
D = duty cycle of waveform ($= \tau/T$)
t_{pw} = pulse width scale on zero duty cycle graph
Z_{JCN} = normalized transient thermal impedance
Z_1 = Z_{JCN} at $D = 0$ and $t_{pw} = T + \tau$
Z_2 = Z_{JCN} at $D = 0$ and $t_{pw} = \tau$
Z_3 = Z_{JCN} at $D = 0$ and $t_{pw} = T$

Program Description and Listing

This program is quite straightforward and has two entry points. The first entry point allows the desired duty cycle to be set once and then retained for a series of calculations. The second entry point allows a whole curve to be generated, without having to re-enter the duty cycle with each iteration through the program.

Program Use

This program requires SIZE on the HP-41CV to be set to at least five.

1. When executing the program from label ZTINT, the prompt "ENTER D" will appear in the display, and the program will stop. Entering a number between 0.0 and 1.0 into the X-register and pressing the R/S key will cause this value to be used as the duty cycle in the calculations.

2. After (1) above or when executing the program from label ZTRAN, the prompt "ENTER TAU" will appear in the display, and the program will stop. Entering the desired pulse width in seconds and pressing the R/S key will cause this value to be saved for use in calculations.

3. The program will now prompt for Z_1, Z_2, and Z_3 in sequence. When the display shows "ENTER ZN, TPW = X.XXX E ± X", where "ZN" = Z_1, Z_2, or Z_3 and "TPW" is shown in the Alpha-ENG 3 format, the program will stop. The point on the zero duty-cycle curve (Z_1, Z_2, or Z_3) at the indicated pulse width (TPW) should be entered into the X-register and the R/S key pressed. This must be done three times.

4. After Z3 has been entered and the R/S key pressed, the program will calculate and then display the normalized transient thermal impedance, Z_{JCN}. The display will show "ZJCN = X.XXX E ± X", where the numerical value is in the Alpha-ENG 3 format and the program will stop. Pressing the R/S key again will cause the program to restart at label ZTRAN.

8

Program Listing

01◆LBL "ZTINT"	20 ARCL X	39 PROMPT
02 ENG 3	21 PROMPT	40 STO 04
03 "ENTER D"	22 STO 02	41 RCL 00
04 PROMPT	23 RCL 01	42 1
05 STO 00	24 "ENTER Z2"	43 RCL 00
06◆LBL "ZTRAN"	25 AVIEW	44 −
07 "ENTER TAU"	26 PSE	45 RCL 02
08 PROMPT	27 "TPW"	46 ★
09 STO 01	28 ARCL X	47 +
10 RCL 00	29 PROMPT	48 RCL 03
11 1/X	30 STO 03	49 +
12 1	31 RCL 01	50 RCL 04
13 +	32 RCL 00	51 −
14 RCL 01	33 /	52 "ZJCN"
15 ★	34 "ENTER Z3"	53 ARCL X
16 "ENTER Z1"	35 AVIEW	54 PROMPT
17 AVIEW	36 PSE	55 GTO "ZTRAN"
18 PSE	37 "TPW ="	56 END
19 "TPW ="	38 ARCL X	

Register Map

R_0 = D
R_1 = τ
R_2 = Z_1
R_3 = Z_2
R_4 = Z_3

Reference

Phillips Application Note #68, April 1975, "SOAR—The Basis for Reliable Power Circuit Design," page 13.

A.3 Table of Symbols
(Preferred Usage)

BV_{CEO}	Breakdown Voltage, collector-to-emitter, base open	$I_{D(on)}$	Current, drain-to-source, ON state
BV_{CES}	Breakdown Voltage, collector-to-emitter, base shorted	I_{DS}	Current, drain-to-source (same as I_D)
BV_{DSS}	Breakdown Voltage, drain-to-source, gate shorted	I_{DSS}	Current, drain, with gate shorted to source
BJT	Bipolar Junction Transistor	I_{DZ}	Current, drain, zero-temperature coefficient
C_{ds}	Capacitance, drain-to-source	I_{GSS}	Current, gate (leakage), source tied to drain
C_{gd}	Capacitance, gate-to-drain		
C_{gn}^+	Capacitance, gate to n^+ diffusion	J	Current density
C_{gp}	Capacitance, gate to p-type body diffusion	k	Boltzmann's constant
		L	Drain-to-source separation (channel length)
C_{gs}	Capacitance, gate-to-source		
C_{iss}	Capacitance, common-source input	L_D	Intrinsic Debye length
C_{oss}	Capacitance, common-source output	L'	Effective channel length
C_{ox}	Gate oxide capacitance per unit area	M	Carrier multiplication
C_{rss}	Capacitance, reverse transfer (same as C_{gd})	MTBF	Mean Time Between Failure
		n	Electron concentration per unit volume
CMRR	Common-Mode Rejection Ratio	p	Hole concentration per unit volume
DMOS	Double-diffused MOS	P_C	Conduction loss
D/CMOS	CMOS logic integrated with DMOS monolithically	P_D	Conduction loss, reverse diode
		P_L	Power loss due to I_{DSS} (enchancement mode only)
E_{chan}	Electric field, lateral (across channel)		
E_g	Energy, band-gap (of silicon = 1.12 eV)	Q	Charge
		Q_{SS}	Fixed positive charge density per unit area at the silicon-silicon dioxide interface
E_{ox}	Electric field within the gate oxide layer		
g_o	Conductance, output	R_B	Resistance, body
$I_{D(off)}$	Current, drain, gate potential below threshold	RBSB	Reverse Bias Second Breakdown

8

R_{CHAN}	Resistance, conducting channel
$R_{DD(on)}$	Resistance, drain-to-drain (bilateral switch)
$r_{DS(on)}$	Resistance, ON, drain-to-source
R_L	Resistance, load
$R_{\Theta ja}$	Resistance, thermal, junction-to-ambient
$R_{\Theta jc}$	Resistance, thermal, junction-to-case
$R_{\Theta cs}$	Resistance, thermal, case-to-sink
$R_{\Theta sa}$	Resistance, thermal, sink-to-ambient
R_p	Resistance, p diffusion
$r(t)$	Transient thermal impedance, normalized
SOA	Safe Operating Area
t	Transit time
T	Temperature
T_A or T_a	Temperature, ambient
T_C	Temperature, case
$t_{d(off)}$	Turn OFF delay time
t_d or $t_{d(on)}$	Turn ON delay time
t_f	Turn OFF fall time
T_j	Temperature, junction
T_{ox}	Thickness of the insulating gate oxide
t_p	Pulse width of power pulse
t_r	Turn ON rise time
t_{rr}	Turn OFF time, reverse recovery (diode)
v	Velocity, drift
V_B	Voltage, breakdown, junction
V_{BE}	Voltage, base-emitter
VDMOS	Vertical DMOS

$V_{DS(sat)}$ or V_{SAT}	Voltage, saturated, drain-to-source
V_{GS}	Voltage, gate-to-source
$V_{GS(on)}$	Voltage, gate-to-source, greater than V_T
VMOS	Vertical MOS (DMOS)
V_o	Voltage, contact (potential)
V_{ox}	Voltage across oxide layer
V_R	Voltage, reverse, applied
V_T or V_{th}	Voltage, threshold [also $V_{gs(th)}$]
VVMOS	Vertical V-groove MOS
W	Channel width
W_d	Depletion width in epi layer

Greek Symbols

β	Current gain of a bipolar transistor (dI_c/dI_b)
ϵ_{ox}	Dielectric constant of the oxide insulating layer
ϵ_o	Permittivity of free space (8.86×10^{-14} F/cm)
μ	Mobility, electron
μ_n	Mobility of electrons in n region
μ_p	Mobility of holes in p region
ρ	Space charge density per unit volume
σ_c	Conductivity
τ	Relaxation time
τ_t	Time, transit
τ_s	Time, storage
ϕ	Electrostatic potential
ϕ_F	Fermi potential
ϕ_B	Potential, diode (for silicon 0.7 V)

A.4 Glossary of Terms

Acceptor. Also known as a "*p*-type" dopant. An element containing three electrons in its outer shell. Combined with silicon (a semiconductor), it produces a region deficient in conduction electrons leaving holes available for conduction. Elements commonly used as acceptor dopants include boron and aluminum. In the periodic table they are classified as Group or Column III-A.

Allotrope. The existence of two or more crystalline structural forms of an element.

Anisotropic etch. Exhibits etching properties dependent upon the crystal orientation. This was commonly used in the original V-groove power MOS technology. Now it is generally obsolete for DMOS technology. (See Double-Diffused.)

Bilateral. The ability to pass an electric current in either of two directions with equal, or near-equal, affinity.

BJT. Bipolar Junction Transistor.

Breakdown. Phenomenon by which, increasing the reverse bias voltage applied to a *pn* junction, causes it to go from a high resistance to a lower resistance region, with a corresponding increase on the reverse current.

Carriers. Moving charges within a semiconductor, either conduction electrons or holes.

Depletion layer. The region of a semiconductor device where essentially all charge carriers are swept out by an electric field. This electric field can result either from an externally applied electric charge, viz., a voltage, or it may simply be that field resulting from the close proximity of a chemically doped region that, of itself, contains a charge, either positive (*p*-doped) or negative (*n*-doped).

Diffusion. The motion of particles away from regions of high concentration. In semiconductors, this process is used to obtain a dopant profile that is graded through the semiconductor material with the heaviest concentration being at the surface.

Donor. The "mirror image" of **Acceptor** above. Popular donor dopant elements include phosphorous, arsenic and antimony. These are classified within the periodic table as Group or Column V-A.

Dopant. An element that, when diffused into a semiconductor, alters the conductivity of the semiconductor by contributing either a hole or a conduction electron. For silicon (a Group IV material), the dopants are found within either Group IIIA (*p*-dopant) or Group V-A (*n*-dopant).

Double-diffused. Two superimposed diffusions using a single mask separated by time and temperature. Commonly used in the fabrication of power MOSFETs where the first diffusion becomes the body, and the second (or last) diffusion creates the source.

Enhancement. In a normally-OFF MOSFET, the collection of carriers under the gate oxide until majority-carrier conduction occurs.

8

Epitaxial. The growth of a single crystal semiconductor film upon a single-crystal substrate. The epitaxial layer can have a vastly different dopant concentration from the substrate.

Hole. Electron vacancy in the structure of a semiconductor which acts like a positive electronic charge.

Intrinsic. A pure, undoped semiconductor crystal.

Impurities. Atoms of a different chemical element which are foreign to the perfect crystal structure.

Inversion. An electric field-induced phenomenon whereby a doped semiconductor surface changes its predominant charge characteristic. For example, a type p substrate is inverted to type n.

Ion-implantation. A procedure for doping a semiconductor by bombarding the surface with high-energy dopant ions. Known for its precise control.

Junction. Area of transition between semiconductor regions of different electrical characteristics.

Latch-back. In MOSPOWER transistors, a breakdown voltage phenomenon resulting from the sudden interaction of a parasitic element (usually taking the form of a bipolar transistor) whose characteristic breakdown voltage is lower than that of its host.

Majority carriers. Charge carriers constituting more than one half the total charge-carrier concentration.

Latch-up. In CMOS circuits and other semiconductors having four layers of alternating conductivity type, a current flow phenomenon where, during an abnormal conduction period, a parasitic thyristor action takes over and normal control is lost. Burnout generally results if drastic action is not taken quickly.

Minority-carriers. The non-predominant mobile charge carriers in a semiconductor. For example, the injected base current in a bipolar transistor.

Photolithography. A technique using light and selective masking to transfer a pattern to a surface.

Polycrystalline. A material composed of many small crystals with random orientations.

Secondary breakdown. In contrast to avalanche or first breakdown, generally results in irreversible damage to the transistor. It is the primary cause of failure in bipolar transistors when used to turn OFF an inductive load. Although several theories have been advanced, the mechanism has never been fully understood.

Thermal resistance. The temperature rise per unit power dissipation of a junction with respect to the temperature of an external reference point under conditions of thermal equilibrium.

A.5 Additional References

Adler, M.S. "A Comparison Between BiMOS Device Types," *PESC'82 Record,* IEEE Power Electronics Specialists Conference, pp. 371-377.

Adler, M.S., et al. "Theory and Breakdown Voltage for Planar Devices with a Single Field Limiting Ring," *IEEE Transactions on Electron Devices,* Vol. ED-24 (1977), p. 107.

Adler, M.S., and S.R. Westbrook. "Power Semiconductor Switching Devices—A Comparison Based on Inductive Switching," *IEEE Transactions on Electron Devices,* Vol. 29, No. 6 (June 1982), pp. 947-952

Ahiya, B.K. "Fabrication and Modeling of VMOS Devices," *M. Eng. Thesis,* Carelton University, Ottawa, April 1976.

Ahiya, B.K. and A.R. Boothroyd. "Modeling of V-Channel MOS Transistor," *Technical Digest,* International Electron Devices Meeting, 1976, pp. 573-577.

Alexander, Mark. "Boost OP-AMP Output Power with Complementary Power MOSFETs," *Siliconix Application Note AN83-5,* Santa Clara, California: Siliconix incorporated, 1983.

Alexander, Mark. "Frequency Response Analysis of the MOSFET Source-Follower," *Siliconix Application Note AN83-8,* Santa Clara, California: Siliconix incorporated, 1983.

Appels, J.A. and H.M.J. Vaes. "High Voltage Thin Layer Devices (Resurf Devices)," *Technical Digest,* International Electron Devices Meeting, 1979, p. 238.

Armstrong, G.A. and J.A. Magowan. "Modelling of Short-Channel M.O.S. Transistors," *Electronics Letters,* Vol. 6 (1970), p. 313.

Awane, K., et al. "High-Voltage DSA-MOS Transistor for Electroluminescent Display," *Digest of Technical Papers,* IEEE International Solid State Circuits Conference, 1978, p. 224.

Baker, W.E., Jr. "The Effects of Radiation on Characteristics of Power MOSFETs," *Proceedings,* POWERCON 7, 1980, pp. D3.1-D3.6.

Baliga, B. Jayant. "MOS Controlled Bipolar Devices," *Session Record 7,* IEEE ELECTRO/83.

Baliga, B. Jayant. "Power MOSFET Integral Diode Reverse Recovery Tailoring Using Electron Irradiation," *Proceedings,* 40th Annual Device Research Conference, June 21-23, 1982, IVA-8.

Baliga, B. Jayant and John P. Walden. "Improving the Reverse Recovery of Power MOSFET Integral Diodes by Electron Irradiation," *Solid-State Electronics,* 26, No. 12 (December 1983), pp. 1133-1141.

Barlage, F. Michael. "Exploit VMOS FET's Advantage to Drive Bipolar Power Transistors," *EDN,* November 5, 1978, pp. 93-98.

Barlage, F.M. "A New Switched-Mode Converter Technique Combines VMOS and Bipolar," *Proceedings,* POWERCON 5, 1978, pp. D2.1-D2.8.

Barnes, J.J., S.N. Shalide, and F.B. Jenne. "The Buried Source VMOS Dynamic RAM Device," *Technical Digest,* International Electron Devices Meeting, 1977, pp. 272-276.

Baum, B. and H. Beneking. "Drift Velocity Saturation in MOS Transistors," *IEEE Transactions on Electron Devices,* Vol. 17 (1970), p. 481.

Berger, J. and K. Lisiak. "Electron Mobility in Silicon Surface Inversion Layers," presented at 1977 Semiconductor Interface Specialists Conference, Miami, Florida.

Bhatti, I.S., E. Fuller, and F. Jenne. "VMOS Electrostatic Protection," *16th International Reliability Physics Symposium,* 1978, pp. 140-145.

Bhatti, I.S. and R.Y. Yau. "Geometry Effects in VMOS Transistors," *Technical Digest,* International Electron Devices Meeting, 1978, pp. 30-34.

Bhatti, I.S., T.J. Rodgers, and J.R. Edwards. "Minimization of Parasitic Capacitances in VMOS Transistors," *Technical Digest,* International Electron Devices Meeting, 1976, pp. 565-569.

Bierhenke, H. and H. Herbst. "Comparison of the Switch-On Behavior of MOS-Transistors of Different Technologies," *AEU Band 31,* Heft 1, West Germany, 1977.

Blanchard, Dick and Mark Alexander. "Using MOSPOWER as Synchronous Rectifiers in Switch-Mode Power Supplies," *Powerconversion International,* Vol. 9, No. 4 (April 1983), pp. 16-26.

Blanchard, Dr. Richard. "Status and Emerging Directions of MOSPOWER Technology," *Proceedings,* 7th Power Conversion International Conference, 1983, pp. 162-174.

8

Blanchard, R. and R. Severns. "MOSFETs Move In On Low Voltage Rectification," *Technical Article TA84-2*, Santa Clara, California: Siliconix incorporated, 1984.

Blanchard, Richard. "MOSFETs in Arrays and Integrated Circuits," *Session Record 7*, IEEE ELECTRO/83.

Blanchard, Richard. "MOSPOWER Devices Are Coming On Strong," *Electronic Products*, July 2, 1984, pp. 71-76.

Blanchard, Richard. "Power Control with Integrated CMOS Logic and DMOS Output," *Session Record 25*, IEEE ELECTRO/84.

Blanchard, Richard and Rudy Severns. "Designing Switched-Mode Power Converters for Very Low Temperature Operation," *Proceedings*, POWERCON 10, 1983, pp. D2.1-D2.11.

Blanchard, Richard and Rudy Severns. "MOSFETs, Schottky Diodes Vie for Low-Voltage Supply Designs," *EDN*, June 28, 1984, pp. 197-206.

Blanchard, Richard and Sheldon Haynie. "Power MOS Transistors: Structure and Performance," *Powerconversion International*, Vol.6, No. 2 (March/April 1980), pp. 43-48.

Blanchard, Richard A. "Optimization of Discrete High Power MOS Transistors," PhD. Dissertation, Stanford University, December 1981.

Blanchard, Richard A. "The Use of MOS Power Transistors in Hybrid Circuits," *The International Journal for Hybrid Microelectronics*, Vol. 5, No. 2 (November 1982), pp. 130-137.

Bloom, Gordon (Ed) and Rudy Severns. "The Generalized Use of Integrated Magnetics and Zero-Ripple Techniques in Switchmode Power Converters," *PESC'84 Record*, Power Electronics Specialists Conference, pp. 15-33.

Bloom, Gordon (Ed) and Rudy Severns. "Magnetic Integration Methods for Transformer-Isolated Buck and Boost DC-DC Converters," *Proceedings*, POWERCON 11, 1984, pp. A2.1-A2.15.

Brown, Steve and Bob Ghent. "Protecting MOSFETs Against Electrostatic Discharge," *Electronic Products*, 26, No. 16 (June 14, 1984), pp. 77-83.

Buhler, Otto R. "Use Equations to Parallel Transistors," *Electronic Design*, Vol. 25, No. 4 (February 15, 1977), pp. 84-86.

Chang, M.F., et al. "25A, 500V Insulated Gate Transistor," *Technical Digest*, IEEE Electron Devices International Meeting, 1983, pp. 83-86.

Chen, Dan Y. and Chandrasekaran. "FET-Gated High Voltage Bipolar Transistor," *Conference Record*, Industrial Applications Society, Mexico City, 1983, pp. 713-720.

Chen, Dan Y. and Shaoan Chin. "Design Considerations for FET-Gated Power Transistors," *PESC'83 Record*, Power Electronics Specialists Conference, pp. 144-149.

Cheng, Hwa and A.G. Milnes. "Power MOSFET Characteristics with Modified SPICE Modeling," *Solid State Electronics*, Volume 25 (December 1982), pp. 1209-1212.

Chi, Min-hua and Chenming Hu. "An Intrinsic Power MOSFET Model for Circuit Design and Analysis," *Proceedings*, POWERCON 10, 1983, pp. H2.1-H2.9.

Clark, Lowell E. "Ambimodel DMOS Devices," *Proceedings*, 40th Annual Device Research Conference, June 21-23, 1982, IVA-9.

Clemente, S. "An Introduction to International Rectifier P-Channel HEXFETs," *International Rectifier Application Note 940*, El Segundo, California: International Rectifier.

Clemente, S. and B. Pelly. "A Chopper for Motor Speed Control Using Parallel Connected Power HEXFETs," *International Rectifier Application Note 941*, El Segundo, California: International Rectifier Semiconductor Division, 1981.

Clemente, S.M. and B.R. Pelly. "Analysis and Characterization of Power MOSFET Switching-Interval Performance in Power Converters," *Proceedings*, POWERCON 8, 1981, pp. H2.1-H2.11.

Clemente, S., B. Pelly, and R. Ruttonsha. "A Universal 100 kHz Power Supply Using a Single HEXFET," *International Rectifier Application Note 939*, El Segundo, California: International Rectifier Semiconductor Division, 1981.

Clemente, S., B.R. Pelly, and R. Ruttonsha. "Current Ratings, Safe Operating Area, and High Frequency Switching Performance of Power HEXFETs," *International Rectifier Application Note 939*, El Segundo, California: International Rectifier Semiconductor Division, 1981.

Clemente, S., B.R. Pelly, and B. Smith. "The HEXFET's Integral Body Diode -- Its Characteristics and Limitations," *International Rectifier Application Note 934A*, El Segundo, California: International Rectifier.

Clemente, S., B.R. Pelly, and A. Sisdori. "Understanding HEXFET Switching Performance," *International Rectifier Application Note 947*, El Segundo, California: International Rectifier.

Clemente, Steve. "Gate Drive Characteristics and Requirements for Power HEXFETs," *International Rectifier Application Note 937*, El Segundo, California: International Rectifier.

Cobbold, Richard S.C. *Theory and Application of Field Effect Transistors*. New York: John Wiley & Sons, Inc., 1970.

Coe, D.J. and H.E. Brockman. "Corner Breakdown in MOS Transistors with Lightly-Doped Drains," *Solid State Electronics*, Vol. 22 (1979), p. 444.

Colak, S., B. Singer, and E.H. Stupp. "Design of High-Density Power Lateral DMOS Transistors," *PESC'80 Record*, Power Electronics Specialists Conference, p. 164.

Colak, S., B. Singer, and E.H. Stupp. "Lateral DMOS Power Transistors Design," *IEEE Electron Device Letters*, Vol. 1 (1980), p. 51.

Colak, S. and E.H. Stupp. "Reverse Avalanche Breakdown in Gated Diodes," *Solid State Electronics*, Vol. 23 (1980), p. 467.

Compeers, J., H.J. DeMan, and W.M.C. Sansen. "A Process and Layout Oriented Short-Channel MOST Model for Circuit-Analysis Programs," *IEEE Transactions on Electron Devices*, Vol. ED-24 (1977), p. 739.

Connolly, Arthur P. "The FREDFET -- A New MOSFET for Motor Speed Control Circuits," *Proceedings*, 8th Power Conversion International Conference, April 1984, pp. 105-109.

Dance, Mike. "More Power from VMOS" (Interview with Dick Lee), *Electronics Industry*, March 1978.

D'Avanzo, D.C., S.R. Combs, and R. W. Dutton. "Characterization and Modeling of Simultaneously Fabricated DMOS and VMOS Transistors," *Technical Digest*, International Electron Devices Meeting, 1976, pp. 569-573.

D'Avanzo, D.C., S.R. Combs, and R.W. Dutton. "Effects of the Diffused Impurity Profile on the DC Characteristics of VMOS and DMOS Devices," *IEEE Journal of Solid State Circuits*, Vol. SC-12 (1977), p. 356.

Declercq, M. and J. Plummer. "Avalanche Breakdown in High-Voltage DMOS Devices," *IEEE Transactions on Electron Devices*, 23 (1976), pp. 1-4.

Dewsbury, P. "MOSPOWER FETs and Support ICs," *New Electronics (UK)*, Vol. 16, No. 4 (February 22, 1983), pp. 30-32.

Dobray, E. and P. Freundel. "A New Power MOSFET with a Fast-Recovery Inverse Diode," *Proceedings*, Power Conversion International Conference, April 1983, pp. 152-161.

Draper, D.A., J.J. Barnes, and F.B. Jenne. "Fabrication and Characterization of a VMOS EPROM," *Technical Digest, International Electron Devices Meeting,* 1977, pp. 277-283.

Edwards, J.R., et al. "VMOS Reliability," *Proceedings,* 16th International Reliability Physics Symposium, 1978, pp. 23-25.

Edwards, John R., Inderjit S. Bhatti, and Earl Fuller. "VMOS Reliability," *IEEE Transactions on Electron Devices,* Vol. ED-26, No. 1 (January 1979), pp. 43-48.

Erickson, R., et al. "Characterization and Implementation of Power MOSFETs in Switching Converters," *Proceedings,* POWERCON 7, 1980, pp. D4.1-D4.17.

Evans, Arthur D., et al. "Higher Power Ratings Extend VMOS FETs' Dominion," *Electronics,* 51 (June 27, 1978), pp. 105-112.

Fang, Z.D., et al. "Designing a High Frequency Snubberless FET Power Inverter," *Proceedings,* POWERCON 11, 1984, pp. D1-4.1 - D1-4.10.

Farzan, B. and C.A.T. Salama. "Depletion V-Groove MOS (VMOS) Power Transistors," *Solid State Electronics,* Vol. 19 (April 1976), pp. 297-306.

Fabian, M. "Performance in Power F.E.T.s," *New Electronics* (UK), Vol. 9, No. 4 (February 1976).

Fay, Gary. "CMOS Integrates with High Power Devices in New ICs," *Session Record 25,* IEEE ELECTRO/84.

Fazi, C. and S. Rattner. "Radiation Effects on Power VMOS Devices," *Sessions Record 23,* IEEE MIDCON/79.

Fong, Edison. *Design Considerations for MOS Amplifiers,* Signetics Corporation, 1979, 368 pp.

Forsythe, James. "Techniques for Controlling Dynamic Current Balance in Parallel Power MOSFET Configurations," *Proceedings,* POWERCON 8, 1981, pp. G3.1-G3.11.

Fortier, Tim. "Power MOSFETs - Look Beyond the Data Sheet," *Electronic Products,* October 24, 1983, pp. 125-128.

Fragale, W., B. Pelly, and B. Smith. "The HEXFET's Integral Reverse Rectifier," *Powerconversion International,* Vol. 6, No. 2 (March/April 1980), pp. 17-36.

Frank, W.E. and C.F. Der. "Solid-State RF Generators for Induction Heating Applications," *Conference Record,* Industrial Applications Society, Mexico City, 1983, pp. 939-944.

Freire, J. Costa and Pedro Teixeira. "Modelling of Epidrain Effects in VMOS Power Transistors for CAD," *Proceedings,* European Conference on Electron Design Automation, Brighton, UK, 1981.

Frey, G. "VMOS Power Amplifiers," *EDN,* September 5, 1977, pp. 83-85.

Frohman-Benchkowsky, D. and A. S. Grove. "Conductance of MOS Transistors in Saturation," *IEEE Transactions on Electron Devices,* Vol. ED-17 (1969), p. 108.

Fuoss, Dennis and Verma Krishna. "A Fully Implanted V-Groove Power MOSFET," *Technical Digest,* International Electron Devices Meeting, Washington D.C., 1978, pp. 657-660.

Gauen, Kim. "Match Power-MOSFET Parameters for Optimum Parallel Operation," *EDN,* February 23, 1984, pp. 249-267.

Gauen, Kim. "Paralleling Power MOSFETs in Linear Applications," *Proceedings,* 8th Power Conversion International Conference, April 1984, pp. 144-151.

Gauen, Kim. "Power MOSFET Variant Excells at High Loads," *Electronic Design,* April 5, 1984, pp. 103-110.

Gauen, Kim. "Special Design Considerations for High Power Multiple Die MOSFET Circuits," *Powerconversion International,* Vol. 9, No. 10 (November/December 1983), pp. 24-30.

Ghandhi, S.K. *Semiconductor Power Devices.* New York: Wiley-Interscience, 1977.

Giandomenico, David. "Analysis and Prevention of Power MOSFET Anomalous Oscillations," *Proceedings,* POWERCON 11, 1984, pp. H.4-1 - H.4-9.

Giandomenico, David. "Anomalous Oscillations and Turn-Off Behavior in a Vertical Power MOSFET," MS *Thesis,* University of California, Berkeley, December 1983.

Goodenough, Frank. "Power Control," *Electronic Design,* Vol. 32 January 12, 1984, pp. 244-258.

Goodman, A.M., et al. "Improved COMFETs with Fast Switching Speed and High-Current Capability," *Technical Digest,* IEEE Electron Devices International Meeting, 1983, pp. 79-82.

Greenreich, E. "Theoretical Considerations on the Effects of Bulk Charge on VMOST Characteristics," *IEEE Transactions on Electron Devices,* May 1979, pp. 807-810.

Grove, A.S., D. Leistiko, and W. W. Hooper. "Effect of Surface Fields on the Breakdown Voltage of Planar Silicon p-n Junctions," *IEEE Transactions on Electron Devices,* Vol. ED-14 (1976), p. 157.

Gyma, Dennis, John Hyde, and Daniel Schwartz. "The Power MOSFET as a Switch From a Circuit Designer's Perspective," *Proceedings,* POWERCON 7, 1980, pp. D1-D16.

Harper, D.H. and R.E. Thomas. "Diffusion Current Effects in DMOS Transistors," *Technical Digest,* International Electron Devices Meeting, 1978, p. 34.

Hayward, W. "A VMOS FET Transmitter for 10 Meters CW," *QST,* May 1979, pp. 23-26.

Hebenstreit, E. "Driving the SIPMOS Field-Effect Transistor as a Fast Power Switch," *Siemens Forsch u. Entwickl-Ber. Bd 9,* Nr. 4 (1980).

Heinzer, Walt. "Don't Trade Off Analog-Switch Specs," *Electronic Design,* Vol. 25, No. 15 (July 19, 1977), pp. 56-61.

Heng, T.M.S. and H.C. Nathanson. "Vertical MOS Transistor Geometry for Power Amplification at Gigahertz Frequencies," *Electronics Letters,* Vol. 10 (November 1974), pp. 490-492.

Heng, T.M.S., et al. "Vertical Channel Metal-Oxide-Silicon Field Effect Transistor," *Final Report,* Westinghouse Research and Development Center, Pittsburgh, Pennsylvania, Contract N00014-74-C-0012, November 1, 1976.

Heng, T.M.S., J.G. Oakes, and R.A. Wickstrom. "Experimental Investigations of the Intrinsic Performance Limitations of Short Channel Non-Planar MOS Transistors," *Technical Digest,* International Electron Devices Meeting, 1976, p. 7.

Heron, Charles J. "A New Triac Device Utilizing Integral FET Drive," *Proceedings,* POWERCON 8, 1981, pp. N2.3.1-N2.3.4.

Hoang, L.H., et al. "Power Transistor Base Drive Circuits Using Power MOSFETs," *Conference Record,* Industrial Applications Society, 1982, pp. 774-781.

Hoffman, D. "Designing a VMOS 250 Watt Off-Line Inverter," *Proceedings,* POWERCON 5, 1978, pp. G1.1-G1.5.

Hoffman, D. "A Microprocessor-Controlled VMOS Power Supply," *Proceedings,* POWERCON 6, 1979, pp. H1.1-H1.8.

Hoffman, David. "VMOS -- A Key to the Advancement of SMPS Technology," *Powerconversion International,* Vol. 6, No. 2 (March/April 1980), pp. 37-42.

Hoffman, K. and R. Losehand. "VMOS Technology Applied to Dynamic RAMs," *IEEE Journal of Solid State Circuits,* Vol. SC-13, No. 5 (October 1978).

Hoffman, K., R. Losehand, and K. Zapf. "VMOS Dynamic RAM," *Digest of Technical Papers,* IEEE International Solid State Circuits Conference, 1978, pp. 156-157.

Hofstein, S.R. and G. Warfield. "Carrier Mobility and Saturation Velocities in Silicon Inversion Layers," *Journal of Applied Physics,* Vol. 41 (1970), p. 1825.

8

Holmes, F.E. and C.A.T. Salama. "VMOS - A New MOS Integrated Circuit Technology," *Solid State Electronics*, Vol. 17 (1974), pp. 791-797.

Hower, P.L. and M.J. Geisler. "Comparison of Various Source-Gate Geometries for Power MOSFETs," *IEEE Transactions on Electron Devices*, 28, No. 9 (September 1981), pp. 1098-1101.

Hower, P.L., T.M.S. Heng, and C. Huang. "Optimum Design of Power MOSFETs," *Technical Digest*, IEEE Electron Devices International Meeting, Washington D. C., 1983, pp. 91-94.

Hower, Philip L. "Bipolar vs. MOSFET -- Seeing Where the Power Lies," *Electronics*, December 18, 1980, pp. 106-110.

Hu, Chenming. "A Parametric Study of Power MOSFETs," *PESC'79 Record*, Power Electronics Specialists Conference, Paper 5.7.

Hu, Chenming and M.H. Chi. "Second Breakdown of Vertical Power MOSFETs," *IEEE Transactions on Electron Devices*, 30, No. 8 (August 1982), pp. 1287-1293.

Jambotkar, C.G. and P.P. Wang. "Power FET and Manufacturing Method," *IBM Technical Disclosure Bulletin*, Vol. 23, No. 7A (December 1980), pp. 2861-2862.

Jansson, L.E. and J.A. Houldsworth. "Cost-Effective Design for Mains-Input SMPS Using POWERMOS," *Proceedings*, 8th Power Conversion International Conference, April 1984, pp. 110-120.

Jenkins, J.D.M. "MOSPOWER: The Challenge to Power Bipolars," *Microelectronics and Reliability*, Vol. 16, No. 5 (1977), pp. 607-616.

Jenne, F.B. "Grooves Add New Dimension to VMOS Structure and Performance," *Electronics*, August 18, 1977, pp. 100-106.

Jenne, F.B., J.J. Barnes, and T.J. Rodgers. "A Theoretical and Experimental Analysis of the Buried-Source VMOS Dynamic RAM Cell, "*IEEE Transactions on Electron Devices*, Vol. ED-25, No. 10 (October 1978), p. 1204.

Jenne, F.B., et al. "VMOS EPROM and Buried Source Cells," *Technical Digest*, International Electron Devices Meeting, Late News Paper, December 1976.

Johnson, Robert J. and Helge Granberg. "Design, Construction and Performance of High Power RF VMOS Devices," *Technical Digest*, International Electron Devices Meeting, Washington, D.C., 1979, pp. 93-96.

Kay, Steeve. "Development and Fabrication of Low ON Resistance High Current Vertical VMOS Power FETs," NASA Lewis Research Center, Contract No. NAS 3-21034.

Kay, Steeve, C.T. Trieu, and Bing H. Yeh. "A New VMOS Power FET," *Technical Digest*, International Electron Devices Meeting, 1979, pp. 97-101.

Keller, J.H., D.R. Kerr, and J.R. Winnard. "Plasma/RF Annealing of FET MOS Devices," *IBM Technical Disclosure Bulletin*, Vol. 25, No. 7A (December 1982), p. 3589.

Kennery, D.P. and P.C. Murley. "Steady State Mathematical Theory for the Insulated Gate Field Effect Transistor," *IBM Journal of Research and Development*, Vol. 17 (1973), p. 2.

Kerr, John W. "High Side Switching with n-Channel MOSFETs," *Electronic Products*, February 7, 1984, pp. 115-118.

Klaassen, F.M. and W.C.J. deGroot. "Modelling of Scaled-Down MOS Transistors," *Solid State Electronics*, Vol. 23 (1980), p. 237.

Koehler, Charles. "Monolithic Analog Multiplexers for High Voltage Applications," *Session Record 25*, ELECTRO/84.

Kuo, D.S.C., Hu, and M.H. Chi. "dV/dt Breakdown in Power MOSFETs," *IEEE Electron Device Letters*, Vol. EDL-4, No. 1 (January 1983), pp. 1-2.

Lazarus, M.J. "The Short Channel IGFET," *IEEE Transactions on Electron Devices*, 22 (1975), p. 351.

Leighton, L., et al. "HF Power Amplifier Design Using VMOS Power FETs," *RF Design*, January 1980, pp. 32-37.

Leighton, Larry. "Two Meter Transverter Using Power FETs," *Ham Radio*, September 1976, pp. 32-34.

Leupold, L. and J. Tihanyi, "Experimental Study of a High Blocking Voltage Power MOSFET with Integrated Input Amplifier," *Technical Digest*, International Electron Devices Meeting, 1983, pp. 428-431.

Lidow, A., T. Herman, and H. Collins. "Power MOSFET Technology," *Technical Digest*, International Electron Devices Meeting, 1979, pp. 79-83.

Lidow, Alexander and William Collins. "Solid-State Power Relays Enter the IC Era," *Electronics*, December 29, 1982, pp. 59-62.

Lin, H.C. and W.N. Jones. "Computer Analysis of the Double-Diffused MOS Transistor of Integrated Circuits," *IEEE Transactions on Electron Devices*, Vol. ED-20 (1973), p. 275.

Lisiak, K. and J. Burger. "Optimization of Nonplanar Power MOS Transistors," *IEEE Transactions on Electron Devices*, 25 (1978), pp. 1229-1234.

Lisiak, K., guest ed. "Special Issue on Power MOS Transistors," *IEEE Transactions on Electron Devices*, Vol. ED-27 (1980), p. 321.

Lorenz, Ing. L., "Study of the Switching Performance of a MOSFET Circuit," *PESC'84 Record*, Power Electronics Specialists Conference, pp. 215-224.

Ludvik, Stephan. "Vertical Geometry is Boosting FETs into Power Uses at Radio Frequencies," *Electronics*, Vol. 51, No. 5 (March 2, 1978), pp. 105-108.

Marshall, Trevor G. "Switched Power Amplification at Low and Medium Frequencies," *Master's Thesis*, Western Australia Institute of Technology, 1978.

Masuhara, T. and R.S. Muller. "Analytical Technique for the Design of DMOS Transistors," *Japanese Journal of Applied Physics*, Suppl. 16-1 (1977), p. 173.

McNulty, T.C. "Power MOSFETs - What the Designer Needs to Know," *Electronic Products*, February 7, 1984, pp. 133-137.

Meador, Jim and Nathan Zomer. "Using Bipolar-MOSFET Combinations to Optimize the Switching Transistor Function," *Proceedings*, POWERCON 8, 1981, pp. F4.1-F4.6.

Miller, Bill. "V-Channel FETs Threaten Bipolar Devices in the Power and Speed Applications," *Canadian Electronics Engineering*, May 1978.

Minisian, R.A., "Computer Aided Simulation of Power MOSFET Switch-Mode Converters," *Electronic Letters*, Vol. 20, No. 11, 24 May 1984.

Mok, T.D. and C.A.T. Salama. "A V-Groove Power Junction Field Effect Transistor for VHF Applications," *Technical Digest*, International Electron Devices Meeting, 1976 Supplement, pp. 6-7. (Also published in *IEEE Transactions on Electron Devices*)

Mok, T.D. and C.A.T. Salama. "A V-Groove Schottky-Barrier FET for UHF Applications," *IEEE Transactions on Electron Devices*, Vol. ED-25, No. 10 (October 1978), p. 1235.

Mosita, Y., H. Takahashi, and H. Matayoshi. "Si UHF MOS High Power FET," *IEEE Transactions on Electron Devices*, Vol. ED-21 (1974), p. 733.

Morenza, J.L. and D. Esteve. "Entirely Diffused Vertical Channel JFET: Theory and Experiment," *Solid State Electronics*, Vol. 21, No. 5, (May 1978), pp. 739-746.

Moyer, Curtis D. "Using Power MOSFETs in Stepping Motor Control," *Application Note AN-876*. Motorola Semiconductor Products, Inc.

Nagata, M. "High-Power MOSFETs," *Inst. Phys. Conf. Ser.*, Vol. 32 (1977), p. 101.

Nagata, Minoru. "Power Handling Capability of MOSFET," *Proceedings*, 8th International Conference on Solid State Devices, Tokyo, 1976.

Nakagiri, M. and K. Iida. "Damage Introduced by Second Breakdown in n-Channel MOS Devices," *Japanese Journal of Applied Physics*, 16, No. 7 (July 1977), pp. 1187-1193.

Navon, D.H. "Numerical Modeling of Power MOSFETs," *Solid State Electronics*, Vol. 26, No. 4 (1983), pp. 287-290.

National CSS, Inc. *ISPICE Short Channel MOSFET Model Note*, Norwalk, Connecticut, 1977.

Neumark, G.F. "Theory of the Influence of Hot Electron Effects on Insulated Gate Field Effect Transistors," *Solid State Electronics*, Vol. 10 (1967), p. 169.

Ng, K.K. and G.W. Taylor. "Effects of Hot-Carrier Trapping in n- and p-Channel MOSFETs," *IEEE Transactions on Electron Devices*, 30, No. 8 (August 1983), pp. 871-876.

Nienhaus, H.A. and J.C. Bowers. "A High Power MOSFET Computer Model," *Powerconversion International*, January 1982, pp. 65-73.

Oakes, James, et al. "A Power Silicon Microwave MOS Transistor," *IEEE Transactions on Microwave Theory and Techniques*, Vol. MTT-24 (June 1976), pp. 305-311.

Ohata, Yu. "New MOSFETs for High Power Switching Applications," *Proceedings*, POWERCON11, 1984, C5.1 - C5.11.

Omura, Y. and K. Ohwada. "Threshold Voltage Theory for a Short-Channel MOSFET Using a Surface Potential Distribution Model," *Solid State Electronics*, 22 (1979), pp. 1045-1051.

Ou-Yang, P. "Double Implanted VMOS," *Technical Digest*, International Electron Devices Meeting, 1976, pp. 577-578.

Ou-Yang, Paul. "Double Ion Implanted VMOS Technology," *IEEE Journal of Solid State Circuits*, Vol. SC-12, No. 1 (February 1977), pp. 3-10.

Overstraeten, R.V. and H. DeMan. "Measurement of the Ionization Rates in Diffused Silicon p-n Junctions," *Solid State Electronics*, Vol. 13 (1970), p. 583.

Oxner, E.S. "MOSPOWER Semiconductors, An Historical Perspective," *Powerconversion International*, Vol. 8, No. 6 (June 1982), p. 14ff.

Oxner, E.S. "MOSPOWER Semiconductors in Industrial Applications," *Powerconversion International*, Vol. 8, No. 7 (July/August 1982), p. 25ff.

Oxner, E.S. "MOSPOWER Semiconductors New Application Frontiers," *Powerconversion International*, Vol. 8, No. 8 (September 1982), pp. 50-53.

Oxner, Ed. "Build a Broadband Ultralinear VMOS Amplifier," *QST*, May 1979, pp. 23-26.

Oxner, Ed. "Characterization and Application of MOSPOWER VHF FETs in High Frequency Communications," 21st Annual West Coast (Amateur) VHF/UHF Conference, May 1976.

Oxner, Ed. "High-Frequency Performance of the Short-Channel MOS," *Session Record 15*, IEEE SOUTHCON/81.

Oxner, Ed. "Integrated Circuits," *1979 Yearbook of Science and Technology*, New York: McGraw-Hill Book Co. (1979), pp. 233-235.

Oxner, Ed. "Meet the VMOSFET Model," *RF Design*, January/February 1979, pp. 16-22.

Oxner, Ed. "MOSPOWER FET as a Broadband Amplifier," *Ham Radio*, December 1976, pp. 32-34.

Oxner, Ed. "MOSPOWER: Will it be the Universal Replacement for High Frequency Power Bipolars?" *Record*, 27th Annual IEEE Vehicular Technology Group Conference, March 1977, pp. 185-189.

Oxner, Ed. "A New Technology: Application of MOSPOWER FETs for High Frequency Communication," *Sessions Record*, IEEE Mid-America Electronics Conference, November 1976.

Oxner, Ed. "Power FETs," *Wireless World* (taken from Ed Oxner's News You Can Use, April 1976), May 1977, pp. 74-76.

Oxner, Ed. "Using Power MOSFETs as High-Efficiency Synchronous and Bridge Rectifiers in Switch-Mode Power Supplies," *Session Record 24*, IEEE MIDCON/82.

Oxner, Ed. "VMOS: A New Technology Takes on HF Power Bipolars," *MSN*, Vol. 6 (October/November 1976), pp. 107-110.

Oxner, Ed. "Will VMOS Replace Bipolar in RF Systems?" *EDN*, June 20, 1977, p. 71.

Oxner, Edwin S. *Power FETs and Their Applications*. New Jersey: Prentice-Hall, 1982.

Oxner, Edwin S. "Static and Dynamic dV/dt Characteristics of Power MOSFETs," *Proceedings*, 8th Power Conversion International Conference, April 1984, pp. 144-151.

Ozawa, Osamu, Hiroshi Iwasaki, and Kenichi Muramoto. "A Vertical Channel JFET Fabricated Using Silicon Planar Technology," *IEEE Journal of Solid State Circuits*, Vol. SC-11, No. 4 (August 1976), pp. 511-518.

Pao, H.C. and C.T. Sah. "Effects of Diffusion Current on Characteristics of Metal-Oxide (Insulator) - Semiconductor Transistors," *Solid State Electronics*, Vol. 9 (1966), p. 927.

Park, H.S. and A. van der Ziel, "Noise Measurements in Ion Implanted MOSFETs," *Solid State Electronics*, Vol. 26, No. 8 (1983), pp. 747-751.

Parks, C.M. and C.A.T. Salama. "V-Groove (VMOS) Conductively Connected Charge Coupled Devices," *Solid State Electronics*, Vol. 18 (1975), pp. 1061-1067.

Pelly, Brian. "Optically Coupled Power MOSFET Technology: A Monolithic Replacement for Electromagnetic Relays," *Sessions Record 7*, IEEE ELECTRO/83.

Pelly, Brian R. "Applying International Rectifier's Power MOSFETs," *International Rectifier Application Note 930*, El Segundo, California: International Rectifier.

Pelly, Brian R. "The Do's and Don'ts of Using Power MOSFETs," *International Rectifier Application Note 936*, El Segundo, California: International Rectifier.

Pelly, Brian R. and Zoltan Zansky. "Using High Voltage Power MOSFETs in Off-Line Converter Applications," *Proceedings*, POWERCON 6, 1979, pp. C2-1—C2-13.

Pelly, B.R. "A New Gate Charge Factor Leads to Easy Drive Design for Power MOSFET Circuits," *International Rectifier Application Note 944*, El Segundo, California: International Rectifier.

Pocha, Michael D. and Robert W. Dutton. "A Computer-Aided Design Model for High Voltage Double Diffused MOS (DMOS) Transistors," *IEEE Journal of Solid State Circuits*, 11 (1976), pp. 718-726.

8

Pocha, Michael D., J.D. Plummer, and J.D. Meindl. "Tradeoff between Threshold Voltage and Breakdown in High Voltage Double Diffused MOS Transistors," *IEEE Transactions on Electron Devices*, 25 (1978), pp. 1325-1327.

Pshaenich, Al. "The MOS SCR, A New Thyristor Technology," *Motorola Engineering Bulletin EB-103*, Phoenix, Arizona: Motorola Semiconductor Products, Inc., 1982.

Redl, Richard, Bela Molnar, and Nathan O. Sokal. "Class-E Resonant Regulated DC/DC Power Converters: Analysis of Operation and Experimental Results at 1.5 MHz," *PESC '83 Record*, Power Electronics Specialists Conference.

Redl, Richard, Bela Molnar, and Nathan O. Sokal. "Small Signal Dynamic Analysis of Regulated Class E DC/DC Converters," *PESC '84 Record*, Power Electronics Specialists Conference, pp. 62-71.

Regan, Phillip. "New VMOS Technology Threatens Bipolar Supremacy," *Electronic Engineering*, Vol. 49, No. 588 (February 1977), pp. 40-41.

Regan, Phillip. "Working with VMOS Power FETs," *Electronics Weekly* (UK), (a six-part, weekly series), March 22, 29, April 5, 12, 19, and 26, 1978.

Richman, Paul. *MOS Field-Effect Transistors and Integrated Circuits*. New York: John Wiley & Sons, Inc., 1973.

Risch, L. "Electron Mobility in Short-Channel MOSFETs with Series Resistance," *IEEE Transactions on Electron Devices*, 30, No. 8 (August 1983), pp. 959-960.

Rodgers, T.J. "Low Capacitance V-Groove MOS NOR Gate and Method of Manufacture," U. S. Patent 3,924,265, December 2, 1975.

Rodgers, T.J. and J.D. Meindl. "Epitaxial V-Groove Bipolar Integrated Circuit Process," *Digest of Technical Papers*, IEEE Transactions on Electron Devices, Vol. ED-20 (March 1973), pp. 226-232.

Rodgers, T.J. and J.D. Meindl. "Short Channel V-Groove MOS (VMOS) Logic," *Digest of Technical Papers*, IEEE International Solid State Circuits Conference, Paper THAM 10.4, 1974, pp. 112-113.

Rodgers, T.J., R. Hiltpold, and J.W. Zimmer. "VMOS ROM," *IEEE Journal of Solid State Circuits*, Vol. SC-11 (October 1976), p. 614.

Rodgers, T.J., et al. "VMOS Memory Technology," *IEEE Journal of Solid State Circuits*, Vol. SC-12, No. 15 (October 1977), pp. 515-524.

Rodgers, T.J., et al. "An Experimental and Theoretical Analysis of Double-Diffused MOS Transistors," *IEEE Journal of Solid State Circuits*, Vol. SC-10 (1975), p. 322.

Rodgers, T.J. and J.D. Meindl. "VMOS: High Speed TTL Compatible MOS Logic," *IEEE Journal of Solid State Circuits*, Vol. SC-19 (October 1974), pp. 239-249.

Rodriguez, Edward T. "Optically Coupled Power MOSFET Technology: A Monolithic Replacement for Electromechanical Relays," *Session Record 7*, IEEE ELECTRO/83.

Roehr, Bill. "The Autobias Amplifier," *Journal* Audio Engineering Society, Vol. 30, No. 4 (April 1982), pp. 208-216.

Roehr, Bill. "High Voltage MOSFET and Bipolar Power Switches, Part I," *Powerconversion International*, January 1984, p. 15ff.

Roehr, Bill. "High Voltage MOSFET and Bipolar Power Switches, Part II," *Powerconversion International*, February 1984, p. 56ff.

Roehr, Bill. "Mounting Techniques for Power Semiconductors," *Motorola Application Note AN-778*. Phoenix, Arizona: Motorola Semiconductor Products Inc., 1978.

Roehr. W. "VMOS - A Giant Step Toward the Ideal," *Canadian Electronics Engineering*, May 1979, pp. 38-40.

Ronan, Harold R. and C. Frank Wheatley. "Power MOSFET Switching Waveforms: A New Insight," *Proceedings*, POWERCON 11, 1984, pp. C3.1 - C3.10.

Russell, J.P., et al. "The COMFET - A New High Conductance MOS-Gated Device," *IEEE Electron Device Letters*, Vol. EDL-4 (March 1983), pp. 63-65.

Saladin, Herb and Al Pshaenich. "Paralleling Power MOSFETs in a Very Fast, High Voltage High Current Switch," *Motorola Technical Design Tip TDT-104*, Phoenix, Arizona: Motorola Semiconductor Products, Inc.

Salama, C.A.T. "V-Groove Power Field Effect Transistors," *Technical Digest*, International Electron Devices Meeting, 1977, pp. 412-417.

Salama, C. and J. Oakes. "Nonplanar Power Field-Effect Transistors," *IEEE Transactions on Electron Devices*, 25 (1978), pp. 1222-1228.

Sampei, Tohru and Shin-ichi Ohashi, "100 Watt Super Audio Amplifier Using New MOS Devices," *IEEE Transactions on Consumer Electronics*, CE-23, No. 3 (August 1977), pp. 409-417.

Sampei, Tohru, et al. "Highest Efficiency and Super Quality Audio Amplifier Using MOS Power FETs in Class G Operation," *IEEE Transactions on Consumer Electronics*, CE-24, No. 3 (August 1978), pp. 300-307.

Scharf, B.W. and J.D. Plummer. "Insulated-Gate Planar Thyristors: II-Quantitative Modeling," *IEEE Transactions on Electron Devices*, Vol. ED-27 (1980), p. 387.

Schultz, Warren. "High Current FETs -- A New Level of Performance," *Powerconversion International*, Vol. 10, No. 3 (March 1984), pp. 43-46.

Schultz, Warren. "Multichip-Power MOSFETs Beat Bipolars at High-Current Switching," *Electronic Design*, Vol. 32, No. 12 (June 14, 1984), pp. 223-232.

Sekigawa, T. and Y. Hayasi. "An Effect of Semiconductor Surface Shape on Gate Breakdown Voltage of a VMOS Transistor," *Solid-State Electronics*, Vol. 26, No. 9 (September 1983), pp. 925-927.

Selberherr, S. "Modeling Static and Dynamic Behavior of Power Devices," *Technical Digest*, International Electron Devices Meeting, 1983.

Severns, R. "dV/dt Effects in MOSFET and Bipolar Junction Transistor Switches," *PESC'81 Record*, Power Electronics Specialists Conference.

Severns, R. "Simplified HEXFET Power Dissipation and Junction Temperature Calculation Speeds Heatsink Design," *International Rectifier Application Note 942*, El Segundo, California: International Rectifier.

Severns, R. "Using the Power MOSFET as a Switch," *Intersil Application Bulletin A036*, Cupertino, California: Intersil, Inc., 1981.

Severns, Rudolf. "The Design of Switchmode Converters Above 100 kHz," *Intersil Application Bulletin A034*, Cupertino, California: Intersil, Inc., 1980.

Severns, Rudy. "dV_{DS}/dt Turn-on In MOSFETs," *Siliconix Technical Article TA84-4*, Santa Clara, California: Siliconix incorporated, 1984.

Severns, Rudy. "Power MOSFETs and Radiation Environments," *Siliconix Technical Article TA84-3*, Santa Clara, California: Siliconix incorporated, 1984.

Severns, Rudy. "The Power MOSFET as a Rectifier," *Power-conversion International*, Vol. 6, No. 2 (March/April 1980), pp. 49-50.

Severns, Rudy. "Safe Operating Area and Thermal Design for MOSPOWER Transistors," *Siliconix Application Note AN83-10*, Santa Clara, California: Siliconix incorporated, 1983.

Severns, Rudy. "Using the Power MOSFET as a Switch," *Electronic Product Design*, May 1981, pp. 49-53.

Shaeffer, Lee. "Improving Converter Performance and Operating Frequency with a New Power FET," *Proceedings, POWERCON 4*, 1977, p. C2.

Shaeffer, Lee. "Use FETs to Switch High Currents," *Electronic Design*, Vol. 24, No. 9 (April 26, 1976), pp. 66-72.

Shaeffer, Lee. "Vertical MOSFETs (VMOS) in High Quality Audio Power Amplifiers," *Audio Engineering Society Preparation for 54th Convention*, Preprint 1106 (F-8), 1976.

Shimada, Yuuki, K. Kato, and T. Sakai. "High Efficiency MOSFET Rectifier Device," *PESC'83 Record*, Power Electronics Specialists Conference, pp. 129-136.

Siegal, Bernard, "Power MOSFETs Thermal Resistance Measurement," *Powerconversion International*, Vol. 6, No. 2 (March/April 1980), pp. 10-14.

Sloane, T.H., H.A. Owen, Jr., and T.G. Wilson. "Switching Transients in High-Frequency High-Power Converters Using Power MOSFETs," *PESC'79 Record*, Power Electronics Specialists Conference, San Diego, California.

Slusark, Laurie, Neilson and Finn. "Catastrophic Burn Out in Power MOSFET Field-Effect Transistors," *IEEE Reliability Physics Symposium*, 1983, pp. 173-177.

Smith, Marvin W. "Applications of Insulated Gate Transistors," *Proceedings, 8th Power Conversion International Conference*, April 1984, pp. 121-131.

Smith, Marvin W. and Sebald R. Korn. "The Dynamics of Power MOS Design," *Proceedings, POWERCON 10*, 1983, pp. C3.1-C3.13.

Smith, Marvin, William Sahm, and Sridhar Babu. "Insulated-Gate Transistors Simplify AC-Motor Speed Control," *EDN*, February 9,1984, pp. 181-201.

Special Issue on Power MOS Devices, *IEEE Transactions on Electron Devices*, 27 (1980), pp. 321-400.

Stone, Robert T. and Howard M. Berlin. *Design of VMOS Circuits with Experiments*. Indianapolis, Indiana: Howard W. Sams & Company, Inc., 1980.

Sun, S.C. and J.D. Plummer. "Modeling of the On-Resistance of LDMOS, VDMOS, and VMOS Power Transistors," *IEEE Transactions on Electron Devices*, Vol. ED-27 (1980), p. 356.

Takesuye, Jack and Gary Fay. "MOSFETs Replace Bipolars and Circuits Are Better for It," *Electronic Design*, December 20,1980, pp. 93-97.

Tamar, A.A., K. Rauch, and J.L. Moll. "Numerical Comparison of DMOS, VMOS and UMOS Power Transistors," *IEEE Transactions on Electron Devices*, 30, No. 1 (January 1983), pp. 73-76.

Tarasewicz, S. and C.A.T. Salama. "A High Voltage UMOS Transistor," *Solid-State Electronics*, Vol. 24, No. 5 (May 1981), pp. 435-443.

Tarasewicz, S.W. and C.A.T. Salama. "Transconductance Degradation in VVMOS Power Transistors Due to Thermal and Field Effects," *Solid-State Electronics*, Vol. 25, No. 12 (December 1982), pp. 1165-1170.

Temple, Victor A.K. "Ideal FET Doping Profile," *IEEE Transactions on Electron Devices*, 30, No. 6 (June 1983), pp. 619-625.

Temple, V. and P. Gray. "Theoretical Comparison of DMOS and VMOS Structures for Voltage and ON Resistance," *Technical Digest*, IEEE Electron Devices International Meeting, 1979, pp. 88-92.

Temple, V.A.K. and R.P. Love. "A 600 Volt MOSFET with Near Ideal On Resistance," *Technical Digest*, International Electron Devices Meeting, 1977, pp. 664-666.

Toyable, Toru, Ken Yamaguchi, and Shojiro Asai. "A Numerical Model of Avalanche Breakdown in MOSFETs," *IEEE Transactions on Electron Devices*, 25 (1978), pp. 825-832.

Tihanigi, Jenoe, et al. "MIS Field-Effect Transistor Having a Short Channel Length," U. S. Patent 4,190,850, February 26, 1980.

van der Kooi, Marvin and Larry Ragle. "MOS Moves into Higher Power Applications," *Electronics*, Vol. 49, No. 13 (June 24, 1976), pp. 98-103.

Wang, Paul P. "Device Characteristics of Short-Channel and Narrow-Width MOSFETs," *IEEE Transactions on Electron Devices*, 25 (1978), pp. 779-786.

Westbrook, Scot Rossman. "The Switching Behavior of the Power MOSFET," *Master's Thesis*, Massachusetts Institute of Technology, 1980.

Wheatley, C. Frank and Harold R. Rowan, Jr. "Switching Waveforms of the L^{2}FET: A 5-Volt Gate-Drive Power MOSFET," *PESC'84 Record*, Power Electronics Specialists Conference, pp. 238-246.

Wildi, Eric J., T. Paul Chow, and Mike E. Cornell. "A 500 V Junction-Isolated BiMOS High-Voltage IC," *Session Record 25*, IEEE ELECTRO/84.

Wilson, Peter. "Linear Power Amplifier Using Complementary HEXFETs," *International Rectifier Application Note 948*, El Segundo, California: International Rectifier.

Wood, P. "Transformer-Isolated HEXFET Driver Provides Very Large Duty Cycle Ratios," *International Rectifier Application Note 950*, El Segundo, California: International Rectifier.

Yoshida, I., T. Okabe, and N. Hashimoto. "A Composite Gate Structure for Enhanced-Performance Power MOSFETs," *Transactions* of the Institute of Electronics and Communication Engineers of Japan, Section E, Vol. E64, No. 12 (December 1981), pp. 824ff.

Yoshida, Isao. "1600V Power MOSFET with 20nS Switching Speed," *Technical Digest*, IEEE Electron Devices International Meeting, 1983, pp. 91-94.

Yoshida, Isao, et al. "Thermal Stability and Secondary Breakdown in Planar Power MOSFETs," *IEEE Transactions on Electron Devices*, 27 (1980), pp. 395-398.

Yoshida, Isao, Masaharu Kubo, and Shikayaki Ochi. "A High Power MOSFET with a Vertical Drain Electrode and a Meshed Gate Structure," *IEEE Journal of Solid State Circuits*, Vol. SC-11, No. 4 (August 1976), pp. 472-477.

Yoshida, I., et al. "A High Power MOSFET with a Vertical Drain and Meshed Gate Structure," *Technical Digest*, International Electron Devices Meeting, 1975, p. 159.

Yoshida, I., et al. "A High Power MOSFET with a Vertical Drain and Meshed Gate Structure," *Japanese Journal of Applied Physics Suppl.*, Vol. 15 (1976), p. 179.

Yoshida, I., et al. "A High Power MOSFET with a Vertical Drain Electrode and a Meshed Gate Structure," *IEEE Journal of Solid State Circuits*, August 1967, pp. 472-477.

Zommer, Nathan. "Designing the Power-Handling Capabilities of MOS Power Devices," *IEEE Transactions on Electron Devices*, 27 (1980), pp. 1290-1296.

8

A.6 MOSPOWER Selector Guide

N-Channel MOSPOWER

Device	Breakdown Voltage (Volts)	rDS (on) (ohms)	ID Continuous (Amps)	Power Dissipation (Watts)	Part Number
	650	0.550	11.700	176	VNT013A
	650	0.750	9.300	150	VNT012A
	650	1.500	5.700	125	VNT008A
	650	2.000	5.000	125	VNT009A
	600	0.550	11.700	176	VNS013A
	600	0.750	9.300	150	VNS012A
	600	1.500	5.700	125	VNS008A
	600	2.000	5.000	125	VNS009A
	500	0.300	20.000	250	VNP006A
	500	0.400	13.000	150	IRF450
	500	0.400	12.000	150	2N6770
	500	0.500	12.000	150	IRF452
	500	0.600	9.600	125	BUZ45
	500	0.850	8.000	125	IRF440
	500	1.100	7.000	125	IRF442
	500	1.500	6.500	175	BUP71
	500	1.500	6.500	75	IRF430
	500	1.500	4.500	75	2N6762
	500	1.500	4.500	100	BUP66
	500	1.500	4.500	100	VN5001A
	500	1.500	4.500	175	VNP002A
	500	2.000	4.200	78	BUZ46
	500	2.000	4.000	100	BUP67
	500	2.000	4.000	75	IRF432
	500	2.000	4.000	100	VN5002A
	500	3.000	2.500	40	IRF420
	500	4.000	2.000	40	IRF422
	450	0.300	20.000	250	VNN006A
	450	0.400	13.000	150	IRF451
	450	0.500	11.000	150	2N6769
	450	0.500	12.000	150	IRF453
	450	0.850	8.000	125	IRF441
	450	1.100	7.000	125	IRF443
	450	1.500	6.500	175	BUP70
TO-3	450	1.500	6.500	75	IRF431
	450	1.500	4.500	100	BUP64
	450	1.500	4.500	100	VN4501A
	450	1.500	4.500	175	VNN002A
	450	2.000	4.000	75	2N6761
	450	2.000	4.000	100	BUP65
	450	2.000	4.000	75	IRF433
	450	2.000	4.000	175	VN4502A
	450	3.000	2.500	40	IRF421
	450	4.000	2.000	40	IRF423
	400	0.200	25.000	250	VNM005A
	400	0.300	15.000	150	IRF350
	400	0.300	14.000	150	2N6768
	400	0.400	13.000	150	IRF352
	400	0.400	10.500	125	BUZ64
	400	0.550	10.000	125	IRF340
	400	0.800	8.000	125	IRF342
	400	1.000	8.000	175	BUP69
	400	1.000	8.000	75	IRF330
	400	1.000	6.000	125	BUP62
	400	1.000	6.000	125	VN4000A
	400	1.000	5.500	75	2N6760
	400	1.000	5.900	75	BUZ63
	400	1.000	5.500	175	VNM001A
	400	1.500	5.000	125	BUP63
	400	1.500	5.000	75	IRF332
	400	1.500	4.500	125	VN4001A
	400	1.800	3.000	40	IRF320
	400	2.500	2.500	40	IRF322
	350	0.200	25.000	250	VNL005A
	350	0.300	15.000	150	IRF351
	350	0.400	13.000	150	IRF353
	350	0.400	12.000	150	2N6767
	350	0.550	10.000	125	IRF341

N-Channel MOSPOWER (Cont'd)

Device	Breakdown Voltage (Volts)	rDS (on) (ohms)	ID Continuous (Amps)	Power Dissipation (Watts)	Part Number
	350	0.800	8.000	125	IRF343
	350	1.000	8.000	175	BUP68
	350	1.000	8.000	75	IRF331
	350	1.000	6.000	125	BUP60
	350	1.000	6.000	125	VN3500A
	350	1.000	5.500	175	VNL001A
	350	1.500	5.000	125	BUP61
	350	1.500	5.000	75	IRF333
	350	1.500	4.500	75	2N6759
	350	1.500	4.500	125	VN3501A
	350	1.800	3.000	40	IRF321
	350	2.500	2.500	40	IRF323
	200	0.060	47.000	250	VNJ004A
	200	0.085	30.000	150	2N6766
	200	0.085	30.000	150	IRF250
	200	0.120	25.000	150	IRF252
	200	0.120	22.000	125	BUZ36
	200	0.180	18.000	125	IRF240
	200	0.220	16.000	125	IRF242
	200	0.400	9.900	78	BUZ35
	200	0.400	9.000	75	2N6758
	200	0.400	9.000	75	IRF230
	200	0.600	8.000	75	IRF232
	200	0.800	5.000	40	IRF220
	200	1.200	4.000	40	IRF222
	150	0.060	47.000	250	VNG004A
	150	0.085	30.000	150	IRF251
	150	0.120	25.000	150	2N6765
	150	0.120	25.000	150	IRF253
	150	0.180	18.000	125	IRF241
	150	0.220	16.000	125	IRF243
TO-3	150	0.400	9.000	75	IRF231
	150	0.600	8.000	75	2N6757
	150	0.600	8.000	75	IRF233
	150	0.800	5.000	40	IRF221
	150	1.200	4.000	40	IRF223
	120	0.180	14.000	75	VN1200A
	120	0.250	12.000	100	VN1201A
	100	0.035	65.000	250	VNE003A
	100	0.055	40.000	150	IRF150
	100	0.055	38.000	150	2N6764
	100	0.060	32.000	125	BUZ24
	100	0.080	27.000	125	IRF140
	100	0.085	33.000	150	IRF152
	100	0.110	24.000	125	IRF142
	100	0.180	14.000	75	IRF130
	100	0.180	14.000	100	VN1000A
	100	0.180	14.000	75	2N6756
	100	0.200	10.000	78	BUZ23
	100	0.250	12.000	75	IRF132
	100	0.250	12.000	100	VN1001A
	100	0.300	8.000	40	IRF120
	100	0.400	7.000	40	IRF122
	90	4.000	1.900	25	2N6658
	90	4.500	1.800	25	VN99AA
	90	5.000	1.700	25	VN90AA
	80	0.180	14.000	100	VN0800A
	80	0.250	12.000	100	VN0801A
	60	0.035	60.000	250	VNC003A
	60	0.055	40.000	150	IRF151
	60	0.080	33.000	150	IRF153
	60	0.080	31.000	150	2N6763
	60	0.085	27.000	125	IRF141
	60	0.110	24.000	125	IRF143
	60	0.120	18.000	100	VN0600A
	60	0.150	16.000	100	VN0601A
	60	0.180	14.000	75	IRF131

8

N-Channel MOSPOWER (Cont'd)

Device	Breakdown Voltage (Volts)	rDS (on) (ohms)	ID Continuous (Amps)	Power Dissipation (Watts)	Part Number
TO-3	60	0.250	12.000	75	2N6755
	60	0.250	12.000	75	IRF133
	60	0.300	8.000	40	IRF121
	60	0.400	10.000	80	VN64GA
	60	0.400	7.000	40	IRF123
	60	3.000	2.000	25	2N6657
	60	3.500	2.000	25	VN67AA
	40	0.120	18.000	100	VN0400A
	40	0.150	16.000	100	VN0401A
	35	1.800	2.000	25	2N6656
	35	2.500	2.000	25	VN35AA
TO-220	650	1.500	6.000	125	VNT008D
	650	2.000	5.000	125	VNT009D
	600	1.500	5.700	125	VNS008D
	600	2.000	5.000	125	VNS009D
	500	0.850	8.000	125	IRF840
	500	1.100	7.000	125	IRF842
	500	1.500	4.500	75	IRF830
	500	1.500	4.500	75	VN5001D
	500	2.000	4.000	75	BUZ42
	500	2.000	4.000	75	IRF832
	500	2.000	4.000	75	VN5002D
	500	3.000	2.500	40	IRF820
	500	3.000	2.400	40	BUZ74
	500	4.000	2.000	40	IRF822
	450	0.850	8.000	125	IRF841
	450	1.100	7.000	125	IRF843
	450	1.500	4.500	75	IRF831
	450	1.500	4.500	75	VN4501D
	450	2.000	4.000	75	IRF833
	450	2.000	4.000	75	VN4502D
	450	3.000	2.500	40	IRF821
	450	4.000	2.000	40	IRF823
	400	0.550	10.000	125	IRF740
	400	0.800	8.000	125	IRF742
	400	1.000	6.000	75	VN4000D
	400	1.000	5.500	75	BUZ60
	400	1.000	5.500	75	IRF730
	400	1.500	5.000	75	VN4001D
	400	1.500	4.500	75	IRF732
	400	1.800	3.000	40	BUZ76
	400	1.800	3.000	40	IRF720
	400	2.500	2.500	40	IRF722
	400	3.600	1.500	20	IRF710
	400	5.000	1.300	20	IRF712
	350	0.550	10.000	125	IRF741
	350	0.800	8.000	125	IRF743
	350	1.000	6.000	75	VN3500D
	350	1.000	5.500	75	IRF731
	350	1.500	5.000	75	VN3501D
	350	1.500	4.500	75	IRF733
	350	1.800	3.000	40	IRF721
	350	2.500	2.500	40	IRF723
	350	3.600	1.500	20	IRF711
	350	5.000	1.300	20	IRF713
	240	6.000	1.400	20	VN2406D
	200	0.180	18.000	125	IRF640
	200	0.220	16.000	125	IRF642
	200	0.400	9.500	75	BUZ32
	200	0.400	9.000	75	IRF630
	200	0.600	8.000	75	IRF632
	200	0.800	5.000	40	IRF620
	200	1.200	4.000	40	IRF622
	200	1.500	2.500	20	IRF610
	200	2.400	2.000	20	IRF612

MOSPOWER Selector Guide (Cont'd)

N-Channel MOSPOWER (Cont'd)

Device	Breakdown Voltage (Volts)	rDS (on) (ohms)	ID Continuous (Amps)	Power Dissipation (Watts)	Part Number
	170	6.000	1.400	20	VN1706D
	150	0.180	18.000	125	IRF641
	150	0.220	16.000	125	IRF643
	150	0.400	9.000	75	IRF631
	150	0.600	8.000	75	IRF633
	150	0.800	5.000	40	IRF621
	150	1.200	4.000	40	IRF623
	150	1.500	2.500	20	IRF611
	150	2.400	2.000	20	IRF613
	120	0.180	14.000	45	VN1200D
	120	0.250	12.000	75	VN1201D
	120	6.000	1.400	20	VN1206D
	100	0.085	27.000	125	IRF540
	100	0.110	24.000	125	IRF542
	100	0.180	14.000	75	IRF530
	100	0.180	14.000	75	VN1000D
	100	0.200	12.000	75	BUZ20
	100	0.250	12.000	75	IRF532
	100	0.250	12.000	75	VN1001D
	100	0.250	9.000	40	BUZ72A
	100	0.300	8.000	40	IRF520
	100	0.400	7.000	40	IRF522
	100	0.600	2.500	20	IRF510
	100	0.800	2.000	20	IRF512
TO-220	80	0.180	14.000	75	VN0800D
	80	0.250	12.000	75	VN0801D
	80	4.000	1.700	20	BSR82
	80	4.000	1.700	20	VN88AD
	80	4.500	1.600	20	VN89AD
	60	0.085	27.000	125	IRF541
	60	0.110	24.000	125	IRF543
	60	0.120	18.000	75	VN0600D
	60	0.150	16.000	75	VN0601D
	60	0.180	14.000	75	IRF531
	60	0.250	12.000	75	IRF533
	60	0.300	8.000	40	IRF521
	60	0.400	7.000	40	IRF523
	60	0.600	2.500	20	IRF511
	60	0.800	2.000	20	IRF513
	60	3.000	1.900	20	VN66AD
	60	3.500	1.800	20	VN67AD
	50	0.100	19.300	75	BUZ10
	50	0.100	12.000	40	BUZ71
	40	0.120	18.000	75	VN0400D
	40	0.150	16.000	75	VN0401D
	40	3.000	1.900	20	BSR80
	40	3.000	1.900	20	VN46AD
	40	5.000	1.500	20	VN40AD
	30	1.200	2.500	20	VN0300D
	80	4.000	1.500	15	VN88AF
	80	4.500	1.400	15	VN89AF
	80	5.000	1.300	15	VN80AF
	60	3.000	1.700	15	VN66AF
	60	3.500	1.600	15	VN67AF
TO-202	40	3.000	1.600	15	VN46AF
	40	5.000	1.300	15	VN40AF

8

N-Channel MOSPOWER (Cont'd)

Device	Breakdown Voltage (Volts)	r_{DS} (on) (ohms)	I_D Continuous (Amps)	Power Dissipation (Watts)	Part Number
TO-52	60	5.000	0.200	0.315	VN10KE
	60	5.000	0.200	0.315	VN10LE
	500	1.500	2.500	25	2N6802
	500	1.500	2.500	75	IRFF430
	500	2.000	2.230	75	IRFF432
	500	3.000	1.600	40	IRFF420
	500	3.000	1.500	20	2N6794
	500	4.000	1.400	40	IRFF422
	450	1.500	2.500	25	2N6801
	450	1.500	2.500	75	IRFF431
	450	2.000	2.230	75	IRFF433
	450	3.000	1.600	40	IRFF421
	450	3.000	1.500	20	2N6793
	450	4.000	1.400	40	IRFF423
	400	1.000	3.370	25	IRFF330
	400	1.000	3.000	25	2N6800
	400	1.500	2.750	25	IRFF332
	400	1.800	2.090	20	IRFF320
	400	1.800	2.000	20	2N6792
	400	2.500	1.770	20	IRFF322
	400	3.600	1.350	15	IRFF310
	400	3.600	1.250	15	2N6786
	400	5.000	1.150	15	IRFF312
	350	1.000	3.370	25	IRFF331
	350	1.000	3.000	25	2N6799
	350	1.500	2.750	25	IRFF333
	350	1.800	2.090	20	IRFF321
	350	1.800	2.000	20	2N6791
	350	2.500	1.770	20	IRFF323
	350	3.600	1.350	15	IRFF311
	350	3.600	1.250	15	2N6785
TO-39	350	5.000	1.150	15	IRFF313
	240	6.000	0.800	6.25	VN2406B
	200	0.400	5.500	25	2N6798
	200	0.400	5.500	25	IRFF230
	200	0.600	4.500	25	IRFF232
	200	0.800	3.500	20	2N6790
	200	0.800	3.500	20	IRFF220
	200	1.200	2.800	20	IRFF222
	200	1.500	2.250	15	2N6784
	200	1.500	2.200	15	IRFF210
	200	2.400	1.800	15	IRFF212
	170	6.000	0.800	6.25	VN1706B
	150	0.400	5.500	25	2N6797
	150	0.400	5.500	25	IRFF231
	150	0.600	4.500	25	IRFF233
	150	0.800	3.500	20	2N6789
	150	0.800	3.500	20	IRFF221
	150	1.200	2.800	20	IRFF223
	150	1.500	2.250	15	2N6783
	150	1.500	2.200	15	IRFF211
	150	2.400	1.800	15	IRFF213
	150	4.500	0.500	15	NOS100B
	120	4.500	0.500	15	NOS101B
	120	6.000	0.800	6.25	VN1206B
	100	0.180	8.000	25	2N6796
	100	0.180	8.000	25	IRFF130
	100	0.250	7.000	25	IRFF132
	100	0.300	6.000	20	2N6788

N-Channel MOSPOWER (Cont'd)

Device	Breakdown Voltage (Volts)	rDS (on) (ohms)	ID Continuous (Amps)	Power Dissipation (Watts)	Part Number
TO-39	100	0.300	6.000	20	IRFF120
	100	0.400	5.000	20	IRFF122
	100	0.500	4.000	15	VNE010B
	100	0.500	4.000	15	VNE011B
	100	0.600	3.680	15	IRFF110
	100	0.600	3.500	15	2N6782
	100	0.800	2.010	15	IRFF112
	90	4.000	0.900	75	2N6661
	90	4.500	0.900	6.25	VN99AB
	90	5.000	0.800	6.25	VN90AB
	80	0.500	4.000	15	VND010B
	80	0.500	4.000	15	VND011B
	80	4.500	0.800	15	NOS102B
	60	0.180	8.000	25	2N6795
	60	0.180	8.000	25	IRFF131
	60	0.250	7.000	25	IRFF133
	60	0.300	6.000	20	2N6787
	60	0.300	6.000	20	IRFF121
	60	0.400	5.000	20	IRFF123
	60	0.500	4.000	15	VNC010B
	60	0.500	4.000	15	VNC011B
	60	0.600	3.500	15	2N6781
	60	0.600	3.680	15	IRFF111
	60	0.800	2.010	15	IRFF113
	60	3.000	1.100	6.25	2N6660
	60	3.500	1.000	6.25	VN67AB
	35	1.800	1.400	6.25	2N6659
	35	2.500	1.200	6.25	VN35AB
	30	1.200	0.350	6.25	VN0300B
TO-237	240	6.000	0.300	1.0	VN2406M
	240	10.000	0.250	1.0	BSR76
	240	10.000	0.250	1.0	VN2410M
	170	6.000	0.300	1.0	VN1706M
	170	10.000	0.250	1.0	BSR72
	170	10.000	0.250	1.0	VN1710M
	170	24.000	0.080	1.0	VN1720M
	120	6.000	0.300	1.0	VN1206M
	120	10.000	0.250	1.0	BSR70
	120	10.000	0.250	1.0	VN1210M
	80	4.000	0.350	1.0	BSR67
	80	4.000	0.350	1.0	VN0808M
	60	3.000	0.400	1.0	BSR66
	60	3.000	0.400	1.0	VN0606M
	60	5.000	0.300	1.0	VN10KM
	60	5.000	0.300	1.0	VN10LM
	60	7.500	0.300	1.0	BSR65
	60	7.500	0.250	1.0	BSR64
	60	7.500	0.250	1.0	VN2222KM
	60	7.500	0.250	1.0	VN2222LM
	30	1.200	0.700	1.0	VN0300M
TO-92	240	6.000	0.210	0.4	VN2406L
	240	10.000	0.160	0.4	VN2410L
	240	24.000	0.080	0.4	VN2420L
	200	24.000	0.080	0.4	VN2020L
	200	28.000	0.100	0.4	BS107
	170	6.000	0.210	0.4	VN1706L
	170	10.000	0.160	0.4	VN1710L
	120	5.000	0.230	0.4	VP1008L
	120	6.000	0.210	0.4	VN1206L
	120	10.000	0.160	0.4	VN1210L
	60	5.000	0.200	0.4	VN0610L
	60	5.000	0.200	0.4	VN0610LL
	60	5.000	0.179	0.4	BS170
	60	7.500	0.150	0.4	VN2222L

8

MOSPOWER Selector Guide (Cont'd)

N-Channel MOSPOWER (Cont'd)

Device	Breakdown Voltage (Volts)	rDS (on) (ohms)	ID Continuous (Amps)	Power Dissipation (Watts)	Part Number
TO-92	60 30	7.500 1.200	0.150 0.350	0.4 0.4	VN2222LL VN0300L
14-Pin Dual-In-Line (Side Brazed)	90 60 60 30	4.500 3.500 5.500 1.000	0.400 0.460 0.225 0.850	1.3 1.3 1.3 1.3	VQ1006P VQ1004P VQ1000P VQ1001P
14-Pin Dual-In-Line (Plastic)	90 60 60 30	4.500 3.500 5.500 1.000	0.400 0.460 0.225 0.850	1.3 1.3 1.3 1.3	VQ1006J VQ1004J VQ1000J VQ1001J

P-Channel MOSPOWER

Device	Breakdown Voltage (Volts)	r_{DS} (on) (ohms)	I_D Continuous (Amps)	Power Dissipation (Watts)	Part Number
TO-39	-100	5.0	-0.90	6.25	VP1008B
	-80	5.0	-0.9	6.25	VP0808B
	-30	2.5	-1.3	6.25	VP0300B
TO-237	-100	5.0	-0.37	1.0	VP1008M
	-80	5.0	-0.37	1.0	VP0808M
	-40	2.5	-0.48	1.0	VP0300M
	-30	2.5	-0.48	1.0	BSR78
TO-92	-80	5.0	-0.23	0.4	VP0808L
	-45	14.0	-0.107	0.4	BS250
	-30	2.5	-0.30	0.4	VP0300L
14-Pin Dual-In-Line (Side Brazed)	-90	5.0	-0.4	1.3	VQ2006P
	-60	5.0	-0.4	1.3	VQ2004P
	-40	2.0	-0.6	1.3	VQ2001P
14-Pin Dual-In-Line (Plastic)	-90	5.0	-0.4	1.3	VQ2006J
	-60	5.0	-0.4	1.3	VQ2004J
	-40	2.0	-0.6	1.3	VQ2001J

N- and P-Channel Quad MOSPOWER

Device	Breakdown Voltage (Volts)	r_{DS} (on) (ohms)	I_D Continuous (Amps)	Power Dissipation (Watts)	Part Number
14-Pin Dual-In-Line (Side Brazed)	30	3.0**	N-0.85 P-0.60	1.3	VQ3001P
	20	3.0**	N-2.0 P-2.0	1.75	VQ7254P
14-Pin Dual-In-Line (Plastic)	30	3.0**	N-0.85 P-0.60	1.3	VQ3001J
	20	3.0**	N-2.0 P-2.0	1.75	VQ7254J

**Total (N + P)

8

A.7 Worldwide Sales Offices

United States

Central

Siliconix Incorporated
1327 Butterfield Rd., Suite 616
Downers Grove, IL 60515
(312) 960-0106
Twx: 910-695-3232

Siliconix Incorporated
Two King James South, Suite 143
24650 Center Ridge Road
Westlake, OH 44145
(216) 835-4470
Twx: 810-427-9258

Siliconix Incorporated
3310 Keller Springs Road
Suite 110A
Carrollton, TX 75006
(214) 385-4046/4047
Twx: 910-860-9262

Eastern

Siliconix Incorporated
31 Bailey Avenue
Ridgefield, CT 06877
(203) 431-3535
Twx: 710-467-0660

Siliconix Incorporated
395 Totten Pond Road
Waltham, MA 02154
(617) 890-7180
Twx: 710-324-1783

Siliconix Incorporated
Cranes Roost Office Park
311 Whooping Loop
Altamonte Springs, FL 32701
(305) 831-3644
Twx: 810-853-0320

Northwestern

Siliconix Incorporated
2201 Laurelwood Road
Santa Clara, CA 95054
(408) 988-8000
Twx: 910-338-0227

Siliconix Incorporated
Denver, CO
(303) 771-9068

Southwestern

Siliconix Incorporated
17821 E. 17th Street
Suite 240
Tustin, CA 92680
(714) 544-8378
(714) 544-7275
Twx: 910-595-2643

International

EUROPEAN

FRANCE

Siliconix S.A.R.L.
Centre Commercial de L'Echat
Place de l'Europe
94019 Créteil Cedex
Tel: (1) 377.07.87
Tlx: Siliconx 230389F

WEST GERMANY

Siliconix GmbH
Johannesstrasse 27
D-7024 Filderstadt-1
Postfach 1340
Tel: (0711) 702066
Tlx: 7-255 533

Siliconix GmbH
Kirchfeldstr. 4
D-8025 Unterhaching
Munich
Tel: 089-611-4091
Tlx: 528305

UNITED KINGDOM

Siliconix Ltd.
3 London Rd.
Newbury, Berks
RG13 1JL
Tel: (0635) 30905
Tlx: 849357

FAR EAST

HONG KONG

Siliconix (H.K.) LTD.
5th Floor
Liven House
61-63 King Yip Street
Kwun Tong, Kowloon
Tel: 3-427151
Tlx: 44449SILXHX

JAPAN

Nippon Siliconix Incorporated
101 Daigo Tanaka Bldg.
4-4 Iidabashi 3-Chome
Chiyoda Ku, Tokyo 102
Tel: (03) 264-7905
Tlx: 2322739 NSIXJ

Siliconix

U.S. and Canadian Sales Representatives

U.S. Sales Representatives

ALABAMA
Huntsville (35815)
Rep Inc.
P.O. Box 4889
11535 Gilleland Rd.
(205) 881-9270
Twx: 810-776-2102

ARIZONA
Tempe (85281)
Quatra Associates, Inc.
1801 S. Jen Tilly Lane
Suite B-10
(602) 894-2808
TWX: 910-950-1153

CALIFORNIA
San Diego (92126)
OHM SPUN Electronics
9560 Black Mt. Rd.,
Suite 125
(619) 695-8102
Telex: 821-267

COLORADO
Denver
SILICONIX
(303) 771-9068

CONNECTICUT
Cheshire (06410-0160)
Scientific Components
1185 South Main Street
(203) 272-2963
Twx: 710-455-2078

FLORIDA
Altamonte Springs (32701)
Semtronic Associates
657 Maitland Ave.
(305) 831-8233
Twx: 810-854-0321

Ft. Lauderdale (33309)
Semtronic Associates
3471 Northwest 55th St.
(305) 731-2484

Clearwater (33515)
Semtronic Associates
300 S. Duncan Ave.,
Suite 270
(813) 461-4675

GEORGIA
Tucker (30084)
Rep Inc.
1944 Cooledge Road
(404) 938-4358
Twx: 810-766-0822

ILLINOIS
Des Plaines (60018)
Electron Marketing Corp.
3158 Des Plaines Ave.
Suite 135
(312) 298-2330
Twx: 910-233-0183

INDIANA
Indianapolis (46240)
Wilson Technical Sales, Inc.
P.O. Box 40699
4021 W. 71st Street
(317) 298-3345
Twx: 910-997-8120

IOWA
Cedar Rapids (52403)
Electromec Sales, Inc.
1500 2nd Ave. S.E.
Suite 205
(319) 362-6413
Twx: 910-525-1342

KANSAS
Wichita (67217)
Design Solutions Inc.
4502 Cherry
(316) 529-0114

MARYLAND
Baltimore (21208)
Pro Rep
107 Sudbrook Lane
(301) 653-3600
Twx: 710-862-0862

MASSACHUSETTS
Tyngsborough (01879)
Comp Tech, Inc.
1 Bridgeview Circle
(617) 649-3030
Twx: 710-347-6661

MICHIGAN
Brighton (48116)
A.P. Associates
225 E. Grand River Ave.
(313) 229-6550
Telex: 287310 (APAIUR)

MINNESOTA
Burnsville (55337)
Electromec Sales Inc.
101 W. Burnsville Pkwy.
(612) 894-8200
Twx: 910-576-0232

MISSOURI
Gladstone (64118)
Design Solutions Inc.
1502 N.E. 58th Street
(816) 452-8871

St. Louis (63144)
Design Solutions
1243 Hanley Industrial Ct.
(314) 961-7170
Twx: 910-968-8332

NEW JERSEY
Teaneck (07666)
R.T. Reid Associates
705 Cedar Lane
(201) 692-0200
Twx: 710-990-5086

NEW YORK
Endwell (13760)
Tri-Tech Electronics Inc.
3215 E. Main Street
(607) 754-1094
Twx: 510-252-0891

Fayetteville (13066)
Tri-Tech Electronics, Inc.
6836 E. Genesee Street
(315) 446-2881
Twx: 710-541-0604

Fishkill (12524)
Tri-Tech Electronics, Inc.
14 Westview Drive
(914) 897-5611

E. Rochester (14445)
Tri-Tech Electronics, Inc.
300 Main Street
(716) 385-6500
Twx: 510-253-6356

Melville (11747)
R.T. Reid Associates
20 Broadhollow Rd.
(516) 351-8833
Twx: 990-997-3030

NORTH CAROLINA
Raleigh (27607)
Rep Inc.
7330 Chapel Hill Road
Suite 206A
(919) 851-3007
Twx: 810-726-2102

Charlotte (28212)
Rep Inc.
Independence Office Park
6407 Idlewild Rd., Ste. 226
(704) 563-5554
Twx: 810-726-2102
Telex: 821765

OHIO
Cleveland (44143)
Arthur H. Baier Company
16 Alpha Park
(216) 461-6161
Twx: 810-427-9278

Dayton (45414)
Arthur H. Baier Company
4940 Profit Way
(513) 276-4128
Twx: 810-459-1624

OKLAHOMA
Tulsa (74133)
Electronics Mktg. Assoc.
7917 S. 72nd East Ave.
(918) 492-0390

OREGON
Beaverton (97005)
Blair Hirsh Co., Inc.
9645 S.W. Beaverton Hwy.
(503) 641-1875

TENNESSEE
Jefferson City (37760)
Rep Inc.
P.O. Box 728
113 So. Branner Ave.
(615) 475-4105/9012/9013
Twx: 810-570-4203

TEXAS
Grapevine (76051)
Electronics Marketing Assoc.
P.O. Box 487
403 E. Wall
(817) 481-7502/7503 &
(817) 488-6583/6584
Twx: 910-890-8659

Houston (77099)
Electronics Marketing Assoc.
P.O. Box 42388
11450 Bissonnet,
Suite 309
(713) 498-8120

UTAH
Denver
SILICONIX
(303) 771-9068

WASHINGTON
Lynnwood (98036)
Blair Hirsh Co., Inc.
P.O. Box 2250
19410 36th Avenue West
Suite 106
(206) 774-8151
Twx: 910-977-0131

WISCONSIN
Wauwatosa (53226)
Larsen Associates
10855 West Potter Road
(414) 258-0529
Twx: 910-262-3160

CANADIAN SALES REPS

Islington, Ontario (M9B 6E3)
Pipe Thompson, Ltd.
5468 Dundas St. W., Ste. 206
(416) 236-2355
Twx: 610-492-4367

North Gower, Ontario
Pipe Thompson, Ltd.
(613) 258-4067

8

U.S. Distributors

ALABAMA

Huntsville (35803)
Hamilton/Avnet, #23
4812 Commercial Drive
(205) 837-7210
Twx: 810-726-2162

Huntsville (35805)
Pioneer/Huntsville
4825 University Square
(205) 837-9300
Twx: 810-726-2197

ARIZONA

Phoenix (85021)
Wyle Laboratories-EMG
8155 N. 24th Ave.
(602) 249-2232
Twx: 910-951-4282

Tempe (85281)
Hamilton/Avnet, #04
505 South Madison Dr.
(602) 231-5100
Twx: 910-950-0077

CALIFORNIA

Anaheim (92807)
Zeus West, Inc.
1130 Hawk Circle
(714) 632-6880
Twx: 910-591-1696

Canoga Park (91304)
Marshall Industries
8015 Deering Ave.
(818) 999-5001
Twx: 910-494-4821

Chatsworth (91311)
Hamilton/Avnet, #48
9650 DeSoto Ave.
(818) 700-6500
Twx: 910-321-3639

Chatsworth (91311)
Avnet Electronics #71
20501 Plummer
(213) 700-2600

Costa Mesa (92626)
Avnet Electronics
350 McCormick Ave.
(714) 754-6111
Twx: 910-595-1928

Costa Mesa (92626)
Hamilton Electro Sales, #29
3170 Pullman Street
(714) 641-4100
Twx: 910-595-2638

Culver City (90230)
Hamilton Electro Sales, #01
10912 W. Washington Blvd.
(213) 558-2121 or (714)
522-8200
Twx: 910-340-6364

El Segundo (90245)
Wyle Laboratories-EMG
124 Maryland Street
(213) 322-8100
Twx: 910-348-7140

Irvine (92714)
Marshall Industries
17321 Murphy Ave.
(714) 660-0951

Irvine (92714)
Wyle Laboratories-EMG
17872 Cowan Ave.
(714) 863-9953
Twx: 910-595-1572

Ontario (91764)
Hamilton/Avnet #49
3002 E. 'G' Street
(714) 989-8309
Twx: 910-321-2806

Rancho Cordova (95670)
Wyle Distribution Group
Sacramento Division
11151 Sun Center Dr.
(916) 638-5282

Reseda (91335)
Jan Devices
6925 Canby, Bldg. 109
(213) 708-1100

Sacramento (95348)
Hamilton/Avnet #35
4103 Northgate Blvd.
(916) 920-3150

San Diego (92123)
Hamilton/Avnet, #02
4545 Viewridge Ave.
(619) 571-7510
Twx: 910-335-1216

San Diego (92123)
Wyle Laboratories-EMG
9525 Chesapeake Drive
(619) 565-9171
Twx: 910-335-1590

Santa Clara (95052)
Wyle Distribution Group
3000 Bowers Avenue
(408) 727-2500
Twx: 910-379-6480

Sunnyvale (94086)
Bell Industries
1161 No. Fairoaks Ave.
(408) 734-8570
Twx: 910-339-9378

Sunnyvale (94086)
Hamilton/Avnet, #03
1175 Bordeaux Avenue
(408) 743-3300
Twx: 910-339-9332

Sunnyvale (94086)
Marshall Industries
788 Palomar Ave.
(408) 732-1100
Twx: 910-339-9298

COLORADO

Englewood (80111)
Hamilton/Avnet, #06
8765 E. Orchard Rd., Suite
708
(303) 740-1000
Twx: 910-931-0510

Thornton (80241)
Wyle Distribution Group
451 E. 124th Avenue
(303) 457-9953
Twx: 910-936-0770

Wheatridge (80033)
Bell Industries
8155 W. 48th Avenue
(303) 424-1985
Twx: 910-938-0393

CONNECTICUT

Danbury (06810)
Hamilton/Avnet, #21
Commerce Drive, Commerce
Park
(203) 797-2800
Twx: 710-460-0594

Milford (06460)
Falcon Electronics
5 Higgins Drive
(203) 878-5272
Twx: 710-462-8407

Wallingford (06492)
Marshall Industries
20 Sterling Dr.
Barnes Industrial Park
(203) 265-3822
Twx: 710-465-0747

FLORIDA

**Altamonte Springs
(32701)**
Pioneer Electronics
221 North Lake Blvd.
(305) 834-9090
Twx: 810-850-0177

Ft. Lauderdale (33309)
Hamilton/Avnet, #17
6801 N.W. 15th Way
(305) 971-2900
Twx: 510-956-3097

Ft. Lauderdale (33309)
Pioneer Electronics
1500 N.W. 62nd St.
Suite 506
(305) 771-7520
Twx: 510-955-9653

St. Petersburg (33702)
Hamilton/Avnet, #25
3197 Tech Drive No.
(813) 576-3930
Twx: 810-863-0374

Winter Park (32792)
Hamilton/Avnet
6947 University Blvd.
(305) 628-3888

Winter Park (32789)
Milgray Electronics
1850 Lee Avenue
(305) 647-5747

GEORGIA

Norcross (30092)
Hamilton/Avnet, #15
5825 Peachtree Corners E-D
(404) 447-7500
Twx: 810-766-0432

Norcross (30093)
Marshall Industries
4364B Shakelford Road
(404) 923-5750
Twx: 810-766-3969

ILLINOIS

Bensenville (60106)
Hamilton/Avnet, #10
1130 Thorndale Ave.
(312) 860-7780
Twx: 910-227-0060

Elk Grove Village (60007)
GBL/Goold Electronics
610 Bonnie Lane
(312) 593-3222

Elk Grove Village (60007)
Pioneer/Chicago
1551 Carmen Drive
(312) 437-9680
Twx: 910-222-1834

INDIANA

Carmel (46032)
Hamilton/Avnet, #28
485 Gradle Drive
(317) 844-9333
Twx: 810-260-3966

Indianapolis (46250)
Pioneer/Indiana
6408 Castleplace Drive
(317) 849-7300
Twx: 810-260-1794

KANSAS

Lenexa (66214)
Marshall Industries
8321 Melrose Dr.
(913) 492-3121

Overland Park (66215)
Hamilton/Avnet, #58
9219 Quivira Road
(913) 888-8900
Twx: 910-743-0005

MARYLAND

Columbia (21045)
Hamilton/Avnet, #12
6822 Oak Hall Lane
(301) 995-3500 (MD)
(301) 621-5410 (DC)
Twx: 710-862-1861

Gaithersburg (20760)
Pioneer/Washington
9100 Gaither Road
(301) 921-0660
Twx: 710-828-0545

Gaithersburg (20760)
Marshall Industries
16760 Oakmont Ave.
(301) 840-9450
Twx: 710-828-0223

MASSACHUSETTS

Burlington (01803)
Milgray Electronics
79 Terrace Hall Ave.
(617) 272-6800
Twx: 510-225-3673

Burlington (01803)
Marshall Industries
1 Wilshire Road
(617) 272-8200
Twx: 710-332-6359

Woburn (01801)
Hamilton/Avnet, #18
50 Tower Office Park
(617) 935-9700
Twx: 710-393-0382

MICHIGAN

Grand Rapids (49508)
Hamilton/Avnet, #67
2215 29th St. S.E. A5
(616) 243-8805

Livonia (48150)
Hamilton/Avnet, #66
32487 Schoolcraft
(313) 522-4700
Twx: 810-242-8775

Livonia (48150)
Pioneer/Michigan
13485 Stamford
(313) 525-1800
Twx: 810-242-3271

MINNESOTA

Minneapolis (55435)
Industrial Components
5229 Edina Industrial Blvd.
(612) 831-2666
Twx: 910-576-3153

Minnetonka (55343)
Hamilton/Avnet, #63
10300 Bren Road, East
(612) 932-0600
Twx: 910-576-2720

U.S. Distributors (Cont'd)

Minnetonka (55343)
Pioneer/Twin Cities
10203 Bren Road, East
(612) 935-5444
Twx: 910-576-2738

MISSOURI
Earth City (63045)
Hamilton/Avnet, #05
13743 Shoreline Ct.
(314) 344-1200
Twx: 910-762-0606

St. Louis (63146)
Franklin Electronics
11638 Page Service Dr.
(314) 993-5333

NEW HAMPSHIRE
Manchester (03103)
Hamilton/Avnet
444 E. Industrial Park Dr.
(603) 881-7435

NEW JERSEY
Cherry Hill (08003)
Hamilton/Avnet, #14
One Keystone Avenue
(609) 424-0100
Twx: 710-940-0262

Fairfield (07006)
Hamilton/Avnet, #19
10 Industrial Road
(201) 575-3390
Twx: 710-734-4388

Fairfield (07006)
Marshall Industries
101 Fairfield Road
(201) 882-0320
Twx: 710-989-7052

Mt. Laurel (08057)
Marshall Industries
102 Gaither Dr., Unit 2
(609) 234-9100 (NJ)
(215) 627-1920 (PA)
Twx: 710-941-1361

NEW MEXICO
Albuquerque (87123)
Bell Industries
11728 Linn N.E.
(505) 292-2700
Twx: 910-989-0625

Albuquerque (87119)
Hamilton/Avnet, #22
2524 Baylor Drive S.E.
(505) 765-1500
Twx: 910-989-0614

NEW YORK
Buffalo (14202)
Summit Inc.
916 Main Street
(716) 884-3450
Twx: 710-522-1692

Commack (11720)
Falcon Electronics
2171 Jericho Turnpike
(516) 462-6350

East Syracuse (13057)
Hamilton/Avnet, #08
1600 Corporate Circle
(315) 437-2642
Twx: 710-541-1560

Endwell (13760)
Marshall Industries
10 Hooper Road
(607) 754-1570
Twx: 510-252-0194

Farmingdale (11735)
Milgray Electronics, Inc.
77 Schmitt Blvd.
(516) 420-9800
Twx: 510-2250-3673

Hauppauge (11788)
Hamilton/Avnet #20
933 Motor Pkwy.
(516) 231-9800

Hauppauge (11787)
Marshall Industries
275 Oser Avenue
(516) 273-2424
Twx: 510-224-6109

Port Chester (10523)
Zeus Components, Inc.
100 Midland Ave.
(914) 937-7400
Twx: 710-567-1248

Rochester (14623)
Hamilton/Avnet, #61
333 Metro Park
(716) 475-9130
Twx: 510-253-5470

Rochester (14623)
Marshall Industries
1260 Scottsville Road
(716) 235-7620
Twx: 510-253-5526

Westbury (11590)
Hamilton/Avnet #39
1065 Old Country Rd.
(516) 997-6868

Woodbury (11797)
Pioneer/Long Island
60 Crossways Park W.
(516) 921-8700
Twx: 510-227-9869

NORTH CAROLINA
Charlotte (28210)
Pioneer/NC
9801-A Southern Pine Blvd.
(704) 527-8188
Twx: 810-620-0366

Raleigh (27604)
Hamilton/Avnet, #24
3510 Spring Forrest Rd.
(919) 878-0819 X210
Twx: 510-928-1836

OHIO
Cleveland (44105)
Pioneer/Cleveland
4800 E. 131st Street
(216) 587-3600
Twx: 810-422-2210

Dayton (45459)
Hamilton/Avnet, #64
954 Senate Drive
(513) 433-0610
Twx: 810-450-2531

Dayton (45424)
Pioneer/Dayton
4433 Interpoint Blvd.
(513) 236-9900
Twx: 810-459-1622

Warrensville Heights (44128)
Hamilton/Avnet, #62
4588 Emery Industrial Pkwy.
(216) 831-3500
Twx: 810-427-9452

Westerville (43081)
Hamilton/Avnet #79
777 Brooks Edge Blvd.
(614) 882-7004

OKLAHOMA
Tulsa (74129)
Quality Components
9934 E. 21st Street So.
(918) 664-8812

Hamilton/Avnet

OREGON
Hillsboro (97123)
Wyle Laboratories-EMG
5289 N.E. Elam Young Parkway
Bldg. E-100
(503) 640-6000

Lake Oswego (97034)
Hamilton/Avnet, #27
6024 S.W. Jean Road
Bldg. C. Suite 10
(503) 635-8836
Twx: 910-455-8179

PENNSYLVANIA
Horsham (19044)
Pioneer Electronics
261 Gibraltar Road
(215) 674-4000
Twx: 510-665-6778

Pittsburg (15222)
Hamilton/Avnet
2800 Liberty Ave.
(412) 281-4150

Pittsburg (15238)
Pioneer/Pittsburgh
259 Kappa Drive
(412) 782-2300
Twx: 710-795-3122

TEXAS
Addison (75001)
Quality Components
4257 Kellway Circle
(214) 387-4949
Twx: 910-860-5459

Austin (78758)
Hamilton/Avnet, #26
2401 Rutland Drive
(512) 837-8911
Twx: 910-874-1319

Austin (78758)
Wyle Distribution
2120-F W. Braker Lane
(512) 834-9957

Austin (78758)
Pioneer Electronics
9901 Burnet Rd.
(512) 835-4000

Austin (78758)
Quality Components
2427 Rutland Drive
(512) 835-0220
Twx: 910-874-1377

Dallas (75234)
Pioneer Electronics
13710 Omega Rd.
(214) 263-3168
Twx: 910-860-5563

Dallas (75240)
Zeus Components
14001 Goldmark
(214) 783-7010

Houston (77063)
Hamilton/Avnet, #11
8750 Westpark
(713) 975-3500
Twx: 910-881-5523

Irving (75062)
Hamilton/Avnet, #16
2111 W. Walnut Hill Lane
(214) 659-4151
Twx: 910-860-5929

Richardson (75083)
Wyle Distribution
1810 N. Greenville Ave.
(214) 235-9953
Twx: 310-378-7663

Sugarland (77478)
Quality Components
1005 Industrial Blvd.
At Bournewood
(713) 491-2255

UTAH
Salt Lake City (84119)
Hamilton/Avnet, #09
1585 West 2100 South
(801) 972-2800
Twx: 910-925-4018

Salt Lake City (84104)
Wyle Laboratories-EMG
1959 South 4130 West
(801) 974-9953

WASHINGTON
Bellevue (98005)
Hamilton/Avnet, #07
14212 N.E. 121st Street
(206) 453-5844
Twx: 910-443-2469

Bellevue (98005)
Wyle Distribution Group
1750 132nd Avenue N.E.
(206) 453-8300
Twx: 910-443-2526

WISCONSIN
Milwaukee (53214)
Marsh Electronics, Inc.
1563 South 101st Street
(414) 475-6000
Twx: 910-262-3321

New Berlin (53151)
Hamilton/Avnet, #57
2975 Moorland Road
(414) 784-4510
Twx: 910-262-1182

CHIP DISTRIBUTOR
FLORIDA
Orlando (32807)
Chip Supply, Inc.
1607 Forsyth Rd.
(305) 275-3810
Twx: 810-850-0103

Orlando (32817)
Chip Supply, Inc.
7725 N. Orange Blossom
(305) 298-7100

8

Canadian Distributors

BRITISH COLUMBIA
Burnaby (V5G 4J7)
RAE Industrial Elec. Ltd.
3455 Gardner Court
(604) 291-8866
Twx: 610-929-3065
Tlx: 04-356533

ONTARIO
Downsview (M3J 123)
Future Electronics
82 St. Regis Crescent N.
(416) 638-4771
Twx: 610-491-1470

Mississauga (L4V 1M5)
Hamilton/Avnet, #59
6845 Redwood Drive
(416) 677-7432
Twx: 610-492-8867

Nepean (K2E 7L5)
Hamilton/Avnet, #60
2110 Colonade Road
(613) 226-1700
Twx: 0534971

Ottawa (K2C 3P2)
Future Elec.
Baxter Centre
1050 Baxter Rd.
(613) 820-8313

ALBERTA
Calgary (T2E 6Z2)
Hamilton/Avnet
2816 21st Street N.E.
(403) 230-3586

QUEBEC
Pointe Claire (H9R 4C7)
Future Elec.
237 Humus Blvd.
(514) 694-7710
Twx: 610-421-3251

St. Laurent (H4S 1M2)
Hamilton/Avnet, #65
2670 Sabourin Street
(514) 335-1000
Twx: 610-421-3731

Siliconix

European Representatives/Distributors

AUSTRIA
Ing. Ernst Steiner
Hummelgasse 14
A-1130 Vienna
Tel: 0222/827474/0
Tlx: 135026

BELGIUM
J. P. Lemaire S.A.
Av. Limburg Styrum 243
B-1810 Wemmel
Tel: (02) 460-05-60
Tlx: 24610

DENMARK
Ditz Schweitzer A.S.
Vallensbaekvej 41, Postboks 5
DK-2600 Glostrup
Tel: (02) 45-30-44
Tlx: 33257

FINLAND
Instrumentarium Electronics
P.O. Box 64
SF 02631 Espoo 63
Tel: (358)-0-5281
Tlx: 124426 HAVUL SF

FRANCE
Almex
48 Rue de L'Aubepine
B.P. 102
92164 Anthony Cedex
Tel: (1) 666-21-12
Tlx: 250067

Alrodis
40 Rue Villon
69008 Lyon
Tel: (7) 800-87-12
Tlx: 380636

A. Baltzinger
B.P. 63
67042 Strasbourg Cedex
Tel: (88) 331852
Tlx: 870952F

Composants S.A.
Avenue Gustave Eiffel
B.P. 81
33605 Pessac Cedex
Tel: (56) 36-4040
Tlx: 550696F

Composants S.A.
55 Avenue Louis Breguet
31400 Toulouse
Tel: (61) 20-82-38
Tlx: 530957

Composants S.A.
183 Route de Paris
86000 Poitiers
Tel: (49) 88-60-50
Tlx: 591525

Composants S.A.
57 Rue Manoir de Servigne
ZI, Route de Lorient
B.P. 3209
35013 Rennes Cedex
Tel: (99) 54-01-53
Tlx: 740311

I.T.T. Multicomposant
38 Avenue Henri Barbusse
92220 Bagneux
Tel: (1) 664-16-10
Tlx: 270763

Sanelec Electronique
7 Rue de la Couture
ZI, de la Pilaterie
59700 Marcq-en-Baroeuil
Tel: (20) 98-92-13
Tlx: 160-143F

SCAIB
80 Rue d'Arcueil
94523 Rungis Cedex
Tel: (1) 687-23-13
Tlx: 204-674F

GERMANY
Ditronic GmbH
Julius-Hoelder Str. 42
7000 Stuttgart 70
Tel: (0711) 720010
Tlx: 7255638

EBV Elektronik GmbH
Oberweg 6
D-8025 Unterhaching
Tel: 089-61105-1
Tlx: 524535

EBV Elektronik GmbH
Alexanderstrasse 42
7000 Stuttgart 1
Tel: (0711) 247481
Tlx: 722271

EBV Elektronik GmbH
Ostrasse 129
4000 Dusseldorf
Tel: (0211) 84846/7
Tlx: 8587267

EBV Elektronik GmbH
Kiebitzrain 18
3006 Burgwedel 1/Hannover
Tel: (05139) 5038
Tlx: 923694

EBV Elektronik GmbH
Schenckstr 99
6000 Frankfurt M.90
Tel: 069/785037
Tlx: 413590

Ing. Büro Rainer König
Königsbergerstrasse 16A
1000 Berlin 45
Tel: 030-772-8009
Tlx: 184-707

Ing. Büro K.H. Dreyer
Albert Schweitzer-Ring 36
2000 Hamburg 70
Tel: (040) 669027
Tlx: 2164484

Ing. Büro K.H. Dreyer
Flensburger Strasse 3
2380 Schleswig
Tel: (04621) 24055
Tlx: 02-21334

Ultratronik GmbH
Münchner Strasse 6
8031 Seefeld
Tel: (08152) 7090
Tlx: 05-26459

GREECE
General Electronics Ltd.
209 Thivon Street
Nikaia, Piraeus GR-184 54
Tel: (1) 4904934
Tlx: 212949 GELT GR

ITALY
Dott Ing. Giuseppe De Mico
S.P.A.
20060 Cassina de Pecchi (MI)
Via Vittorio Veneto 8
Milano
Tel: (02) 9520551
Tlx: 330869

NETHERLANDS
Koning en Hartman
Elektrotechniek BV
Postbus 43220
2504 AE The Hague
30 Koperwerf
Tel: 070-211641
Tlx: 31528

NORWAY
A.S. Kjell Bakke
Øvre Raelingsvei 20
P.O. Box 27
N-2001 Lillesstrøm
Tel: (02) 83-02-20
Tlx: 19407

PORTUGAL
Telectra S.A.R.L.
Praceta Av. Dr. Mario
Moutinho, Lote 1258
1400 Lisboa
Tel: (11) 616221
Tlx: 12598

SPAIN
Comercial Espanola de Componentes S.A.
Arzobispo Morcillo
24 Oficina 5
Madrid 28029
Tel: (1) 733-7054/55
Tlx: 47010

Redis Logar SA
Lopez de Hoyos
78 DPDO, Madrid 28002
Tel: (1) 4113561
Tlx: 23967

Redis Logar SA
Aragon 208-210
Barcelona 11
Tel: (3) 2549048

SWEDEN
Komponentbolaget NAXAB
Box 4115
S-171-04 Solna
Tel: 08-985140
Tlx: 17912 KOMP

SWITZERLAND
Abalec A.G.
Landstrasse 78
CH-8116 Würenlos
Tel: 01-730-0455
Tlx: 59834

UNITED KINGDOM
Abercorn Electronics Ltd.
Abercorn House, 3 Pittville St.
Edinburgh, Scotland EH15 2BZ
Tel: 031-669-6479

Barlec-Richfield Ltd.
Foundry Road, Horsham
West Sussex RH13 5PX
Tel: 0403-51881
Tlx: 877222

Dage Eurosem Ltd.
Rabans Lane
Aylesbury
Bucks HP19 3RG
Tel: 0296-32881
Tlx: 83518

Farnell Electronic
Components Ltd.
Canal Rd.
Leeds LS12 2TU
Tel: 0532-636311
Tlx: 55147

Hartech Ltd.
Forum House, Stirling Rd.
Chichester PO19 2EN
West Sussex
Tel: 0243-773511
Tlx: 86230

Macro-Marketing Ltd.
Burnham Lane
Slough, Berks
Tel: (06286) 4422
Tlx: 847945

Micro-Tek Mktg. Ltd.
4 Bellfield Avenue
Harrow Weald
Middlesex
HA3 6SX
Tel: 01-428-3265

Semiconductor Specialists (UK)
Ltd.
Carroll House
159 High Street
West Drayton
Middlesex UB7 7XB
Tel: (08954) 45522/46415
Tlx: 21958

YUGOSLAVIA
Contact: Belram S.A.
83 Avenue des Mimosas
1150 Brussels, Belgium
Tel: 734-33-32/734-26-19
Tlx: 21790

8

Worldwide Sales Offices

International Representatives/Distributors

AUSTRALIA
NSD (Australia) PTY, Ltd.
22 Michellan Ct.
Bayswater, Victoria 3153
P.O. Box 148
Tel: 729-8855
Tlx: AA35443

BRAZIL
Cosele Comercio e Servicios
Electronicos Ltda.
Rua da Consolacao, 867
01310 Sao Paulo
Tel: 255-1733
Tlx: 1130869-CSEL-BR

HONG KONG
Atek Electronics Co. Ltd.
Rm. 1302 Argyle Centre, Phase 1
688 Nathan Rd.
Kowloon
Tel: 3-916833/4
Twx: 37119 ATEK HX

Century Electronic Products Co.
Unit 1111, 11/F., Century Centre
44-46 Hung To Rd.
Kwun Tong
Kowloon, Hong Kong
Tel: 3-420101/2
Tlx: 51226 CEPCO HX
Cable: CENTURYEPC

Gibb, Livingston & Co, Ltd.
53 Wong Chuk Hang Rd
Sungib Industrial Centre
Aberdeen, Hong Kong
Tel: 5-558331
Tlx: HX73470

INDIA
Zenith Electronics
541 Panchratna
Mama Parmanand Marg
Bombay 400004
Tel: 384214
Tlx: 011-3152

Authorized U.S. Agent
Fegu Electronics Inc.
2584 Wyandotte St.
Mtn. View, CA 94043
Tel: (415) 961-2380
Tlx: 345599

ISRAEL
Telsys Ltd.
12 Kehilat Venetsia St.
Tel Aviv
Tel: 69010 494891
Tlx: 032392

INDONESIA
Sinar Electrik
Hayam Wuruk,
Jakarta-Barat
Tel: (021) 623243
Tlx: LIONGKINYAMCO JKT

JAPAN
Teijin Advanced Products Corp.
1-1 Uchisaiwai-Cho.2-Chome
Chiyoda-Ku, Tokyo, 100
Tel: (03) 506-4670
Tlx: J-23548

KOREA
A-Mee Trading Co., Ltd.
155-12 Yumri-Dong
Mapo-Ku
Seoul
Tel: 716-6883
Tlx: K28569 TBTRACO

LATIN AMERICA
Intectra Inc.
2629 Terminal Blvd.
Mt. View, CA 94043
Tel: (415) 967-8818
Tlx: 345545 Intectra MNTV
Cable: INTECTRA

MALAYSIA
Carter Semiconductor (M)
SDN Berhad
Jalan Lapangan Terbang, Ipon
Tel: 514-033
Tlx: MA44050

NEW ZEALAND
S.T.C. (NZ) Ltd.
10 Margot Street
Epsom, Auckland 3
Tel: 500-019
Tlx: NZ21888

PHILIPPINES
Alexan Commercial
812 Elcano Street
Binando, Manila
Tel: 405-952
Tlx: 27484 CEI PH

SINGAPORE
Carter Semiconductor (S)
PTE Ltd.
400 Orchard Road
Orchard Towers No. 09-07
Singapore 0923
Tel: 235-6653
Tlx: 36443 CARSIN

SOUTH AFRICA
Electrolink (PTY) Ltd.
P.O. Box 1020
Capetown 8000
Tel: 215-350
Tlx: 572-7320

TAIWAN
Don Business Corp.
6F, No. 33, Alley 24,
Lane 251, Nanking East Road
Sec. 5, Taipei
Tel 766-4515, 760-7801-3
Tlx: 25641 DONBC
Cable: "DONBC" TAIPEI

THAILAND
Choakchai Electronic Supplies
128/22 Thanon Atsadang
Bangkok 2
Tel: 221-0432
Tlx: 84809 CESLP T-H
Cable: SAHAPIPHAT

VENEZUELA
P. Benavides S.R.L.
Avilanas a Rio Edificio
Rio Caribe, Local 9
Caracas
Tel: 52-92-97
Tlx: 21801 PBTH

Manufacturing Facilities

TAIWAN
Siliconix (Taiwan) Ltd.
Nantze Export Processing Zone
Kaohsiung
Tel: 3615101-4
Tlx: 78571235

HONG KONG
Siliconix (H.K.) Ltd.
5/6/7th Floors
Liven House
61-63 King Yip Street
Kwun Tong, Kowloon
Tel: 3-427151
Tlx: 44449SILXHX

UNITED KINGDOM
Siliconix Ltd.
Morriston, Swansea
SA6 6NE
Tel: (0792) 74681
Tlx: 48197

UNITED STATES
2201 Laurelwood Road
Santa Clara, CA 95054
Tel: (408) 988-8000
Tlx: 910-338-0227

Index